INDUSTRIAL
DYNAMICS

INDUSTRIAL DYNAMICS

JAY W. FORRESTER

 THE M.I.T. PRESS • MASSACHUSETTS INSTITUTE OF TECHNOLOGY

CAMBRIDGE • MASSACHUSETTS

TYPOGRAPHY BY BURTON J. JONES, JR.

LIBRARY OF CONGRESS CATALOG CARD NO. 61-17871

PRINTED IN THE UNITED STATES OF AMERICA

To Susan

Preface

THIS book is intended for the student of management, whether he is in a formal academic program or in business. It treats the central framework underlying industrial activity. The goal is "enterprise design" to create more successful management policies and organizational structures.

Industrial dynamics is a way of studying the behavior of industrial systems to show how policies, decisions, structure, and delays are interrelated to influence growth and stability. It integrates the separate functional areas of management — marketing, investment, research, personnel, production, and accounting. Each of these functions is reduced to a common basis by recognizing that any economic or corporate activity consists of flows of money, orders, materials, personnel, and capital equipment. These five flows are integrated by an information network. Industrial dynamics recognizes the critical importance of this information network in giving the system its own dynamic characteristics.

The approach is one of building models of companies and industries to determine how information and policy create the character of the organization. The "management laboratory" now becomes possible. The first step is to identify the problems and goals of the organization. The second is to formulate a model that shows the interrelationships of the significant factors. Such a model is a systematic way to express our wealth of descriptive knowledge about industrial activity. The model tells us how the behavior of the system results from the interactions of its component parts. These interactions are often more important than the pieces taken separately. Finally, proposed changes can be tried in the model and the best of them used as a guide to better management.

Industrial dynamics now becomes possible as a result of four foundations developed during the last twenty years. The theory of information-feedback systems gives us a basis for understanding the goal-seeking, self-correcting interplay between the parts of a business system. Investigation of the nature of decision making in the context of modern military tactics forms a basis for understanding the place of decision making in industry. The experimental model approach to the design of complex engineering and military systems can be applied to social systems. The digital computer has become a practical, economical tool for the vast amount of computation required. These accomplishments now make it possible to cope with the greater complications that we find in the dynamics of industrial and economic behavior.

At M.I.T. we have found that industrial dynamics can be taught to management students of any age and experience. It can begin in the management curriculum any time from the first undergraduate year through to the special development programs for senior executives. During the 1961-1962 academic year, study projects in industrial dynamics are being extended to a new optional research program for freshmen entering M.I.T.

This volume is intended as a classroom text and also as a guide for practicing managers or management scientists who wish to explore the dynamic interactions within the business system.

vii

The four parts of the book differ greatly in form and pace. The Introduction and Part I treat the background, nature, and objectives of industrial dynamics. Part II gives detailed methodology. Part III applies the methodology to examples of industrial systems. Part IV provides a look at the future. The title pages for the separate parts give suggestions to several classes of readers.

This book presents my own personal view of the management process. It does not purport to be a comprehensive treatment of management science. In this spirit, no effort has been made to include a complete bibliography. References have been limited to those that are especially pertinent to the discussion. Because the book reports on an active research program, new results will probably lead to alterations in some of the views presented here.

A series of publications on industrial dynamics is being planned. Others now being prepared by various authors include a description of the DYNAMO computer program compiler described in Appendix A, problems and assignments given students in the teaching of industrial dynamics, case studies and formal models of industries, and a variety of management and economic situations beyond the examples in Part III.

The book results from the first five-year phase of a research program to develop a coordinating structure for the separate facets of management and economics. It marks a transition in the research program. The first phase dealt primarily with philosophy and methodology and with the "steady-state" dynamics of mature industries. The new phase will deal more with transient situations. Research is already in progress toward the design of policies controlling industry and company growth. During the past academic year, a graduate seminar at M.I.T. began the development of models for the dynamics of product and market growth. A study of the dynamics of economic development is now being started.

The research represented in this book has evolved directly from my own experiences. The cattle ranch operated by my parents, M. M. and Ethel W. Forrester, at Anselmo, Nebraska, provided my first exposure to business and to the nature of commodity markets. The study of electrical engineering at the University of Nebraska laid the foundation for graduate research. My graduate study at M.I.T. was under Professor Gordon S. Brown, who was then starting the Servomechanisms Laboratory and developing the concepts of information-feedback systems in a research project atmosphere that gave leadership experience to graduate students and junior staff. In the late 1940's the challenging environment of the M.I.T. Division of Industrial Cooperation under Mr. Nathaniel McL. Sage, Sr., gave me an opportunity to plan, and to direct with broad managerial responsibility, the construction of Whirlwind I, which was one of the first high-speed electronic digital computers. As head of the Digital Computer Division of the Lincoln Laboratory, I had the opportunity to manage a growing technical organization, to coordinate the early planning of the Air Force's Semi-Automatic Ground Environment (SAGE) system for air defense, and to guide the early stages of industrial company manufacturing to build the needed equipment. Together, these experiences provided a view of management problems at all levels as well as a foundation in the methodology on which the book is based.

ACKNOWLEDGMENTS

The work that led to this book was started in 1956 as a direct result of arrangements by Professors E. P. Brooks, Eli Shapiro (then Dean and Associate Dean of the M.I.T. School of Industrial Management), and Edward L. Bowles. It continues with the encouragement of Professor Howard W. Johnson, now Dean of the School.

PREFACE

The Industrial Dynamics program at M.I.T. was launched with financial support from the Ford Foundation. The grant continues to sustain the basic research and the development of industrial dynamics methodology.

The Sloan Research Fund of the School of Industrial Management, established from grants by the Alfred P. Sloan Foundation, has given substantial assistance.

The encouragement and financial support of American industry are especially significant to the effectiveness of the program. The industrial system and the model discussed in Chapters 17 and 18 are included through the cooperation of the Sprague Electric Company of North Adams, Massachusetts. For several years, the Sprague Electric Company has financed a joint research program in industrial dynamics with the M.I.T. School of Industrial Management. The interest and participation of Mr. Robert C. Sprague, Mr. Ernest L. Ward, Mr. Bruce R. Carlson, and others in the company have made possible the applications of the methods of this book to an actual problem in the management of an industrial system. Mr. Willard R. Fey on the M.I.T. staff has been in charge of the research reported in Chapters 17 and 18. Mr. Wendyl A. Reis, Jr. (now at the Sprague Electric Company), and Mr. Carl V. Swanson have contributed to the work reported in those chapters.

The Digital Equipment Corporation of Maynard, Massachusetts, and the Minute Maid Corporation, Orlando, Florida, are providing financial support for other aspects of the industrial dynamics research.

Digital computer time has been provided by the M.I.T. Computation Center, and in addition, the International Business Machines Corporation has made available IBM 709 computer time for the preparation of the illustrations.

This book was undertaken partly as a result of the favorable reception accorded my July 1958 and March 1959 articles in the *Harvard Business Review,* which have been a basis for Chapters 2 and 16. The encouragement of Professors Edward C. Bursk and John F. Chapman of the *Review* has been most helpful.

Without the DYNAMO compiler, the industrial dynamics research could not have progressed so rapidly. Appendix A identifies the contributions of Mr. Richard K. Bennett, Dr. Phyllis Fox (Mrs. George Sternlieb), Mr. Alexander L. Pugh, III, Mr. Edward B. Roberts, Mrs. Grace Duren, and Mr. David J. Howard.

The comments by those who carefully read the first typescript have made it possible to improve the text. The criticisms by the following were especially complete and helpful: Professors Lynwood S. Bryant, Edward H. Bowman, David Durand, Billy E. Goetz, and Chadwick J. Haberstroh, and Messrs. Robert G. Brown, Willard R. Fey, W. Edwin Jarmain, Alexander L. Pugh, III, Edward B. Roberts, and F. Helmut Weymar.

Messrs. John F. Buoncristiani, Arthur L. Douty, Jr., and Ole C. Nord handled the computer runs, and Mrs. Faith Richards typed manuscript and assisted in numerous details.

Miss Constance D. Boyd of the M.I.T. Press has been especially helpful in editing the text.

JAY W. FORRESTER

Massachusetts Institute of Technology
Cambridge, Massachusetts
August, 1961

Contents

Introduction

MANAGEMENT AND MANAGEMENT SCIENCE   1

    I.1   Management as an Art   1
    I.2   The Manager and Today's Management Science   3
    I.3   The Precedent of Engineering   4
    I.4   The Challenge to Management   6
    I.5   The Manager and Future Management Science   8

PART I

THE MANAGERIAL VIEWPOINT

1 · INDUSTRIAL DYNAMICS   13

    1.1   Information-Feedback Control Theory   14
    1.2   Decision-Making Processes   17
    1.3   Experimental Approach to System Analysis   17
    1.4   Digital Computers   18

2 · AN INDUSTRIAL SYSTEM   21

    2.1   The Approach   21
    2.2   Needed Information   22
    2.3   Simulation Method   23
    2.4   System Experiments   24
    2.5   Adding a Market Sector   36

3 · THE MANAGERIAL USE OF INDUSTRIAL DYNAMICS   43

    3.1   The Management Laboratory   43
    3.2   Steps in Enterprise Design   43
    3.3   Effect on the Manager   45

PART II

DYNAMIC MODELS OF INDUSTRIAL AND ECONOMIC ACTIVITY

4 · MODELS   49

    4.1   Classification of Models   49
    4.2   Models in the Physical Sciences, Engineering, and the Social Sciences   53
    4.3   Models for Controlled Experiments   55
    4.4   Mechanizing the Model   55
    4.5   Scope of Models   55
    4.6   Objectives in Using Mathematical Models   56
    4.7   Sources of Information for Constructing Models   57

CONTENTS

5 • PRINCIPLES FOR FORMULATING DYNAMIC SYSTEM MODELS   60

5.1 What to Include in a Model   60
5.2 Information-Feedback Aspects of Models   61
5.3 Correspondence of Model and Real-System Variables   63
5.4 Dimensional Units of Measure in Equations   64
5.5 Continuous Flows   64
5.6 Stability and Linearity   66

6 • STRUCTURE OF A DYNAMIC SYSTEM MODEL   67

6.1 Basic Structure   67
6.2 Six Interconnected Networks   70

7 • SYSTEM OF EQUATIONS   73

7.1 Computing Sequence   73
7.2 Symbols in Equations   75
7.3 Time Notation in Equations   75
7.4 Classes of Equations   76
7.5 Solution Interval   79
7.6 Redundancy of Equation Type and Time Notation   80
7.7 First-Order versus Higher-Order Integration   80
7.8 Defining All Variables   80

8 • SYMBOLS FOR FLOW DIAGRAMS   81

8.1 Levels   81
8.2 Flows   82
8.3 Decision Functions (Rate Equations)   82
8.4 Sources and Sinks   82
8.5 Information Take-off   83
8.6 Auxiliary Variables   83
8.7 Parameters (Constants)   83
8.8 Variables on Other Diagrams   83
8.9 Delays   83

9 • REPRESENTING DELAYS   86

9.1 Structure of Delays   86
9.2 Characteristics of Delays   87
9.3 Exponential Delays   87
9.4 Time Response of Exponential Delays   89

10 • POLICIES AND DECISIONS   93

10.1 Nature of the Decision Process   95
10.2 Policy   96
10.3 Detecting the Guiding Policy   97
10.4 Overt and Implicit Decisions   102
10.5 Inputs to Decision Functions   103
10.6 Determining the Form of Decision Functions   103
10.7 Noise in Decision Functions   107

CONTENTS

11 · AGGREGATION OF VARIABLES   109

11.1 Using Individual Events to Formulate Aggregate Flow   109
11.2 Aggregation on Basis of Similar Decision Functions   110
11.3 Effect of Aggregation on Time Delays   110

12 · EXOGENOUS VARIABLES   113

13 · JUDGING MODEL VALIDITY   115

13.1 Purpose of Models   115
13.2 Importance of the Specific Objectives   116
13.3 Predicting Results of Design Changes   116
13.4 Model Structure and Detail   117
13.5 Behavior Characteristics of a System   119
13.6 Model of a Proposed System   121
13.7 Comments on Model Testing   122

14 · SUMMARY OF PART II   130

PART III

EXAMPLES OF DYNAMIC SYSTEM MODELS

15 · MODEL OF THE PRODUCTION-DISTRIBUTION SYSTEM OF CHAPTER 2   137

15.1 Objectives   137
15.2 Scope   138
15.3 Factors to Include   139
15.4 Basis for Developing Equations   140
15.5 System Equations   141
    15.5.1 Equations for the Retail Sector   141
    15.5.2 Equations for the Distributor Sector   158
    15.5.3 Equations for the Factory Sector   161
    15.5.4 Initial Conditions   165
    15.5.5 Parameters (Constants) of the System   168
15.6 Philosophy of Selecting Reasonable Parameter Values   171
15.7 Test Runs of Model   172
    15.7.1 Step Increase in Sales   172
    15.7.2 One-Year, Periodic Input   175
    15.7.3 Random Fluctuation in Retail Sales   177
    15.7.4 Limited Factory Production Capacity   180
    15.7.5 Reduction in Clerical Delays   182
    15.7.6 Removal of the Distributor Sector   183
    15.7.7 Rapidity of Inventory Adjustment   186

16 · ADVERTISING IN THE SYSTEM MODEL OF CHAPTER 2   187

16.1 Equations for the Advertising Sector   191
16.2 Initial-Condition Equations   197
16.3 Values of Constants   198

CONTENTS

16.4 Behavior of System with Advertising 199
    16.4.1 Response to Step Input 200
    16.4.2 Random Fluctuations at Retail Sales 206

17 · A CUSTOMER-PRODUCER-EMPLOYMENT CASE STUDY   208

17.1 Description 208
17.2 What Constitutes a System? 210
17.3 Factors to Be Included 211
17.4 Equations Describing System 215
    17.4.1 Order Filling 215
    17.4.2 Inventory Reordering 220
    17.4.3 Manufacturing 223
    17.4.4 Material Ordering 228
    17.4.5 Labor 229
    17.4.6 Delivery-Delay Quotation 235
    17.4.7 Customer Ordering 236
    17.4.8 Cash Flow 240
    17.4.9 Profit and Dividends 245
17.5 Supplementary Output Information 246
17.6 Input Test Functions 248

18 · DYNAMIC CHARACTERISTICS OF A CUSTOMER-PRODUCER-EMPLOYMENT SYSTEM   253

18.1 The Old System 253
    18.1.1 Step Change in Demand 253
    18.1.2 One-Year Period 258
    18.1.3 Two-Year Period 260
    18.1.4 Random Releases from Engineering Department 262
    18.1.5 Cash Flow and Continuous Balance Sheet 262
    18.1.6 Adequacy of the Model 263
18.2 Variations in Parameters of the Old System 268
    18.2.1 Sensitivity Analysis 268
    18.2.2 Rapidity of Labor Adjustment 269
18.3 New Policies 276
    18.3.1 Changes in Inventory Ordering 278
    18.3.2 Changes in Manufacturing Department 280
    18.3.3 Changes in Labor Hiring 282
18.4 Effect of New Policies 284
18.5 Improvements in New Policies 290
18.6 Characteristics of the New System 291
18.7 New System with Large Disturbances 300
18.8 Summary 308

PART IV

FUTURE OF INDUSTRIAL DYNAMICS

19 · BROADER APPLICATIONS OF DYNAMIC MODELS   311

19.1 Market Dynamics 311
19.2 Growth 317
19.3 Commodities 321
19.4 Research and Development Management 324

CONTENTS

19.5   Top-Management Structure   329
19.6   Money and Accounting   335
19.7   Competition   336
19.8   The Future in Decision Making   337
19.9   Models of Entire Industries   340

20 · INDUSTRIAL DYNAMICS AND MANAGEMENT EDUCATION   344

20.1   Industrial Dynamics as an Integrating Structure   345
20.2   Principles of System Structure   347
20.3   Academic Programs in Industrial Dynamics   350
20.4   People   354
20.5   Management Games   357
20.6   Management Research   360

21 · INDUSTRIAL DYNAMICS IN BUSINESS   362

APPENDICES

APPENDIX A	DYNAMO	369
APPENDIX B	Model Tabulations	383
APPENDIX C	Example of Computation Sequence	396
APPENDIX D	Solution Interval for Equations	403
APPENDIX E	Smoothing of Information	406
APPENDIX F	Noise	412
APPENDIX G	Phase and Gain Relationships	415
APPENDIX H	Delays	418
APPENDIX I	Phase Shift and Turning Points	421
APPENDIX J	Value of Information	427
APPENDIX K	Prediction of Time Series	430
APPENDIX L	Forecasting	437
APPENDIX M	Countercyclical Policies	441
APPENDIX N	Self-generated Seasonal Cycles	443
APPENDIX O	Beginners' Difficulties	449
REFERENCES		457
INDEX		459

Management and Management Science

Management of countries and industries has developed over the centuries as an empirical art. During the last half century a management science has begun to develop but is not yet an effective basis for dealing with top-management problems. Just as the merging of physical science and engineering in the last twenty-five years became the basis for the modern upsurge in technology, so will the development of a foundation structure of industrial and economic behavior provide a new dimension in management effectiveness in the next twenty-five years.

THE manager's task is far more difficult and challenging than the normal tasks of the mathematician, the physicist, or the engineer. In management, many more significant factors must be taken into account. The interrelationships of the factors are more complex. The systems are of greater scope. The nonlinear relationships that control the course of events are more significant. Change is more the essence of the manager's environment.

In the past the arts, the sciences, and the traditional professions have been placed on an intellectual pedestal with a status above the study and practice of management. The illusion that the study of management lacked intellectual challenge has arisen, not because the field of management is wanting in unexplored frontiers, but because the intellectual opportunities were not recognized and the problems lay beyond the reach of traditional analysis methods.

Our most challenging intellectual frontier of the next three decades probably lies in the dynamics of social organizations, ranging from growth of the small corporation to development of national economies. As organizations grow more complex, the need for skilled leadership becomes greater. Labor turmoil, bankruptcy, inflation, economic collapse, political unrest, revolution, and war testify that we are not yet expert enough in the design and management of social systems.

I.1 Management as an Art

Management is in transition from an art, based only on experience, to a profession, based on an underlying structure of principles and science.

Any worthwhile human endeavor emerges first as an art. We succeed before we understand why. The practice of medicine or of engineering began as an empirical art representing only the exercise of judgment based on experience. The development of the underlying sciences was motivated by the need to understand better the foundation on which the art rested.

The relationship between the growth of an art and the underlying science is illustrated in Figure I-1. The art develops through empirical experience but in time ceases to grow because of the disorganized state of its knowledge. When the need and necessary foundations coincide, a science develops to explain, organize, and distill experience into a more compact and usable form. As the science grows, it provides a new basis for further extension of the art.

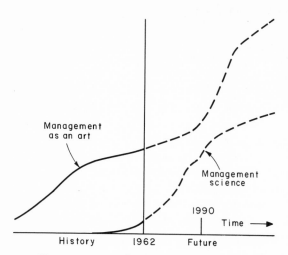

Figure I-1 Management art and science.

Over the centuries, management as an art has progressed by the acquisition and recording of human experience. But as long as there is no orderly underlying scientific base, the experiences remain as special cases. The lessons are poorly transferrable either in time or in space.

The corporate manager today finds little help in experiences recorded in the literature and carried forward from a generation ago. The descriptions are incomplete and lack precision. They arose from circumstances that cannot be properly related to today's events.

Likewise, contemporary management experiences are not so helpful as they could be to other managers. We still find each company and each industry believing its problems are unique. A discussion of present-day experiences in the context of another situation often elicits the rejoinder, "Yes, but my industry is different." Because of the lack of a suitable fundamental viewpoint, we fail to see how industrial experiences all deal with the same material, financial,

and human factors — all representing variations on the same underlying system.

Management education and practice have been highly fragmentized. Manufacturing, finance, distribution, organization, advertising, and research have too often been viewed as separate skills and not as a part of a unified system. Too often management education consists of gathering current industrial practice and presenting it to the student as a sequence of unrelated subjects. Similarly, in his work in industry, the manager specializes within departments where his experience perpetuates the atmosphere of unrelated compartmentalization.

To unify the separate facets of management, selected experiences have been recorded as "cases" to provide a vehicle to discuss management as an interrelated system. This has been the best method available for integrating management knowledge, although it has been far from adequate.

From a discussion of management situations the student has been expected to gain intuitive insight into principles underlying the cases he studies, even though these principles of the industrial system are themselves not specifically formulated. Were engineering still to rest on the same descriptive transmittal of experience, we should not have today's advanced technology. The liberal-arts training through multiple exposure to recorded incidents of the past presumes that the student will distill an intuitive structure of human and social behavior around which to assemble and interpret his own experiences.

The rapid strides of professional progress come when the structure and principles that integrate individual experiences can be identified and taught explicitly rather than by indirection and diffusion. The student can then inherit an intellectual legacy from the past and build his own experiences upward from that level, rather than having to start over again at the point where his predecessors began.

Without an underlying science, advancement of an art eventually reaches a plateau. Management has reached such a plateau. If progress is to continue, an applied science must arise as a foundation to support further development of the art. Such a base of applied science would permit experiences to be translated into a common frame of reference from which they could be transferred from the past to the present or from one location to another, to be effectively applied in new situations by other managers.

I.2   The Manager and Today's Management Science

The search for a scientific base underlying management goes back at least to the beginning of this century. It has progressed through work simplification and statistical quality control. It has expanded into the "operations research" activities which followed World War II. With few exceptions, the attempts toward a management science have dealt with isolated situations at the bottom of the management structure.

Thus far, management science still has not penetrated the inner circle of top management.[1] Partly this is because much of the work in operations research has dealt with problems of operating departments — not the problems of top management and the board of directors. Partly, as shown in Figure I-1, management science is not effective at the top-management level because the science is still embryonic. Management science has not reached a level that is effective as a supplement to the skillful practice of top management as an art. Partly it is because management science has only begun to deal with the time dimension in business — the changes and evolution with time that are the manager's principal concern.

But primarily, management science has failed to assist top management because the philosophy and objectives of management science have often been irrelevant to the manager. Mathe-

matical economics and management science have often been more closely allied to formal mathematics than to economics and management. The difference in viewpoint is evident if one compares the business literature with publications on management science, or descriptive economics books with texts on mathematical economics. In many professional journal articles the attitude is that of an exercise in formal logic rather than that of a search for useful solutions of real problems. In such an article, assumptions having doubtful validity are stated in an introductory paragraph and adopted without justification. On this formal but unrealistic foundation is then constructed an analytic mathematical solution for the behavior of the assumed system.

Another evidence of the bias of much of today's management science toward the mathematical rather than the managerial motivation is seen in a preoccupation with "optimum'" solutions. For most of the great management problems, mathematical methods fall far short of being able to find the "best" solution. The misleading objective of trying only for an optimum solution often results in simplifying the problem until it is devoid of practical interest. The lack of utility does not, however, detract from the elegance of the analysis as an exercise in mathematical logic, nor does it prevent the publication of a paper in a professional journal.

Figure I-2 illustrates the conspicuous dichotomy which has persisted between a Region A, representing practicing managers and economists, and a Region B, representing the mathematical analyst of business and economic phenomena. In business, Region A is inhabited by men responsible for decisions and policy, and Region B by staff specialists. In academic circles, Region A is the home of the descriptive social scientist whose skill is measured by his acuteness in perceiving the motivations and interrelationships in economic and managerial affairs; in contrast, Region B is more apt to include those searching for problems to fit avail-

---

[1] See Peter Drucker's discussion (Reference 1) of the failure of management science to deal thus far with the "risk-making and risk-taking decisions of business enterprise."

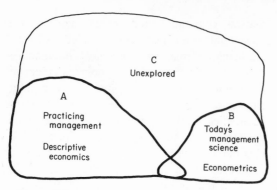

Figure I-2 The management-economics field.

able mathematical tools. In Region A the goal is improvement of real situations; in Region B the goal is often the explicit or the optimum solution to unrealistically simplified hypotheses. In Region A the manager acts on such information as he can obtain; in Region B the analyst often ignores phenomena that he admits are crucial but that cannot be precisely measured. In Region A success is measured by financial results and by economic development; in Region B reputations are based on published papers and mathematical elegance. In Region A the literature includes the business press and the descriptive books on economics and management; in Region B the literature comprises the journals and books on operations research and mathematical economics; and not often do the inhabitants of one region read the literature of the other. In Region A system nonlinearities[2] are recognized as primary causes of important occurrences; in Region B nonlinear behavior is usually ignored. In Region A data come from personal observation and participation in economic and business affairs; in Region B inputs are often limited to those few for which statistical measurements are available. In Region A opinions are more apt to be built up from individual incidents — how the individual person reacts, how the actual production process is designed, how long it takes to build a factory; in

---

[2] Where result is not proportional to cause. Discussed more fully in Section 4.1.

Region B relationships are extracted from averages and aggregates in which the nature of individual action is often lost. In Region A the manager and the government policy maker deal with the dynamic interactions between men, materials, decisions, equipment, and money; in Region B simplified money-flow relationships have often claimed the exclusive attention of the mathematical analyst.

These distinctions between Regions A and B have been sharply etched and somewhat exaggerated. However, the proper impression emerges — the "art" of Region A is still better able to deal with decisions of great consequence than the "science" of Region B. The overlap between the two is slight. The manager has often found that management science did not deal with his most urgent problems. It has not learned to take into account the variables that he knows to be important. It is not cast in a language with which he is familiar.

If the lack of acceptance of management science by management is to change, management science must change. It must attack the major problems of corporations and countries. It must accept the world as it is, not as an idealized abstraction that fails to be meaningful. It must search for improvement, not hold out for the optimum and perfection. It must use the information that is available, all that is pertinent, but, like the manager, it cannot wait for measurement of everything that one might like to know. It must be willing to deal with "intangibles" where these are important. It must speak in the language of the practicing manager.

I.3 The Precedent of Engineering

The separation between management and management science is now closing. Management research is being realigned to coincide with the objectives of practicing managers. The trends that are shaping for the future appear to have a precedent in the recent history of engineering.

In the last twenty-five years the technologi-

4

cal upsurge in our modern society illustrates the changes that can be anticipated in management during the next two or three decades. The changes in the status and world position of engineering since 1935 are of the same nature, and have occurred for the same fundamental reasons, as the changes we can expect in management and economics between now and 1985.

Over the last twenty years, technology has held the spotlight in the center of the world stage, just as management and economics will during the remainder of this century. Furthermore, the fundamental reasons for great advances in management will be essentially the same as those that have thrust science and engineering to a dominant international position.

Empirical Practice and Scientific Base. Before 1935 engineering tended to be an empirical art following handbook procedures, precedent, and experience. In the same way, management today is an empirical art.

Before World War II, basic scientific developments in the world's universities lacked close ties to the practice of engineering. There was no strong applied science intermediate between basic information and its practical application. Industrial research laboratories were the exception rather than the rule. In the same way we now see a developing body of applicable basic science that has not yet found its way into the practice of management.

After 1940, the practice of engineering began to converge with the underlying science. Engineering advancement was founded more closely on mathematics, physics, and chemistry. The methods, instruments, and attitudes of research were no longer foreign to the practical fields of application. Research came to be recognized as an essential part of the technological process. Likewise, managers are now beginning to recognize the importance of an applied science foundation underlying the practice of management.

I believe that developments are now begin-
ning in the theory of industrial organizations that will bear the same relationship to management that physics does to engineering. Both the strengths and weaknesses of this statement are intended. Physics has provided the foundation for a great upsurge in technology. But it is not adequate to specify the "best" design of a space vehicle nor to guarantee our ability to make a roof that does not leak. Physics is a foundation of principle to explain underlying phenomena but not a substitute for invention, perception, and skill in applying the principles.

Organization for Research. Twenty years ago the concept of how to organize for scientific research began to change. Before that, research was most often a one-man activity. Now team research is recognized as essential if resources are to match the tasks adequately.

Until recently research in the social sciences has been largely at the individual level in the form of doctoral theses and university faculty research. (We can scarcely count as research the large group efforts devoted to the collection of statistical data. These have provided information useful to the practice of management as an art, but gathering data is not the creation of an underlying science.)

The attitude toward management research is now changing, and we already see corporations beginning to assign groups for research into the underlying fundamentals of management policies and decisions. Just as we now expect a percentage of the sales dollar to go to product research, so we may come to expect in the future that a certain fraction of the management and white-collar payroll should be devoted to research toward managerial innovation.

System Awareness. Over the last two decades, engineering has developed an articulate recognition of the importance of *systems engineering*. Systems engineering is a formal awareness of the interactions between the parts of a system. A telephone system is not merely wire, amplifiers, relays, and telephone sets to be considered separately. The interconnections, the compatibility, the effect of one upon the other, the objectives of the whole, the relationship of the system to the users, and the economic feasibility must receive even more attention than the parts, if the final result is to be successful.

Until now much of management education and practice has dealt only with components. Accounting, production, marketing, finance, human relations, and economics have been taught and practiced as if they were separate, unrelated subjects. Only in the topmost managerial positions do managers need to integrate the separate functions. Our industrial systems are becoming so large and complex that a knowledge of the parts taken separately is not sufficient. In management as in engineering, we can expect that the interconnections and interactions between the components of the system will often be more important than the separate components themselves.

I.4 The Challenge to Management

The challenge to the future manager is to design improved enterprises. To do this he must devote himself more to the major decisions of strategy and less to the routine decisions and short-term tactics. He must relieve himself of the minor repetitive decisions by putting them under the control of policy so that he is free to push back the management frontiers. This will occur as an acceleration of trends that are already quite evident.

The past evolution of management decision making might be shown as in Figure I-3, which represents the entire scope of management activity. In Region A we already accept as automatic a class of decisions that a century ago would have been a managerial prerogative — calculating payroll and reordering of goods sold. In Region B lie the semiautomatic decisions made by middle management but closely circumscribed by corporate policy — employment levels, inventories, and production sched-

ules. In Region C are today's top-management decisions — capital investment, advertising budgets, and union contract conditions.

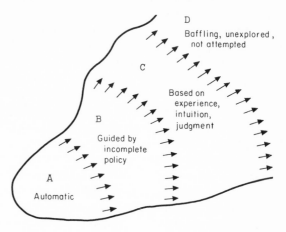

Figure I-3 Scope of management decisions.

The boundaries between regions are moving to the right. Just as the pace of technological change has quickened, so is the pace of managerial change accelerating. As present policy-guided decisions become automatic, and as today's judgment decisions are prescribed by policy, the creative manager will be faced by new problems and opportunities in Region D.

The industrial dynamics discussed herein will help expand the area of automatic decisions in Region A. It should show how to extend Region B by helping design policies to cover more of today's vexing management questions. It should be a tool for providing background against which to make intuitive judgment decisions in Region C. Top management should then have time and opportunity to push out the frontier into Region D and to accept the challenge of unexplored areas in industrial management and economic development.

What are some of the unexplored problems of Region D?

International Economic Development. On the international scene are the problems of the dynamics of industrialization as the newly emerging countries of the world move from agricultural to industrial economies.

The history of economic development in the Western countries is not directly applicable as a guide for today's new societies. In the economic development of Europe and the United States the pace was gradual; education, capital accumulation, and technological change advanced together. Other parts of the world are developing in a different environment. Their elementary economies exist in the presence of nations with a high standard of living. Their people are impatient to reach the economic level of more advanced countries. Capital formation, education, and the aspirations of the people must grow in synchronism if revolution and war are not to overtake economic development.

The manager of the future must assist in this orderly growth while pursuing his own business. He must avoid policies and investments in other countries that are incompatible with the development of those countries and hence with his own long-range welfare.

The manager will find a world commerce far different from the past. Newly industrialized countries will compete for markets. The world's raw materials will be more in demand.

We see already that the international struggle of the 1950's that was based on military technological competition is changing to a struggle to achieve economic strength and sufficient understanding of economic change to form a new basis for world leadership.

Middle Management and Technical Effectiveness. Closer to home, the corporation has a challenge to improve the effective use of its middle-management and technical personnel. In this group, from the 22-year-old college graduate to the vice-president, the average lifetime contribution is a small fraction of that which is potentially possible. Individuals are not developed to their fullest abilities, and existing abilities are not challenged to their greatest contribution. A bureaucracy has evolved in which

the individual has become the "organization man"; and the essence of capitalism, with its competition and its objective basis for reward, has disappeared from the immediate environment of the individual.

As an estimate, I would put this middle-management lifetime effectiveness as low as a tenth of the maximum possible. We should not hope for the maximum; but certainly a manyfold increase in a 10% effectiveness is possible before a point of diminishing returns is reached.

Our problems of international technological competition lie here in the ineffectiveness with which we use our human resources of scientists, engineers, and managers within the corporation — not in attempts to divert more students into technology. We cannot, nor could we afford to, attract enough of the population into middle-management and technical positions to compensate for such low levels of ability development and motivation.

Long-Range Planning. Another future challenge to management is in the whole realm of long-range planning. Here we see much lip service but little real action. Plans, where they exist, are apt to be but wishes — goals of greater sales or higher profit without a plausible sequence of steps for achieving those goals.

Where long-range plans do exist, they seldom have substantial content beyond a five-year span; yet the momentum of our corporations and our economy is such that five years is practically the minimum time in which it is possible to create any real changes. Past decisions on new product research, development of personnel, choice of products and markets, and construction of manufacturing facilities have already determined the essential characteristics of the corporation for the next five years. The challenge lies in how today's decisions will affect the time interval between five and twenty years hence.

Our industrial society has become increasingly dependent on decisions that are made further and further ahead of the period of their maximum impact; yet the social pressures on the individual have diminished the personal emphasis on serious thought about the future. The consequences of important decisions that control economic growth once lay inside the time horizon of the individual's personal planning; now the reverse is more typical.

In a simple agrarian economy, a decision to plant and till affected production and the ability to consume six months or a year later. In early industrial economies, the essential decisions to construct simple factories affected production and consumption two or three years later. As the industrial age emerged, the capital equipment industries appeared wherein decisions on new tools led to their manufacture, which, in turn, made new consumer goods factories possible, which eventually produced consumable products — the interval between primary decisions and economic results had lengthened to ten years or more. Now we find that key decisions relate to research in frontier products that will require the development of materials and tools, then of production facilities, then of markets — the span from primary research decision to the full consequences has risen to twenty years or more.

As the momentum of the industrial society has increased, the delays in its response to key decisions have lengthened. By contrast, the time span of interest to the individual has been shortening.

In personal planning for the future in an elementary agricultural economy, the family expects to spend a lifetime on one farm; a person looks ahead to building an estate for old age and retirement; decisions are made in the light of their future effect on the decider and his children. Likewise, in earlier industrial economies, capitalism meant a direct relationship between success and reward, both financial reward and the feeling of accomplishment. The entrepreneur stayed with his business and committed his own future to the wisdom of his management decisions. In our modern economy

there has been an essential change. The manager of the large, mature corporation is a trustee. He does not stay more than a few years in one position nor reap the long-range successes and failures that follow his decisions. Emphasis is on short-term results; even his own personal future is assured by retirement plans. His business and his position in it are not to be left to his children. As our economy has evolved, the personal-interest time horizon has shrunk from a lifetime to a few years.

This modern discrepancy between the distant consequences of required decisions and the brief tenure of men in the positions where the decisions are made unavoidably reduces responsibility and morale. The man is judged on results determined by his predecessors and makes decisions that will affect primarily his successors.

I.5 The Manager and Future Management Science

Managing is the task of *designing* and *controlling* an industrial system. Management science, if it is to be useful, must evolve effective methods to analyze the principal interactions among all the important components of a company and its external environment. It must be able to synthesize improved industrial systems.

Management now stands at a frontier leading to a new understanding of industrial growth and stability. This new insight into industrial and economic behavior will come from a better grasp of the time-varying interactions between the major facets of our social systems. Management education and practice are, I believe, on the verge of a major breakthrough in understanding how industrial company success depends on the interactions between the flows of information, orders, materials, money, personnel, and capital equipment. The way these six flow systems interlock to amplify one another and to cause change and fluctuation will form a basis for anticipating the effects of decisions, policies, organizational forms, and investment choices.

"Industrial dynamics," as used here, is a method of systems analysis for management. It deals with the time-varying interactions between the parts of the management system.

As a science emerges with the power and scope to deal with today's practical top-management problems, the gap between the science and today's management art will at first narrow. The science will develop more rapidly than it is accepted. As the science reaches maturity, it then becomes a basis for a further development of the art of management.

The task of the manager will become more challenging. His training will become more rigorous. The new professionalism will not bring automatic success. New tools used without proper understanding can lead to disaster. In the hands of those who use them correctly they become a new competitive advantage toward business success.

Management science does not mean automatic management. It means a platform from which to reach further by the exercise of managerial intelligence and judgment.

The following chapters treat the way in which the components of the business enterprise interact on one another. The approach is one of building experimental models of companies and industries to determine the influences of organization and policy.

Although the methods are quantitative and result in a mathematical representation of the business system, the philosophy, objectives, and approaches are more akin to those of the practicing manager than to the management science specialist of recent years. From verbal description, experience, field observations, and such data as are available will be evolved the corresponding mathematical models. I hope to show, however, that mathematical notation can be kept close to the vocabulary of business; that each variable and constant in an equation has individual meaning to the practicing manager; that the successful manager of the future can understand, in fact will help originate, the relationships described by the equations; that the required mathematics is within the reach of almost anyone who can successfully manage a modern corporation.

If a management science useful to the top manager is to be achieved, it must coincide with Region A of Figure I-2, bringing a tool that can enter the circles of responsible line management. Staff specialization in management science that is apart from the responsible manager will in time give way to a professional level of line management, where the manager understands his tools well enough to know their strengths and their limitations. This imposes new demands on the manager and his education. The answers obtained from using any scientific methods are only as pertinent as the questions that are asked. The validity of the results are only as good as the assumptions on which the study is based. The manager must become able to take the responsibility for the questions asked, for the assumptions inserted, and for the interpretation of results.

I

THE MANAGERIAL VIEWPOINT

Part I is a nontechnical introduction to the nature and objectives of industrial dynamics. The three chapters are recommended to all readers.

● Chapter 1 defines and describes what we mean by industrial dynamics. It mentions the steps in an industrial dynamics study (which are further developed in Chapter 3), states several premises underlying industrial dynamics (which are illustrated and supported throughout the book), and reviews the four lines of historical development that make the present book possible.

● Chapter 2 provides an illustration, in managerial terms, of the way the dynamic responses of an industrial system can be subjected to experimental study.

● Chapter 3 discusses the concept of a management laboratory for enterprise design and suggests some changes that can be expected in management practice and education.

CHAPTER 1

Industrial Dynamics

Industrial dynamics is the investigation of the information-feedback character of industrial systems and the use of models for the design of improved organizational form and guiding policy. Industrial dynamics grows out of four lines of earlier development — information-feedback theory, automatizing military tactical decision making, experimental design of complex systems by use of models, and digital computers for low-cost computation. Each of these is reviewed as part of the background from which this book has evolved.

THIS book treats the time-varying (dynamic) behavior of industrial organizations — industrial dynamics.

Industrial dynamics is the study of the information-feedback characteristics of industrial activity to show how organizational structure, amplification (in policies), and time delays (in decisions and actions) interact to influence the success of the enterprise. It treats the interactions between the flows of information, money, orders, materials, personnel, and capital equipment in a company, an industry, or a national economy.

Industrial dynamics provides a single framework for integrating the functional areas of management — marketing, production, accounting, research and development, and capital investment. It is a quantitative and experimental approach for relating organizational structure and corporate policy to industrial growth and stability.

Industrial dynamics should provide a basis for the design of more effective industrial and economic systems. An industrial dynamics approach to enterprise design progresses through several steps:

- Identify a problem.
- Isolate the factors that appear to interact to create the observed symptoms.
- Trace the cause-and-effect information-feedback loops that link decisions to action to resulting information changes and to new decisions.
- Formulate acceptable formal decision policies that describe how decisions result from the available information streams.
- Construct a mathematical model of the decision policies, information sources, and interactions of the system components.
- Generate the behavior through time of the system as described by the model (usually with a digital computer to execute the lengthy calculations).
- Compare results against all pertinent available knowledge about the actual system.
- Revise the model until it is acceptable as a representation of the actual system.
- Redesign, within the model, the organizational relationships and policies which can be altered in the actual system to find the changes which improve system behavior.
- Alter the real system in the directions that model experimentation has shown will lead to improved performance.

Such an approach is based on several premises:

- Decisions in management and economics take place in a framework that belongs to the general class known as information-feedback systems.

- Our intuitive judgment is unreliable about how these systems will change with time, even when we have good knowledge of the individual parts of the system.

- Model experimentation is now possible to fill the gap where our judgment and knowledge are weakest — by showing the way in which the known separate system parts can interact to produce unexpected and troublesome over-all system results.

- Enough information is available for this experimental model-building approach without great expense and delay in further data gathering.

- The "mechanistic" view of decision making implied by such model experiments is true enough so that the main structure of controlling policies and decision streams of an organization can be represented.

- Our industrial systems are constructed internally in such a way that they create for themselves many of the troubles that are often attributed to outside and independent causes.

- Policy and structure changes are feasible that will produce substantial improvement in industrial and economic behavior; and system performance is often so far from what it can be that initial system design changes can improve all factors of interest without a compromise that causes losses in one area in exchange for gains in another.

Why are these premises now a sound basis for developing a better understanding of the behavior of industrial systems? The approach discussed in this book would not have been feasible even a decade ago. The need has long existed for better insights into the problems of management and economics, but the cornerstones for an adequate approach have only recently been laid.

Four foundations on which to construct an improved understanding of the dynamics of social organizations have been built in the United States since 1940, primarily as a by-product of military systems research. These four are:

- The theory of information-feedback systems

- A knowledge of decision-making processes

- The experimental model approach to complex systems

- The digital computer as a means to simulate realistic mathematical models.

Because of the important part these four play in making industrial dynamics possible, each will be reviewed in some detail.

1.1 Information-Feedback Control Theory

The first and most important foundation for industrial dynamics is the concept of servomechanisms (or information-feedback systems) as evolved during and after World War II.[1] Until recently we have been insufficiently aware of the effect of time delays, amplification, and structure on the dynamic behavior of a system. We are coming to realize that the interactions between system components can be more important than the components themselves.

The concepts of information-feedback systems will become a principal basis for an underlying structure to integrate the separate facets of the management process. What is an information-feedback system? I should like to give a broad definition:

An information-feedback system exists whenever the environment leads to a decision that results in action which affects the environment and thereby influences future decisions.

This is a definition that encompasses every

[1] In References 2 and 3 are listed early texts recommended to the beginner as simplified treatments of the mathematical theory of servomechanisms. Brown and Campbell (Reference 4) give a more complete treatment of transient analysis. In these three books will be found references to the early papers of the field. The serious worker in dynamics of industrial systems should acquire as background an understanding of linear servomechanisms analysis. The methods will find little use as such, but they help provide the intuitive "feel" about the way amplification and delays combine to create total system behavior. The books in Reference 5 carry the treatment of information-feedback systems into the effects of noise and intermittent data, give an excellent picture of the scope of linear information-feedback system theory, and contain generous references.

conscious and subconscious decision made by people. It also includes those mechanical decisions made by devices called servomechanisms.

Systems of information-feedback control are fundamental to all life and human endeavor, from the slow pace of biological evolution to the launching of the latest space satellite. To illustrate:

- A thermostat receives temperature information and decides to start the furnace; this raises the temperature, and the furnace is stopped.

- A person senses that he may fall, corrects his balance, and thereby is able to stand erect.

- In business, orders and inventory levels lead to manufacturing decisions that fill orders, correct inventories, and yield new manufacturing decisions.

- A profitable industry attracts competitors until the profit margin is reduced to equilibrium with other economic forces, and competitors cease to enter the field.

- The competitive need for a new product leads to research and development expenditure that produces technological change.

All of these are information-feedback control loops. The regenerative process is continuous, and new results lead to new decisions which keep the system in continuous motion. Such systems are not necessarily well behaved. In fact, a complex information-feedback system designed by happenstance or in accordance with what may be intuitively obvious will usually be unstable or ineffective.

The study of feedback systems deals with the way information is used for the purpose of control. It helps us to understand how the amount of corrective action and the time delays in interconnected components can lead to unstable fluctuation. Driving an automobile provides a good example:

The information and control loop extends from steering wheel, to automobile, to street, to eye, to hand, and back to steering wheel. We accept this complex system thoughtlessly. Consider the effects of making only slight changes in system structure and time delays. Suppose the driver were blindfolded and drove only by instructions from his front-seat companion. The resulting few seconds of increased information delay and the slightly greater information distortion caused by inserting voice and ear between the observing eye and the driver's brain would cause erratic driving.

Still worse, if the blindfolded driver could get instructions only on where he had been from a companion who could see only through the rear window, his driving would be even more erratic. Yet this is the situation in business. Top executives do not see the salesmen calling on customers, do not see the prospective buyers watching a television commercial. They do not attend the board meetings of competitors. They do not have a clear view of the road ahead. The only thing they can tell, and that with only partial certainty, is what happened in the past.

In an information-feedback system it is always the presently available information about the past which is being used as a basis for deciding future action.

Everything we do as an individual, as an industry, or as a society is done in the context of an information-feedback system. The definition of such a system is so all-inclusive as to seem meaningless at first. Yet we are only now becoming sufficiently aware of the tremendous significance of information-feedback system parameters in creating the behavior of these systems.

Information-feedback systems, whether they be mechanical, biological, or social, owe their behavior to three characteristics — structure, delays, and amplification. The structure of a system tells how the parts are related to one another. Delays always exist in the availability of information, in making decisions based on the information, and in taking action on the decisions. Amplification usually exists throughout such systems, especially in the decision policies of our industrial and social systems. Amplification is manifested when an action is more force-

15

ful than might at first seem to be implied by the information inputs to the governing decisions. We are only beginning to realize the way in which structure, time lags, and amplification combine to determine behavior in our social systems.

Why has the fundamental nature and the importance of information-feedback systems escaped notice until the last three decades? I think it is because of the peculiar classes into which these systems fell before 1940. On the one hand, we had the biological information-feedback systems that regulate body temperature, muscular coordination, and other functions. These systems had been so ideally perfected for their purposes, and at the same time we had so completely accepted their shortcomings, that the systems and their information-feedback character went unnoticed. On the other hand, social, economic, and industrial systems have evolved in recent centuries but on so large a scale compared with the individual that the fundamental information-feedback characteristics were most difficult to discern. Furthermore, hosts of other explanations have been advanced for the behavior that arises directly from the information-feedback character of these systems. Explanations have been in terms of the superficial specific manifestations of the particular problem rather than in the more fundamental terms of generalized closed-loop systems.

The theory and concepts of information-feedback systems have developed recently as a result of trying to build simple self-regulating control systems. As control devices developed beyond the early steam-engine governor, greater precision was needed. The systems to be controlled became more complex. The dynamic characteristics and system difficulties became obvious and on a scale small enough for study. Strong commercial and military incentives encouraged attempts to master the theory of information-feedback-system design. Initial simple problems lay within the reach of the available mathematics. The problems, the needs, and the tools were adequately matched for two decades of progress in the dynamics of physical systems.

Our social systems are a great deal more complex than the information-feedback systems that have already been mastered in engineering. Are we ready to tackle them?

Our knowledge of information-feedback systems has been growing in that exponential manner so characteristic of the early phase of any area of human knowledge. The ability to deal with information-feedback systems seems to have been progressing by about a factor of 10 per decade.

In the late 1930's the scientific papers in the field dealt with the dynamic characteristics of very simple control systems described by linear differential equations of two variables. By the early 1940's the field had developed into the concepts of Laplace transforms, frequency response, and vector diagrams.

But as usual, the forefront of mathematics was unable to cope with the problems of major engineering interest. Military necessity exerted pressure. Engineers did not long linger waiting for analytical solutions to information-feedback-system behavior. Linear and nonlinear mathematical models were constructed for solution on analog computing machines. By 1945 systems of 20 variables were easy to handle — a tenfold increase over 1935.

By the end of the second decade, in 1955, new methods and new areas were again being pioneered. The digital computer had appeared, opening the way to the simulation of systems far beyond the capability of analog machines. With the new tools, attention began to center on the information-feedback characteristics of military combat systems incorporating both equipment and people. Systems of 200 variables could be feasibly studied.

The pace has not slackened. We are now entering the 1965 era with another factor of 10 within reach. Models of 2,000 variables, with no restrictions on representing nonlinear phenomena, put us within a vast area of important managerial and economic questions.

1.2 Decision-Making Processes

The second foundation for industrial dynamics is the better understanding of decision making achieved in the 1950's during the automatizing of military tactical operations.

Historically, military necessity has often led not only to new devices like aircraft and digital computers but also to new organizational forms and to a new understanding of social forces. These developments have then been adapted to civilian usage.

Such innovations have appeared in the military command (or management) function. As the pace of warfare has quickened, there has of necessity been a shift of emphasis from the tactical decision (moment-by-moment direction of the battle) to strategic planning (preparing for possible eventualities, establishing policy, and determining in advance how tactical decisions will be made). The battle commander can no longer plot the course of his enemy on a chart and personally calculate the aiming point. In fact, with a ballistic missile he would have no time even to select his defensive weapon.[2]

During World War II, fire-control prediction decisions were made automatically by machine, but before 1950 there was almost no acceptance of automatic threat evaluation, weapon selection, friend and foe identification, alerting of forces, or target assignment. In a mere ten years these automatic decisions were pioneered, accepted, and put into practice. In so doing, it was necessary to interpret the "tactical judgment and experience" of military decision making into formal rules and procedures. This change was forced because the pace of modern military operations exceeded the ability of a hu-

man organization to respond. A decade of time and thousands of people were involved in this interpretation of military decision processes and automatizing the operational policies that are the basis for tactical military decision making. It has been amply demonstrated that carefully selected formal rules can lead to short-term tactical decisions that excel those made by human judgment under the pressure of time, or with men having insufficient experience and practice, or in the rigidity of large organizations.

Men who started to formalize decision-making policy in 1950 in the environment of "you can't make a machine substitute for my military training and command experience" have in ten years seen the same critics accept as better and as commonplace the automatic execution of front-line military "judgment." The resulting body of practical experience in determining the basis for decisions and the content of "judgment" is now becoming available to the study of management systems. Many men from military systems research are now moving into the study of industrial and economic systems.

As in military decisions, we shall see that there is an orderly basis that prescribes much of our present managerial decision making. Decisions are not entirely "free will" but are strongly conditioned by the environment. This being true, we can set down the policies governing such decisions and determine how the policies are affecting industrial and economic behavior.

1.3. Experimental Approach to System Analysis

The third foundation for industrial dynamics is the experimental approach to understanding system behavior.

Mathematical analysis is not powerful enough to yield general analytical solutions to situations as complex as are encountered in business. The alternative is the experimental approach.

[2] There is little useful literature on the subject of automatizing of military decisions, not so much because of military security classification as because the changes in attitudes toward military decision making have received scant philosophical documentation.

A mathematical model of the industrial system is constructed. Such a mathematical model is a detailed description that tells how the conditions at one point in time lead to subsequent conditions at later points in time. The behavior of the model is observed, and experiments are conducted to answer specific questions about the system that is represented by the model.

"Simulation" is a name often applied to this process of conducting experiments on a model instead of attempting the experiments with the real system. During the 1950's simulation was extensively developed in the design of air defense systems[3] and in engineering work. For example:

In planning the development of a river basin, numbers in a digital computer represent water volumes, flow rates, electric demand, and rainfall. A few seconds of computer time can represent a whole day of actual system operation. Dams can be located and designed for an adequate compromise between the conflicting demands of power generation, irrigation, navigation, and flood control.

Likewise, simple simulation studies of parts of a business have been common in the last few years and are reported in the operations research literature. Simulation techniques have now reached a state of development where they can be applied to the top-management problems of industrial organizations.

In business, simulation means setting up in a digital computer the conditions that describe company operations. On the basis of the descriptions and assumptions about the company, the computer then generates the resulting time charts of information concerning finance, manpower, product movement, and so on. Different management policies and market assumptions

can be tested to determine their effects on company success.

Instead of going from the general analytical solution to the particular special case, we have come to appreciate the great utility, if not the mathematical elegance, of the empirical approach. In this we study a number of particular situations, and from these we generalize as far as we dare.

Use of simulation methods will not require great mathematical ability. To be sure, details of setting up a model need to be monitored by experts because there are special skills required and pitfalls to be avoided. However, the job of choosing the situations to be explored, judging the assumptions, and interpreting the results is within the ability of the type of men we now see in management schools and executive development programs.

1.4 Digital Computers

The fourth foundation for industrial dynamics progress is the electronic digital computer that became generally available between 1955 and 1960. Without it, the vast amount of work to obtain specific solutions to the characteristics of complex systems would be prohibitively expensive. In the last fifteen years the cost of arithmetic computation has fallen by a factor of 10,000 or more in those areas where digital computers can be used in their most efficient modes of operation. The simulation of information-feedback models of industrial behavior is such an area of high efficiency. A cost reduction factor of 10,000 or 100,000 creates a totally different research environment than existed even a decade ago.

After World War II the advent of computing machines brought the feasibility of dealing with more complex systems. Analog-type computers, as used in electrical-power-system network analyzers and in differential equation analyzers, had been developed from 1930 to 1950. At first, attempts were made to use analog computing devices for the study of eco-

[3] Simulation studies of military systems have received tens of thousands of man-years of effort in the last decade, both to study system behavior and to provide ways to train men in their duties in the system.

18

nomic systems.[4] However, these analog computers proved inadequate to cope with problems of practical interest. They do not readily deal with nonlinear systems. The upper capacity limit of such machines barely overlaps the minimum size and complexity that must interest us in economic and corporate problems.

The appearance of high-speed electronic digital computers has removed the practical computational barrier. The technical performance of electronic computers increased by a factor of nearly 10 per year over the decade of the 1950's; in almost every year there was a tenfold increase in speed, memory capacity, or reliability. Over-all, this was a technological change greater than that effected in going from chemical to atomic explosives. Society cannot absorb so big a change in a mere ten years. We have a tremendous untapped backlog of potential devices and applications. It is now to be expected that machine progress will stay ahead of conceptual progress in industrial and economic dynamics. Computing machines are now so widely available, and the cost of computation and machine programming is so low relative to other costs, that the former difficulties in activating a simulation model need no longer determine our rate of progress in understanding system dynamics.

[4] A book by Tustin (Reference 6) discusses the similarities between economic systems and electromechanical servomechanisms. In broad concept, the book anticipates present simulation studies; in detail, it is out of date in referring only to analog computers and classical types of simple models of economic behavior. It is recommended, however, as background on the information-feedback-system view of economic behavior.

CHAPTER 2

An Industrial System

Before starting the discussion in Part II of the philosophy and methodology of industrial dynamics, it seems best to give the reader a quick exposure to the kinds of results toward which we are working. In this chapter a simple model of a distribution system will be used to show how the form and the policies of an organization can give rise to characteristic and undesirable modes of behavior. Indication of ways to improve the system will appear, but the process of system redesign will be illustrated in another context in Chapter 18. We begin with a verbal description of the system. The details of actual model construction are postponed until Chapter 15. But using that model, this chapter proceeds to illustrate the kinds of questions that can be explored. It is the intent of this chapter to provide a motivation for further study of the subject. We shall here examine:

- *How small changes in retail sales can lead to large swings in factory production*
- *How reducing clerical delays may fail to improve management decisions significantly*
- *How a factory manager may find himself unable to fill orders although at all times able to produce more goods than are being sold to consumers*
- *How an advertising policy can have a magnifying effect on production variations*

THE general description of industrial dynamics in the first chapter should become more meaningful in the context of a simple example. How do the concepts of information-feedback systems apply to specific business situations? Or, stated differently, how do the delays and amplifications in the circuital flow of information in a company affect its operations? How can we use a model of such a system to discover the way its components affect its over-all behavior?

2.1 The Approach

The first step in a system study is to identify clearly the problem to be explored and the questions to be answered. This initial example must be kept simple. It will be wise, for the sake of clear explanation, to start with a very limited subsystem of an entire company. To maintain this initial simplicity, the questions we ask must deal with problems whose causes lie within a restricted part of the industrial enterprise. Later we can range more broadly over the management field.

The central core of many industrial companies is the process of production and distribution. A recurring problem is to match the production rate to the rate of final consumer sales. It is well known that factory production rate often fluctuates more widely than does the actual consumer purchase rate. It has often been observed that a distribution system of cascaded inventories and ordering procedures seems to amplify small disturbances that occur at the retail level. This example will deal with the structure and policies within a multi-stage distribution system. How does the system create amplification of small retail sales changes? What changes in management policies will affect internally generated oscillations? What will such a system do in response to various assumed sales rate changes at retail?

We shall find that our questions can be explored using only the flows of information, orders, and materials. Of the six flow systems previously enumerated, we shall postpone consideration of capital equipment, money, and personnel.

Even such a restricted system will still be interesting and meaningful. It will contain many of the elements that give rise to baffling behavior in real systems. If we examine the basic internal behavior of the distribution system, even with independently specified customer orders and no interaction between the company and the market, we shall see that typical manufacturing and distribution practices can generate the types of business disturbances which are often blamed on conditions outside the company. Random, meaningless sales fluctuations can be converted into annual, seasonal production cycles. Advertising and price discount policies of an industry can create two- and three-year sales cycles. Factory capacity, even though always exceeding retail sales, can seem to fall short of meeting demand, with the result that production capacity is overexpanded.

2.2 Needed Information

To begin the study of our production-distribution example, we need to know three kinds of information about the system: its organizational structure, the delays in decisions and actions, and the policies governing purchases and inventories.

Organizational Structure. Figure 2-1 shows a typical organizational structure for the production and distribution functions in a hard goods industry like household appliances. The bottom box represents the retail level. Next above are the distributors, and at the left the factory and factory warehouse. The circled lines show the upward flow of orders for goods. The solid lines show the shipment of goods flowing downward. Note that three levels of inventory exist — at factory, distributor, and retailer.

Delays in Decisions and Actions. To be able to determine the dynamic characteristics of this system, we must also know the delays in the flow of orders and goods. The time delays are shown on the diagram in weeks and are reasonable values for a consumer–durable product line.

Delivery of goods to the customer averages a week after the customer places an order.

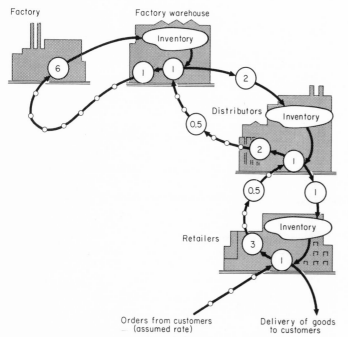

Figure 2-1 Organization of production-distribution system.

At the retail level the accounting and purchasing delays average three weeks between the time of a sale and the time when that sale is reflected in an order sent out to obtain a replacement. Mailing delay for the order is half a week. The distributor takes a week to process the order, and shipment of goods to the retailer takes another week. Similar delays exist between the distributor and the factory warehouse. The factory lead time averages six weeks between a decision to change production rate and the time that factory output approaches the new level.

Policy on Orders and Inventories. To complete the initial description of the example, we need to know the policies that govern the placing of orders and the maintaining of inventory at each distribution level. We shall consider three principal components of orders: (*a*) orders to replace goods sold, (*b*) orders to adjust inventories upward or downward as the level of business activity changes, and (*c*) orders to fill the supply pipelines with in-process orders and shipments. Orders are treated in the following ways:

- After a sales analysis and purchasing delay (three, two, and one weeks for the three levels), orders to the next higher level of the system include replacement for the actual sales made by the ordering level.

- After sufficient time for averaging out short-term sales fluctuations (eight weeks), a gradual upward or downward adjustment is made in inventories as the rate of sales increases or decreases.

- One component of orders in process (orders in the mail, unfilled orders at the supplier, and goods in transit) is necessarily proportional to the average level of business activity and to the length of time required to fill an order. Both an increased sales volume and an increased delivery lead time necessarily result in increased total orders in the supply pipeline. These "pipeline" orders are unavoidable. They are part of the "basic physics" of the system structure. If

not ordered explicitly for the purpose of pipeline filling (and often they will not be), then the pipeline demands will come from a depletion of inventories, and the pipeline orders will be placed unknowingly in the name of inventory adjustment.

The ordering rate will also depend on some assumption about future sales. Prediction methods that amount to extending forward (extrapolating) the present sales trend will in general produce a more unstable and fluctuating system. For our example, however, we shall use the conservative practice of basing the ordering rate on the assumption that sales are most likely to continue at their present level.

2.3   Simulation Method

Before we can determine the implications of the preceding structure, delays, and policies on system behavior, all of the above rather general descriptions of the system must be expressed in explicit quantitative form.

The straightforward but detailed formulation of equations to represent the above relationships will be delayed until Part III.[1] At this point, however, it seems best to assume the availability of a model of the system, and to discuss how such a mathematical model can be used to learn more about the characteristics of the system.

Once an explicit, mathematical description of the system has been established, the next step is to determine how the system as a whole behaves. To do so, we might use some arbitrary pattern of consumer purchases as an input and observe the resulting inventory and production changes. The effect on the production-distribution system can be obtained by "simulation" methods. Simulation consists of tracing through, step by step, the actual flows of orders, goods, and information, and observing the series of new decisions that take place.

---

[1] Chapter 15 develops and explains the seventy-three equations that form the detailed description of the above distribution system.

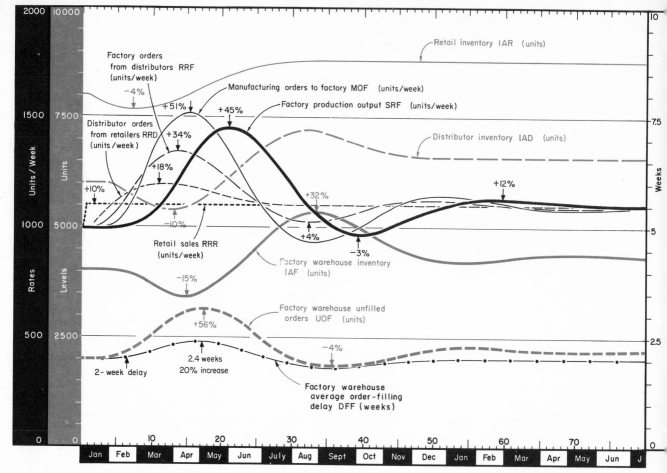

Figure 2-2 Response of production-distribution system to a sudden 10% increase in retail sales.

As an example of simulation, our production-distribution system might be acted out by a group of men around a table, one to represent retailers, another the postal service, another transportation, another the factory, and so on. Five minutes might represent a week,[2] and in each time interval the proper purchase orders and deliveries would be made according to the policies which were described above and which are represented more precisely by the equations in Chapter 15. Alternatively, the whole exercise can be done by one person in tabular form on paper. Better still, the entire sequence can be programmed on a digital computer.[3] Digital-computer simulation was used to obtain the following results.

2.4 System Experiments

It now becomes possible to test how the above production-distribution system would respond to any arbitrarily specified input. We might think that we should choose some actual

[2] For this kind of laborious, manual simulation, larger time steps are usually used than are appropriate if a good representation of system detail is to be generated, as can be done on a computer. The following computer-simulated examples were computed each twentieth of a week. See Appendix D on solution interval.

[3] A description of how these studies are handled with the "DYNAMO" compiler on a digital computer will be found in Appendix A.

sales history as a trial input. This can be done later, but any such complex pattern will be unduly confusing at this introductory stage. It is more informative to take simple, "pure," sequences for preliminary study purposes. Such simple patterns include a *step* (a sudden change from one constant level to another), or a *sinusoid* (a smoothly fluctuating disturbance). A more complex artificial pattern would be some random-noise curve of specified statistical characteristics.

Step Input. A very informative input is the simple step. It provides a single disturbance at the input and allows us to observe how the internal components of the system respond thereafter. Figure 2-2 shows the result of a 10% increase in retail sales introduced in January. The resulting fluctuations are shown in order rates, factory output, factory warehouse inventory, and unfilled orders. (Retail orders are here given as an input that is independent of what happens in the production-distribution system itself. Retail orders are, of course, not actually independent since they are affected by availability of the product and by advertising, a point to be discussed later.)

Because of accounting, purchasing, and mailing delays, the increase in distributors' orders from retailers lags about a month in reaching the 10% level. It is important to note, however, that the rise does not stop at 10%. Instead it reaches a peak of 18% at the 11th week because of the new orders that have been added at the retail level (*a*) to increase inventories somewhat and (*b*) to raise the level of orders and goods in transit in the supply pipeline by 10% to correspond to the 10% increase in sales rate. These inventory and pipeline increments occur as "transient" or nonrepeating additions to the order rate, and when they have been satisfied, the retailers' orders to the distributors drop back to the enduring 10% increase.

The factory warehouse orders from distributors show an even greater swing. This is be-cause the incoming-order level at the distributors persists above retail sales for more than four months and is readily mistaken for an enduring increase in business volume. The distributors' orders to the factory, therefore, include not only the 18% increase in orders they themselves receive but also a corresponding increase for distributor inventories and for orders and goods in transit between distributor and factory. As a result, orders at the factory warehouse reach, at the 14th week, a peak of 34% above the previous December.

Let us turn now to manufacturing orders. They are placed on the basis of the increasing factory warehouse orders and the falling warehouse inventory, which drops 15%. Manufacturing orders rise 51% at the 15th week. As a result, the factory output, delayed by a factory lead time of 6 weeks, reaches a peak in the 21st week, an amount 45% above the previous December. While retail sales are still at 10% above December, the factory increase has become over four times as great.

It is important to note that all of these effects are reversible. As retailers satisfy their inventory requirements, they decrease their order rate. The distributors find they have built up an order rate, an inventory, and a supply-line rate in excess of needs. They take the excess out of current orders to the factory, so that at the 32nd week this order rate is 6% *below* retail sales and only 4% above the previous December. In September and October at the 39th week the factory output drops to 3% below the previous December and 13% below current retail sales.

As a direct result of the typical organizational form shown in Figure 2-1 and of the customary inventory and ordering policies described, over a year is required before all ordering and manufacturing rates stabilize to their proper levels corresponding to the 10% retail sales increase. Ironically, this increase was minor compared to the fluctuations in company operations.

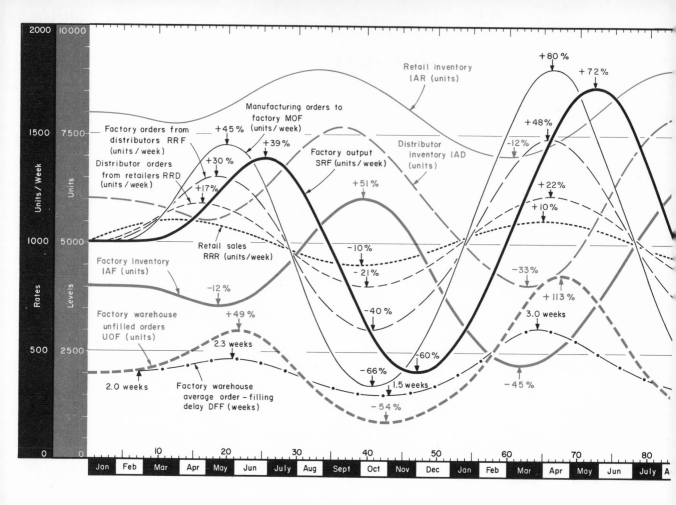

Periodic Variations in Retail Sales. The problem of *unexpected* periodic variations in retail sales will now be examined. Imagine first that our system has operated in the past with constant sales, and then sales rise and fall gradually over a one-year interval.

Figure 2-3 shows the way in which such a variation in retail sales would be accentuated as orders travel toward the factory. Retail sales have been stable at 1,000 per week in the past; therefore, past history has led to no planning for seasonal behavior. In January, sales start rising toward a 10% increase at the end of March. They fall to 10% below "normal" by the end of September and return to normal by the end of December.

The initial upswing in orders and factory output is much like that in Figure 2-2 except that the initial peaks are lower and later. However, at the time the system would naturally rebound from its overproduction state, it is given an additional downward shove by the declining retail sales, which are amplified for the same reasons as previously discussed. As a combined result, factory orders from distributors reach a low of 40% below normal in October, and factory output falls to 60% below normal in November.

In the following year, we see factory production continuing to fluctuate between peaks and valleys that are about 72% above and 60% below normal. Inventories vary over the ranges shown in Table 2-1.

We see here how the periodic disturbance is

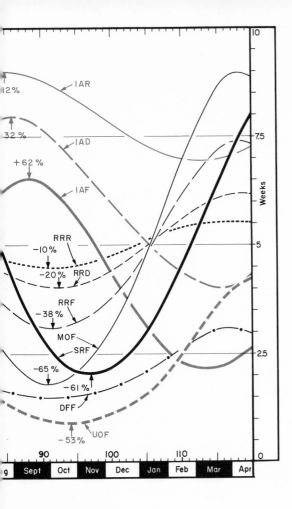

accentuated as it moves from retail toward the factory.

Table 2-1   Ranges of Inventory Variation

Location	Maximum (%)	Minimum (%)
Factory	+62	−45
Distributor	+32	−33
Retail	+12	−12

Random Retail Sales. Note that the curves of Figures 2-2 and 2-3 are smooth and free of the short-term random fluctuations which are seen in most charts of actual industrial orders.

In simulating the performance of an industrial system, it is of course quite impossible to account for all the minor factors that may affect

operations. These residual effects can be represented by "noise" or random disturbances introduced at the decision-making points of the system. Suppose we wish to study the behavior of this production-distribution system, not with a perfectly smooth customer purchasing pattern but with a pattern that fluctuates week by week.

An oscillatory system of the kind shown in Figures 2-2 and 2-3 will respond to the random external disturbances by fluctuating in a manner reflecting the characteristics of the system itself, rather than in a pattern that is easily traceable to the external disturbance.

Even if there is no average change in retail sales level (as in Figure 2-2) and no regular periodic change (as in Figure 2-3), the system will, if it has tendencies toward oscillation, convert random events into upswings and downswings in orders and production. In Figure 2-2 we saw that, in response to a sudden shock, production and inventories tend to fluctuate with peaks some eight or nine months apart. Anyone familiar with the characteristics of information-feedback systems would expect to see similar intervals of fluctuation appearing in response to random input disturbances.

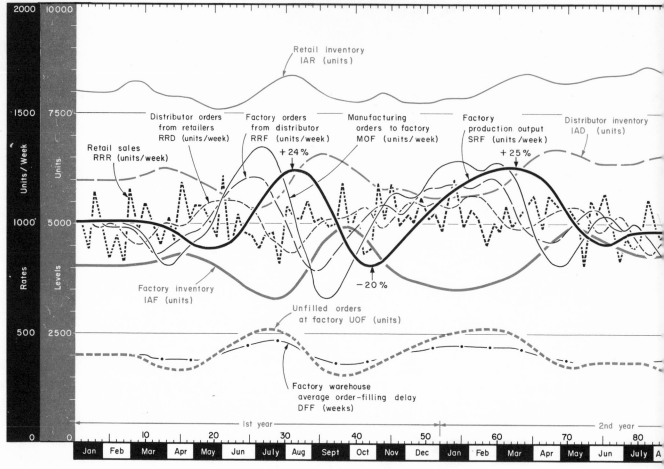

Figure 2-4 Effect of random deviation at retail sales.

Figure 2-4 shows more than three years of week-by-week consumer purchases containing random deviations from the average but no meaningful long-term changes in sales rate. Also shown are distributor orders from retailers and the conditions at the factory.

The system, by virtue of its policies, organization, and delays, tends to amplify those retail sales changes to which the system is sensitive. Conversely, other frequencies of disturbance will be strongly suppressed. Assumed retail sales fluctuate from week to week over a range having a normal deviation of 10% of average sales. Factory output rises and falls over periods of several months, with amplitudes of 25% or more away from the average.

We can see evidence of a 5- to 10-month interval between successive peaks in factory production. These are not regular or predictable but arise from the random input disturbance which is selectively amplified by the characteristic response of the system itself.

At the end of each of the first two years there is some indication of an upswing of incoming factory orders. These might easily lead the un-skilled observer to conclude that a seasonal sales pattern is present. Actually, any tendency here toward annual repetition is occurring only because this production-distribution system has internal inclinations toward oscillation with a period of fluctuation of about a year. This is not an uncommon circumstance because the

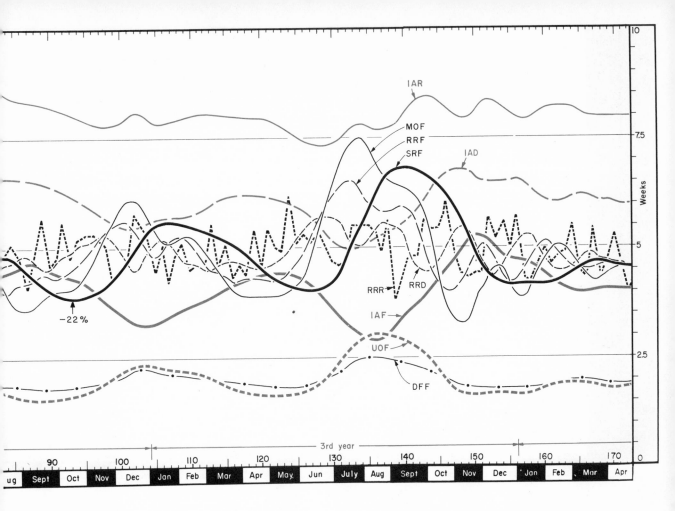

usual combinations of human decision-making reaction times, communications and transportation delays, and factory lead times will often yield systems that tend to show periodicity in the range of six months to two years. Under these circumstances, truly annual influences of even minor magnitude can serve to synchronize the system fluctuations to the calendar. Such "pulling" of system fluctuation into step with the calendar could be caused by reduced ordering or production during the vacation season, annual budgeting for advertising and capital equipment purchase, annual model design changes, and of course by minor but real seasonality in the actual demand for and use of the product.

Management policies adopted in recognition of such apparent seasonal disturbances may often accentuate rather than alleviate the difficulties. We find many company situations in which such an erroneous conclusion about seasonal sales had led to the establishment of employment, inventory, and advertising policies that in succeeding years *caused* a seasonal manufacturing pattern and thereby confirmed the original error.[4] The likelihood of self-generated seasonal sales patterns must be carefully considered in the design of management policies.

[4] This is discussed further in Appendix N.

29

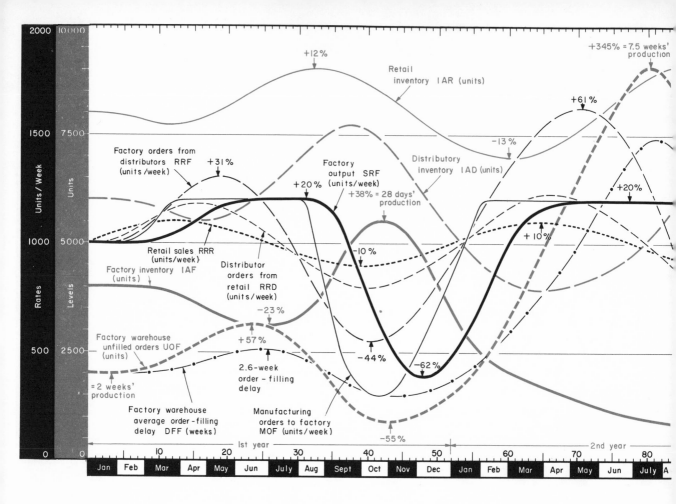

Limited Factory Capacity. So far we have considered only simple situations. Additional realities of company operation can be added as needed. In the preceding exhibits we assumed that the factory was able to produce at whatever level was desired. In a more realistic situation, with limited factory capacity, some surprising new effects develop.

Figure 2-5 shows the effect of a maximum factory capacity limit that is 20% above the average sales level. As before, the system is completely stabilized at the beginning of the first year; then it is assumed that retail sales rise and fall 10% during each year.

Retail sales never reach production capacity. Yet because of the inventory and pipeline effects, distributor orders to the factory *do* exceed its capacity. Furthermore, as factory deliveries become slower, distributors begin to order further in advance of needs, and still more orders are put into the system. As a result the factory operates at full capacity for three months during the first year.

Then inventory demands are filled at a time that coincides with falling retail sales. In the last half of the year, falling retail sales and inventory liquidation combine with the reduction of in-process orders which results from improving delivery. In the third quarter we see a combination of rapidly falling factory orders, a swiftly declining backlog of unfilled orders, and a suddenly rising factory inventory. The natural result is to curtail production from its maximum level, here falling to 62% *below* normal.

Unlike the first year, the system now enters

30

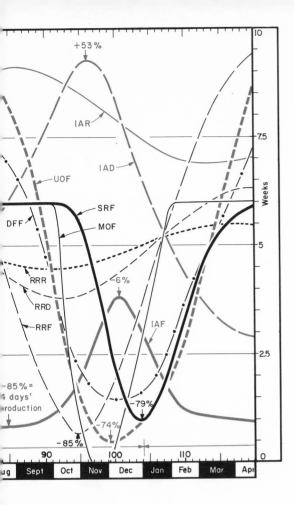

Figure 2-5 Effect of fluctuating retail sales on factory with manufacturing capacity limited to 20% above average sales level.

only 13% and that actually there is little likelihood of losing retail sales. Retail inventories happen to reach their peak when factory inventory is at a minimum.

Note that in the 81st week shipments equal incoming orders; inventory is constant. This is therefore the point at which unfilled orders reach a peak and start down. Since the factory has produced in excess of sales for 8 months and because the long-run average production must necessarily equal retail sales, manufacturing must be sharply curtailed in the remaining 4 months. Factory output is almost suspended and drops to 79% below normal and produces at a level below average retail sales for a period of 17 weeks.

In succeeding years, unless operating policies are changed, the system will act much as it did in the second year.

We see in Figure 2-5 an occurrence that may lead a company to overexpand production capacity. For the entire second year, inventory was below the desired level; for 9 months from November until August the order backlog was rising. Viewed from the circumstances at the factory, this condition might easily lead management to undertake expansion plans — even though retail sales never reached production capacity.

Explaining the behavior of an industrial organization is only the first step. After adequate representation of the current operations of a particular company or industry, the next step is to determine ways to improve management control for company success.

the second year in a state of imbalance with production below average sales, orders rising, and inventory falling. The result in the second year is an accentuation of the first-year performance. Factory orders from distributors rise to 61% above normal and show a broadened peak that is sustained by the distributors' tendency to order ahead when deliveries become slow. Factory production runs at full capacity for 6 months to meet demand from the distributors. Unfilled orders at the factory rise to 345% above normal and represent a backlog which is nearly 6 weeks of production higher than normal. In the meantime, factory inventory has been depleted from the normal 4 weeks' production to less than a week.

In spite of low factory inventory and slow deliveries, note that retail inventories deviate

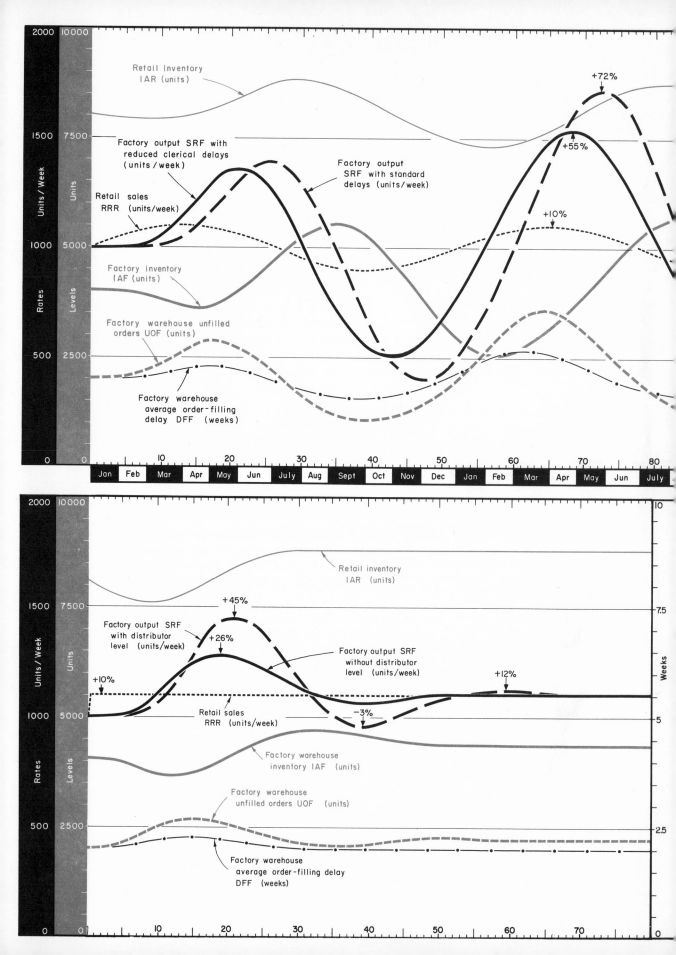

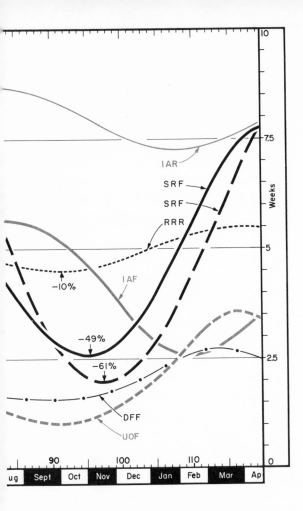

Figure 2-6 labels visible on chart: IAR, SRF, SRF, RRR, IAF, DFF, UOF, −10%, −49%, −61%

← **Figure 2-7** Distributor level eliminated.

Figure 2-6 Effect of reducing clerical delays.

analysis and purchasing time) have been reduced to one-third their previous values. But the effect is only slight improvement: the production peaks are reduced from six times to five times the retail sales changes. Because clerical delays are such a small factor in the operation of the system as a whole, no amount of speed-up can radically change over-all performance. In fact, speeding the flow from previously used information sources is often not the best way to improve operations; entirely different bases for reaching decisions, using different and perhaps already available information, will often produce greater improvement at lower cost.

Eliminating Distributor Level. What about changes in ordering procedures? One possibility would be to place retailers' orders directly with the factory. The result is shown in Figure 2-7. When the inventory accumulation, fluctuation of in-process orders, and delays of one of the three stages are removed, production overshoot in response to a 10% step input change is 26% instead of the 45% in Figure 2-2, which is copied for comparison.

The foregoing raises an interesting question about industries having more than three distribution levels. For example, in the textile industry, which shows marked instability, there are often four or five distribution levels from yarn manufacture to the final consumer. May not a good deal of instability be caused by the existence of so many levels?

Faster Order Handling. To improve industrial stability and to make the company less vulnerable to external influences, executives may consider various changes in company operation. For instance, they might wish to explore the effect of reduced clerical and data-processing delays, which is often suggested as a quick and easy step toward better management control. However, companies often find the results disappointing, for reasons illustrated in Figure 2-6. Here, the clerical delays assumed in Figure 2-1 (the 3, 2, and 1 weeks of sales

33

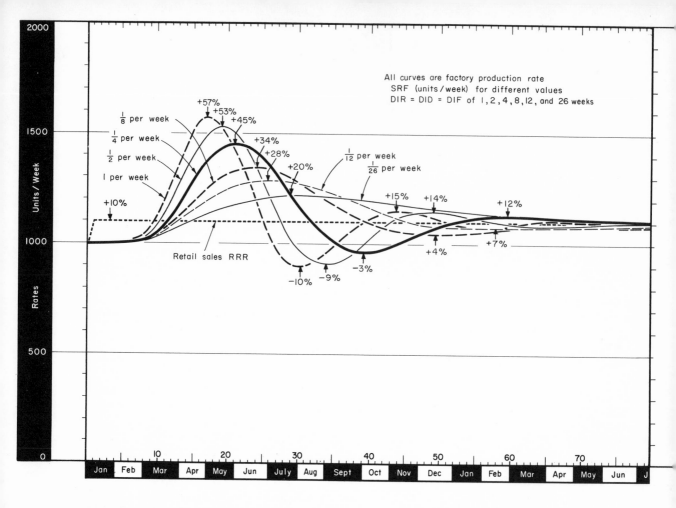

All curves are factory production rate
SRF (units/week) for different values
DIR = DID = DIF of 1, 2, 4, 8, 12, and 26 weeks

Changing Inventory Policy. The behavior of a simple production-distribution system like that shown in the figures is probably more affected by the practices followed in adjusting inventories and in-process orders than by any other single characteristic. The amount of inventory change and its timing are both important. These points deserve special emphasis and should be considered with care by decision-making executives.

To illustrate, Figure 2-8 is the same production-distribution system as before but assumes that management has changed the rapidity with which the inventory corrections are made (but not the total amount of adjustments made). In all the examples thus far, policy has specified that an inventory correction should be based on sales smoothed over the previous 8 weeks

(exponential smoothing with an 8-week time constant). This continues to be true here, so that in no case can the adjustment follow a sales change more rapidly than the delay introduced by the smoothing. Beyond this, there is the question of how quickly to respond in placing orders after a desired inventory correction is recognized. When inventories are lower than desired for the current level of business activity, how much of the difference or imbalance should be added to the orders sent to the factory in the next week?

The factory output curves portray the effect of different speeds of correcting inventory discrepancies. Each curve is based on a different correction rate — a different fraction per week of any inventory discrepancy which is to be corrected through adjustment of the ordering rate

3 4

← **Figure 2-8** Changing the time to make inventory and in-process order corrections.

ally without undershoot as the production rate approaches the 10% higher level.

All other examples in this chapter use the 4-week inventory correction rate so that the "$\frac{1}{4}$ per week" curve of Figure 2-8 is the same as the factory production curve in Figure 2-2.

In Figure 2-8 the more gradual inventory correction to business level changes leads to improved stability. Further, the reduction of manufacturing fluctuation is achieved without increasing the inventory extremes.

Table 2-2 Effect of Inventory Ordering Correction Time on Minimum Inventory Levels

Adjust-ment Time	Retail (%)	Dis-tributor (%)	Factory (%)	Total (%)
1 week	−3.6	−9.4	−16.0	−6.6
2 weeks	−3.9	−9.8	−15.8	−6.9
4 weeks	−4.2	−9.9	−14.8	−7.0
8 weeks	−4.6	−9.7	−13.0	−7.0
12 weeks	−4.8	−9.6	−11.9	−7.1
26 weeks	−5.2	−9.2	−10.2	−7.1

to the next higher supply level. This correction rate varies from the top curve, in which orders for any imbalance in inventory and in-process orders are fully placed in the following week, to the lowest curve, in which only $\frac{1}{26}$ of any *remaining* imbalance is corrected in the following week. The lowest curve leads to about 60% of an initial imbalance being corrected in 26 weeks and 85% corrected in 1 year.

We see that the successive production extremes for the 1-week correction (the top curve) reach 57% above, 10% below, and 15% above the initial value as a result of a sudden 10% rise in retail sales. There is an interval of 27 weeks between successive peaks.

At the other extreme, the 26-week correction time (the lowest curve) leads to a peak of only 20% above the initial value and declines gradu-

Table 2-2 presents, for various correction times, the minimum inventory at each level in the system and the minimum for the sum of all inventories in the system, compared to the initial values. The total inventory changes are less than the sum of the separate levels because the extremes do not occur at the same time.

It should be noted that many sales forecasting methods tend to accelerate the inventory reactions to changes in sales levels. Such forecasting will therefore tend to make the operation of the system less stable.[5] This occurs because the inventory and pipeline corrections are concentrated in high peaks and low valleys and tend to be self-regenerative. The inventory correction rates at one level are taken as *bona fide* changes in business activity at the next

[5] See Appendix L.

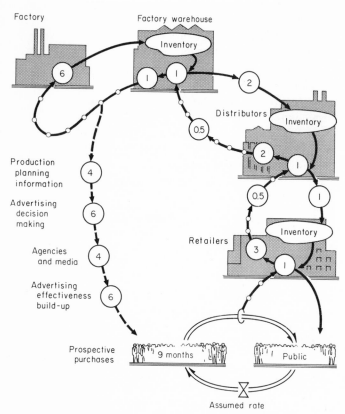

Factory

Factory warehouse

Inventory

6

1 1

2

Distributors

Inventory

0.5

2

1

0.5

1

Production
planning
information

4

Advertising
decision
making

6

Retailers

Inventory

Agencies
and media

4

3

1

Advertising
effectiveness
build-up

6

Prospective
purchases

9 months

Public

Assumed rate

Figure 2-9 Advertising and consumer market.

higher level and there become the basis for still greater inventory adjustments.

2.5 Adding a Market Sector

The production-distribution system discussed thus far is the oversimplified core of a typical industrial enterprise. Later chapters will discuss how to extend system models to other corporate functions. Merely to illustrate the method of extension of the analysis, we shall here add to the previous system one of the many effects of advertising.

Advertising and Market. First we need a schematic diagram of the advertising and mar-

keting behavior which is to be considered. Figure 2-9 will serve our present purposes. It includes the production-distribution organization set forth in Figure 2-1 and adds (1) an advertising function and (2) a special segment of the market.

The figure shows an "organization" or "structure" of the relationship between advertising, the market, and the distribution system. The planned production level is the basis for making advertising decisions. The resulting advertising is delayed by the time to make and execute the decisions. The consumer gradually responds to the advertising by a change in purchasing delay; this change in delay affects the retail purchasing rate that is propagated through the distribution system to change the demands on the factory. The "outer" information-feedback loop is completed when the new production schedules begin to affect new advertising decisions.

The advertising expenditure decision is here assumed to be based on expected sales as indicated by the factory production schedule. Although many managers criticize the practice, advertising expenditure often does rise and fall with industrial activity. Some defend this increase in selling effort when sales increase as "hunting when the hunting is good." We shall see some of the *possible* consequences.

The market characteristic of "deferrability of purchase" has been incorporated. In the earlier examples, an arbitrarily and independently specified rate of retail sales was imposed on the system and the resulting production and distribution flows were generated. Now we introduce somewhat greater reality by including one small part, but far from all, of the dynamics of the consumer market. Instead of arbitrarily specifying retail sales rate, we shall instead move a step further back into the market by arbitrarily specifying one of the several more basic flows that generate the actual retail sales.

The model has therefore been extended so that the independent input is the flow of "prospective purchases." A prospective purchaser is one who has some awareness of need for the product, who expects to purchase, but has not yet done so. For products like household major appliances, this rate would be determined by such factors as the rate of wear-out of existing appliances and by new family formation and new housing construction.

The prospective purchasers are here *defined* as that group who *will* eventually buy and who already have some awareness of their plans and need to purchase. For a mature, well-known, generally accepted product such as household refrigerators, much of the market fits this description. For the industry as a whole, without yet considering the competitive aspects, clearly a large fraction of sales is generated independently of advertising and sales efforts.

What, then, can be the effect of selling effort on these prospective customers? It might influence the average time that the prospective customer waits between first awareness of need and the actual purchase. Could this have any influence on system behavior? How much effect on customer purchasing practices must be created to change the character of the distribution system responses as seen in the previous figures?

· We are here dealing with a situation where the purchase is not mandatory at a particular moment; it is deferrable; and presumably the actual moment of buying can be somewhat influenced by outside persuasion. If advertising has *any* effect on the timing of the purchase, it should be to shorten the contemplation interval as the advertising effort is increased. Customers who are persuaded that prices are now favorable or that the features of new models are desirable, or customers who simply have their attention directed toward acting on a purchase that was already imminent, may pur-chase somewhat sooner than they would have with a lower selling effort.

Now look more closely at the relationships shown in Figure 2-9:

- Production planning information is averaged over a period of 4 weeks, and the average is the basis for advertising decisions.
- A time interval of 6 weeks is assumed for analysis of sales conditions, waiting to be sure of business trends, and reaching a decision to change the advertising expenditure rate.
- Acting on advertising decisions by agencies and media is taken to require 4 weeks.
- Full buyer awareness of the advertising is assumed to develop over a period of 6 weeks after the appearance of a new level of advertising.
- The average time that prospective buyers contemplate purchase and during which they are subject to advertising influence is assumed to be about 9 months in the presence of an "average" level of advertising.
- A 10% change in the advertising level is taken to shorten or lengthen by about 2 weeks the 9-month period during which the average prospective buyer considers his purchase.

This postpones, to a future model extension, the further details of a consumer market — the use of advertising to introduce new products, or the place of advertising in the competitive relationship between companies. The interaction of advertising and sales here proposed is most plausible if the production-distribution-advertising system is thought of as representing an entire industry rather than a single company. The flows are then total industry market, sales, and advertising.

Production Cycle from Advertising Interactions. We can now see what the interaction between advertising, sales, and consumer market might do to the production-distribution processes already studied. What are some possible implications of gearing the advertising budget to the incoming sales dollar?

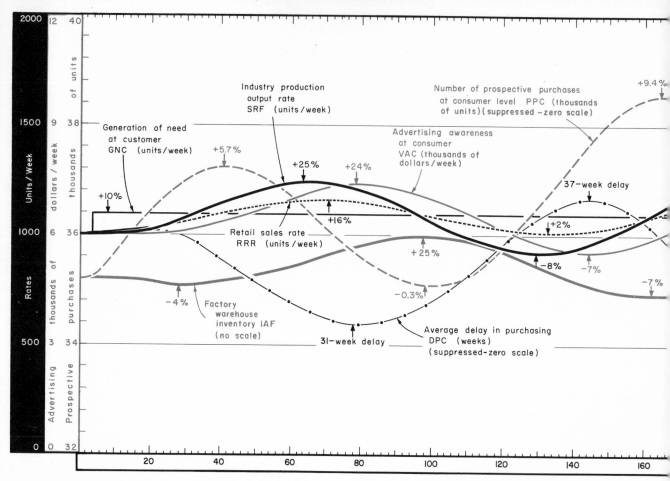

Figure 2-10 Effect of advertising that changes with sales.

Figure 2-10 shows the behavior of the system illustrated in Figure 2-9, when management's policy is to let advertising be proportional to expected sales, as represented by scheduled factory production. To introduce a test disturbance into the system and thus allow observation of the propagation of a change, we here let the rate of appearance of *prospective* purchases (generation of need) suddenly increase 10%. This starts to build up the pool of prospective purchases, and gradually over 20 weeks the rate of retail sales increases. Beginning with the 20th week, industry production starts to rise. Because of the factors earlier described, it exceeds the increase in retail sales.

The decision that leads to increased factory output also leads to a higher advertising commitment. This influences the pool of prospective

buyers and further increases the retail sales between the 20th and 70th weeks. However, this retail sales rate, which rises at the 70th week to 16% above the start, is 6% above the new rate of generation of prospective buyers. After the 42nd week, when retail sales surpass the 10% increase in generating new prospective purchases, the pool of prospective purchases starts to decrease. The decreasing pool of prospective purchases becomes the controlling factor after 70 weeks, and retail sales start to fall despite a still rising advertising awareness.

We then see falling retail sales and a more rapidly falling production rate. The consequent fall in advertising commitments and awareness lets retail sales sink to a still lower level. Retail sales fall for over a full year, from the 70th to the 134th weeks. During this period,

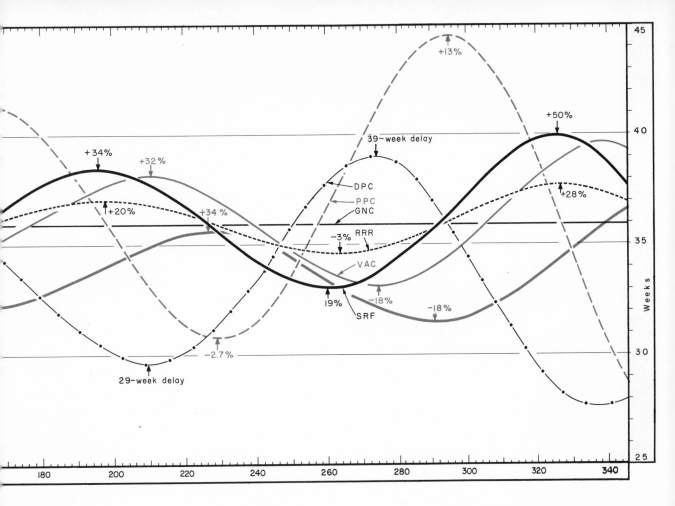

retail sales are below average demand after the 99th week because buyers are reverting to a slightly longer average wait before purchasing (since advertising has been cut back in response to the lower level of sales).

The pool of prospective purchases starts to rise at the 98th week; by the 134th it overpowers the falling advertising awareness, and retail sales start to rise again. The average delay in consumer purchasing falls to 31 weeks at the 78th week and rises to 37 weeks at the 144th week.

Because of the interaction of production swings that are greater than retail sales swings, advertising that is proportional to production plans, and a small advertising influence on how soon the average prospective buyer makes his purchase, there is created a 2.5-year production

cycle that rises and falls some 30% above and below the new 10% increase in average demand. Furthermore, the amplitude of the production swings is gradually increasing and will do so until the swings are curbed by manufacturing limitations, lost sales due to delays in shipments, and other factors.

This pattern, with over two years elapsing between production peaks, reminds us of occurrences in household appliances and some other industries. It indicates that advertising is a *possible* cause of long cyclic disturbances. It would be premature to assume that advertising is *the* cause, however, until many other factors (such as consumer credit, sales forecasting methods, price discounting, and growth and saturation characteristics of the market) are also studied.

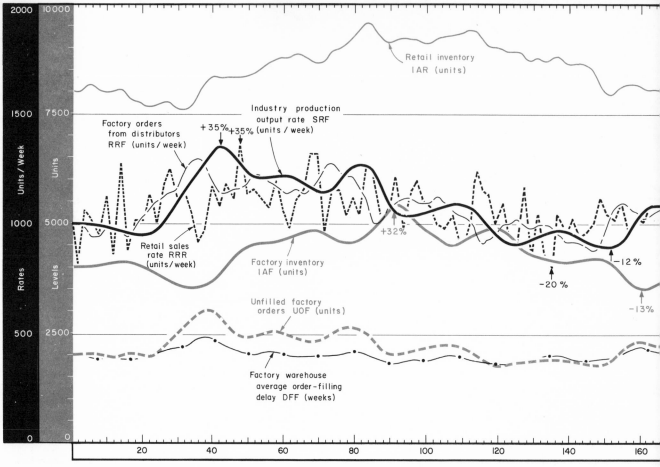

Figure 2-11   Industry instability from advertising interactions with deferred purchase at market, factory limit 40% above initial retail sales, 10% increase in generation of need, random variations in retail sales.

A Boom and Recession Industry. To make our industry model show more realistic behavior, we can combine some of the factors already examined separately.

Figure 2-11 portrays the result when we add to the system of Figure 2-10

- A random pattern to consumer purchases while keeping the advertising influences already discussed in Figure 2-10. This random pattern superimposed on retail sales does not affect long-run average sales but does affect week-by-week sales. It represents the variety of miscellaneous influences that continuously disturb economic behavior.

- A factory production capacity limit that is 40% above the initial sales level.

As in Figure 2-10, a sudden, permanent, 10% increase in the rate of appearance of prospective customers is followed by long fluctuations of over two years' duration. Superimposed on this broad pattern are shorter disturbances that are caused by the system's response to the random retail fluctuations.

At 330 weeks, the production rate is flattened by encountering the maximum production capacity of the industry. This maximum rate persists for 20 weeks even though it is far in excess of the average retail demand. Such circumstances lead to competitive pressures to expand production still more. In this type of unstable system, the counterpart of which seems to exist in several important industries, the in-

40

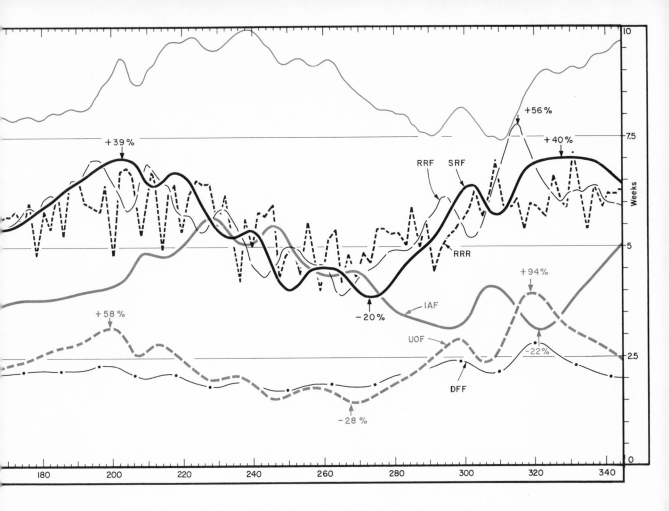

crease of production capacity leads to heightened production peaks, and this in turn deepens the recessions in the following periods. The recovery from the recession results in still more capacity expansion.

The preceding discussion has shown how deferrability of purchase can join with other influences (advertising in the above) to produce fluctuating conditions in an industry. Economic conditions and price discounts can also interact with purchase deferrability. For example, the farm equipment industry sells to a market that can defer purchases for several years. Sales are then concentrated, depending not only on years of farm prosperity but also on the history of machine usage and the existing inventory in the hands of the customer.

How can this "roller-coaster" effect be reduced? What can flatten out the cycles? Different management policies can have dramatic effects. Changes in advertising policy, inventory management, sources of information used in decisions, and the decision and action delays in the system will all affect stability of operations.

The behavior of the system will be influenced by plant expansion policies, pricing changes, competitive practices, labor contract terms, and by the factors that arise from the technology of the industry. In practice, management would want to work toward including all these elements in a dynamic model of industrial behavior.

For instance, by adding money flow, including the factors determining the cost of carrying

41

inventory, cost of changing production levels, and losses from special inventory clearance sales, executives would be able to get direct profit comparisons between various methods of operation. The model could include the restrictions imposed by available labor supply, maximum hiring and training rate, overtime, and policies regarding work-force reduction.

To set up a dynamic model for simulating company or industry behavior, we must adequately describe the real system that it represents. Gathering the information will sometimes be difficult. However, the kinds of information needed involve the essential characteristics of the company. We need factory lead time, inventory policies, construction time for new plants, union contract provisions, and advertising policy. These are the same items that we consider and try to interrelate in the everyday management of a business.

In an adequately complete company model, more intangible items must also be specified, such as the likelihood of results from research expenditures, consumer response to advertising, and market behavior. Many of these quantities will be unknown at first; a few can be measured; some can be closely estimated; others will be guesses. Today these same factors are combined intuitively, and sometimes quite effectively, but the systems-analysis approach should help increase our understanding of the systems that we are attempting to manage.

CHAPTER 3

The Managerial Use of Industrial Dynamics

The information-feedback concepts of system behavior, mathematical models of dynamic interrelationships, and the digital computer to simulate system interactions make experimental industrial system design possible. The manager can have a design laboratory to assist him in creating improved control policies and the information flows on which they depend. This will lead to greater depth in management education and to changes in the qualifications of managers. It can be expected to alter the present staff versus line management duties in companies.

IN Chapter 2 the example of an industrial system has provided an elementary demonstration of a way to study the effect of policy and organizational structure. The example has been based on very simplified flows of information, orders, and materials, but even so, realistic and informative results have emerged. What then is the place of such analysis in the quest for better methods of management?

It is only through costly experience and errors that managers have been able to develop effective intuitive judgment. We need to expedite this learning process. Other professions in similar circumstances have turned to laboratory experiments.

3.1 The Management Laboratory

Controlled laboratory experiments on industrial and economic situations are now possible with computers to do the work required by mathematical models that simulate the system being studied. Unlike real life, all conditions but one can be held constant and a particular time-history repeated to see the effect of the one condition that was changed. Circumstances can be studied that might seldom be encountered in the real world. Daring changes that might seem too risky to try with an actual company can be investigated. The manager, like the engineer, can now have a laboratory in which he can learn quickly and at low cost the answers that would seldom be obtainable from trials on real organizations.

The controlled laboratory experiment is a powerful tool — when used properly. The wind-tunnel model of an airplane illustrates the strengths and dangers of this technique. Today's aircraft would be impossible without laboratory model experiments; but the erroneous design of a model or mistaken interpretation of laboratory results can lead to disaster in the final airplane. In the same way the manager must come to understand his new tools, their dangers, and their effective use.

3.2 Steps in Enterprise Design

The management laboratory approach will follow the same steps common to other design laboratories. These have been mentioned in Chapter 1. The goals must be defined, the significant factors identified, a model constructed, the model tested, and the results interpreted.

The Goals. First come the goals and the questions to be answered. To be productive, laboratory design must be addressed to an important goal. The questions guide the work that follows. The problems must be worthy and significant. This seems obvious, but too often management research and academic thesis studies have been addressed to questions whose answers could have but little significance to the improvement of management. The important problems are often no harder to subdue than the trivial ones; the results are more rewarding.

The Description of the Situation. Next, the factors that bear on the answers must be visualized, interrelated, and described. This should not be a formal step of mere statistical procedure but rather the point where intuition and insight have their greatest opportunity. This is a step for the philosophical, sensitive, perceptive observer. Here, with respect to the past and present, we deal with what are indeed the actual practices — not merely with formal organization charts, or profit-maximizing concepts, or "economic man," but with the social, psychological, and organizational forces that influence decisions.

Here we need experience, alertness, and a strong intuitive feel for the nature of information-feedback systems. We need to look in the proper places for the policies, delays, and information sources that determine dynamic system behavior.

This step should result in an unambiguous, clear, verbal description of the factors and their interrelationships that bear on the questions to be answered.

The Mathematical Model. Even the best verbal description will lack clarity; the spoken language is notoriously vulnerable to misunderstanding. The next step then is to clarify the verbal statement by translating it into a less ambiguous form and into a form that will allow

us to experiment with the implications of the statements already made. Any truly complete and unambiguous verbal statement (whether correct or not is immaterial) can be cast into mathematical notation. The task here is routine — if the verbal statement was indeed clear. The job is essentially no different from the translation from one verbal language to another. Indeed, mathematics is but another language form — one with more rigid rules than English for controlling its definitions, syntax, and logic.[1] The difficulties that will often arise in this conversion to mathematical language arise because of ill-conceived questions in the first step above, or an inadequate and incomplete description in the second step.

This third step gives us a model that contains the mechanisms of interaction that have been visualized between the parts of the system as described in the second step. In general, this model will constitute a straightforward, understandable description of a managerial situation that is, nevertheless, far too complex in detail to permit "optimum" (best possible) mathematical solutions. It does, however, make possible the generation of a specific time history of behavior that would result if the system (as described) were started with a specified initial state and with any specified behavior of the external environment.

Simulation. This tracing of a specific time history is called *simulation*. The model takes the place of the real system and simulates its operation under circumstances that are as realistic as was the original description of the system. This is the counterpart of trying a new policy or organizational structure in the real system, but here cost is insignificant compared with the cost of a real-life experiment.[2] Furthermore, a great deal more is learned because the

---

[1] Part II that follows discusses a mathematical language that is adequate for this translation.
[2] Around $30 per experiment at present commercial computing costs. Such economic feasibility has existed only since 1956.

experimental conditions are fully known, controllable, and reproducible, so that changes in system behavior can be traced directly to their causes. Examples of such time histories were seen in the preceding chapter.

The simulation step requires a vast amount of arithmetical drudgery, and here the electronic computer comes into its own. The computer can take the mathematical statements from the preceding step, automatically do the detailed computer programming, generate a time history showing the implications of the system description when combined with the input conditions as specified, and prepare the requested tabular data and graphical curves.

Interpretation. After a simulation run comes interpretation of the results. Did it turn out as expected? If not, why? As the experiment is examined, new questions arise. Perhaps a totally implausible result discloses a defect in the system description of step two, or in the mathematical translation of step three. Perhaps it even becomes evident that the wrong questions had been asked in the first step.

System Revision. If the goal is improvement of an existing industrial system, the first model will usually be built to represent the system as it has been. When the results adequately represent the important behavior characteristics of the past, the next step in search of improvement is the redesign of system structure and policies. This is a process of invention and trial like the design of a physical system. The skilled designer will obtain better results than the novice or the unimaginative.

Repeated Experimentation. At each step in the above sequence, the prior steps may need to be reviewed and revised. Each simulation result teaches, and it also prompts additional questions. The procedure is analogous to the repeated design and test of an experimental machine until performance has been improved and the difficulties have been reduced to a point where the resulting design can be put into use.

3.3   Effect on the Manager

The ultimate goal of industrial dynamics and of the management laboratory is "enterprise design." But the effect on the manager often becomes evident before the effect on the enterprise.

One of the first effects is to develop in the manager a better intuitive feel for the time-varying behavior of industrial and economic systems. The study of particular situations, as in the previous chapter, improves his judgment about the factors influencing company success. A background in experimental system design helps the developing manager to enhance the benefit from his work experience. His operating experience becomes more meaningful when he better understands how his immediate environment is related to other company functions.

At the college level, system simulation (simulation as used here does not mean management games, see Section 20.5) allows the student to observe the nature of the industrial enterprise. He can for the first time experience the result of policy changes. He can "establish" controlling *policies* for "his" company and observe the results. He can acquire a better appreciation for the underlying similarities between seemingly diverse kinds of organizations, and begin to sense what makes them appear on the surface to be different.

What will these changes demand of a manager in training and ability?

Just as automation requires new skills at the worker level, so will improved methods require new abilities at the management level. The top executive of the future will be concerned less with actual operating decisions and more with the policy *basis* that should guide operating decisions.[3] He will be concerned not so much with day-to-day crises as with the establishment

[3] See Leavitt and Whisler, "Management in the 1980's" (Reference 7), for additional views on the changing nature of policy making and middle management.

of policies and plans that minimize emergencies.

We can expect changes in various management responsibilities. There will be a merging of many line-and-staff functions. Today the kind of systems studies discussed here would be the province of staff specialists, yet the results determine the future success of the company. Therefore the methods of enterprise design must in time become well known to the top line executive.

In the past, power has been concentrated at the top in order to improve over-all integration and control. We can expect that improved definition of objectives and more pertinent standards for the measurement of managerial success will permit managers at the lower levels to take on more responsibility in a form which can be more effectively discharged.

This will lead to a more decisive separation of policy making from operations, with the dividing line much lower in the organization than at present. We can expect a reduction in the size of the middle-management hierarchy. The upper levels will become more exclusively concerned with the design of policies to integrate the separate functions of the business and less with actual decision making. At the lower management levels those decisions that impinge on other departments and on the outside environment will be removed to a point where the guiding policy can be set with regard to the system as a whole. Instead, at these lower levels the responsibilities will become better matched to the opportunities and authorities that can there exist. Stress will increase on innovation, efficiency, and personnel morale within a better-defined local framework.

As with any major social change, a new approach to the management process will take time — time to develop ideas, time to test them, time to train a new generation of managers, and time for these men to reach positions of responsibility.

Industrial dynamics is in an early stage of development. At a few isolated places, in both companies and universities, research is under way as an outgrowth of military or operations research projects. But the full impact will not be felt for a decade or more.

Progress will come in two stages. It is likely that the five years from 1960 to 1965 will be devoted to the development of basic analytical techniques, demonstration by success in specific industrial situations, and establishment of new academic programs for training future managers.

After that will come general recognition of the advantage enjoyed by the pioneering managements who have been the first to improve their understanding of the interrelationships between separate company functions and between the company and its markets, its industry, and the national economy. Competitive pressures will then lead other managements to seek the same advantage.

Many companies are already developing the background skills that will be needed. Computers that are now being acquired for other purposes can be used as systems simulators. Operations research departments can expand into the broader activity of studying management systems. Simulation techniques, feedback-control-system theory, and analysis of decision processes are now being developed in the engineering and product research departments of many firms.

Already some companies are taking the lead. For them, research on the management process is a part of the long-range planning of company improvement.

II

DYNAMIC MODELS OF INDUSTRIAL

AND ECONOMIC ACTIVITY

Part II is more technical than Part I. It treats the nature of mathematical models, the principles of formulating industrial dynamics models, and the specific methodology to be used in Part III.

- *Chapter 4 classifies models, to help place the models of this book into the context of models as found in engineering, the physical sciences, and the social sciences.*
- *Chapter 5 discusses the basis of model formulation, stressing the view that a model is for a purpose. The objective comes first to guide the model content.*
- *Chapter 6 gives a simple model structure of alternating levels and flow rates into which can be cast the broad managerial systems investigations outlined in Part I.*
- *Chapter 7 describes the system of equations that is used within the model structure of Chapter 6.*
- *Chapter 8 introduces flow-diagram symbols to illustrate the relationships between system components.*
- *Chapter 9 discusses the nature and the mathematical representation of delays which exist in all system actions and which are essential to an understanding of system dynamics.*
- *Chapter 10 on policies and decisions treats the formal nature of decision making. It establishes the basis whereby the policy statements in a model are formulated to generate the system decisions through time.*
- *Chapter 11 gives a basis for aggregating the separate items and events of a system into the essential main channels.*
- *Chapter 12 describes the use of independent (exogenous) variables in dynamic-model experimentation.*
- *Chapter 13 discusses the several viewpoints from which models must be judged in establishing their validity and in determining their utility.*
- *Chapter 14 summarizes the principal points made in Part II.*

In recommending a reading procedure for Part II, we shall recognize three groups of readers:

● *Group A includes the manager who is not taking time for the more technical material, nor for the specific model-formulation detail in Part III. Chapter 4 should be read; unfamiliar terminology can be ignored. The preview paragraphs at the beginning of the other chapters will provide the central thought of each chapter. Chapters 5 and 10 should be examined and read insofar as they are of interest. Chapter 14 gives a summary of Part II. All managers who can be persuaded to place themselves in Group B should do so.*

● *Group B includes those who are interested in how models of industrial systems can be set up but who do not at this reading wish to spend time on underlying model-building philosophy. Chapters 4 through 9 should be read, but sections that seem obscure can be left after a single reading and can be re-examined as necessary when reading Part III. The first part of Chapter 10, the preview paragraphs of Chapters 11 through 13, and all of Chapter 14 should be read.*

● *Group C includes the serious student of the book who wants to take time to get the most out of it. All of Part II should be read carefully. Experience has shown, however, that this material falls short of its intended purpose until after the reader has seen the material applied. The Group C reader should study Part II carefully but need not be concerned about those parts that do not seem to fall readily into place. He should then continue to a study of Part III, which contains two complete examples of dynamic models. He will want to refer back to Part II while studying Part III. After Part III he should reread Part II. It is only on this rereading, after seeing examples, that most students achieve a good grasp of the intended content of Part II.*

CHAPTER 4

Models

Models can be a basis for experimental investigations at lower cost and in less time than trying changes in actual systems. Social science models need to be models of systems, not merely of isolated components of an information-feedback system. Our descriptive knowledge provides a wealth of material from which to formulate dynamic models.

MODELS have become widely accepted as a means for studying complex phenomena. A model is a substitute for some real equipment or system. The value of a model arises from its improving our understanding of obscure behavior characteristics more effectively than could be done by observing the real system. A model, compared to the real system it represents, can yield information at lower cost. Knowledge can be obtained more quickly and for conditions not observable in real life.

4.1 Classification of Models

Models might be classified in many ways. Figure 4-1 shows models subdivided into the categories of interest here.

Physical or Abstract. First, models can be distinguished as being either physical models or abstract models.

Physical models are the most easily understood. They are usually physical replicas, often on a reduced scale, of objects under study. Static physical models, such as architectural models, help us to visualize floor plans and space relationships. Dynamic physical models are used as in wind tunnels to show the aerodynamic characteristics of proposed aircraft designs.

An abstract model is one in which symbols, rather than physical devices, constitute the

model. The abstract model is much more common than the physical model but is less often recognized for what it is. The symbolism used

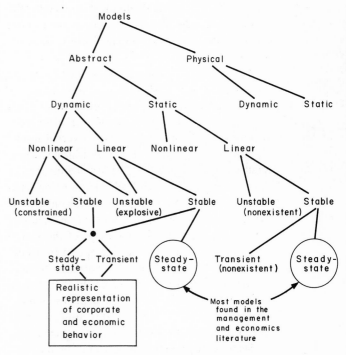

Figure 4-1 Classification of models.

can be a written language or a thought process.

A mental image or a verbal description in English can form a model of corporate organization and its processes. The manager deals continuously with these mental and verbal models of the corporation. They are not the real corporation. They are not necessarily correct. They are models to substitute in our thinking for the real system that is being represented.

A mathematical model is a special subdivision of abstract models. The mathematical model is written in the "language" of mathematical symbols. A mathematical model, like other abstract models, is a *description* of the system it represents. Mathematical models are in common use but are less easy to comprehend than physical models and are less frequent in everyday life than verbal models. An equation relating the lengths and weights on each side of a playground seesaw is a static mathematical model. The equations for stress in a structure make up a static mathematical model of the girders and supports. The equations of motion of the planets around the sun form a dynamic mathematical model of the solar system.

Mathematical notation is a more specific language than English. It is less ambiguous. A mathematical model is therefore a description having greater clarity than most verbal models. In model building, we start with a verbal model and then refine it until it can be translated into mathematical language. The translation is not inherently difficult. The problems in going from the verbal to mathematical statements arise when the initial verbal model is not an adequate description, and the shortcomings of the verbal model are revealed in the attempt to translate.

The mathematical model is valuable because it can be manipulated more easily than verbal or physical models. Its logical structure is more explicit. It can be more readily used to trace assumptions to their resulting consequences.

Static or Dynamic. Models may or may not represent situations that change with time. A static model describes a relationship that does not vary with time. A dynamic model deals with time-varying interactions.

Linear or Nonlinear. Systems represented by a model may be "linear" or "nonlinear," and the models can be similarly classified.

In a linear system, external effects on the system are purely additive.[1] A linear representation of a factory would be one in which a doubling of the incoming-order rate would, at every future moment of time, produce exactly ten times the changes that would come from a 10% increase in orders. In such a factory model, production capacity limits would not be permissible; man-hour productivity would not decrease as employment began to crowd the available equipment; large changes in capacity would take no longer to accomplish than small changes. Labor, equipment, and materials would each make its own contribution to production rate entirely independently of the state of the other two, implying, for example, that labor and equipment could produce a product even if materials were zero. Linear models are adequate in much of the work in the physical sciences but fail to represent essential characteristics of industrial and social processes.

In obtaining explicit mathematical solutions, linear models are much simpler than nonlinear. With negligible exceptions, mathematical analysis is unable to deal with the general solutions to nonlinear systems. As a consequence, linear models have often been used to approximate phenomena that are admittedly nonlinear.

[1] A linear model is one in which the concept of "superposition" holds. In a linear system the response to every disturbance runs its course independently of preceding or succeeding inputs to the system; the total result is no more nor less than the sum of the separate components of system response. The response to an input is independent of when the input occurs in the case of a linear system with constant coefficients (not for a linear system having time-varying coefficients). Only damped or sustained oscillations can exist in an actual linear system; an oscillation that grows is not bounded and must become explosively larger. These are not descriptions of real industrial and economic systems. Nonlinear phenomena are the causes of much of the system behavior that we shall wish to study.

As a result, the nonlinear characteristics have been lost.[2]

When we no longer insist that we must obtain a general solution that describes, in one neat package, all possible behavior characteristics of the system, the difference in difficulty between linear and nonlinear systems vanishes. Simulation methods that obtain only a particular solution to each separately specified set of circumstances can deal as readily with nonlinear as with linear systems.

Stable or Unstable. Dynamic models, in which conditions change with time, can be subdivided into stable and unstable models. Likewise, the actual systems they represent are characterized as being stable or unstable.

A stable system is one that tends to return to its initial condition after being disturbed. It may overshoot and oscillate (like a simple pendulum that is set in motion), but the disturbances decline and die out.

In an unstable system that starts at rest, an initial disturbance is amplified, leading to growth or to oscillations whose amplitude increases. A nonlinear system that is unstable under normal conditions can exhibit fluctuations that grow until restrained by the appearance of nonlinear influences (labor shortage, production capacity, declining availability of materials). The sustained fluctuation might then be thought of as having reached a stable amplitude of peak-and-valley excursion. Clearly, in economic systems, upper levels of activity are limited by resources, and the lower levels are at least bounded by zero activity.

The indications are that the industrial and economic systems of greatest interest will often be of this type wherein small disturbances grow in an unstable manner until restrained by nonlinearities.

Steady-State or Transient. Models (and systems) can be further subdivided according to whether their behavior is primarily steady-state or transient.

A steady-state pattern is one that is repetitive with time and in which the behavior in one time period is of the same nature as any other period. (For some purposes, a model of a nongrowing national economy that shows business-cycle patterns could be considered a steady-state fluctuation, even though never repeating identically any particular sequence of events. Likewise, the long, mature portion of a product life cycle, as now illustrated by automobiles, might be considered a steady-state dynamic model for the answering of certain questions.) In business systems, steady-state behavior is a restricted, special case. (The system discussed in Chapter 2 is a steady-state, dynamic model.)

Transient behavior describes those changes where the *character* of the system changes with time. A system that exhibits growth would show transient behavior. Transient responses are "one-time" phenomena that cannot repeat. Many of the important management problems are transient in character — company growth, new plant construction, and market development.

Open or Closed. In addition to the classification of Figure 4-1, models may be "open" or "closed." The distinction is not as sharp as the words would indicate. Different degrees of "openness" can exist.

The closed dynamic model is one that functions without connection to externally supplied (exogenous) variables that are generated outside the model. A closed model is one that internally generates the values of variables through time by the interaction of the variables, one on another. The closed model can exhibit interesting and informative behavior without receiving an input variable from an external source.[3]

Information-feedback systems are essentially

---

[2] For an interesting, descriptive, nonmathematical discussion of nonlinearity, see Reference 8.

[3] See Chapter 12 on exogenous inputs to models.

closed systems. They are self-regulating, and the characteristics of principal interest are those that arise from the internal structure and interactions rather than those responses that reflect merely the externally supplied inputs.

The models of interest to us here can be operated as closed systems. The internal dynamic interactions are of primary interest. We shall not always choose to study such models in completely closed form. It is often informative to depart from strictly closed operation enough to permit a test input that serves as excitation of the internal responses of the system. Impulses, steps, sinusoids, trends, and noise (random disturbances) are common test inputs. Such external (exogenous) inputs are valid only under conditions where we are willing to assume that the external inputs are themselves entirely independent of the resulting response within the system.

Models of Industrial Systems. Most of the mathematical models found thus far in the management and economics literature belong in one of the two circles in the figure. Nearly all are steady-state, stable, and linear. Some are static and some dynamic. The practical utility of these models in dealing with economic systems has not been notable. The models of industrial situations in the field of operations research have often repaid their cost manyfold, but even so they have not dealt with the major problems of corporate top management.

To deal with practical management and economic problems of pressing importance, a mathematical model must be able to include all of the categories leading to the square in Figure 4-1. Corporate management must cope with the transients of growth and the steady state of normal business fluctuation and uncertainty. Stable industrial systems may exist in mature product lines. Systems that are unstable and are restrained only by their nonlinearities are to be expected in capital goods industries, commodities, and probably in our economic system as a whole. The nonlinearities of maximum fac-

tory capacity, labor and credit shortage, and the dependence of decisions on complex relationships between variables, all compellingly insist on being included in a usefully realistic model of the industrial enterprise. Since time and changes with time are the essence of the manager's task, a useful model must be dynamic and capable of adequately generating its own evolution through time.

Consequently, we are speaking here of mathematical models that can be used to simulate the time-sequential operation of *dynamic* systems, *linear* or *nonlinear, stable* or *unstable, steady-state* or *transient*. The model must be able to accept our descriptions of *organizational form, policy,* and the *tangible* and *intangible* factors that determine how the system evolves with time. Such models will be far too complex (tens, hundreds, or thousands of variables) to yield analytical solutions. In fact, for nonlinear systems modern mathematics can achieve analytical solutions to only the most trivial of problems. The models considered here are instead to be used to simulate (that is, to trace through time) a particular course of action that results from a specific set of starting conditions coupled with one specific combination of noise and other inputs which are introduced. This is an experimental, empirical approach in search of better knowledge, and thereby better results, but not promising "optimum" solutions to any question.

In the management science and economics literature, the term "mathematical model" has customarily been used to mean any mathematical relationship between the inputs and the output of a part of a system. In the terminology of engineering systems, this output reaction of a system component to one or more inputs is commonly called a "transfer function." The transfer function specifies how conditions at the input will be transferred to the output. In this text, a simple mathematical relationship describing the response of a component of the system to its immediate environment will not be called a "model" but instead will be referred to synonymously as "transfer function," "functional relation," "decision equation," or "rate equation." In contrast, a "model" defines a system consisting of an interacting set of "decision equations."

4.2 Models in the Physical Sciences, Engineering, and the Social Sciences

Mathematical models in the social sciences have often been compared with the simple models in the physical and biological sciences. This may have been misleading.

Useful models of the solar system, the atom, Newton's laws of motion, and heredity are much simpler than models that will be helpful in industrial and economic systems. Linear analysis is much more widely applicable in such physical science systems. Most physical science systems that have been represented by successful models have contained much smaller amounts of noise (uncertainty) than in our social systems. Physical science models have been deduced to explain phenomena that can be observed but usually not altered. Statistical inference methods that succeed in relating unidirectional cause and effect in biological heredity are not necessarily sound in studying social systems where effect reacts on cause.

The attitudes toward the source and purpose of physical science and social science models have been similar, to the detriment of progress in models of social systems.

Models in the engineering and military worlds are so different in degree from the physical science models that we can fairly say they differ in principle. They arise in a different way, to be used for a different purpose.

Models in engineering and military usage provide a much better precedent for the social sciences than do physical science and biology models. Economics and management, like engineering, deal with aggregate systems above the level of the individual elementary events that are the subject of many physical science models. Unlike systems that are commonly modeled in the physical sciences, engineering systems have complexity approaching the complexity of social systems. Both engineering and social systems have a continuous gradation (from the obviously important, through the doubtful, into the negligible) of influences that affect each action and decision; by contrast, the physical science systems have often been different, with a substantial gap in importance between the few factors that must be included in a model and nearly insignificant ones that can be omitted. Social systems are strongly characterized by their closed-loop (information-feedback) structure, like many engineering systems that have been modeled but unlike most models in the basic physical sciences. In models of social systems, as in engineering but unlike simple physical science models, we must be interested in transient, noncyclic, nonrepeating phenomena.

Dynamic models have proved indispensable in designing physical systems. They are used in aircraft engineering, the planning of military command systems, and in studying communications networks. They have included both equipment and people; therefore, they take on aspects of social systems. Today's advanced technology would be impossible without the knowledge that has resulted from mathematical models.

The same cannot be said for the impact of mathematical models on business and economic decisions. Economic models have enjoyed a long history of research but little general acceptance as a tool to aid top management of a company or a country.

Many of the past failures in economic model building can be traced to unsound methods and to attempts to reach unachievable objectives. We need new attitudes toward the construction and use of models of social systems.

Objectives. The past contrast in usefulness between engineering and economic dynamic models can be partly traced to the way the tools of model building have been employed. The difference between the two uses of models seems to arise from a different emphasis on end objective. In engineering, models have been used for *designing* new systems; in economics, a common use is to *explain* existing systems. A model that is useful for design must also explain. But it appears that models that have been undertaken only for explanation have often had their goals set so low that they fail not only for design but also for their explanatory purposes.

Basis for a Model. In engineering systems, models have been built upward from available knowledge about the separate components. Designing a system model upward from identifiable and observable pieces is a sound procedure with a history of success.

In economics, models have often been constructed working backward from observed total-system results. Even as a theoretical goal, there is no evident reason to believe that the inverse process of going from total-system behavior to the characteristics of the parts is possible in the kinds of complicated, noisy systems that are encountered in business and economics.

The attempt merely to reproduce an existing economic system in a model has led to models that are statistically derived from observed time series of past behavior. It is most unlikely that the internal causal mechanisms of a complex, nonlinear information-feedback system can be derived from a sequence of external observations of its normal performance. By contrast, the use of models for the *design* of physical systems has emphasized models of systems that do not yet exist, models that could not possibly be constructed backward from observed results. A wind-tunnel model of an airplane is not constructed to reproduce merely observable overall behavior of a known airplane; it is not designed merely to match the aggregate average of all airplanes designed in the past. It is designed part by part to represent a proposed new aircraft so that the interaction of the parts and the performance of the airplane as a whole can be studied.

In formulating a model of a system, we should rely less exclusively on statistics and formal data and make better use of our vast store of descriptive information.

Validity of a Model. The test of the *adequacy* of a model also differs between engineering and economic usage. In technical and military circles, models have been judged by their ability to exhibit the dynamic characteristics of systems such as amplification, bandwidth,[4] and transient response. In economics, models have often been judged by their ability to predict the *specific state* of the system at some future time; and the models have customarily failed to pass this prediction test.

In using a model, we should look less for prediction of *specific actions* in the future and more for enhancing our understanding of the inherent *characteristics* of the system. There seem to be good reasons why models cannot be expected to predict specific system condition far enough into the future to be particularly significant. If so, prediction of a specific sequence of actions is not a useful test of a model.[5] Instead a model should be judged by its ability to reproduce or to predict the *behavior characteristics* of the system — stability, oscillation, growth, average periods between peaks, general time relationships between changing variables, and tendency to amplify or attenuate externally imposed disturbances.

Similarity of Model and System. In engineering, mathematical models have shown a greater correspondence to the structural and operational details of the real systems they represent than appears in the classical economic models. The communications barrier has been nearly impenetrable between the mathematical models in the social sciences, and the industrial and governmental executives. This is accentuated because models of social systems, unlike models of physical systems, have not been cast in the terms commonly employed by the active prac-

[4] An indication of the resistance to disturbance of any existing trends and cycles.
[5] See Chapter 13 on model validity and also Appendix K.

titioners of the art. The difference in terminology may arise from the differing initial viewpoint. The manager deals with the components of his organization just as the engineer does with the components of his airplane; the manager does not use abstract coefficients that cannot be tied to specific sources in the real system. The modelmaker, who derives relationships from statistical analysis, is apt to leave his coefficients as abstract, empirical results that are not identified with particular features of the real system.

In the chapters that follow, we shall attempt to make every variable and every constant have individual significance in the context of everyday managerial practice. It should be possible to discuss the individual plausibility of the value of any constant because that constant will, in its own right, have physical or conceptual meaning.

4.3 Models for Controlled Experiments

The mathematical model makes controlled experiments possible. The effects of different assumptions and environmental factors can be tested. In the model system, unlike real systems, the effect of changing one factor can be observed while all other factors are held unchanged. Such experimentation will yield new insights into the characteristics of the system that the model represents. By using a model of a complex system, more can be learned about internal interactions than would ever be possible through manipulation of the real system. Internally, the model provides complete control of the system organizational structure, its policies, and its sensitivities to various events. Externally, a wider range of circumstances can be generated than are apt to be observable in real life.

In the model, observations can be made of variables that are unmeasurable in the real system. An adequate model must include any "intangibles" that we believe contribute importantly to the behavior being studied. In the model, the intangibles, and our assumptions about them, become tangible and observable. We then have a means for tracing the implications of our assumptions.

4.4 Mechanizing the Model

A dynamic mathematical model is a description of how to generate the actions that are to be taken progressively through time. To be useful, the model must be mechanized by providing some way of carrying out the specified actions.

The actions called for by the model could be executed by a group of people playing the separate parts of the real system that is being simulated. Decisions and actions would generate results that would in turn become the inputs for the decisions and actions that follow. Such simulation, using groups of people, has been employed in studies of real systems. It is a good technique for classroom demonstration of basic principles. For the study of large systems it is burdensome and expensive.

A digital computer can be instructed to execute the same procedures that would be followed by the group. The cost is less than one-thousandth of the cost of the same clerical operations executed by a group of persons. The task is ideally suited to the unique characteristics of the electronic digital computer.

4.5 Scope of Models

In recent years it has become possible to formulate dynamic models of industrial behavior with sufficient reality to cope with the interactions of production, distribution, advertising, research, investment, and the consumer market. Within such a formulation, both physical and psychological factors can be included. Model-building technique and computational cost no longer limit the systems that can be studied. Instead, progress will be set by the rate at which our knowledge of the industrial world can be sifted, refined, and reorganized into an explicit form.

The immediate goal is to take our literature and knowledge of "descriptive management" and "descriptive economics" and formalize what we believe about the separate parts. It then will become possible to improve our understanding of how these parts interact. In discussing the formulation of dynamic models, no distinction is made in this book between corporations, industries, and complete economies. There should be no difference in approach or arbitrary distinction between microeconomics and macroeconomics. The same principles control. The same theoretical considerations will guide the way in which aggregation can be accomplished. The opportunities for improving our understanding are similar, with the same restrictions on achievable objectives. The comments in this book are intended to apply equally at all levels, from dynamic behavior of the individual firm to international economics.

4.6 Objectives in Using Mathematical Models

A mathematical model of an industrial enterprise should aid in understanding that enterprise. It should be a useful guide to judgment and intuitive decisions. It should help establish desirable policies. Using a model implies the following:

- We have some knowledge about the detailed characteristics of the system.

- These known and assumed facts interact to influence the way in which the system will evolve with time.

- Our intuitive ability to visualize the interaction of the parts is less reliable than our knowledge of the parts individually.

- By constructing the model and watching the interplay of the factors within it, we shall come to a better understanding of the system with which we are dealing.

These assumptions form the same basis on which we construct models of floor plans and of equipment. The model of a company is justified to the extent that it will allow us to manage the company better. There is no implication that the results need be perfect to be beneficial. A model can be useful in determining the degree to which the industrial system is sensitive to changes in a policy or in system structure. It can help determine the relative value of information of differing kind, accuracy, and timeliness. It can show the extent to which the system amplifies or attenuates disturbances impressed by the outside environment. It is a tool for determining vulnerability to fluctuation, overexpansion, and collapse. A model can point the way to policies that yield more favorable performance. In short, mathematical models should serve as tools for "enterprise engineering," that is, for the design of an industrial organization to meet better the desired objectives.

The preceding comments imply that a useful model of a real system should be able to represent the *nature* of the system; it should show how changes in policies or structure will produce better or worse behavior. It should show the kinds of external disturbance to which the system is vulnerable. It is a guide to improving management effectiveness.

But note especially that quantitative prediction of *specific events* at *particular future times* has not been included in the objectives of a model. It has often been erroneously taken as self-evident that a useful dynamic model must forecast the specific condition of the system at some future time.[6] This may be desirable, but the usefulness of models need not rest on their ability to predict a specific path in the future. This is fortunate because there is ample reason to believe that such a prediction will not be achieved in the foreseeable future.[7]

---

[6] Least-squares tests of the period-by-period differences between model variables and real-system variables imply such an expectation that the model should predict the specific future configuration of the system. See Chapter 13 on model validity.

[7] See Appendix K for an example of system *nature* versus future system *condition* prediction.

4.7 Sources of Information for Constructing Models

Many persons discount the potential utility of models of industrial operations on the assumption that we lack adequate data on which to base a model. They believe that the first step must be extensive collecting of statistical data. Exactly the reverse is true.

We usually start already equipped with enough descriptive information to begin the construction of a highly useful model. A model should come first. And one of the first uses of the model should be to determine what formal data need to be collected. We see all around us the laborious collection of data whose value does not equal the cost. At the same time highly crucial and easily available information is neither sought nor used.

The routine, clerical collection of numerical data is unlikely to expose new concepts or previously unknown but significant variables. Extensive data collecting is not apt to shed new light on the *general nature* of the variables. Some of the most important sources for a realistic dynamic model do not exist as "data" in the usual sense of tabulated statistical information.

What is the relative importance of the many different variables? How accurately is the information needed? What will be the consequences of incorrect data? These questions should be answered before much time or money is expended in data gathering.

Actually we use models of corporate and economic systems continuously with only the data that we have at hand. A word picture or description is a model; our mental picture of how the organization functions is a model. A verbal model and a mathematical model are close kin. Both are abstract descriptions of the real system. The mathematical model is the more orderly; it tends to dispel the hazy inconsistencies that can exist in a verbal description. The mathematical model is more "precise." By precise is meant "specific," "sharply defined," and "not vague." The mathematical model is not necessarily more "accurate" than the verbal model, where by accuracy we mean the degree of correspondence to the real world. A mathematical model could "precisely" represent our verbal description and yet be totally "inaccurate."

Much of the value of the mathematical model comes from its "precision" and not from its "accuracy." The act of constructing a mathematical model enforces precision. It requires a specific statement of what we mean. Constructing a model implies nothing one way or the other about the accuracy of what is being precisely stated.

There seems to be a general misunderstanding to the effect that a mathematical model cannot be undertaken until every constant and functional relationship is known to high accuracy. This often leads to the omission of admittedly highly significant factors (most of the "intangible" influences on decisions) because these are unmeasured or unmeasurable. To omit such variables is equivalent to saying they have zero effect — probably the only value that is known to be wrong!

Different attitudes toward data and their accuracy can be traced to the different goals and objectives of models already discussed.

If the only useful and acceptable model is one that fully explains the real system and predicts its specific future condition, then precision is not sufficient; it must also be accurate. Lacking such accuracy, the endeavor flounders.

If, on the other hand, our objective is to enhance understanding and to clarify our thinking about the system, a model can be useful if it represents only what we *believe* to be the nature of the system under study. Such a model will impart precision to our thinking; vagueness must be eliminated in the process of constructing a mathematical model; we are forced to commit ourselves on what we believe is the relative importance of various factors. We shall discover inconsistencies in our basic assumptions. We shall often find that our assumptions about the separate components cannot lead to our expected consequences. Our verbal model, when converted to precise mathematical form, may be inconsistent with the *qualitative* nature of the real world we observe around us. We may find that cherished prejudicies cannot, by any plausible combination of assumptions, be shown to have validity. Through any of these we learn.

In these ways we use a model as does the engineer or military strategist. What would the situation be like if the real system corresponds to our basic assumptions? What would a *proposed* system be like if we designed it to agree with the model? What changes in the model would give it more nearly the characteristics of the existing system that it presumably represents? These are questions that can be asked of a closed, or closable, model and are significant when the system is so complex that the correct answers are not evident by inspection.

A model must start with a "structure," meaning the general nature of the interrelationships within it. Assumptions about structure must be made before we can collect data from the real system. Having a reasonable structure that fits our descriptive knowledge of the system, we can take the next step and assign plausible numerical values to coefficients, since the coefficients should represent identifiable and describable characteristics of the real system. We can then proceed to alter the model and the real system to eliminate disagreement and move both toward a more desirable level of performance.

This is the attitude of the manager toward the verbal image that he uses as a model of the company he directs. He strives to grasp the implications of the separate factors he observes about him. He attempts to relate individual policies and characteristics of the system to the consequences that they imply. He tries to estimate the result of changing those parts of the system over which he has control.

As a model is detailed toward an approximation of a real or proposed system, we can use the model itself to study the significance of various assumptions that have gone into it. With respect to every numerical value that we have been forced to estimate arbitrarily, there is some range within which we are practically certain that the "true" value must fall. It will often happen that the model is relatively insensitive to changes in value within this range; refining our estimate would then be unjustified.[8]

On the other hand, the entire qualitative behavior of the system may depend significantly on some different numerical value that has been assumed.[9] We are then alerted to the critical nature of this assumption. When the vulnerability to an error in numerical value is demonstrated, we must then choose between

- Measuring the value with adequate accuracy
- Controlling the value to a desired range
- Redesigning the system and the model to make the value less important

A mathematical model should be based on the best information that is readily available, but the design of a model should not be postponed until all pertinent parameters have been accurately measured. That day will never come. Values should be estimated where necessary, so that we can get on with the many things that can be learned while data gathering is proceeding. In general sufficient information exists in the descriptive knowledge possessed by the active practitioners of the management and economics arts to serve the model builder in all his initial efforts. He will find that he is more in danger from being insensitive to and unperceiving of important variables than from lack of information, once the variables have been exposed and defined. Searching questions, asked at points throughout the organization under study by one skilled in knowing what is critical in system

[8] See Figure 2-6 on the system insensitivity to clerical delays.
[9] See Figure 2-8 showing sensitivity to rapidity of inventory adjustment.

dynamics, can divulge far more useful information than is apt to exist in recorded data.

These comments are not to discourage the proper use of the data that are available nor the making of measurements that are shown to be justified; they are to challenge the common opinion that measurement comes first and foremost. Lord Kelvin's famed quotation, that we do not really understand until we can measure, still stands. But before we measure, we should name the quantity, select a scale of measurement, and in the interests of efficiency we should have a reason for wanting to know. Even in the context of basic research that is presumed to seek information for its own sake, the world has limited resources, and the researcher should have conviction that his investigation promises a high probability of important results.

To some, this attitude toward the data on which to base a model will seem highhanded and will be repugnant. To others, it will seem the practical and necessary avenue along which to attack a difficult problem.

One important use for a model is to explore system behavior outside the normal and historical ranges of operation. These ranges will be outside the region of any data that could have been collected in the past. We are dependent on our insight into the separate parts of the system to establish how they would respond to new circumstances. Fortunately, this is usually possible. In fact, we may be more certain about the extreme limiting circumstances of human behavior, the likely decisions, and the technological nature of production and inventories than we are about the "normal" range. These limiting conditions are part of our body of descriptive knowledge. Incorporating the full possible range of functional relationships in a model makes it feasible to study wider ranges of system operation. It also improves the accuracy of model representation over normal ranges because incorporating the known extreme values helps to bound and determine many characteristics in the normal ranges.

Useful results can be expected from models constructed as herein discussed — by building upward from the characteristics of the separate components and by incorporating and estimating the values of all factors that our descriptive familiarity with the system tells us are important. Such models will communicate easily with the practicing manager because they arise from the same sources and in the same terminology as his own experience.

CHAPTER 5

Principles for Formulating Dynamic System Models

The questions to be answered precede model design. The closed-loop system structure must be reflected in a model. Time delays, amplification, and information distortion must be adequately represented. All constants and variables of the model can and should be counterparts of corresponding quantities and concepts in the actual system. Dimensional units of measure of model quantities must be meticulously consistent. The preferred practice is to begin with the continuous (nonstochastic) structure of system decisions and later add randomness and periodic influences. Model formulation methods should not presuppose system linearity or stability.

BEFORE proceeding with the specific details of equations and a suitable mathematical structure for a dynamic model, we shall discuss some general principles that should guide the construction of a model of an industrial system.

5.1   What to Include in a Model

In practice there will be <u>no such thing as *the* model of a social system</u>, any more than there is *the* model of an aircraft. An airplane is represented by *several* aerodynamic wind-tunnel models for various purposes, plus cockpit arrangement mock-ups, models for maximum stress loading, etc. In designing a dynamic simulation model of a company or economy, the factors that must be included arise directly from the questions that are to be answered.

In the absence of an all-inclusive model, which we are unlikely to achieve, <u>there may well be different models for different classes of questions about a particular system.</u> And a particular model will be altered and extended as each new question is explored.

The skill of the person who undertakes to use a model is tested immediately — his first decision is to ask pertinent questions having important answers. Trivial questions can but lead to trivial answers. Questions that are too general — such as how to be more successful — fail to define an area of attack. Questions that are too restrictive may confine the investigation to an area that does not contain the answer. Questions that are impossible to answer can lead only to disappointment.

<u>Questions to be answered control the content of a model</u>. But how? Again the perception and judgment of the investigator are taxed. He must select, on the basis of his knowledge of the situation, the factors that he believes are pertinent. Here skill and practice in working with the dynamic behavior of systems are important. What the novice often considers paramount may prove to be insignificant. Some factors that are stressed in static analysis may not even exist as essential concepts in a dynamic model. Factors that are omitted, both in static analysis and in ordinary descriptive debates about a problem, may prove to be crucial.

Here, "art" re-enters the picture as a guide to proper use of the tools of science. For the present we can do no better than to discuss general principles and later supply specific examples that will help the beginner to start developing his own skills.

Since the objective is to include those factors that influence the answers sought, the basis of model building cannot be limited to any one narrow classification of intellectual discipline. We must feel free to include technical, legal, managerial, economic, psychological, organizational, monetary, and historical factors. All these must find their proper places in our formulation of system component interactions.

In formulating a quantitative model, we must have courage to include all the facets that we should consider essential to a verbal description of the phenomena under study. In the past, so long as mathematical models were restricted to those which could be solved analytically, the models could not accept the wealth of concepts that exist in descriptive knowledge. Simulation models and electronic computers change this picture.

As a generalization, significant models for the answering of top-management questions range in size between 30 and 3,000 variables. The lower limit is about the minimum that will exhibit the kinds of behavior of interest to the manager; the upper limit will bound for some time our ability to conceive of a system and its meaningful interrelationships. Already there have been several preliminary explorations with both steady-state and transient growth models incorporating up to a few hundred variables.

5.2   Information-Feedback Aspects of Models

Economic and industrial activities are closed-loop, information-feedback systems. Models of such systems must preserve the closed-loop structure that gives rise to so much of the interesting behavior.

In an information-feedback system, condi-tions are converted to information that is a basis for decisions that control action to alter the surrounding conditions. The cycle is continuous. We cannot properly speak of any beginning or end of the chain. It is a closed loop.

This is a broad definition covering most human, social, and technical activities. Economic fluctuations are one manifestation of the time-varying interactions that occur in information-feedback loops. At the company level, rising sales exceed plant capacity, leading to expansion plans and construction that restore the balance of demand and output. Falling sales and rising inventories may lead to greater marketing effort to increase sales to equal production. The need for product improvement leads to research expenditure, to technological progress, to new competition, to the need for still newer products, and to more research. These and all other management decisions are made in the framework of an information-feedback system where the decision ultimately affects the environment that caused the decision.

The general concepts of information-feedback systems are essential because such systems exhibit behavior as a whole which is not evident from examination of the parts separately. The pattern of system interconnection, the amplification caused by decisions and policy, the delays in actions, and the distortion in information flows combine to determine stability and growth.[1]

As we have seen in Chapter 2, the interconnection of entirely ordinary corporate actions

---

[1] We are not dealing here with the simple concept of the servomechanism which is often treated in engineering literature as having a single "error function" (difference between actual and desired results) and a single control mechanism. Instead our economic systems have "distributed error functions" represented by the individual goals of many participating persons. The control function is likewise dispersed, so that it exists in part at each decision point in the system. Formal servomechanisms theory is unlikely to find direct use in economic models but is an excellent background to guide the selection of factors that should enter such models.

can lead to production fluctuations, unemployment, and excess plant capacity. As one action feeds into another and eventually back to the first, it causes instability that is the counterpart of "hunting" in mechanical servomechanisms. Careful attention must be given to represent properly the time delays, information distortion, and the basis for reaching decisions.

Time Relationships. The behavior of information-feedback systems is intimately associated with the time-sequence relationships between the actions in the system. Time delays arise in every stage of system activity — in decisions, in transportation, in averaging data, and in inventories and stocks of all kinds.

Variable accumulations of all types must be carefully noted and must be represented if the time magnitudes associated with them are judged to be important. Points of accumulation permit, or intentionally introduce, time delays between input and output rates of flow. The fundamental purpose of an inventory is to uncouple somewhat the inflow and outflow rates so that the incoming receipt of goods need not match exactly at all times the rate of delivery from the inventory. The level of the inventory fluctuates to make up the difference in the flow rates.

We must include reservoirs in information and order channels as well as in physical goods. Such reservoirs contain unfilled-order backlogs, in-transit orders in the mail, decisions that have been made but not executed, and data that are being gathered and processed but not yet used. Reservoirs of materials include in-process and finished inventories and goods in transit. Money flow reservoirs include bank balances and loans.[2] Personnel reservoirs include employees, various categories of unemployed, and the classes of consumers and investors. Capital equipment reservoirs include plant and machines, segregated as necessary by age, type, productivity index, and life expectancy.

Although time delays are important, and without delays our systems would not exhibit their usual characteristics toward instability, it is far from true that *all* delays are bad in a system. Nor is it true that the practical solution to system improvement always lies in reducing delays.[3] There are occasions when the introduction of time lags at properly selected points may be the most economical way to achieve a desired objective.[4]

The method of model formulation and the types of equations used should permit time delays to be represented both in average duration and in transient characteristics, in close agreement with our knowledge of how the delays are actually created.[5] Neither the solution interval of the model equations nor the intervals at which data on the actual system may have been collected should dictate the way in which delays are generated in the model.

Amplification. Amplification is a most important characteristic determining the behavior of information-feedback systems. "Amplification" is used here as a nontechnical term implying a response from some part of a system which is greater than would at first seem to be justified by the causes of that response. For example, it is often observed that fluctuations in factory production rate greatly exceed the magnitude of retail sales rate changes.

Amplification occurs in many places in our social system. Amplification arises, as to be discussed later, in the policies that define decisions that control rates of flow. Policies, and the decisions that result from them, must be carefully examined, since they are the source of amplification in social systems.

---

[2] In the context of many situations these will be considered as information rather than money. Accounts payable and receivable and capital equipment and depreciation accounts ordinarily are not "money" in the cash flow sense and will usually appear in the information flow channels.

[3] For instance, in Figure 2-6, reducing the clerical delays produces only slight improvement.

[4] See Figure 2-8, where increasing the delay in inventory adjustment reduces production fluctuation.

[5] See Chapter 9.

Amplification occurs in many places. Orders for goods do not merely reproduce the rate of sales; orders are added for inventory accumulation, to fill supply pipelines, and for speculation. Amplification occurs in many forecasting and predicting procedures, as when growth rates are extrapolated and lead to overinvestment in factory capacity. Amplification arises from the tendency to order ahead when deliveries slow down.[6] Ordering ahead during price rises and delayed ordering during price declines produce amplification. All these are of the greatest importance in representing the factors that control industrial growth and stability.

Information Distortion. Information is the input to decisions, and the decisions are affected by all the influences that act on the information flows. Information can be distorted in other ways than by delays and amplification. Information is modified by averaging procedures and by summarizing individual business transactions into composite data for executive decision. Information is interpreted differently by different people and organizations. Prejudices, past history, integrity, hope, and the internal political environment of an organization all bias information flows. Information contains errors and random noise and unknown perturbations from external sources.

Since information is the raw material of decisions, information distortion must be included if we are to represent decisions properly. Particular attention must be given to determining what information is actually available and used at each point in the system—much information is not available, much of what is available is not actually used, what is used is not necessarily the most effective of that available.

5.3 Correspondence of Model and Real-System Variables

In the proper formulation of a dynamic simulation model, the model variables should correspond to those in the system being represented.

[6] As seen in Figure 2-5.

Model variables should be measured in the same units as the real variables. At first glance this may seem obvious and elementary, but this principle has rather consistently been violated in model building.

Sufficiently close correspondence of model and real-system variables is obtained only through constant alertness and re-examination of whether the decision functions adequately represent the concepts, social pressures, and sources of information that control the actual decisions.

Flows of goods should be measured in physical units, not equivalent dollars. Money flow should be separate. Prices relate the two. Goods cannot be represented in terms of equivalent dollars, or else we lose the significance of prices and of the fact that dollars do not flow in synchronism with goods but usually with a delayed relationship. Orders for goods are not goods, goods delivered are different from accounts receivable, and accounts receivable are not money. We sometimes see equation systems built in dollar-equivalent terms, as if refrigerators were made of dollar bills instead of *from* materials, *by* labor, *using* only existing capital equipment. The concentration on money flow using the money equivalent of labor and materials (and the complete omission of information as a separate quantity) has obscured the character of the industrial system in many analytical attempts.

Time-sequential relationships exist between the actual variables of the real system. These must be represented by our analysis methods if the true dynamic behavior of the information-feedback system is to be preserved. For example, demand (information) leads to shortages (material) that lead to construction (flow of materials and labor) that increases productive capacity (capital equipment) that requires payment for the capital equipment (money flow).

We must distinguish between "actual" and "desired" quantities, and between the "true" value of a variable and what our available information may give as the "indicated" value.

In an economic system model we should use actual money prices, not prices deflated by multiplying with a price index. The actual money prices and their changes produce important psychological effects, as in wage negotiations. Price changes would not create the same illusions if calculated in dollars of constant purchasing power.

In many models it will be necessary to deal separately with length of work week, number of employees, productivity per man-hour, and wages per man-hour, all of which may be system variables that interact. Each has its separate nonlinearities and restrictions that must be recognized in decisions to change production.

Effort to achieve correspondence between the model and real-system variables will be rewarded in many ways. Important system time delays often appear at the conversion points between flows, such as from orders to goods, men to research results, and plans to factories. As the model variables correspond more closely to life, our vast body of descriptive knowledge becomes more directly applicable. Clarity in identifying the real-system variables will make us more sensitive to the relationships that must be incorporated between them.

5.4 Dimensional Units of Measure in Equations

In writing equations, meticulous attention must be given to the correctness of the dimensions (units of measure) associated with each term. Dimensional analysis plays an important part in guiding equation writing in engineering and the physical sciences. Inconsistency in dimensions often divulges an incorrect equation formulation. The units of measure of all variables *and* constants should be precisely stated and should be checked for consistency in each equation that is written. Carelessness on this point can lead to much needless confusion.

A special caution should be inserted here against the practice of arbitrarily inserting co-efficients in equations *merely* to make the dimensions correct. Each coefficient should appear because it has a meaning in its own right; then the units of measure should be checked to see that all terms in the equations are consistent. The formulation of the model should be suspect until each coefficient, and its dimensions of measure, can be defended as having individual meaning and reality and has been selected in such terms that its numerical value can be discussed in the context of the industrial practice that it represents. This may at first seem easy and obvious, but it is one of the major pitfalls of the beginner.

5.5 Continuous Flows

In formulating a model of an industrial operation, we suggest that the system be treated, at least initially, on the basis of continuous flows and interactions of the variables.[7] Discreteness of events is entirely compatible with the concept of information-feedback systems, but we must be on guard against unnecessarily cluttering our formulation with the detail of discrete events that only obscure the momentum and continuity exhibited by our industrial systems.

In beginning, decisions should be formulated in the model as if they were continuously (but not implying instantaneously) responsive to the factors on which they are based. This means that decisions will not be formulated for intermittent reconsideration each week, month, or year. For example, factory production capacity would vary continuously, not by discrete additions. Ordering would go on continuously, not monthly when the stock records are reviewed.

There are several reasons for recommending the initial formulation of a continuous model:

- Real systems are more nearly continuous than is commonly supposed. There may be an annual budget, but it is frequently modified by new

[7] Here "continuous" and "discontinuous" do not refer to the common distinction between differential and difference equations but to the intent of the verbal description of the phenomena being represented.

conditions. Even the signing of a particular contract, which may seem a discontinuous step, is usually preceded by a period of negotiation, by a continuously rising expectation of completing the agreement, and by actions to prepare oneself for the obligations that it will impose. The signing is followed by further actions to consolidate the plans that have been under way. In factory capacity, an old plant will be pushed to new limits while waiting for added facilities; temporary expediencies can fill a production gap; and the new facility probably starts producing gradually, and its full capacity may not be needed at first. Even major executive decisions represent a rather continuous process; they are reached after a period of consideration; advance actions may be taken in anticipation of the probable outcome; action is not taken immediately after the decision; and the decision is interpreted and "smoothed" and produces gradual change as it overcomes resistance and the inertia of other persons in the organization.

- There will usually be considerable "aggregation" (that is, grouping of individual events into classes) in our models. Orders arrive as separate pieces of paper, but we represent them as a continuous order flow, and our interest in the model, like the executive's interest in the company, is from a viewpoint above the separate individual transactions.

- A continuous-flow system is usually an effective first approximation even where repetitive but discrete decisions and events do occur. It provides a good starting point from which to add later the greater reality of separated actions where such a representation is necessary.

- There is a natural tendency of model builders and executives to overstress the discontinuities of real situations. This is partially compensated by stressing here the continuously varying nature of all flows in the system.

- A continuous-flow model helps to concentrate attention on the central framework of the system. This framework is more orderly and unchanging than usually expected. Diversion of attention toward separate isolated events tends to obscure the central structure of the system that we are trying to define.

- As a starting point, the dynamics of the continuous-flow model are usually easier to understand and should be explored before complications of discontinuities and noise are added. For this reason, it seems better to start with a continuous representation of the system than with a stochastic model (one that generates each decision on a random basis from some controlled probability distribution). Noise (random disturbances) can later be added to the decision functions in the equations. Very often such noise will adequately generate the effects on the system that would be created by the discontinuous generation of decisions at spaced points in time.

- A discontinuous model, which is evaluated at infrequent intervals, such as an economic model solved for a new set of values annually, should never be justified by the fact that data in the real system have been collected at such infrequent intervals. The model should represent the continuously interacting forces in the system being studied.[8] The frequency with which measurements on the real system may happen to have been taken is not relevant to the frequency with which internal dynamic performance must be calculated.[9]

These comments must not be construed as suggesting that the model builder should lack interest in the microscopic separate events that occur in a continuous-flow channel. The course of the continuous flow is the course of the separate events in it. By studying individual events we get a picture of how decisions are made and how the flows are delayed. The study of individual events is one of our richest sources of information about the way the flow chan-

[8] Nor can we safely assume that long- and short-range phenomena can be separated into different models that would allow the computation of long-term growth factors at widely spaced time intervals. In a nonlinear system the separate phenomena may not be superposable, and the long-term characteristics may depend on the nature of the superposed short-term fluctuations.

[9] The theory of sampled-data servomechanisms deals with the intervals that are permissible between solutions, relative to the bandwidth characteristics of the system.

nels of the model should be constructed. When a decision is actually being made regularly on a periodic basis, like once a month, the continuous-flow equivalent channel should contain a delay of half the interval; this represents the average delay encountered by information in the channel.

The preceding comments do not imply that discreteness is difficult to represent, nor that it should forever be excluded from a model. At times it will become significant. For example, it may create a disturbance that will cause system fluctuations that can be mistakenly interpreted as externally generated cycles (such as where annual design changes, or summer employee vacations, or yearly budgeting procedures create effects that resemble a seasonal demand in consumer sales). When a model has progressed to the point where such refinements are justified, and there is reason to believe that discreteness has a significant influence on system behavior, discontinuous variables should then be explored to determine their effect on the model.

5.6 Stability and Linearity

In formulating a model, we should not base our work on the assumption that the system being represented will necessarily be stable.[10] Many models that appear in the management and economics literature have assumed that the system being represented is linear. It has then been necessary, in consequence, to assume a stable system. This excludes the study of a class of systems that appears very important in

economics and industry — the system which is bounded by nonlinear influences and is stable only for large amplitudes of disturbance.

There is strong indication that among real systems of importance some are unstable in the usual mathematical sense. They do not tend toward a state of static equilibrium (even in the absence of randomness or external disturbance). They are unstable, tending toward increasing amplitudes of oscillation that are contained by a continuously shifting balance of forces among the system nonlinearities. Our social systems are highly nonlinear and most of the time are operating against limitations of overemployment, politically unacceptable unemployment, money shortage, pressures to overcome inflation or recession, or inadequacy of capital equipment. It seems likely that such nonlinearities, coupled with the unstable tendencies caused by amplifications and time delays, create the characteristic modes of behavior that we see in free-enterprise economic systems.

"Small-signal" linear analysis is not justified for nonlinear, unstable systems. Such an analysis of small disturbances from equilibrium necessarily assumes a static equilibrium point and a system that tends to return to such a point. Instead, results from working with dynamic models[11] and the observed behavior of actual systems both suggest that the important manifestations are often "large-signal" and nonlinear phenomena.

Formulating a model in a way that presupposes stability may exclude from consideration some of the most interesting and significant system characteristics.

---

[10] By stable is meant a tendency toward a state of equilibrium, either by following a nonoscillatory or a damped-oscillation approach to equilibrium.

[11] For example, see Figure 2-11.

CHAPTER 6

Structure of a Dynamic System Model

A basic structure of alternating levels and flow rates seems to represent the nature of industrial management systems. The levels determine the decisions that control flow rates. The flow rates cause changes in the levels. These levels and rates make up six interconnected networks that constitute industrial activity. Five of these represent materials, orders, money, capital equipment, and personnel. The sixth, the information network, is the connecting tissue that interrelates the other five.

WE are now ready to discuss a type of model structure that is amenable to the objectives and principles outlined in earlier chapters. A very simple model framework will suffice. A particular model may become complex because of its size and wealth of detail, but its fundamental nature will still be one of alternating "levels" and "decisions," as will be outlined in this chapter.

The form of a model should be such as to achieve several objectives. The model should have the following characteristics:

- Be able to describe any statement of cause-effect relationships that we may wish to include

- Be simple in mathematical nature

- Be closely synonymous in nomenclature to industrial, economic, and social terminology

- Be extendable to large numbers of variables (thousands) without exceeding the practical limits of digital computers, and

- Be able to handle "continuous" interactions in the sense that any artificial discontinuities introduced by solution-time intervals will not affect the results. It should, however, be able to generate discontinuous changes in decisions when these are needed.

6.1 Basic Structure

The preceding requirements can all be met by an alternating structure of reservoirs or levels interconnected by controlled flows, as shown in Figure 6-1.

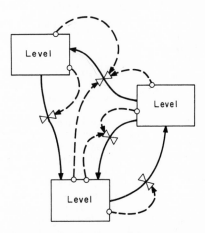

Figure 6-1 contains four essential features that will later be discussed separately:

Figure 6-1 Basic model structure.

- Several *levels*
- Flows that transport the contents of one level to another
- Decision functions (drawn as valves) that control the *rates* of flow between levels
- Information channels that connect the decision functions to the levels

This is the basic structure that will be used here, even with industrial and economic models that at first glance may seem vastly more complex. If the reader acquires a good grasp of this structure and the basic equations of the next chapter that go with it, he should have no difficulty with even extensive models. At each separate point the larger systems will still correspond to one of the above four features in the fundamental structure of Figure 6-1.[1]

The following sections will discuss each of the four features of the basic model structure.

Levels. The levels are the accumulations within the system. They are the inventories, goods in transit, bank balances, factory space, and number of employees. Levels are the present values of those variables that have resulted from the accumulated difference between inflows and outflows.[2] Levels exist in all of the six networks to be discussed later: information, materials, orders, money, personnel, and capital equipment.

It is very important to note that the units of measure of a variable do not serve to distinguish

levels from rates. Some levels have the dimension of units over time (as units per week). This may cause confusion until the basic distinction between levels and rates is clear.

A good test to determine whether a variable is a level or a rate is to consider whether or not the variable would continue to exist and to have meaning in a system that had been brought to rest. If all activity, in the form of flows, were to cease, the levels would still exist. Stopping the receiving and shipping of goods does not affect the continued existence of inventory that is in a warehouse. If all activity in a system is momentarily stopped, rates are unobservable. There is no movement to be detected, but the levels continue to exist. The levels of physical quantities such as goods, money, and personnel would be countable in the stationary system.

Levels exist in the information network as well as in the physical network of material, etc. "Awareness levels" exist in the mental attitudes that influence decisions. Levels of satisfaction, of optimism, and of recollection of a past disastrous depression, all influence economic behavior. The level of awareness of inflation trends influences investment. In a more prosaic context, the level of recent business activity influences ordering and inventory decisions. All memory and continuity from the past to the future exist in the levels of the system.

In the information network especially, we are apt to find levels that carry the dimensional units usually thought of as belonging to rates. Take, for example, the *average* sales *level* for the last year. It may be measured in items per month or dollars per year. It is not a rate. It is not a flow that indicates the moment-by-moment, present transfer rate between any two levels. In a continuously rising trend, the average sales rate that is computed at any given moment would never be equal to the actual current sales rate at that same moment. Average sales are often spoken of in business practice as a level — the level of sales or the level of

---

[1] The reader who is already familiar with the mathematical organization of a problem for solution on a differential analyzer will note the identity of Figure 6-1 to a system of integrators and function generators. The equations of the next chapter are a set of first-order difference equations equivalent to the integrations ordinarily executed by the differential analyzer. As already noted in Chapter 1, the analog-type differential analyzer is inadequate because it does not readily handle problems of the size or degree of nonlinearity that we shall deal with in the field of management. The digital computer can handle the same class of problems to perhaps 100 times the size that has been done successfully on analog machines.

[2] Levels are the time integrals of the net flow rates. A level may have any number of inflow or outflow channels.

business. The average sales *level* is obtained by integrating (summing) the moment-by-moment actual sales *rate* over some period of time like a year; it is therefore of the same mathematical equation form that describes the other levels in the system. By the test of bringing the system to rest, we see that average sales for the last year is a level. We can stop all present sales and shipping activity without destroying the concept and numerical value of average sales for the preceding year.

Flow Rates. From the preceding discussions of levels, the nature of flow rates has emerged. Rates define the *present,* instantaneous flows between the levels in the system. The rates correspond to *activity,* while the levels measure the resulting *state* to which the system has been brought by the activity. Rates as well as levels exist in all six networks that may constitute a system — materials, orders, money, personnel, capital equipment, and information.

The rates of flow are determined by the levels of the system according to rules defined by the decision functions. In turn the rates determine the levels. The levels determining a particular flow rate will usually include the level from which the flow itself comes.

Decision Functions. The decision functions (called rate equations in some later sections) are the statements of policy that determine how the available information about levels leads to the decisions (current rates). All decisions pertain to impending action and are expressible as flow rates (generation of orders, construction of equipment, hiring of people). As will be discussed in Section 10.4, the decision functions pertain both to managerial decisions and to those actions that are inherent results of the physical state of the system.

A decision function may appear as a simple equation that determines, in some elementary way, a flow in response to the condition of one or two levels (like the output of a transportation system that may often be adequately represented by the goods in transit, which is a level,

and the average transportation delay, a constant). On the other hand, a decision function may be described by a long and elaborate sequence of computations that progresses through the evaluation of a number of intermediate concepts (for example, a personnel hiring decision might involve the following levels: present employees, average order rate, employees in training, employee requisitions already initiated, backlog of unfilled orders, present inventory levels, available capital equipment, available materials, etc.).

Information as a Basis for Decisions. In Figure 6-1 it is shown that the decision functions determining the rates are dependent only on information about the levels. Rates are not determined by other rates. This is always true in principle.[3]

In principle, present, instantaneous rates are not available as inputs to the making of other decisions. In fact, present rates are in general unmeasurable. If we shorten sufficiently what we mean by the "present" time, we have no knowledge of other action rates occurring at the same time. We do not know the sales rate of our company at precisely this moment; in fact it would be meaningless to us in view of the normal short-run fluctuations that occur. What we usually mean by "present" rates are really averages (levels) such as average sales for the week, or the month, or last year.

By making the time span sufficiently short, we can in principle establish the impossibility of having a particular decision dependent on some other *present* decision (or rate) that is being determined at the same instant in another part of the system. (Another decision stream might depend on the same input levels.) The decision to attempt to hire more employees in New York is not conditioned by the corresponding decision at the same moment by a

---

[3] At times the modelmaker may for simplicity take liberties with this principle by letting a generated instantaneous rate be used instead of a very short run average.

competing firm in Chicago. The two are eventually interconnected by resulting levels such as available unemployed workers and inventories of products.[4]

The principle of independence of decisions is applicable in practice and is effective as a cornerstone of model formulation. It does not lead to oppressively short time intervals for computation in a model. It makes possible a formulation that is free of simultaneous algebraic equations with their computationally expensive requirements for matrix inversion.

6.2 Six Interconnected Networks

The basic model structure in Figure 6-1 shows only one network, with a rudimentary set of information ties from levels to rates. Several interconnected networks are necessary to represent industrial activity.

In Figure 6-1 it should be noted that flow rates transport the contents of one level to another. Therefore, the levels within one network must all have the same kind of content. Inflows and outflows connecting to a level must transport the same kind of items that are stored in the level. For example, the network for material deals only with material and accounts for the transport of the material from one inventory to another. Items of one type must not flow into levels that store another type.

It will be convenient to identify six networks to represent the grossly different types of variables that will be encountered — orders, materials, money, personnel, and capital equipment, all interconnected by information. This particular subdivision into six networks is arbitrary. Any one of the networks may appear in several separate segments; the *kind* of materials in one segment of the materials network may be dif-

ferent and not mixable, for our purposes, with another kind, and therefore the levels of one segment cannot connect by flow rates with the levels of another segment.

The information network can extend from a level in any one of the six networks to a rate in the same or any other network. The information network is therefore in a unique and superior position relative to the other five. This should be contrasted with the superior position that has often been assigned previously to money in economic analysis. We must stretch the concept of money beyond any reasonable meaning to make it serve the purpose here achieved by a distinct information network.

Materials Network. Here we include all rates of flow and stocks of physical goods, whether raw materials, in-process inventories, or finished products.

Orders Network. Here are orders for goods, requisitions for new employees, and contracts for new plant space. Orders are the result of decisions that have not been executed into flows in one of the other networks. In general they form a link between overt decisions[5] and results in the form of implicit decisions in the materials, money, personnel, or capital equipment networks.

Orders may at times be difficult to distinguish from concepts in the information network. However, orders customarily have a sufficiently clear identity and important position in the system that a separate recognition seems warranted. The orders network will usually occur in fragments that connect the order-generating decision to the order-filling action.

Money Network. Money is used here in the cash sense. Money flow is the actual transmittal of payments between money levels.

From the standpoint of a model of company operations, the bank balance is a money level. Accounts receivable would not be money, even though they are salable like an inventory of

[4] As in the actual economic world, production and consumption rate equations in a model should not be instantaneously equal. They are separately and independently reached but are interacting in time through their separate effects on inventories, which are an input level both to the *desire* to produce and to the physical *ability* to purchase.

[5] Overt and implicit decisions are defined in Section 10.4.

goods; they can be carried in the information network as information telling us our rights to the receipt of payment. Price is also information, and not part of the money network.

At the point of filling orders, three flows are created. First is the flow of completed orders out of the unfilled-order backlog. Second is the flow of physical goods from supplier to customer. Third is the flow of information to accounts receivable, which is the product of price multiplied by the goods flow rate. The *level* of accounts payable at the customer, along with invoice payment delay and his ability to pay, determine the *rate* of money flow from the customer's bank account to the supplier. This flow rate increases the supplier's bank account level, and information about the rate decreases the level of accounts receivable.

Different points of view may lead to different treatment of information and money. The bank account, treated by the company as a money level, is to the bank an account payable. The account contains no money but is information concerning the right of the company to make a withdrawal. Representation of money, credit, and information will require more subtle treatment as the refinements of a banking and national monetary system are incorporated into more comprehensive dynamic models.

Personnel Network. Many important dynamic effects in the company arise because of the policies and behavior patterns surrounding the acquisition and utilization of people. Company policies, labor union contracts, and availability of people establish a framework that affects the changing of employment levels. The timing of such changes can interact with other parts of the company to produce unexpected results. In Part III, Chapters 17 and 18, is an example of the way the interaction of the manufacturer's labor policy with the customers' ordering policy creates instability in an industry.

In the personnel network we deal with people as countable individuals, not as man-hours of labor. Man-hours per week is a product of men multiplied by the length of work week. Different considerations control changes in numbers of men from those that circumscribe the changes in length of work week. In most situations we shall need to distinguish men in the personnel network from the variables that are in the information network, such as length of work week and the productivity per man-hour.

Capital Equipment Network. The capital equipment network includes factory space, tools, and equipment necessary to the production of goods. It describes the way that factories and machines come into existence, the stock of existing capital equipment, what part (to measure the level of productivity) of our capital equipment stock is in use at any instant, and the discard rate of capital equipment.

Within a capital goods producing sector, it should be noted that the capital equipment output is a result, among other things, of material that flowed into the capital equipment production facility. The capital equipment producer has his own stock of equipment used for the production process.

Interconnecting Information Network. The information network is itself a sequence of alternating rates and levels. In this book it is raised to a position superior to the other networks because it is the interconnecting tissue between all of them.

In general, the information network starts at the levels and rates in the other five networks and ends at the rate-generating decision functions in those five. It transfers level information to decision points, and rate information in the other networks to the levels in the information network. Levels and rates of flow exist within the information network itself. For example, *information* about the actual, current *rate* of incoming orders is averaged to produce the *level* of average incoming-order rate. This is a level in the information network and will usually be one of the inputs to an ordering decision in the order network.

The value of an information variable must

not be confused with the true variable it represents. *Orders* usually go into an unfilled-order file; *information* about the level of average order rate and about the level of unfilled orders may go many places in the company. Information will often be delayed, as will flows in the other five networks. As discussed in Section 5.2, information may contain noise and bias. Information is not necessarily identical in value to the "true" variable that it represents.

The information network will contain the generation of various "concepts" that are inputs to decision making, such as desired inventory level, projected size of necessary plant, forecast of sales, desired employment level, and knowledge from research results.

The major fraction of a model will usually be within the information network. Information is the essence of decision making. It is the linkage that causes the other five networks to act on one another. In many economic models, and in our attitudes toward accounting, there is often an effort to visualize money in this central, interlocking relationship to the other parts of industrial activity. The money network does not provide adequate inputs for creating actual managerial and economic decisions. The money network constitutes a summary of past transactions and acts as a restraint on future decisions but is not a sufficient guide to the making of those decisions.

Careful observation of the information network of an organization is essential to understanding its true character. Its dynamic behavior cannot be represented without a proper treatment of the flows, levels, delays, and distortions that lie in the information channels that interweave the separate organizational components.

CHAPTER 7

System of Equations

The model structure described in Chapter 6 leads to a simple system of equations that suffices for representing information-feedback systems. The equations tell how to generate the system conditions for a new point in time, given the conditions known from the previous point in time. The equations of the model are evaluated repeatedly to generate a sequence of steps equally spaced in time. Level equations and rate equations generate the levels and rates of the basic model structure. In addition, auxiliary, supplementary, and initial-value equations are used. The interval of time between solutions must be relatively short, determined by the dynamic characteristics of the real system that is being modeled. Simple first-order integration of level equations is satisfactory.

IN the preceding chapter an arrangement for the basic structure of a model was illustrated in terms of levels (reservoirs) interconnected by rates of flow. To describe this general structure, we must have a suitable system of equations. The equations should be compatible with the five objectives of a model listed at the beginning of Chapter 6.

The system of equations should be adequate to describe the situations, concepts, interactions, and decision processes that constitute the world of management and economics. The model should be able to fit our concepts of reality rather than requiring violent and unacceptable simplifications to bring those concepts into the bounds of a model structure.

The equations to be described here form a basic system going with, and elaborating slightly on, the model structure already described. Classes of equations are discussed in this chapter but not the special forms that particular equations of these classes may assume.[1]

[1] In particular, this discussion omits various intermittent operations like periodic sampling and "boxcar" delays that are useful in segregating and saving historical information, as in bringing forward annual sales curves from past years and separating capital equipment by age and productivity.

Basically, the equation system consists of two types of equations corresponding to the levels and the rates discussed in the preceding chapters. In a following section will be introduced other incidental types of equations which add convenience and clarity to complex systems but which are not fundamental to the model structure. Before doing that, the fundamental time sequence of computation can be described in terms of levels and rates.

7.1 Computing Sequence

A system of equations is written in the context of certain conventions that state how the equations are to be evaluated. We are dealing here with a system of equations that controls the changing interactions of a set of variables as time advances. This system evolution implies that the equations will be computed periodically to yield the successive new states of the system.

At each point in time, there may be a particular sequence of computation implied by the equation system. The sequence to be used here is shown in Figure 7-1. The continuous advance of time is broken into small intervals of equal length DT (Standing for "delta time," or time increment). By definition this interval must be short enough so that we are willing to accept constant rates of flow over the interval as a

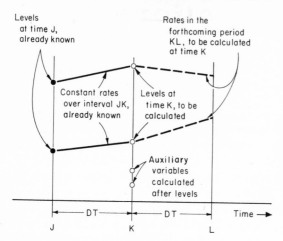

Figure 7-1 Calculations at time K.

satisfactory approximation to continuously varying rates in the actual system. This means that decisions made at the beginning of the interval will not be affected by any changes that occur during the interval. At the end of the interval new values of levels are calculated, and from these new rates (decisions) are determined for the next interval.

It is clear that in principle we can select time intervals spaced closely enough so that straight-line segments over the intervals will approximate any curve as closely as we wish. Figure 7-2 shows such a straight-line approximation.

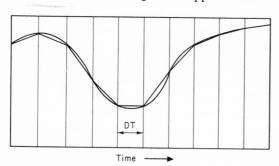

Figure 7-2 Straight-line approximation to a variable level.

Shorter and more numerous intervals would yield a still closer approximation to the curve. In practice, we shall be able to achieve as small an interval as needed and still have it long enough to keep the amount of computation within the capacity of present-day computing machines.[2]

Now let us return to Figure 7-1, where successive points in time are given the designations, J, K, and L. The instant of time K is used to denote the "present." The interval JK has just passed, and information about it and earlier times is in principle available for use. No information from a time later than K, like the interval KL, or the time L, or beyond, can ever be available for use in an equation being evaluated at present time K.

There are no exceptions to this unavailability of future information. Forecasts are not future information; they are *present* concepts *about* the future, based on information available in the *present* and *past*.

For the purposes of numerical evaluation, the basic equations of a model are here separated into two groups — the level equations in one group and the rate equations in another. For each time step, the level equations are evaluated first, and the results become available for use in the rate equations. (Auxiliary equations, to be discussed later, are an incidental convenience and are evaluated between the level and rate groups.)

The equations are to be evaluated at the moments of time that are separated by the solution intervals DT. The equations are written in terms of the generalized time steps J, K, and L, using the arbitrary convention that K represents the "present" point in time at which the equations are being evaluated. In other words, we assume that the progress of the solution has

[2] The criteria for selecting this interval are discussed in Section 7.5 and Appendix D. Theoretical, sampled-data, information-feedback-system considerations demand a much smaller solution interval than that which has been attempted in many dynamic economic models.

just reached time K, but that the equations have not yet been solved for levels at time K, nor for rates over the interval KL.

The level equations show how to obtain levels at time K, based on levels at time J, and on rates over the interval JK. At the time K, when the level equations are evaluated, all necessary information is available and has been carried forward from the preceding time step.

The rate equations are evaluated at the present time K *after* the level equations have been evaluated. The rate equations can, therefore, have available as inputs the present values of levels at K.[3] The values determined by the rate (decision) equations determine the rates that represent the actions that *will be* taken over the *forthcoming* interval KL. Constant rates imply a constant rate of change in levels during a time interval. The slopes of the straight lines in Figure 7-1 are proportional to the rates and connect the values of the levels at the J, K, and L points in time.

After evaluation of the levels at time K and the rates for the interval KL, time is "indexed." That is, the J, K, L positions of Figure 7-1 are moved one time interval to the right. The K levels just calculated are relabeled as J levels. The KL rates become the JK rates. Time K, "the present," is thus advanced by one interval of time of DT length. The entire computation sequence can then be repeated to obtain a new state of the system at a time that is one DT later than the previous state. The model traces the course of the system through time as the environment (levels) leads to decisions and actions (rates) that in turn affect the environment. Thus the interactions within the system are made to follow the "description" that has been set down in the equations of the model.[4]

---

[3] The rates from the JK interval are also available and may sometimes be used as an expediency, but in principle are not properly or necessarily involved in the computation of new rates.

[4] In Appendix C will be found an example of the computational steps in evaluating a simple set of equations.

7.2 Symbols in Equations

Symbols used to represent quantities in the equations of the model should be chosen to have as much mnemonic significance as possible, that is, to remind us of the commonly used terminology of everyday affairs. Partly to agree with common usage, partly to recognize the restricted choice of characters available on digital computer output printers, we shall use only groups of capital English letters and Arabic numerals to signify the variables and constants in a model.[5] Thus the number of employees at Plant A might be represented as EPLTA; the inventory in warehouse number 5 could be INVW5; the inventory desired at the retail level could be IDR. Manufacturing output rate from the factory could be MOF.

Because of the limitations of computer printing equipment, no subscripts or superscripts will be used.

7.3 Time Notation in Equations

A convention must be adopted for indicating time notation so that we can specify the moment of time at which a numerical value applies. Often in the literature, time has been indicated by small subscripts below the line of printing; this is not readily accomplished on a typewriter and not at all on many computing-machine printers.

To designate time, we shall use one or two capital letters following a variable and separated by a period.

Thus in the previous examples, the *level* of employees at time J would be EPLTA.J, the level at time K would be written as EPLTA.K. Note that the single letter is used as a time suffix because the values of levels are calculated at, and are correct only at, the separate instants

---

[5] To agree with the specifications of the DYNAMO computing system (see Appendix A), which has been designed especially for this equation system, we shall use groups of five characters or less, of which the first will be alphabetic.

of time J and K, respectively. Levels (and auxiliary variables to be discussed later) will have a single time-notation letter.

On the other hand, rates will be shown by a double letter. For example, the manufacturing rate which existed *during* the interval from J to K is written as MOF.JK, and the rate which *will* exist in the succeeding interval is MOF.KL.[6]

Constants will have no time notation. They do not change from one interval to the next. The constant delay in shipping to retail could be indicated by DSR.

7.4 Classes of Equations

Level and rate equations have already been discussed in describing the fundamental nature of the dynamic model structure to be used here. Reference has been made to other equation types which are convenient but which do not contribute new dynamic characteristics to the model. These are the auxiliary, supplementary, and initial-value equations. All of the equation forms will now be discussed.

Level Equations. Levels are the varying contents of the reservoirs (stocks, inventories, balances) of the system. They would exist (for example, an inventory) even though the system were brought to rest and no flows existed. New values of levels are calculated at each of the closely spaced solution intervals. Levels are assumed to change at a constant rate between solution times, but no values are calculated between such times.

---

[6] It should be noted that no values earlier than those at time J are used or need to be used in the equations. This is a practice unlike that in many difference-equation systems, in which a past time series has been saved by using a time notation extending prior to the last interval. The same effect can be obtained by defining a variable that is the *present* knowledge about a certain past time, for example, our *present* information about the level of sales the year before last. Here exponential delays and smoothing will carry over earlier values, and when these are not suitable, boxcar delays (see Section 8.9) will be employed.

The following is an example of a typical level equation:

$$IAR.K = IAR.J + (DT)(SRR.JK - SSR.JK) \qquad 7\text{-}1,\ L$$

The symbols represent variables as follows, given with the dimensions of measurement:

IAR   Inventory Actual at Retail (units), "actual" being used to distinguish from "desired" and other inventory concepts

DT   Delta Time (weeks), the solution-time interval between evaluations of the set of equations

SRR   Shipments Received at Retail (units/week)

SSR   Shipments Sent from Retail (units/week)

The indication "7-1, L" on the right shows that this is the first equation in Chapter 7 (all equations are given a numerical identification) and that it is a level equation "L."[7]

The equation states the straightforward accounting relationship that the present value of inventory IAR at time K will equal the previously computed value IAR.J plus the difference between the inflow rate SRR.JK during the last time interval and the outflow rate SSR.JK, the difference in rates multiplied by the length of time DT during which the rates persisted. In short, what we have equals what we had plus what we received less what we sent away.[8]

Note that the "dimensions" of each term in the equation are "units" of goods. In the parentheses on the right, "units" result from multiply-

---

[7] This is not to be confused with the instant of time L that will appear in rate equations as a KL time-step indicator.

[8] Notice that level equations are "integrating" equations that accumulate the net result of past rate of change in the level. If we were using a differential-equation formulation, these equations would appear as integrals:

$$IAR = IAR_{t=0} + \int_0^{t_1}(SRR - SSR)\ dt$$

where $IAR_{t=0}$ is the initial value of inventory at the beginning of the model run. Since a digital computer operates with difference equations, and since they seem easier than differential equations for most persons to visualize, we shall formulate models directly in the difference equations that are to be computed.

ing time in fractions of a week by flow rates in units/week.

The flow rates are always measured in some customary unit of time such as day, week, or month, not in terms of the solution interval DT. But the time units of the rates and the time units of DT must be the same, for example, all in weeks or months. The equation is valid and independent of the solution interval DT so long as the interval does not exceed a maximum value to be discussed later. The solution interval can be changed without requiring any change in the equation formulation or in any constants that may have appeared. By letting the interval DT appear explicitly in the equation, we can follow the procedure of using the normal customary units of measure found in the real system that is being modeled.

Level equations are independent of one another. Each depends only on information before time K. It does not matter in what order level equations are evaluated. At the evaluation time K, no level equation uses information from other level equations at the same time. A level at time K depends on its previous value at time J and on rates of flow during the JK interval.

Variables that should be classed as levels may carry units of measure, like "items per week," that at first may seem to indicate rates. The "stationary-system" test as discussed in Section 6.1 should be applied. As explained there, *average rates* are by nature levels, not rates.

Rate Equations (Decision Functions). The rate equations define the rates of flow between the levels of the system. The rate equations are the "decision functions" that will be discussed further in Chapter 10. A rate equation is evaluated from the presently existing values of levels in the system, very often including the level from which the rate comes and the one into which it goes. The rates in turn cause the changes in the levels. The rate equations may be either of the "overt" or "implicit" [9] decision

[9] See Section 10.4.

types; there is no structural or evident difference in the equations themselves.

The rate equations, being decision equations in the broad sense to be discussed later, are best thought of as controlling what is to happen next in the system. A rate equation is evaluated at time K to determine the decision governing the rate of flow over the forthcoming interval KL. Rate equations in principle depend only on the value of levels at time K.[10] (In practice, *rates* from the last time interval JK may occasionally be used as a short cut in place of the *level* of an *average* rate where a very short averaging time would have been used.)

Rate equations are evaluated independently of one another within any particular time step, just as level equations are. Interaction occurs by their ensuing effect on levels that then influence other rates at later times. A rate equation determines an immediately forthcoming action. If this action is sufficiently immediate (that is, the length of the solution interval DT is sufficiently short), it is evident that the decision cannot itself be affected by other decisions being made at the same time instant in other parts of the system.[11] The rate equations are therefore independent of one another and can be evaluated in any sequence. Since they do depend on the values of the levels, the group of rate equations is evaluated after the level equations.

An example of a rate equation is provided by the outflow rate of a first-order exponential delay. The justification of the equation will be given in Chapter 9; its form will be examined here.

[10] Auxiliary variables appear in rate equations merely as subdivisions of the rate equation and so are not a violation of this principle.

[11] A sufficiently short time interval precludes intercommunication of information between decisions within one time step. Meeting this requirement does not lead to excessively short solution intervals. A decision can depend on as many information inputs as are thought significant. The value of any one level may be used by several separate decision points at any one time step.

$$OUT.KL = \frac{STORE.K}{DELAY} \qquad \text{7-2, R}$$

in which the letter groups are defined as

OUT   the OUTflow rate   (units/week)

STORE   the present amount STOREd in the delay (units)

DELAY   a constant, the average length of time to traverse the DELAY   (weeks)

The equation is our second equation and is a rate equation, "7-2, R." The equation defines the rate "OUT" and tells us the value over the next time interval KL. The rate is to be equal to the value of a level "STORE" at the present time K, divided by a constant that is called "DELAY" (no time notation because it is a constant). At the time of evaluating the equation, there would of course be available numerical values for STORE and DELAY.

Auxiliary Equations. Often a rate equation will become very complex if it is actually formulated only from levels, as thus far stated. Furthermore, a rate may often be best defined in terms of one or more concepts which have independent meaning and, in turn, arise from the levels of the system. It is often convenient to break down a rate equation into component equations that we shall call "auxiliary equations." The auxiliary equation is a great help in keeping the model formulation in close correspondence with the actual system, since it can be used to define separately the many factors that enter decision making.

As the name implies, auxiliary equations assist but are incidental. They can be substituted forward into one another (if there are several "layers" of auxiliary equations) and thence into the rate equations.[12] By algebraic substitution, auxiliary variables can be made to disappear at the expense of increasing the complexity of the rate equations, and probably at the same

time losing the simplicity and obscuring the meaning of the equations of the model.

The auxiliary equations are evaluated at time K but *after* the level equations for time K because, like the rates of which they are a part, they make use of present values of levels. They must be evaluated *before* the rate equations because their values are obtained for substitution into rate equations.

Unlike the level and rate equations, the auxiliary equations cannot be evaluated in an arbitrary order. Some auxiliary equations may be components of others. Two or more auxiliary equations may form "chains" that must be evaluated in the proper order so that one can be used in the next. A possible substitution sequence must exist if the equation formulation is correct. A closed loop of auxiliary equations where two variables depend on the values of each other indicates a set of simultaneous equations and represents a forbidden and unnecessary equation formulation.[13]

The following is a chain of two auxiliary equations between two levels and a rate equation:

$$IDR.K = (AIR)(RSR.K) \qquad \text{7-3, A}$$

where RSR is a level, and AIR is a constant

$$DFR.K = DHR + DUR\frac{IDR.K}{IAR.K} \qquad \text{7-4, A}$$

where IAR is a level, and DHR and DUR are constants

$$SSR.KL = \frac{UOR.K}{DFR.K} \qquad \text{7-5, R}$$

where UOR is a level.

Note that in Equation 7-3, A (for auxiliary) the level RSR at time K is used as an input to the auxiliary variable IDR at time K. The terms AIR, DHR, and DUR are constants. Also at time K, IDR is an input, along with another level, to the auxiliary variable DFR. In turn, DFR is used,

---

[12] This substitution does not raise the "order" of the difference equations, since the resulting rate equation still contains only information from the levels at time K.

[13] The DYNAMO automatic coding system checks for this and many other erroneous conditions, and also rearranges the equations into proper order.

along with another level, in the rate equation, 7-5, R, to define the rate SSR.

Note that Equation 7-3 can be substituted into Equation 7-4 and the result into Equation 7-5 to give

$$SSR.KL = \frac{UOR.K}{DHR + DUR\ \dfrac{(AIR)(RSR.K)}{IAR.K}} \qquad 7\text{-}6,\ R$$

The auxiliary equations have disappeared, leaving the rate SSR dependent only on levels and constants.

In Chapter 15, Equations 7-3, 7-4, and 7-5 are discussed in the context of an industrial situation. Each of the auxiliary equations defines a variable that has important conceptual meaning. These separate concepts would be hopelessly obscured if we actually made the substitution indicated by Equation 7-6.

An auxiliary variable, in principle, depends only on levels that are already known and on other auxiliary variables that can be computed before being needed. As noted for rate equations, values of rates from the preceding time interval JK may sometimes be used, improperly but as good approximations to short-term averages, in auxiliary equations.

Supplementary Equations. Supplementary equations are used to define variables which are not actually part of the model structure but which arise in printing and plotting values of interest about the model behavior. We may wish to compile information (like the sum of all inventories in the entire system) that is not used in any of the decision processes of the model. The notation "S" indicates a supplementary equation.

Initial-Value Equations. Initial-value equations are used to define initial values of all levels (and some rates) that must be given before the first cycle of model equation computation can begin. They also are used initially to compute values of some constants from other constants. The initial-value equations are evaluated only once before the start of each model run. The notation "N" indicates an initial-value equation.

7.5 Solution Interval

The solution interval DT must be short enough so that its value does not seriously affect the computed results. It should be as long as permissible to avoid using unnecessary digital-computer time.

The general requirement that limits the length of the interval arises from the organization of the system of equations. Levels determine rates and rates determine levels; but the equation system is "open," meaning that the information-feedback channels are uncoupled during the solution interval DT. Therefore the interval must be short enough so that the changes in levels between solution times do not lead to unacceptable discontinuities in the rates. Difficulties from too long an interval will be discussed in Appendix D.

In most of our systems, the exponential delays (to be discussed in Chapter 9) will determine the permissible computing interval. As will be seen, the interval *must* be less than the length of any first-order-delay element and *should* be less than half the delay. Since third-order delays are common and are equivalent to three cascaded first-order sections each containing one-third the delay, the solution interval should be less than one-sixth of the total length of the shortest third-order delay.

The above is a rule of thumb. The most expedient test will be to vary the length of the solution interval, to observe its effect on computed results.[14]

Another, and separate, criterion controlling the maximum permissible solution interval is the relationship between the values of levels that will be experienced in the system and the rates of flow to and from these levels. The solution interval must be short enough so that the net flow to or from a level will not make a major change in the content of the level in one solution interval. For example, if a high rate of flow

[14] See Appendix D.

is possible out of a small inventory, the solution interval should be short enough so that only a fraction of the inventory can be depleted in one solution interval. Clearly, if the interval is so long that more than the content of the inventory is removed, leading to a negative inventory, an absurd answer has resulted.

There is another, and more fundamental, consideration that theoretically influences the solution interval. Sampling theory dealing with intermittent flows in information-feedback systems relates the sampling interval (here the solution interval) to the bandwidth characteristics of interest in the system. (Bandwidth can be thought of as related to the abruptness with which a system can change its condition of operation.) The solution interval must be substantially less than the period of the shortest-period components that are to be properly generated by the computation. It is believed that the above rule-of-thumb guide, applied to the separate components of the model, will always yield an interval that is short enough to represent properly the individual components and that is also less than the maximum permitted by the total-system characteristics.

7.6 Redundancy of Equation Type and Time Notation

The time notation appended to the variables in the equations conveys some of the same information indicated by the equation type (that is the L, R, A, etc.). Level (L) equations evaluate the variable at time K based on variables on the right side of the equation from time J and from the interval JK. Auxiliary (A) equations for time K use information from levels at time K and other auxiliary variables at time K (also, as an occasional expediency, rate values from interval JK). Rate (R) equations yield a value of the rate for the KL interval based on level and auxiliary variables from time K (and also as an occasional expediency, rates from the previous interval JK).

Even though there is some redundancy between the information conveyed by the equa-

tion type and by the time notation, experience has shown that confusion is readily possible in defining equation types and in handling time notation. Therefore, for greater clarity both should be used.

7.7 First-Order versus Higher-Order Integration

It will be noted from examining the form of the level equations[15] that they are difference equations using first-order, step-by-step integration for generating levels from rates. In accurate scientific computation a higher-order integration method is often used. In our work the more elaborate numerical methods do not seem justified, and in fact they have serious disadvantages.

We are not working for high numerical accuracy. The information-feedback character of the systems themselves make the solutions insensitive to round-off and truncation errors. We shall even be intentionally inserting additional noise in the rates that are to be integrated. The solution interval can be empirically adjusted to observe whether or not the solutions are sensitive to the simplified numerical methods that are being used.

To use more complex numerical methods would make the equation formulation less understandable by the manager and economist who does not have ready facility with mathematical techniques. It seems that the simplicity advantages of the more direct formulation far exceed any small improvements in accuracy that might come from more subtle numerical methods.

7.8 Defining All Variables

Each equation defines one variable in terms of constants and other variables that will be already available when needed. There must be as many equations as variables[16] (including a generating equation or source of values for any external test inputs).

---

[15] Section 7.4.
[16] The DYNAMO system checks that all variables are defined once and only once. See Appendix A.

CHAPTER 8

Symbols for Flow Diagrams

The model structure of Chapter 6 and the interrelationships of the equations of Chapter 7 can be displayed diagrammatically. Such a flow diagram helps to prevent confusion and forms another means of communicating the nature of the model to those who are not proficient in "reading" mathematical equations. This chapter explains the symbols that will be used in Chapters 15 to 18.

EXPERIENCE from teaching the formulation of industrial dynamics models has demonstrated that a pictorial representation of an equation system is highly desirable. A diagram that displays the interrelationships between equations helps to lend clarity to the system formulation. Many people visualize the interrelationships better when these are shown in a flow diagram than they do from a mere listing of equations. A detailed flow diagram supplements a set of equations. It gives much of the same information but in a different manner. A properly constructed flow diagram is better than a set of equations for communicating the structure of a system to many practicing managers. A diagram represents an intermediate transition between a verbal description and a set of equations.

A flow diagram of the system interrelationships should be developed *simultaneously* with the development of the equations of a system. Many beginners attempt to develop equations first and then summarize these later in a diagram, if the diagram is used at all. Such neglect of a diagram until a later stage loses the clarifying advantage that a diagram can provide in the initial stages of equation formulation.

This chapter will describe a set of standard symbols for flow diagrams of dynamic models.[1] Symbolism for diagrammatic presentation is based on arbitrary choices that are selected to emphasize and clarify particular aspects of a situation.

The system of symbols used here shows the existence of the interrelationships in the system. It distinguishes levels from rates. It separates the six flow systems of information, materials, orders, money, personnel, and equipment from one another. It discloses what factors enter into each decision (rate) function. But the diagram does not reveal what the functional relationships are within the decision functions. For the specific nature of the interactions between the factors entering into a decision, the diagram carries the equation numbers, and we are thereby referred to the pertinent equation. The diagram corresponds point by point to the equations.

8.1 Levels

A level is shown by a rectangle, as in Figure 8-1. In the upper-left corner is the symbol group (IAR) that denotes this particular level variable.

[1] In Chapter 15 these symbols will be applied to a specific situation.

In the lower-right corner is the equation number to tie the diagram to the equations.

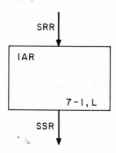

Figure 8-1   Levels.

No distinction need be made in the rectangle for different flow systems because the entering and leaving flow lines identify the kind of flow (here the solid line is for material flow). Arrowheads show the direction of flow to and from a level. The symbols identifying the flow rates are given beside the flow lines (unless otherwise shown nearby in the controlling decision function).

8.2   Flows

A flow occurs into and out of a level. The symbol for a flow line distinguishes the six types of flow systems, as shown in Figure 8-2. The

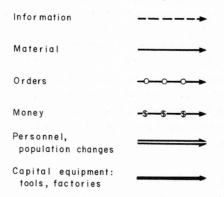

Figure 8-2   Flow symbols.

kinds of lines have been selected either to suggest the type of flow represented or to facilitate drawing.

8.3   Decision Functions (Rate Equations)

Decision functions determine the rate of flow. They act as valves in the flow channels, as shown in Figure 8-3. Two equivalent forms are

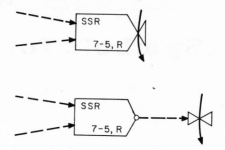

Figure 8-3   Decision functions (rate equations).

displayed. The symbol shows the flow that is being controlled and the information inputs (only information leads into a decision function, or rate equation) that determine the flow rate. The equation number that defines the rate is given.

8.4   Sources and Sinks

Often a rate is to be controlled whose source or destination is considered to lie outside the considerations of the model. For example, an order flow must start from somewhere, but clarity of flow system terminology does not permit a mere extension of information lines into the line symbol for orders. Orders are properly thought of as starting from a supply of blank paper that we need not treat in the model dynamics. Likewise, orders that have been filled

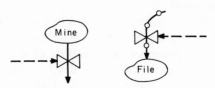

Figure 8-4   Sources and sinks.

must be discarded from the system into a completed-order file which usually has no significant dynamic characteristics. In a manufacturing process we may sometimes be justified in assuming that materials are readily available and that the characteristics of the source do not enter into system behavior.

In such circumstances the controlled flow is from a source or to a sink that is given no further consideration in the system, as shown in Figure 8-4.

8.5 Information Take-off

Information flows interconnect many variables in the system. The take-off of an information flow does not affect the variable from which the information is taken. As in Figure 8-5, an

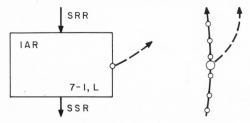

Figure 8-5   Information take-off from levels and rates.

information take-off is shown by a small circle at the source and, of course, by a dashed information line.

8.6 Auxiliary Variables

Auxiliary variables are concepts that have been subdivided out of the decision functions because they have independent meaning. They lie in the information flow channels between levels and the decision functions that control rates. They can be algebraically substituted into the rate equations.

Auxiliary variables are shown by circles, as in Figure 8-6. Within the circle are the name of

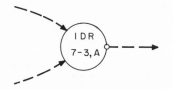

Figure 8-6   Auxiliary variable.

the variable and the equation number that defines it. The incoming information lines show the variables on which the auxiliary depends (levels or other auxiliary variables). The outflow is always an information take-off. The auxiliary variable is not an integration as is a level. No numerical value need be saved from one computational time step to the next. Any number of information lines can enter or leave.

8.7 Parameters (Constants)

Many numerical values that describe the characteristics of a system are considered constant, at least for the duration of computation

Figure 8-7   Parameters (constants).

of a single model run. These are shown by a line above or below the symbol of the constant, with an information take-off as in Figure 8-7.

8.8 Variables on Other Diagrams

Very often the diagram of a system is divided into sections. Figure 8-8 shows the handling

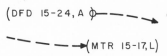

Figure 8-8   Variables appearing on other diagrams.

of sources and terminations of flow lines that lie on other sheets. The name of the variable and its equation number are given; the page number of the other diagram might be shown in addition.

8.9 Delays

Exponential delays (to be discussed in Chapter 9) can be represented by a combination of levels and rates of flow. However, delays are encountered so frequently that an abbreviated

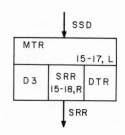

SSD	the input rate
MTR	quantity (level) in transit
15-17, L	equation for the level in transit
D3	order of the delay
15-18, R	equation defining the output rate
DTR	time constant of the delay
SRR	the output rate

Figure 8-9 Symbol for exponential delay.

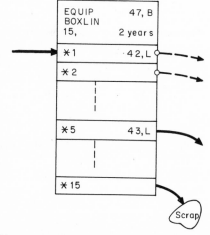

EQUIP	generic name given to the entire train
47, B	equation number that specifies the characteristics of the boxcar train
BOXLIN	specifies a linear progression with discard of last box
15	number of boxes in train
2 years	stepping interval

Equipment flow is shown into the first boxcar and is accumulated by level Equation 42.

Equipment leaves at Box 5 as controlled by a rate that is integrated by level Equation 43.

Discard is automatic from Box 15.

Information take-offs are shown from Boxes 1 and 2.

Individual boxes are identified by EQUIP∗1, EQUIP∗2, etc.

Figure 8-10 Linear boxcar train.

symbol is needed, as shown in Figure 8-9. This symbol takes the place of three levels with interconnecting rates (for a third-order exponential delay). In the box, D3 indicates a third-order delay; D1 would indicate a first-order delay.

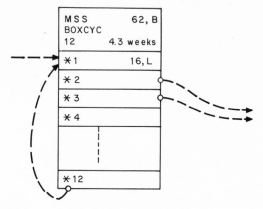

MSS	generic name of the entire train
62, B	equation number that sets up the characteristics of the boxcar train
BOXCYC	specifies recirculation of the last box into the first
12	number of boxes in the loop
4.3 weeks	stepping interval

Information flows into the first box (typical in a calculation of averages).

Information take-offs come from Boxes 2 and 3.

Individual boxes are designated in equations by MSS∗1, MSS∗2, etc.

Figure 8-11 Cyclic boxcar train.

For many purposes involving historical situations, we wish to segregate past information. For this, either a linear or a cyclic boxcar train is effective. The term "boxcar train" connotes segregated boxes that move with time. Here the movement is by stepping the contents of the boxes from one to the next at specified intervals.[2]

Figure 8-10 shows a linear boxcar train in which the contents of the boxes are moved

[2] The processes necessary to accomplish the boxcar train operations are automatically provided by the DYNAMO computer program compiler. See Appendix A.

down one step at the specified intervals and the contents of the last box are discarded. Each box has a designation that can be used in equations in the same manner as any other variable. The example shown might represent the thirty-year life of capital equipment, segregated by two-year age blocks. A parallel information boxcar train could hold separate productivity indices for each age of equipment.

Figure 8-11 shows a cyclic boxcar train in which the contents of the last box are reinserted into the first box. The example shown might be used to generate the average seasonal sales month by month by averaging corresponding months of past years. Such monthly values might enter into decisions that are based on seasonal considerations. Appendix N shows a cyclic boxcar train for seasonal sales analysis.

CHAPTER 9

Representing Delays

Delays are crucial in creating the dynamic characteristics of information-feedback systems. Delays are themselves composed of the standard level and rate equations of Chapter 7. Some forms of delays — exponential and pipeline — are discussed in the present chapter. Because of the prevalence of delays, a "shorthand" notation is introduced to stand for the actual level and rate equations that are to be used in representing a delay.

PRECEDING chapters have stressed the important contribution of delays to creating the characteristics of an information-feedback system. This chapter discusses methods of representing in mathematical models the kinds of delays encountered in our industrial and economic processes.

In principle, delays exist in all flow channels. However, to introduce a time delay in *every* flow would lead to a vast amount of model detail, much of which would contribute little to system behavior. Two kinds of simplifications will always be used to reduce the number of points at which delays must be introduced into a model formulation. First, many system delays will be judged to be so short that their effect is negligible compared with the other longer or more significantly located delays that are to be incorporated. Second, delays that arise from separate, actual processes which are cascaded one after the other can often be combined into a single delay representation. In addition, delays in parallel branches entering a common channel may often be combined by shifting them into the common channel.

9.1 Structure of Delays

A delay is essentially a conversion process that accepts a given inflow rate and delivers a resulting flow rate at the output. The outflow may differ instant by instant from the inflow rate under dynamic circumstances where the rates are changing in value. This necessarily implies that the delay contains a variable amount of the quantity in transit. The content of the delay increases whenever the inflow exceeds the outflow, and vice versa.

A delay is a special, simplified category of the general concept of inventories or levels. All levels exist to permit the inflow rates to differ, over limited intervals, from the outflow rates. In the general concept of a level, there is no restriction on the factors that might control the outflow. For example, the outflow from an inventory can be influenced both by the level of the inventory and by the level of unfilled orders. By contrast, a delay is here thought of as a special class of level wherein the outflow is determined only by the internal level stored in the delay (and by certain descriptive constants). For example, a transportation system can often be adequately represented merely as a delay. If so, the internal level of goods in transit is the only variable. A constant describes the average delay, and a specified computing process is applied to the internal level to generate the appropriate type of transient relationships between the internal level and the output rate.

Time delays in the flow of physical or informational quantities can be created by combinations of the level and rate equations already discussed in Chapters 6 and 7.

Time delays are therefore represented by "packages" consisting of a combination of rate and level equations that are inserted in a flow channel. They modify the time relationships between the input flow which is given and the resulting output flow which is generated by the delay.

9.2 Characteristics of Delays

Two characteristics of a delay are of immediate interest. One is the length of time expressing the average delay D. This fully determines the "steady-state" effect of the delay. By steady state we mean that inflow and outflow rates and the level between them are constant. Under these unvarying circumstances the inflow and outflow rates must be equal. In steady state the flow rate multiplied by the average delay gives the quantity in transit in the delay.

The second characteristic of a delay describes its "transient response." The transient response tells how the time shape of the outflow is related to the time shape of the inflow when the inflow rate is changing with time.

Different delays can have the same average delay D with very different transient responses to changes in input rate. We must take care in selecting the transient response of the delay to be used. If the transient response is grossly wrong, it might have an important effect on the qualitative behavior of a dynamic model.[1]

9.3 Exponential Delays

Various computational processes might be used to create a delay in a flow channel within a mathematical model. We shall consider here delay functions of only a single class — exponential delays. There is no need to exclude other kinds of functions that could be used to create

[1] For additional discussion of delays, see Appendices D and H.

delays in a flow; however, the exponential delays are simple in form, and they have adequate scope to fit our usual degree of knowledge about the actual systems to be represented.

A "first-order" exponential delay, Figure 9-1a, consists of a simple level (which absorbs

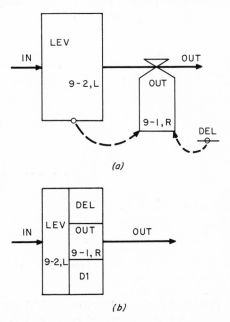

(a)

(b)

Figure 9-1 First-order exponential delay.

the difference between the inflow and outflow rates) and a rate of outflow that depends on the level and on the average delay (a constant). The inflow rate is determined by some other part of the system.

The outflow rate, by the definition of this class of delay, is equal to the level divided by the average delay:

$$\text{OUT.KL} = \frac{\text{LEV.K}}{\text{DEL}} \qquad \text{9-1, R}$$

where the letter groups have the following meaning:

OUT the OUTflow rate   (units/time)
LEV the LEVel stored in the delay   (units)
DEL the DELay constant representing the average time required to traverse the delay   (time)

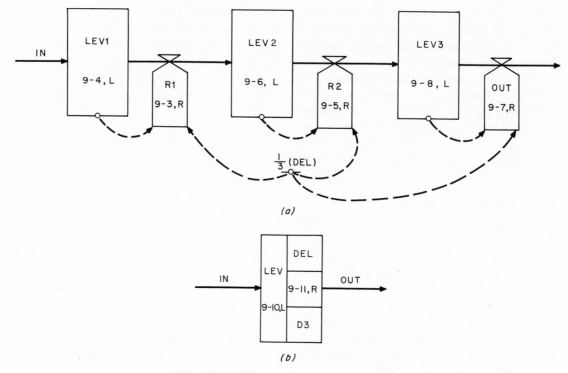

Figure 9-2 Third-order exponential delay.

Equation 9-1 is a rate equation and represents an "implicit" decision[2] because it is not normally considered to be the result of a managerial decision but results instead from the presently existing state of the system as manifested by the variable level LEV.

The representation of the delay is not complete unless there is an equation to generate the internal level of units in transit LEV. The level LEV stored in the delay is accumulated as the difference between inflow and outflow:

LEV.K = LEV.J + (DT)(IN.JK − OUT.JK)		9-2, L
LEV	the LEVel stored in the delay (units)	
DT	the solution interval between successive evaluations of the equation (time)	
IN	the INflow rate, given by some other equation in the system (units/time)	
OUT	the OUTflow rate (units/time)	

The rate equation and the level equation con-

[2] As discussed in Chapter 10.

vert the inflow rate IN into the delayed rate OUT.

For conciseness, and because delays will be such common building blocks of dynamic models, we shall use the symbol of Figure 9-1b in place of the more detailed information flow and decision diagram of Figure 9-1a. In Figure 9-1b, the half-box on the input side gives the name of the level variable and the equation defining the level. The three boxes on the output side give the symbol of the delay constant, the equation number defining the output rate, and the "order" of the delay (D1 for first-order).

"Higher-order" exponential delays are obtained by feeding the flow through two or more cascaded first-order delays. A first-order delay and a delay of higher order can have the same total average delay D but will differ in the transient response to *changes* in flow rate.

Figure 9-2a shows a third-order delay of

total over-all delay DEL. This total delay is distributed one-third in each first-order section. The third-order exponential delay is defined by three pairs of equations, like 9-1, R and 9-2, L, which relate input rate IN to outflow rate OUT:

$$R1.KL = \frac{LEV1.K}{(DEL)/3} \qquad 9\text{-}3, R$$

$$LEV1.K = LEV1.J + (DT)(IN.JK - R1.JK) \qquad 9\text{-}4, L$$

$$R2.KL = \frac{LEV2.K}{(DEL)/3} \qquad 9\text{-}5, R$$

$$LEV2.K = LEV2.J + (DT)(R1.JK - R2.JK) \qquad 9\text{-}6, L$$

$$OUT.KL = \frac{LEV3.K}{(DEL)/3} \qquad 9\text{-}7, R$$

$$LEV3.K = LEV3.J + (DT)(R2.JK - OUT.JK) \qquad 9\text{-}8, L$$

We are more apt to be interested in the total quantity LEV stored in transit in the delay than in the sections separately. If so,

$$LEV.K = LEV1.K + LEV2.K + LEV3.K \qquad 9\text{-}9, A$$

It would be laborious and entirely repetitive to write equations like 9-3 through 9-9 and to make a diagram like Figure 9-2a each time a third-order delay was to be represented.[3] We therefore adopt a shorthand for this frequently employed package. We shall usually be interested in the total in-transit quantity in the delay which is given by Equation 9-9 above but which can also be determined directly by accumulating the difference between the input IN and the output OUT as follows:

$$LEV.K = LEV.J + (DT)(IN.JK - OUT.JK) \qquad 9\text{-}10, L$$

which is the same as Equation 9-2, L.

To provide a concise abbreviation for Equations 9-3 through 9-8, we can use the following functional notation:

$$OUT.KL = DELAY3 \; (IN.JK, \; DEL) \qquad 9\text{-}11, R$$

---

[3] The equations for the third-order delay can be written in various forms, one being Equations 9-3 through 9-8, and another form being given in Appendix H. A digital-computer compiler program like DYNAMO can automatically insert the required group of equations in response to instructions of the form of Equation 9-11.

This is not a true equation but simply an indication that the necessary package of equations for a third-order delay is intended. Here "OUT" is the symbol group designating the name of the output flow; "DELAY3" indicates that a third-order delay is to be inserted in the flow "IN," and the average delay is "DEL."

Two equations of the form of 9-10, L and 9-11, R suffice to designate a third-order delay. Equation 9-10, L can be omitted if the total internal in-transit level is not required elsewhere in the model. As before, the abbreviated symbol in Figure 9-2b will be used in later chapters to designate a third-order delay.

9.4   Time Response of Exponential Delays

Now that we have seen the mathematical form of exponential delays, we shall here in this section examine their behavior. The transient response of an exponential delay changes as the number of first-order sections in it are increased.

As a vehicle for discussing delay characteristics, consider the particular example of a delay in shipping goods from a factory to distributors. The transient response of the actual physical shipping process is best visualized by thinking of shipping a large number of orders simultaneously by many different methods of transportation to a number of different distributors located in various places. This is an "impulse" input, where suddenly a large number of items are inserted into the transportation system at one moment, and we wish to consider their rate of arrival at their destinations.

For this example, average delay could be estimated fairly readily and might be a number of days, depending on the types of transportation and distances involved. Anticipating the proper transient response, that is, the shape of the output rate-of-arrival curve, will require a more thoughtful consideration.

To represent properly the transient response of the example, we could compare the expected

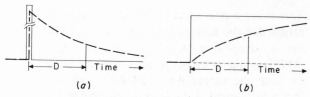

Figure 9-3 First-order exponential delay.

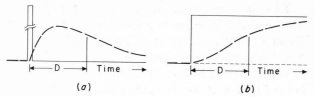

Figure 9-4 Second-order exponential delay.

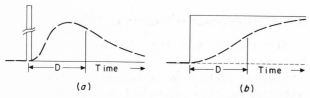

Figure 9-5 Third-order exponential delay.

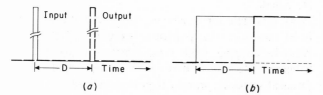

Figure 9-6 Discrete or pipeline delay.

behavior of the actual system with the different exponential delays that might be selected. Figures 9-3, 9-4, 9-5, and 9-6 show selections from the exponential family.

In the figures the solid line represents the input to the delay; the dotted line shows the output. Time moves to the right along the horizontal axis. In each figure the left-hand diagram is for an *impulse input;* that is, a *quantity* is inserted into the delay in a negligibly short time, and the dotted line shows the *rate* of arrival at the output. In the right-hand diagram, the input is a sudden step increase in *input rate,* and the dotted line again shows the resulting *output rate.*

Figure 9-3 shows a <u>first-order exponential</u> delay. In Figure 9-3*a* the maximum rate of output occurs immediately after an impulse input, and the output rate declines exponentially thereafter. <u>This is clearly not representative of the shipping delay,</u> since Figure 9-3*a* shows that the maximum rate of arrival at the destinations occurs at the instant the multiplicity of shipments is made. Certainly the maximum *rate* of arrival does not occur at the moment of shipment.

Figure 9-3*b* represents a step change (solid line) in input *rate* and the resulting exponential rise in output *rate* from a first-order delay. The area between the solid and dotted curves is a measure of the quantity that accumulates in the in-transit level in the delay. So long as there is a flow through the delay, the total amount delivered at the output is less (by the amount in transit) than the total amount that has been put in.

Figure 9-4 is the output response of a second-order exponential delay. The second-order delay is the equivalent of two first-order delays cascaded one after the other so that the output of the first is the input to the second. In Figure 9-4*a* the initial output rate in response to an impulse input is zero, and the output curve has its maximum slope at the origin. <u>This is also qualitatively unacceptable as a shipping delay because we should not expect the rate of arrivals to start rising rapidly at the moment that shipments leave the factory.</u>

In Figure 9-5 is a <u>third-order exponential delay.</u> This shape of output response is the first of the sequence that <u>satisfies the more obvious characteristics of the actual shipping process.</u> In Figure 9-5*a* the output response to an impulse input is initially zero. Also the initial slope of the output curve is zero. The curve begins to rise slowly, reaches a maximum slope and then a peak value, and falls off. In Figure 9-5*b* is the output following a step change in the input rate.

The third-order delay satisfies the major intuitive requirements that we should place on the delay function for the above example of shipping delay. To refine the delay function further would require a careful study in the actual system of item-by-item delays and their time distribution. It is unlikely that any further refinement will have appreciable effect on system behavior.

As the exponential delay of constant over-all length is broken into more-and-more, smaller-and-smaller, cascaded, first-order sections, the initial delay in response to an impulse is increased before the output begins to respond. The rise is steeper when once started, and the falling section is steeper and approaches the zero output rate sooner. The ultimate end member of this family is the hypothetical infinite-order delay.[4] This is sometimes called the "discrete delay" or the "pipeline delay." Figure 9-6 shows the infinite-order exponential delay wherein nothing happens at the output until after the elapsed delay D, at which time the input is reproduced exactly. Figure 9-6a shows an impulse of a certain quantity put into the delay, with the resulting impulse output appearing at time D. Figure 9-6b shows the response to a step-input rate. The input rate rises suddenly from zero to its final value, and the output does likewise D days later. This is clearly not a realistic representation of the shipping delay of the above example, since it implies that all shipments that were started at the same instant would be delivered at exactly the same instant D days later, no matter how far they had to travel.

In a situation different from the above shipping delay, for example the consumer buying rate response to a single advertisement placed in a daily newspaper by a department store, we might select a different shape of delay response. Perhaps the first-order exponential delay would seem better on the basis that the buying response is greatest immediately and falls off with time.

We may occasionally, as in representing the lead-time delay in establishing a revised factory manufacturing rate, want to have a more pronounced initial delay than is given by the third-order exponential. Cascading exponential delays increases the initial delay and steepens the slope of the rising part of the curve; therefore a sixth-order (two cascaded third-order delays) could be used.

Once a functional form is found that exhibits a qualitative shape that matches our existing empirical knowledge of the facts, we shall usually find neither the necessity nor adequate data to refine the function further. This is a specific example of the general comments in Section 4.7 on sources of information for constructing models. So long as a functional relationship violates knowledge that is available to us, we should strive for greater realism. Once the known asymptotes, intercepts, and directions of curvature are satisfied for the transient as well as the steady-state behavior, it is probably best to proceed with other parts of the model until such time as model tests themselves reveal critical sensitivity to some of the uncertain assumptions.[5]

Figure 9-7 shows the accurately plotted output responses of first-, second-, third-, sixth-, and infinite-order exponential delays when an "impulse" is the input. That is, a quantity is put in at zero time, nothing in thereafter, and we observe the output rate. Along the time axis, time is indicated in terms of the total average delay D. The delay is defined in such a way that in a *steady-state flow* the rate multiplied by the delay gives the quantity stored in the delay. In other words, all of the curves are normalized in such a way that for a delay of D, and for a *uniform* flow through the delay of

---

[4] As a practical matter this cannot be very closely approximated by actually using first-order sections; too many are required. It can be approximated by a different type of high-order difference equations or by the "boxcar delays" used in the DYNAMO computing system.

[5] Appendix D shows that a system may actually be only slightly affected by changing the delay characteristics over the extreme range from first-order to infinite-order delays.

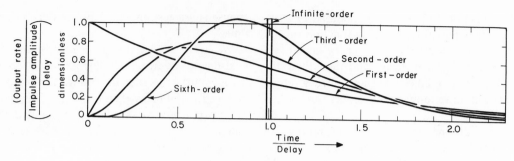

Figure 9-7 Exponential-delay responses to a unit impulse.

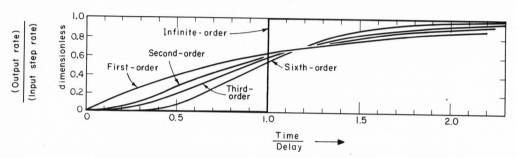

Figure 9-8 Exponential-delay responses to a unit-step rate.

R items per unit of time, the number of items in transit in the delay is equal to (R)(D).

Mathematically speaking, an nth-order delay is equivalent to n cascaded single-order delays, with each single-order delay having a delay time of D/n. In steady-state flow each first-order delay has the same flow rate and consequently $1/n$ of the total units in the delay stored in it.

The vertical scale of Figure 9-7 shows the output rate as a ratio to the initial rate of a first-order delay that is given by I/D, where I is the impulse quantity inserted initially, and D is the average delay. The ratio I/D has the proper dimensions of units/time.

Figure 9-8 shows the output from first, second, third, sixth, and infinite-order delays when

a step function of input *rate* occurs. A sudden input flow rate is impressed at zero time, and the resulting output rates are plotted. The vertical scale shows the ratio, at any instant, of the output rate to the suddenly applied input rate.[6]

[6] It will be noted that the exponential delays, except the extreme infinite-order delay, have a "tail" that approaches but does not reach the value toward which the curve is tending. Mathematically speaking, the responses to impulses (Figure 9-7) never fully empty the delay nor reduce to zero output rate. Correspondingly, the response to a step-rate input (Figure 9-8) never reaches the level of the input rate. This might seem an unrealistic approximation to actual delays, but the magnitude of the effect is very small; it might even be defended on the basis of lost items in shipments, and it will be negligible compared with noise in the flow channels of the system.

CHAPTER 10

Policies and Decisions

The rate equations, as described in Chapters 6 and 7, contain the statements of policy that govern system action. This action takes the form of streams of decisions that cause and control the system flow rates. The formulation of a model is based on an explicit statement of the policy (or rules) that govern the making of decisions in accordance with any condition to which the system may have evolved. The decision-making process consists of three parts — the formation of a set of concepts indicating the conditions that are desired, the observation of what appears to be the actual conditions, and the generation of corrective action to bring apparent conditions toward desired conditions. Distorted and delayed information about actual conditions forms the basis for creating the values of desired and also of apparent conditions. Corrective action will in turn be delayed and distorted by the system before having its influence on actual and then on apparent conditions. All the literature of society is rich in information about the nature of decision-making policy; the literature of management is no exception. Adequate knowledge exists from which to build the central framework of the managerial policy structure.

MANAGEMENT is the process of converting information into action. The conversion process we call decision making. Decision making is in turn controlled by various explicit and implicit policies of behavior.

As used here, a "policy" is a *rule* that states how the day-by-day operating decisions are made. "Decisions" are the actions taken at any particular time and are a result of applying the policy rules to the particular conditions that prevail at the moment.

If management is the process of converting information into action, then it is clear that management success depends primarily on what information is chosen and how the conversion is executed. The difference between a good manager and a poor manager lies at this point. Every person has available a large number of information sources. But each of us selects and uses only a small fraction of the available infor-

mation. Even then, we make only incomplete and erratic use of that information.

The manager sets the stage for his accomplishments by his choice of which information sources to take seriously and which to ignore. When he has chosen certain classes of information and certain information sources to carry the highest priority, the manager's success depends on what use is made of this information. How quickly or slowly is it converted to action? What is the relative weight given to different information sources in the light of the desired objectives? How are these desired objectives created from the information available?

In this book we shall look upon the manager as an information converter. He is a person to whom information flows and from whom come streams of decisions that control actions within the organization. Much human behavior might be properly viewed as the conversion of information into physical action. The manager, however, is not paid his premium salary in recognition of physical effort exerted. He is primarily an information converter at his own particular control point in the organization. He receives incoming information flows and combines these into streams of managerial instructions.

Viewing the manager in this way shows us immediately why we are interested in decision making and information flow. An industrial organization is a complex interlocking network of information channels. These channels emerge at various points to control physical processes such as the hiring of employees, the building of factories, and the production of goods. Every action point in the system is backed up by a local decision point whose information sources reach out into other parts of the organization and the surrounding environment.

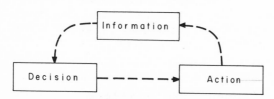

Figure 10-1 Decisions and information feedback.

Figure 10-1 shows a decision stream in the simplest framework of an information-feedback system. Information is the input to a decision-making point that controls actions yielding new information. The diagram shows the structural relationship. In each of these boxes there are delays. Information about actions is not immediately available. The decisions do not respond instantaneously to available information. Time is required for executing the actions indicated by a decision stream. Likewise, each of the boxes contains amplification which I use here in all of its positive, negative, and nonlinear senses. In other words, the output of a box may be either greater or less than is seemingly indicated by the inputs. Likewise, the output may be noisy or distorted. The amplification, the attenuation, and the distortion at each point in the system can make the system more sensitive to certain kinds of disturbing influences than to others.

The industrial system is of course not the simple information-feedback loop as shown in Figure 10-1. Instead, it is a very complex multiple-loop and interconnected system, as implied by Figure 10-2. Decisions are made at multiple

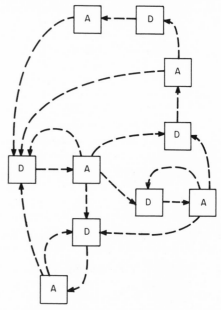

Figure 10-2 Multiloop decision-making system.

points throughout the system. Each resulting action generates information that may be used at several but not at all decision points. This structure of cascaded and interconnected information-feedback loops, when taken together, describes the industrial system. Within a company, these decision points extend from the shipping room and the stock clerk to the board of directors. In our national economy, they extend from the aggregate decisions of consumers about the purchase of automobiles to the discount rate of the Federal Reserve Board.

10.1 Nature of the Decision Process

We should now examine in finer detail the decision process whereby information is converted to action. Figure 10-3 shows the system

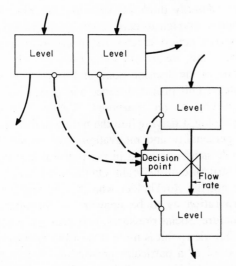

Figure 10-3 Decision making in the system structure.

structure as it surrounds the decision point. A decision is based on the state of the system, which is here shown by the condition of various levels. Some levels describe the present instantaneous condition of the system and others our presumed knowledge about the system. A level may be an inventory, the number of employees, the average sales for last month, the accomplishment we believe has been made to date in a research project, the degree of optimism about the economic future, the size of bank balance, and so forth. These are the inputs to decisions. The output from a decision point controls the rate at which the system levels will change. I am using decision here in a very broad sense. This includes the filling of orders from an existing inventory, the placing of purchase orders for new replacement goods, the authorization of factory construction, the hiring of research sci-

entists, and the authorization of advertising expenditure.

A very important part of the concept of this organizational structure is the directional relationship between the parts shown in Figure 10-3. The levels are the inputs to the flow of decisions. Decisions control flow rates between the levels. The flow rates between levels cause changes in the levels. But flow rates themselves are not inputs to the decisions. Instantaneous present rates of flow are in general unmeasurable and unknown and cannot affect present instantaneous decision making.

In an industrial organization a particular person may be primarily responsible for the control of one particular flow rate, as for example the replacement orders for the maintenance of an inventory. On the other hand, a particular person may embody several separate decision points controlling several separate flow rates. If so, we should look upon these separately as lying within different parts of the information and action network of the system.

A somewhat finer structure of the decision process is of interest to us as shown in Figure 10-4. Decisions fundamentally involve three

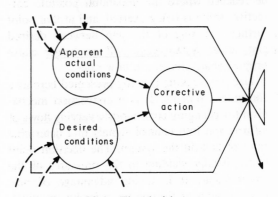

Figure 10-4 The decision process.

things. First is the creation of a concept of a *desired* state of affairs. What should we like to have the condition of the system be? What are we striving for? What are the goals and objectives of this particular decision point? Second, there is the *apparent* state of *actual* conditions. In other words, our available information leads us to certain observations that we believe represent the present state of the system. These apparent conditions may be either close or far removed from the actual present state, depending on the information flows that are being used and the amount of time lag and distortion in these information sources. The third part of the decision process is the generation of the kinds of *action* that will be taken in accordance with any discrepancy which can be detected between the apparent and the desired conditions. In general, the greater the discrepancy, the greater the resulting action, although this entire process of forming a concept of desired conditions, detecting actual conditions, and creating from these a course of action is highly nonlinear and noisy. Very small discrepancies between apparent and desired conditions may seem of no consequence and create little action. A mounting discrepancy may lead to more and more decisive attempts to correct actual conditions toward desired conditions. However, a level of discrepancy can be reached where the maximum possible corrective action is being exerted, and at that point further widening of the gap between desired and actual system state will no longer cause proportionate changes in flow rates.

Decision making is being presented here as a continuous process. It is a conversion mechanism for changing continuously varying flows of information into control signals that determine rates of flow in the system. The decision point is continually yielding to the pressures of the environment. It is taking advantage of new developments as they occur. It is always adjusting to the state of affairs. It is treading a narrow path between too much action and too little. It is always attempting to adjust toward the desired goals. The amount of action is some function of the discrepancy between goals and observed system status.

We note that we are viewing the decision process from a very particular distance. We are not close enough to be concerned with the mechanisms of human thought. We are not even close enough to see each separate decision as we ordinarily think of decisions. We may not be close enough to care whether one person or a group action creates the decision. On the other hand, we are not so far away as to be unaware of the decision point and its place in the system. This proper distance and perspective is important to our purposes. We are not the psychologist delving into the nature and sources of personality and motivation, nor are we the biophysicist interested in the physical and logical structure of the brain. On the other hand, we are not the stockholder who is so far from the corporation as to be unaware of the internal structure, social pressures, and decision points.

Our viewpoint is more that of the managerial superior of a particular person who is charged with certain responsibilities. The superior is close enough to know how desired goals are established. He is in a position to observe and probably provide the information sources to be used by the subordinate to determine his concept of actual conditions. He knows in general the guiding policies and the manner in which the subordinate decision maker would respond to various kinds of circumstances.

10.2 Policy

We are now led to what is here called policy. The word "policy" is used as a broad term to describe how the decision process converts information into action. What actions will result from certain information inputs? What is the conversion relationship between information sources and the stream of resulting decisions?

We shall first define what is here meant by policy. We shall later turn our attention to whether or not such policy does exist and whether or not its form can be determined.

Policy is a formal statement giving the relationship between information sources and resulting decision flows. It is what has often been referred to in the literature as a decision rule. In physical systems, particularly in the field of servomechanisms, the corresponding term is "transfer function." The transfer function tells how the output of a particular box depends on the stream of inputs. The transfer function does not necessarily deal with the particular physical way whereby the conversion is accomplished. We are satisfied if the transfer function tells us adequately well, for a particular purpose, the present resulting action as a function of present and past inputs to the box.

Most of the literature of economics deals with what we are here calling policy. How will individuals and groups respond to various circumstances and pressures? If conditions change in a certain direction, what will be the direction of the response? In industrial organizations, some policy is very formal. It has been reduced to writing for the guidance of the decision makers in the system. Most of the guiding policy is informal but every bit as influential. It depends on habit, conformity, social pressures, ingrained concepts of goals, awareness of power centers within the organization, and personal interest.

We should here note that we have progressed through three different levels of decision-making abstraction. At the lowest level, is random unreasoned action, which does not depend on inputs and which has no basis. At the second level, are unrationalized intuitive reactions, which in fact result from the available flows of information but where there is no comprehension by the participant about the structure and the basis of his actions. We can presume that this represents the thinking and the decisions of the lower animals. There are a basis and a reason for their decisions and actions, but they presumably are unaware of such a basis and the logical structure. At the third level of abstraction, there is an awareness of the formal reasons for decisions. Not only are decisions made, but we have self-awareness of why we make certain decisions, and we are aware and

able to anticipate with some reliability the kinds of reactions that others will exhibit in response to changes in the state of their environment.

The formal awareness for the basis of decisions, which I am here calling the guiding policy, certainly goes back as far as the written record of our civilization. Man is most conspicuously separated from the lower animals by this self-awareness of why he acts. In other words, much of history and literature devotes itself to the basis or policy that causes the human decision maker to react in reasonable and expected ways to his environment. When we say that there is a reasonable and expected reaction, we are in essence describing the policy whereby information will become a certain kind of action.

10.3 Detecting the Guiding Policy

We should now consider whether or not we can detect the nature of the guiding policy with sufficient accuracy so that we can use it to understand better the behavior of the industrial and social systems of which we are a part. Clearly, people are of two minds on this question. Most of the literature on decision making implies great difficulty and subtlety in the subject. The social scientist makes very tentative simple experiments with groups of three or four people, in an effort to determine how their decisions are arrived at in reacting to one another. When we raise the question of understanding the human decision-making process, the frequent answer from scientists is that not even a good beginning has yet been made. Yet the historian, the novelist, the manager, and every one of us in his everyday life has been much bolder. We all discuss why "so-and-so" acted in a particular way. In doing this we are discussing his guiding policy. We are discussing how he did or how he should have responded to information available to him.

The dichotomy in our thinking is illustrated by two recent encounters which I had with two different colleagues. One flatly stated that it was clearly impossible to introduce the actions of the Federal Reserve Board into a formal model of national economic behavior. The impossibility was argued on the basis that we do not know the process by which such decisions are reached. They are too subtle. They are subjective, intuitive decisions for which we know not the guiding policy. The other incident took place in a doctoral oral examination. Another colleague, as a routine matter, casually asked the candidate to describe the factors that would lead the Federal Reserve System to make adjustments in various directions in its discount rate and open-market policies. In other words, the economics doctoral candidate was expected to know the essential nature of the policy that would guide the stream of Federal Reserve Board decisions. To be sure, there may be a high noise content that can cause timing variations and uncertainty in the extent of a response. However, the broad underlying outlines of guiding policy were expected to be within the understanding and comprehension of the student.

This contradiction in the opinions we hold about the process of decision making is very similar to that which we observe in our thinking about the process of invention. There is great argument and little agreement in any discussion of how new ideas are generated and how invention and research results are achieved. Yet we are almost in 100% agreement in acknowledging that more intelligent and more experienced people, and greater research budget expenditure, and greater motivation, and greater need for the results will all enhance the probability of a successful outcome. This agreement on the nature of the conversion function that couples financial and manpower inputs to scientific output is the basis for Congressional action and military department appropriations.

In short, our whole civilization is founded not only on the presumption that a basis exists for the guidance of human action but also on the conviction that we know a great deal about the specific nature and extent of this guiding policy.

We saw in Figure I-3 in the Introduction how decisions could be classified according to the form of the controlling policy. Even in the Region C of judgment decisions, there is a strong presumption concerning what constitutes proper action. In general, it is well known what the direction of effect of various changes in the system status will do to the resulting decisions. The current management press, such as *Business Week,* the *Wall Street Journal,* and *Forbes,* is filled with the rationale for management decisions. Much of the printed material is devoted to a discussion of the pressures of the current state of affairs and the effects these will have on decision makers.

Many people seem to believe that a sharp break exists between the automatic decisions that are completely formal and the other regions of management decisions. Such persons are unwilling to accept the possibility of even the existence of formal policies that could describe the major aspects of management in the other decision-making regions. There is an interesting contradiction in the attitude of many managers toward this matter of understanding the formal basis of the decision process. Any manager must of necessity admit the existence of the region of automatic decisions, since these are common practice. The majority of managers will argue that the region of the intuitive-judgment decision is so subtle that no reasonable approximation can be made to it through formal decision rules. Yet those same managers, when faced with a decision that they recognize as lying beyond the capabilities of their intuitive judgment, will once more fall back on formal decision-making procedures. I here refer to the whole field of sales and market and economic forecasting when done on the basis of statistical analysis of past data or by routine procedures of collecting and tabulating sales department estimates. Forecasting is essentially a decision-making process. It consists of taking past and presently available information and converting this into results that indicate a course of action. I am not a supporter of the wisdom and validity of the majority of this type of forecasting, but I simply point out the conflict in attitudes. There are those who relegate the simple decisions to automatic procedures. They rely, for lack of anything better, on certain formal statistical decision procedures with respect to some of the most subtle and difficult decisions. But they reserve the middle ground as a region for judgment, which they assert is untouchable by formal decision rule.

It seems to me that there is now ample evidence to indicate that this middle region is not the obscure and subtle jungle that it has so often been pictured. Men are not good calculators of the dynamic behavior of complicated systems. The number of variables that they can in fact properly relate to one another is very limited. The intuitive judgment of even a skilled investigator is quite unreliable in anticipating the dynamic behavior of a simple information-feedback system of perhaps five or six variables. This is true even when the complete structure and all the parameters of the system are fully known to him. The verbal model and the mental model that we carry around with us to explain the dynamics of industrial and economic system behavior probably do not rank in effective dynamic complexity beyond a fourth- or fifth-order differential equation. We think that we give consideration to a much larger number of variables, but I doubt that these are properly related to one another in groups larger than five or six at a time. In dealing with the dynamics of information-feedback systems, the human is not a subtle and powerful problem solver.

Management science has so far been very ineffective in dealing with this formulation and use of decision-making policy. The past difficulties might be gathered into three categories.

First, there is the matter of perspective, or viewing distance, mentioned earlier. The social scientist has tended to look at the individual man with emphasis on psychology and individual motivation. Many of the attempts at laboratory experiments have been with small groups assembled in an artificial environment for brief periods of time. The results have not been conditioned by the strong social forces of precedent, conformity, and the attempt to behave as the man believes his superior would want him to. Study of the individual man, especially over short periods of time in an artificial environment, tends to accentuate the feeling that decisions are capricious, infrequent, disconnected, and isolated.

At the other extreme, the economist has viewed the corporation from too great a distance. He has often looked upon the market as maximizing its utility, whether or not it has the information available to do so. He looks upon the entrepreneur as a man who maximizes his profit without asking whether or not he has the available information sources and the mental computing capacity to find a maximum. This viewing from too great a distance tends to overstress the importance of the top-management decisions compared to those made in the lower and middle structure of the corporation. A directive from top management does not change the prejudices, habits, and self-interest objectives of the middle-level decision makers. For example, the public press has well documented the futility felt by the successive Secretaries of Defense who have tried their hands at changing the character, attitudes, and practices of our military department. It is a long, slow process. We tend to be misled by various "upheaval" incidents. A proxy fight followed by a complete change of top management and the firing of half of the middle-management structure will indeed change the attitudes and traditions of an organization. But this drastic surgery is not common.

For an understanding of the corporate information-feedback system, it is very essential that we look neither at the individual person isolated from his environment nor at the exterior of the system. It is at the intermediate viewpoint from which we can know men and groups of men in their working environment that we can capture the true character of the operation.

Second, I think that the understanding of decision making has been greatly handicapped by the presumption that it is a more subtle and more skillful process than it actually is. We have been too heavily influenced by the fact that the highest speed computers cannot yet play chess as well as a person. This is not a typical example. Full and exact information is available to the man. It is a problem of visualizing spatial relationships, which a man does quite well and today's machines do poorly. There are other situations, such as the time histories of Chapter 2, where a computing machine can determine in five minutes what the consequences of a set of policies will be and wherein a group of men could argue inconclusively for a year about what follows from a given set of assumptions.

It is my feeling that in a dynamic information-feedback system the human decision maker is usually using a great deal less than the total amount of information available to him. Furthermore, the information available to him is a great deal less than that commonly presumed. In general, his actions with respect to any given decision stream will be almost entirely conditioned by less than ten information inputs. What he does with these few information sources is apt to be rather stereotyped. Some of them will be used to create a concept of desired objectives. Others will serve to form his impressions of the true state of affairs. From the difference will result reasonably straightforward and obvious actions. What seems obvious may not be best. Some of the biggest improvements that can be made in the dynamics of our industrial systems will come from following pro-

cedures that our traditions and our management folklore have led us to believe are precisely in the wrong direction. Our understanding of the dynamics of complicated information-feedback systems is so inadequate that our intuitive judgment can often not be trusted to tell us whether improvement or degradation will result from a given direction of policy change.

The third difficulty into which many seem to have fallen in attempting to understand decision-making policy has resulted from trying to skip one of the evolutionary steps in the hierarchy of decision-making abstraction mentioned earlier. It was suggested earlier that the lowest level of decision making would be that where actions were random and irrational. The second level would be one where actions are reasoned and rational but where there is no self-awareness of what the governing policies might be. Man was characterized in the third level as having, throughout recorded history, at least a verbal descriptive model of a rational policy that creates the stream of individual decisions. This is already a major fraction of the step toward the ability to formulate explicit, quantitative, policy-governing decisions. This next level would be one in which the art and the intuitive judgment are applied to the development of a better understanding of policy.

Art, judgment, and intuition are at this new level no longer applied to the individual separate decisions but to the definition of a policy that governs the stream of individual decisions. This is the level of abstraction that is just emerging. This is the one wherein we can point to numerous successful examples but where there is not general agreement on method. At this level of abstraction, there is not a descriptive literature on what constitutes the practice of the art of detecting decision-making policy.

In spite of the gap in the art that exists at the policy-detection level, many economists have attempted to skip to the next level of the abstraction hierarchy. They have been attempting to develop statistical methods that can be routinely applied to extract from quantitative data about a system the governing decision policies. This is still another level of abstraction in which intuitive art and judgment are applied to setting up rigid rules whereby the formal decision policies can be derived. I feel we are not ready to attempt this last level of abstraction until we have achieved acknowledged success in applying art and judgment and intuition to the extraction of the decision policies themselves. After this process is well understood, it may then be possible to reduce this method of analysis of an organization to a rigid and orderly procedure. In general, the precedents seem to indicate that we must take these levels of abstraction one at a time. At each point in time, art and judgment are devoted to establishing the rules whereby the lower levels can be automatized.

An example has been the history of computing-machine programming. Ten years ago one wrote the specific machine code for the solution of a particular problem. The next level of abstraction was to write a program of logical instructions to tell the machine how to create its own running program for a specific problem formulation. The hierarchy of abstraction in problem programming is now deepening. At the new level, man develops concepts that allow the computer to formulate the specific statement of the problem which another computer program will in turn convert into machine language. This is philosophically the equivalent of the above hierarchy of orderly rules related to decision making.

To begin to deal with the dynamic characteristics of social systems, we must be able to represent at least the central essential skeleton of the decision-making structure. To do this, we must be able to approximate the controlling policy at each significant decision point in the system. This understanding of policy can be accomplished if

- We have the proper concept of what a decision

is and of the significance of the policy that describes the decision process.

- We have the proper structure relating system status to policy, to decisions, and to action.
- We realize that the process is noisy and we shall not get and do not need high accuracy of decision-making representation.
- We use to best advantage the extensive body of experience and descriptive information that probably contains 98% of the essential information on decision making. The other 2% will come from formal statistical and numerical data.
- We realize that a formal quantitative statement of policy carries with it no implications one way or the other regarding absolute accuracy. We can make a formal quantitative statement corresponding to any statement that can be made in descriptive English. Lack of descriptive accuracy does not prevent quantifying our ideas about decision policy. Assigning a number does not enhance the accuracy of the original statement. The common belief that we cannot quantify a decision rule because we do not know it with high accuracy is mixing two quite separate considerations. We can quantify regardless of accuracy. After that, we deal with the question of what is sufficient accuracy.

I feel it has been adequately demonstrated that these things can be done. We have the tremendous examples of corresponding accomplishments in the understanding of military control systems in the last decade. We have preliminary examples of the application of the same approaches to industrial systems.

Actual, effective decision functions in a company or in an economy go much further than the formal policy that is set down in executive memoranda and in laws. The "effective policy" is the framework for reaching decisions and is established by the environment, the sources of information that are in fact available, the success measures and rewards that affect people at each decision point, the priority order of food and shelter and luxuries, the mores of the society, and the prejudices and habits impressed by past experience. When decisions are examined in the light of this circumscribing framework, we find that they are far from the unpredictable actions that are sometimes supposed. Even for a particular individual, we can assume a certain degree of consistency and can meaningfully discuss probable effects of various pressures on him. For a class of persons in similar environments, the likely average response to changes in specific environmental forces can be estimated with even higher confidence.

A principal use of a dynamic model is to study the influence of policies on system behavior. In formulating a model, we must extend the concept of policy beyond its usual meaning. All decisions in the model come under the complete control of policy. The policy represents the basis for controlling flows at all points in the actual system. We are forced to explore this basis in depth to see how decisions are generated from the various circumstances that arise. The concept of policies governing decisions goes well beyond that of the human, managerial decision. A model must also make "decisions" that are of a physical nature — for example, how many unfilled orders can be filled depending on the state of inventories?

10.4 Overt and Implicit Decisions

It is sometimes helpful in our thinking to divide decision functions into two categories, depending on whether they are ordinarily conscious, "free" human decisions or whether they arise inexorably from the physical condition of the system. The dividing line is not sharp. *Overt decisions* are here defined as the conscious decisions by people as part of the management and economic process. In the overt decisions will be found all executive decisions, consumer purchase decisions, and all psychological factors that are to be included. *Implicit decisions* are the unavoidable result of the state of the system. Included here are usually (1) present ability to deliver, depending on the present status of inventories, (2) the output rate (goods delivered)

from a transportation system, resulting from the input rates, the goods in transit, and the transport delays, and (3) taxes due as a result of profits.

Factory production serves to emphasize the distinction between overt and implicit decisions. Actual, present production rate is usually the result of an *implicit* decision function that shows how production rate is a consequence of employment, available equipment, and materials. It is not usually an overt decision; it is not possible to decide arbitrarily on a production rate and thereby have that rate exist immediately and with certainty. The accompanying *overt* managerial decisions are the decisions to *attempt* to hire people and to *order* equipment and materials. Whether or not people are in fact hired as a result of the overt decision depends on the implicit decision functions within the "physical" state of the system which recognize such factors as the supply of workers, the wages being offered, etc. Whether or not materials and equipment result from the orders depends on the available supplies.

Overt and implicit decisions need not be distinguished in the manner in which they are handled in the model; but thinking of them separately helps us to avoid omission of important steps in the flows of information and the resulting decisions and actions. Including both the overt and implicit decisions makes it possible, in the model, to deal with the *wish* as well as the *actuality*. Conditions lead to a *desire* for a change; the desire interacts with the state and resources of the system to determine what, if anything, is to happen.

Introducing the concepts of both overt and implicit decisions is one factor that here eliminates the necessity of simultaneous equations that arise in some models, such as in making production decisions continuously equal to consumption decisions, etc. Such decisions are actually made separately and independently, and are eventually coupled through inventories, prices, and various information flows. The overt decisions to *want* or to *attempt* are a result of *information* available to the decision maker. The implicit decisions, which create the actions, recognize the existing true state of the system as well as the desires.

10.5   Inputs to Decision Functions

In formulating decision functions (the rate equations) we must take care that the decisions are generated from the variables that would actually be available at the decision point. In general, the information available for overt decision making is not identical to the primary variables that it represents. The information may be late, biased, and noisy. Here again, a distinction between implicit and overt decisions may arise. Overt decisions are usually based on information (which may be perturbed) *about* the primary variables. The more mechanistic implicit decisions control the routine flows that depend on the actual state of the system and, therefore, on the true values of the variables in the model.

The distinction between the *true* value of a variable and the value of information *about* the variable can be illustrated by an inventory. Ability to deliver an item from an inventory, being dependent on whether or not the item is present, can usually be thought of as an implicit decision controlled by the actual, present, true state of the inventory. This present, true inventory is one of the variables in the model. The overt decision function that controls reordering of material for inventory may depend on *information about* the inventory, such information being subject to delay and inaccuracy. What the inventory is *believed* to be may, in some models for some purposes, need to be included as a second and different variable. In addition, we should often deal with the concept of "desired inventory," which is often different from either of the preceding; this would then become a third variable with a still different value relating to the same inventory.

As a second example, a model of an economic system would continuously generate an actual, instantaneous, gross national production rate; but this rate should not be available as an input to any overt decisions (such as planning plant expansions). The available information, to agree with actual conditions, must be delayed and should contain a degree of error and uncertainty even with regard to what the true values were before the delay.

A model, like the real world, must often generate both "true" values of variables and also associated variables that represent the values that are measured or conceived for decision-making use.

10.6   Determining the Form of Decision Functions

A model that can create proper dynamic system behavior requires formal expressions to indicate how decisions are made. The flow of information is continuously converted into decisions and action. No plea about the inadequacy of our understanding of the decision-making process can excuse us from estimating the decision-making criteria. To omit decision making is to deny its presence — a mistake of far greater magnitude than any errors in our best estimate of the process.

Can estimated decision functions be good enough to be useful? In general, it appears that they can. Perceptive observation, searching discussions with persons making the decisions, study of already existing data, and the examination of specific examples of decisions and actions will all illuminate the principal factors that influence decisions. In considering a factor influencing a decision, we progress through four stages:

- What factors are significant enough to include?
Then for each:
- What is the direction of its effect?
- What is the magnitude of its effect?
- What nonlinearities should be recognized?

Factors to Include. In deciding how to formulate a particular decision function (policy) in a model, the first step is to list those factors that are important influences on the decision. The answer is often obscure. What at first may appear to be a most significant factor will sometimes be found to have little influence on the model behavior or on the actual system. A factor that is ordinarily overlooked in everyday management practice may turn out to hold the key to important characteristics of the total system.

The choice of factors that affect a decision must be made from the viewpoint of what affects the characteristics of an information-feedback system. Very few persons have a good intuitive judgment about such systems. Working with system models helps develop sound judgment and intuition. The decisive test of the significance of a factor in a decision function in a model is to observe model performance with and without the factor. In this way the model itself can be used to help determine what it should contain.

The extent of the direct influence of the factor on the decision is not the only consideration. We must also consider the degree of feedback or repercussion of the decision on the factor entering into the decision, and the timing of such feedback. Relatively slight influences on decisions can be important in "positive feedback" conditions, where the variable factor influences a decision and the decision affects the input factor to create still more change in the decision.[1] This is seen in many places. For example, when customers order further ahead in response to increasing delay in the delivery of goods, increased ordering rate increases the backlog of unfilled orders, the rising backlog means that each new order is delayed even more

than those before it, and the increasing delay leads to still more advance ordering.[2]

Direction of Effect. There will usually be little doubt about the direction in which a decision will be influenced by changes in a particular factor affecting the decision. However, we must be alert to represent properly the "worse-before-better" sequences that often arise. The short-term and long-term influences on a decision by a particular factor are often in opposite directions, and the dynamic behavior of a model can be seriously affected if only the long-range effects are included.

Several examples will illustrate the kinds of factors that can have short-term effect (often overlooked) that is opposite to the long-term effects that are ordinarily considered. It is commonly assumed that higher prices are an incentive to greater output of a product, but in the short run this sometimes will not be true. In the production of beef the first step in increasing production is to withhold cattle from market to build up breeding herds, thus reducing sales for two or three years; also, rising prices mean an appreciation in value of live "inventory," which results in lengthening the fattening interval in the feed lot and also reduces sales rates for a period of several months. In some types of mining a rise in prices makes it economically feasible to process lower-grade ore; fixed-tonnage-rate equipment is then applied to a lower-yielding raw material, a procedure which can result in reduced total output until marginal mines that were not previously operating can be opened. In an expanding research activity more people may be needed to accelerate the completion of a project, but the first effect may be reduced progress while these people are trained and absorbed into the organization. In a full-employment national economy, the de-

---

[1] Figure 2-10 is an example where the cumulative cycle of sales increases advertising and advertising increases sales until some other effect (depletion of prospective customer pool) terminates the regenerative process.

[2] This appeared in Figure 2-5. Also, see Chapters 17 and 18 for discussion of a system in which a small percentage change in a factor in a policy can be the essential link in creating an unstable market-to-factory-to-employment interaction.

mand for more goods may require diverting workers from producing goods into the manufacture of additional factories and machinery; the first step to achieve more long-term output tends initially to reduce production (this effect may of course be counterbalanced by other factors such as a longer work week).

Magnitude of Effects in Decision Functions. The dynamic behavior of information-feedback systems is determined by the way in which *changes* in one variable cause *changes* in another. This might lead one to expect a high system sensitivity to the exact values of parameters[3] in the decision functions, but such will usually not be true.

If the model is properly constructed to represent the actual information-feedback structure of our social systems, it will have the same self-correcting adaptability that exists in real-life situations. In the preferred model formulation all parameters that must be estimated for decision functions are acted on by the values of levels to obtain the rates of flow that are controlled by the decisions. The levels are in turn adjusted by the resulting decisions. An inaccurate decision-function parameter can then lead to compensating readjustments of levels in the model until the rates of flow are properly related to one another.

Some examples will illustrate this internal readjustment. In estimating a parameter determining the *delay* in repaying accounts receivable, one might choose too high a value; this would cause the level of accounts receivable to run slightly high, but repayment *rate* is still coupled to the *rate* at which new obligations are being incurred. Or too high a labor mobility in describing a labor market could produce levels of unemployment that run lower in the model than in the actual economy, while still retaining the proper effects of policy changes on model

dynamics; the model would readjust or warp internally, as does the real economy, so that a closer approach to the limit of full employment would counterbalance any built-in tendencies to create excessive employing rates. In a model, too low a consumer propensity to buy automobiles would cause the consumer stock of automobiles to decline and progressively decrease transportation service until the automobile purchase rate increases. Changes in the level of consumer stocks of automobiles will help to compensate for an inaccurate purchase-rate decision function, while still leaving the dynamics of purchase-rate *changes* qualitatively correct with respect to other variables in the model.

We should be more concerned with what the model tells us about the factors that will cause *changes* in the rates and levels than about the accuracy in determining the average magnitudes of the rates and levels.

A properly constructed model is often surprisingly unaffected by plausible changes in most parameter values — even sometimes changes of severalfold. In a model, the sensitivity to selected values of parameters should be no greater than the sensitivity of the real system to the corresponding factors. It seems obvious that our actual industrial and economic activities must not be highly sensitive to their fundamental parameters and that these parameters do not change rapidly. This must be true because the significant *characteristics* of our organizations persist for long times. The successful company tends to remain so for extended periods — a success that is founded in basic organization and policies (including the essential aspects of its leadership). Our national economy has exhibited surprisingly similar economic cycles throughout its history, in spite of vast changes in technology, monetary structure, transportation and communication speed, relative importance of industry and agriculture, and magnitude of government activity.

Nonlinear Decision Functions. Nonlinear

---

[3] A parameter is a constant for a given model run whose value has been assigned in constructing the model. It can of course be changed between one simulation run and another.

models have been mentioned in Section 4.1. The nonlinearities will occur in the model in the decision functions that determine the rates of flow. A linear relationship is one in which the input factors combine by simple addition or subtraction to determine the result. Suppose that the rate R is dependent on the variable factors X, Y, and Z, as described by the following linear function:

$$R = aX + bY - cZ$$

Here X, Y, and Z contribute individually their effect to R. For example, Y and Z do not influence the contribution of X to the resulting R. Furthermore, each contribution to R is proportional to the particular input variable, regardless of the value it may assume. Linear decisions will not suffice to describe the relationships with which we must deal.

By contrast, a nonlinear decision function might assume a vast variety of forms of which the following is an example:

$$R = aX^2 + b(Y)(Z)$$

Here we see two sources of nonlinearity. In the term aX^2, the contribution is no longer proportional to the changes in X. As X goes from 0 to 1, the term increases by the value of a; as X increases from 1 to 2, the value of the term increases by three times the value of a. In the term $b(Y)(Z)$, the contribution of Y and Z each depends on the value of the other. The larger Z is, the more will be the effect of a given change of Y; if one of them is zero, there will be no contribution from the other, regardless of its value.

Nonlinearities of these two types are essential to the proper representation of corporate and economic behavior. Some examples may help.

The first form of nonlinearity was when the influence of a factor that affects a decision is not simply proportional to the factor. For example, the available stock of goods for sale affects the delivery rate of goods. When stocks are low, unavailability of goods reduces the ability to deliver; in the range of "normal" inventories, delivery rate will be very little affected by inventory changes. We can expect that most variable inputs to decision functions will be nonlinear and show increasing or decreasing importance as the ranges of the variables change.

The second source of nonlinearity in decision functions is when the decision is not independently responsive to two or more causative input variables but to a product or other interrelationship of the variables. In the previous example, the delivery of goods is not independently and separately responsive to the stocks of goods and to the unfilled orders that have been received for goods. We may not simply add the two separate contributions. If there are no orders, stocks are immaterial in determining delivery; if there are no stocks to deliver from, orders cannot produce delivery.

The two types of nonlinearity will often occur together. Consider the production rate capability of a factory as it depends on present employment level and on the available capital equipment. Figure 10-5 shows how production

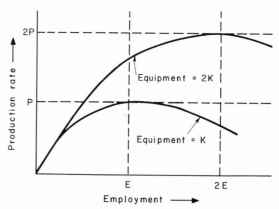

Figure 10-5 Production rate as a function of employment and equipment.

rate might rise as employees are added. At first, as employees are added, each can have access to any equipment needed, the man-hour efficiency is high, and the curve of total production rises steeply. As the maximum capacity of the equipment is approached, the additional production per worker decreases. Still more employees finally result in the maximum possible production from the equipment. Beyond this level of employment, additional men would create only congestion and confusion and loss of total production. We see then that for a fixed amount of equipment the production is not proportional to the employment and is a nonlinear relationship. Furthermore, the contribution to production of any given change of employment depends on the amount of equipment, so that the two inputs interact on one another. At low employment levels, it is unimportant where the amount of equipment might be in the range of K to 2K. At higher levels of employment, the contribution of additional men is increasingly affected by whether or not equipment is added.

Linear approximations to these nonlinear relationships are usually not satisfactory. Normal operations will vary over ranges wide enough so that the nonlinearities are highly significant. Very often, the *approach* to one limit becomes the input signal to some compensating action (in the above example, the lowering man-hour efficiency caused by crowding is one of the inputs to the decision to order more equipment).

The models we formulate should be valid over wide ranges of the variables. This is desirable for several reasons. We shall want to explore wide ranges of conditions. We may not know how far some variables will be called upon to move. We shall want the model to be useful outside the operating ranges that may have been already encountered by the real system because the design of new systems implies operation outside of historical practices.

In building a model we should use all the pertinent information that we have about the system that is being represented. An indispensable class of information is our knowledge of what to expect under extreme conditions of operation. Very often we know more about the extreme limiting conditions of a relationship between variables than we do about the normal range of operation. Very often we know the value or the slope that a curve relating two variables must assume as the input variable approaches zero or some absurdly high value. By selecting functional relationships that match all our bits of knowledge, we enhance our chances of obtaining a model that acts properly.

In regard to correct representation of derivatives (slope, rate of change of slope, etc.), the danger of using broken-line approximations to functions is apparent. Most actual limiting conditions exert their influence in a progressive way as the limit is approached. It is poor practice, often with serious consequences, to approximate such a function by a linear section that is "clipped" to stop its travel suddenly. At the clipping point, all derivatives of the function are in extreme error.

Correctly curved functional relationships facilitate internal self-adjustment as already discussed, wherein the model seeks values that balance one another. Realistic behavior is easier to achieve in a nonlinear than in a linear model because the nonlinear model formulation leads us more quickly to the factors on which actual system behavior depends.

10.7   Noise in Decision Functions

The decision functions of a model will necessarily incorporate only the more important of the factors that influence decisions. Beyond these will be a host of minor influences that will unavoidably be omitted. The omissions can represent two quite different categories.

In the first category of omission are the slight influences from those variables that are a part of the system and of the model under consideration. These omissions are actually the elimination of some of the feedback paths between the model variables. This will often happen because of the necessity for simplification, but the dangers are those discussed in Section 10.6. Omitting a variable within the model from a decision function is the omission of an input that may be correlated in time with the decisions created by the decision function. We cannot substitute for this type of omission by a random variable incorporated into the decision process of the model.

A second class of omission from the decision functions has a very different character. These omitted factors are the ones that are not themselves affected by each other or by the other variables of the model. Their source is outside of and independent of the real system being represented. An example is the uncertain influence of weather[4] — not only its obvious influence on agriculture but also its effect on Easter sales, sports equipment, and day-by-day department store business. Also we might classify as random the incidents of the local, national, and international political news that may not be completely independent of business and economic affairs but that are often sufficiently loosely related to be looked upon as uncorrelated happenings. Here also would be the purchasing agent's vacation or the production manager's illness and the effect they have on the smooth flow of business. This flood of "noise" of a random and unpredictable form adds its

contribution to all decision points in our actual social systems. We can approximate these by including "noise" variation in the decision functions in the model.

The theoretical treatment of noise inputs and how to design them comprise a complex subject to be taken up elsewhere. The practical question of what noise characteristics should be included in a model will, like many other inputs, be decided primarily by such knowledge as we have of the system being represented.[5]

Noise inputs to decision functions can be used to represent the influence of the above-mentioned second class of omissions from the decision functions in a model, where the omitted factors are unrelated to the system being modeled. Noise cannot substitute for the first class of omissions which constitute a simplification and omission of parts of the information-feed-back structure of the system.

---

[4] Until such time as weather is controlled as part of the economic system.

[5] It should be noted that neither the "random numbers" of statistics nor the "white noise" of the engineer can be employed without modification. Due regard must be given to sampling interval, the high and low cutoff frequencies contained in the noise sequence, and to the way the noise is inserted at the decision point with respect to the closed-loop channels of which it becomes a part. In Appendix F is some discussion of the filtering of the outputs of "pseudo-random" noise generators to obtain a noise form that meets desired frequency and amplitude specifications.

CHAPTER 11

Aggregation of Variables

Much of the value of a model comes from distinguishing the important factors in a system and setting them apart from the unimportant. Part of this highlighting of the essence of a system is accomplished by combining similar factors into a single aggregate. Aggregation is desirable and essential and yet a hazard. Permissibility of aggregation is determined largely by examining the decision-making policies of the separate factors that might be aggregated. Items controlled by sufficiently similar policies that depend on sufficiently equivalent information sources may be combined into a single channel. Allowable aggregation, as with other aspects of a model, depends on the purpose of the model.

$\mathbf{I}$T is obvious that a model of a company or of an economy cannot possibly represent every individual decision and transaction taking place in the system. In fact, we should not want to do so, any more than we should want equations that account for each molecule of water in calculating pressures and flows in a water supply system. If individual actions are properly grouped according to similarity of circumstances, the average behavior can be more accurately described than we could hope to do for any individual incident.

How this grouping, or "aggregating," is done is of the greatest importance. If there is insufficient aggregation, the model will be cluttered by unnecessary and confusing detail. If aggregation is too sweeping or accomplished by combining the wrong things, we shall lose elements of dynamic behavior that we wish to observe. Some general guides may help determine the basis for aggregation.

11.1 Using Individual Events to Formulate Aggregate Flow

Even though we speak of aggregating large groups of individual physical items and deci-

sions, the aggregate flows must still traverse the same channels followed by individual items. As an analogy, if we were developing a model of a city water works, we should include individual reservoirs, pumping stations, major distribution lines, and groups of water consumers segregated by type and by geographical location. We should make no effort to identify a particular drop of water, but we should devise a diagram that would represent the circumstances encountered by an individual drop as it passes from wells to reservoir and through piping and pumps until it is used.

Likewise we must formulate our industrial system model by examining particular individual decisions and the sequences in their execution.

A flow diagram is first developed to represent the sequence of individual actions. For example, we might trace the path of an order, its formation, its clerical delays, its transmittal, its waiting in a backlog of unfilled orders, the shipment of goods in response to the order, and the transportation of the goods. Each item in the flow encounters these same circumstances. The channel is easiest to find and to visualize in terms of the route of a single item of business. Having established the channel of one item, we then wish to aggregate into this channel as many separate items as possible.

11.2 Aggregation on Basis of Similar Decision Functions

The decision functions that control rates of flow provide the best clue to permissible aggregation.

We can aggregate into one channel any two items or classes of items if the following conditions exist:

- The items can be assumed to be controlled by the same identical decision-function.
- The controlled outputs are assumed to be used for identical purposes elsewhere in the model.

For example, we can often aggregate even diverse catalogue items being ordered by distributors from a factory. The first condition is met if inventories of the separate items are to be controlled on a similar basis, and if orders are generated in the same manner with the same delays. The second condition is met if the questions of interest at the factory and at the distributor depend only on the combined aggregate flow of orders. This would be true if there is enough production interchangeability so that space, employment, and other factory considerations are adequately described by the combined flow; and if the "mix" proportions of the items are sufficiently constant that total aggregate inventories describe the average ability to fill orders.

Aggregation on the basis of similarity of decision functions will at various times lead to different results, depending on the nature of the particular questions that the model is to answer. Questions of total capital budget requirements may need to be based only on total aggregate production, but the specific equipment to be bought will depend on product lines or the sales volume in particular catalogue items.

By aggregating only when decision functions are similar, we shall combine only those items that have the same time-response characteristics to the factors on which they depend. For example, in a model of an economic system, consumer purchases would usually be aggregated only into classes of similar time response to income and savings. For many purposes it would be possible to aggregate all items of food; and into another category soft goods, such as clothing; into another hard goods, including appliances and automobiles; and into another housing. These major categories must be kept separate from each other because they represent major dynamic differences in the motivation for purchase, the deferrability of purchase, and the life of the inventory of the item held by the consumer. Likewise we might aggregate machine tools and industrial factory construction, on the basis that changes in these occur with approximately the same timing, and on the basis of similar decisions.

Aggregating only similar decision functions will preserve the important nonlinearities of the system which appear in the decision functions. To illustrate, in the previous example of aggregated factory conditions, one product line may be near peak production capacity while another is lightly loaded. The facilities may not be convertible between products. Consequently, rising aggregate sales compared with total aggregate production capacity will not tell us that one line may need expansion while there is still spare capacity in the other. The two may need to be handled separately. The resolution of this problem is determined in part by the use the model builder wishes to make of the model (see Section 5.1).

11.3 Effect of Aggregation on Time Delays

It should be noted that aggregation of diverse items into a common flow channel will broaden the distribution of time intervals taken by items in traversing delays. A particular letter in the mail takes a definite number of hours for delivery. This precise delay should not be used for each and every letter. Some will take longer, some shorter. There will be an average delay and a distribution around that average. The delays in aggregate flow channels should reflect this time distribution of output.

The greater the diversity of the items in the channel, the wider will be the range of individual delays and the less sharply defined [1] will be the delay response.

----

[1] From Section 9.4 on time response of exponential delays, the third-order delay is seen to be less sharply defined in its output response to an input than are higher-order delays.

CHAPTER 12

Exogenous Variables

Variables that are generated independently (exogenous variables) of the system described by the model equations can properly be used as test signals for a model. As such, they allow us to see how the internal system of the model reacts to a particular kind of behavior pattern in the environment external to the system. An independent variable is used as if the response of the system under study had no coupling back to the independent variable; only those questions can be explored for which the assumption of independence is sufficiently valid. Seldom will more than a single exogenous variable be justified (other than random-noise variables, which can be considered as internal to the decision policies of the model). Using more than one exogenous variable implies a rather unlikely condition that the exogenous variables are interlocked by control mechanisms between themselves but all are free of interlocking ties with the variables of the system under study. The purpose of an exogenous variable is not to create more realistic model behavior but to set arbitrarily the external conditions under which the system is to be observed.

OPEN and closed models have been mentioned in Section 4.1. The closed model represents a self-contained system without connection to external (i.e., exogenous, or independent) inputs. Exogenous variables are those whose values arise independently of the internal variables of the model.

In exploring the dynamic characteristics of a model of a system, we shall often use an exogenous input as a test input.[1] In so doing, we assume that the test input is itself independent of, and not affected by, what occurs within the variables that are represented in the model itself. In using the model, we can then validly explore only those questions for which the assumption of independence of the test input is correct. Where there is assumed but not actual independence, the model will fail to exhibit modes of operation that involve closed-loop coupling from the model variables to the exogenous variables and back into the model system of equations.[2] As discussed in Section 10.6, coupling that appears very slight can sometimes produce dramatic effects on system performance if the proper time-phase, direction, and amplifying relationships exist.

It has sometimes been the practice in formulating econometric models[3] to use several time

---

[1] For example, the step function or the seasonal fluctuation or the noise was used as an assumed test input in the illustrations of Chapter 2.

[2] For example, in Figures 2-2 through 2-8 in Chapter 2, we assumed that retail sales were independent of what might happen within the production-distribution system. Clearly this assumption is of limited validity, and the explorations had to be restricted to certain of the internal characteristics of the distribution chain itself. When in Figures 2-10 and 2-11 retail sales became affected by internal conditions in the distribution chain, new modes of system behavior appeared and could be explored.

[3] See Klein (Reference 9), pages 2, 3, 38.

series of values taken from the real economic world and to insert these into the model. These are called "exogenous inputs" and are variables which are used in the model equations but which are not generated by the equations of the model. Such a procedure is fraught with dangers from the viewpoint of dynamic behavior. If the system being modeled is actually sensitive to the variables that are *assumed* to be exogenous and in turn affects those variables, then the failure to define the mechanisms of the exogenous variables so that they may become endogenous (that is, as dependent variables created within the model by appropriate equations) destroys the feedback linkages that may actually exist between the assumed exogenous variables and those of the model. This situation is equivalent to assuming that the model has dynamic meaning independent of the exogenous inputs and that the exogenous inputs are being taken only as a test input to the model. Furthermore, the exogenous inputs are being taken "ensemble" as a single test input. It is implicitly being assumed that the time relationships of the exogenous variables to each other are significant and interlocked, even though the exogenous variables are as a body all assumed independent of the variables being generated in the model. This procedure is apt to be most misleading. Inputs that are improperly assumed to be exogenous may be so strongly linked into the equations of the model that they force the entire response of the model. The model may lack the internal freedom of action from which arises the behavior of the real system. If so, the model results are merely the reflection of the numerous and fully controlling inputs. The dynamic character of the actual system is obscured in the model.

It seems quite incorrect to use taxes, money supply, or government expenditure as exogenous variables in a model of an economic system. These clearly have important coupling both from and back into other economic variables such as profits, price level, and the social and political pressures of unemployment.

In this kind of situation it appears that we should fare much better to represent in the model (no matter how poorly) the policies and political forces that we believe are operating in the actual system than to try to use actual time series of historical or current data from the real system as a model input. The assumed decision functions for the governmental sector may, of course, be highly simplified in a model, just as are all the other decision functions. However, including these simplified policies within the model at least would retain the major closed flow loops that connect governmental expenditure and credit control into the economy of the country, so that the dynamic interactions of the two could be explored.

The use of exogenous inputs seems contrary to sound practice in handling dynamic models except where the external input is completely independent of and unaffected by any of the variables generated in the model.[4] As discussed in Section 10.7, noise of specified characteristics can be inserted in decision functions to represent the independent factors in decisions that are not being generated within the model structure.

The inherent incorrectness of interconnecting real-life variables with model variables in a model run is closely related to the discussion in the next chapter regarding the limitations on the ability of models to predict the values of real-life variables. A model should be considered qualitatively excellent if both it and the real world generated growth and oscillatory modes of *approximately* the same time constants, frequency, and damping.[5] If, however, the frequencies and growth rates are not *exactly* the same, or if they are excited into changing phase owing to differ-

---

[4] In Koopmans (Reference 10), see pages 56 and 393-394.

[5] Damping refers to the tendency of the oscillation to change in amplitude as time progresses.

ent noise in the model and the real systems, the variable from the real system does not possess a plausible relationship to the instantaneous values of other variables in the model. Its inclusion can be a completely extraneous act; the "true" value from the real system is irrelevant to the state of the model variables.

Deciding what variables must be internally generated in the model and what may be taken as independent (exogenous) is a matter for the model builder's judgment and depends on the purpose of the model. Take, for example, an industrial company model of a supplier of components that go into military equipment. Here orders for components from equipment manufacturers cannot be considered an exogenous variable except for some studies of very limited interest on internal interactions at the component maker, such as interrelating inventory, employment, and sales rate. The ordering rate from equipment manufacturer to the component maker is actually very sensitive to what happens within the component-maker sector. It is affected by the ability of the component maker to supply. Inability to supply shifts orders to competitors; also, delay in supplying causes ordering further in advance in anticipation of need. Interactions between component supplier and equipment maker may create the important dynamic behavior of the industry.[6] On the other hand, we probably can assume that military department ordering from equipment manufacturers is an exogenous input so far as the component maker is concerned. Ordering of military equipment will not usually be done with awareness of the supply condition of ordinary components. If our component maker does not meet the demand, other sources will be developed. The actions of the component

[6] See Chapters 17 and 18 for a model of this industrial system.

maker probably have negligible effect on the placing of government contracts for equipment.

Here it is worth noting that as a practical matter we are usually limited to one exogenous, nonnoise, test input. It is unlikely that two or more exogenous inputs will be totally independent of the variables in the model and still have strong causal coupling between themselves. If the two or more exogenous variables have no cause for interrelationship, then every time-phase combination of the two or more constitute a separate test input to be explored. Given several exogenous inputs that presumably have meaningful nonnoise time sequences, the combinations of these in different time-phase relationships to one another soon become prohibitively numerous.

At the other extreme, where there are many noise inputs,[7] we cannot expect to generate more than an infinitesimal fraction of the time-sequence relationships that might exist amongst the noise inputs. However, after a model run that is several times as long as the period of the longest disturbances existing in the system dynamics, the general qualitative nature of the system combined with its noise will become evident.

In summary, exogenous inputs in dynamic models should be thought of as "test inputs." They are themselves independent of what happens in the model. They should not be used to produce greater reality in model behavior; the reverse would usually occur. They are inputs from the independent external environment used to see how the system under study would respond to *assumed* changes in that environment.

[7] Probably most persons would look upon noise as an exogenous input, but we might consider it endogenous (internal) to any decision function, since we specify its qualitative characteristics and its method of generation.

CHAPTER 13

Judging Model Validity

The significance of a model depends on how well it serves its purpose. The purpose of industrial dynamics models is to aid in designing better management systems. The final test of satisfying this purpose must await the evaluation of the better management. In the meantime the significance of models should be judged by the importance of the objectives to which they are addressed and their ability to predict the results of system design changes. The effectiveness of a model will depend first on the system boundaries it encompasses, second on the pertinence of selected variables, and last on the numerical values of parameters. The defense of a model rests primarily on the individual defense of each detail of structure and policy, all confirmed when the total behavior of the model system shows the performance characteristics associated with the real system. The ability of a model to predict the state of the real system at some specific future time is not a sound test of model usefulness.

THE validity (or significance) of a model should be judged by its suitability for a particular purpose. A model is sound and defendable if it accomplishes what is expected of it. This means that validity, as an abstract concept divorced from purpose, has no useful meaning. What may be an excellent model for one purpose may be misleading and therefore worse than useless for another purpose. What then are the standards for judging the models discussed in this book?

13.1 Purpose of Models

The purpose of industrial dynamics models is to aid in the design of improved industrial and economic systems. How are we to judge if the models are suitable for this purpose? The ultimate test is whether or not better systems result from investigations based on model experimentation. By this criterion, the validity of industrial dynamics models is not separable from the effectiveness of industrial dynamics as a total viewpoint and discipline. The pertinent test is that of utility in improving management practice. Any new approach to management problems will be on probation until this test is passed.

The evaluation of improved managerial effectiveness will almost certainly rest on a subjective judgment rendered by managers in regard to the help they have received. Objective, noncontroversial proof of the effectiveness of an experimental system design using models is most difficult to conceive. Who is to say unequivocally that a superior manager is more successful because of a methodology and not because of the judgment and skill which he must also possess?

Although industrial dynamics in its entirety will ultimately be judged by the management change it creates, this does not serve the interim purpose of judging a particular investigation as it progresses. The ultimate test is too remote to be a guide for current work. We then attempt to find ways that can substantiate the work in progress. In the early phases we can expect to establish only a degree of confidence that is less than complete. There can be no certain proof of final success until after that success has been achieved. We must then consider on what our interim confidence is to rest.

13.2 Importance of the Specific Objectives

If the purpose of a series of models is to assist in the design of an improved industrial system, the first concern should be whether or not the particular undertaking is addressed to important questions and problems. The absolute worth of a model can be no greater than the worth of its objectives.

The value of the objective transcends all other considerations in determining the utility of a model. An elaborate and accurate model can do little if it relates to questions and behavior that are of no consequence to the success of the organization. On the other hand, a simple and even inaccurate model may still be tremendously valuable if it yields only a little better understanding of the reasons for major success and failure. Our mental models are examples. The simple and rather inaccurate dynamic models in the minds of skilled managers have been far more effective in carrying our industrial civilization to its present heights than have the formal mathematical models thus far used in management and economic science. The manager's working models in the form of verbal descriptions and thought processes are more attuned to important objectives of the future and more perceptive of the behavior mechanisms of actual organizations than have been the abstract models developed for explanation of the past.

How can we determine whether or not solutions to significant problems are being sought? Is there any formal test to show how promising and how feasible the goals are toward which we are working? For the present at least, the answer is negative. The ultimate significance of

our effort to design a better industrial system rests on the investigator's judgment in selecting a specific objective. Choosing significant objectives depends on keen perception of symptoms and ability to relate them to the probable causes. We must then conclude that the first and most important subsidiary test — the value of the objectives of a specific industrial dynamics study — is still a judgment decision for which there is now no objective criterion.

Can we be more specific? Is there a more concrete operational statement of what makes a useful model?

13.3 Predicting Results of Design Changes

The operational use of the models discussed in this book is to predict the result that will ensue from a *change* in organizational form or policy. We are most interested in the *direction* of the *major changes* in system *performance*[1] that will result from altering a structural relationship or a policy in the system. Second, we wish to know the approximate *extent* of the system improvements that will follow. Improvements that are slight in magnitude are apt to be less certain and also may not be worth the effort required.

How can we determine the model's ability to predict the effect of design changes in the actual system? We can discuss some of the requirements.

If a model is to indicate the effect of real-system changes, there must be a correspondence between the parameters and structure that might be changed in the model and the actual parameters and structure of the system. A proposed model redesign must be meaningful in terms of changes that can be accomplished in the real system. The mechanisms of the model must represent the mechanisms of actuality. Further-

---

[1] By "performance" we mean here such measures of a system as profitability, employment stability, market penetration, product cost, company growth, price fluctuation, investment required, and cash position variations — those characteristics that indicate the "desirability" of system operation.

more, a model must generate the *nature* of the dynamic characteristics that are of interest; otherwise it is not a vehicle for detecting how those characteristics can be changed.

As before, we lack an objective test of whether or not the model will predict design changes. The persuasive test is like that for model purpose — do real-system behavior changes actually occur as predicted by the model? An actual test is possible only after the design change has been made and there is some measurement or observation of the performance changes in the real system. Eventually such evidence will accumulate, but in the meantime on what are we to base our faith to substantiate work in progress?

The *presumption* of model significance rests on two foundations. Primarily, confidence depends on how acceptable the model is as a representation of the separate organizational and decision-making details of the actual system. Secondarily, confidence is confirmed by the correspondence of total model behavior to that of the actual system.

13.4 Model Structure and Detail

During the course of an industrial dynamics study, the defense of any model rests primarily on the defense of the details of its design. This means the evidence and the arguments to justify not only the form of each equation but also the selection of system boundaries, system variables, and assumed system interactions between variables.

The importance of justifying model detail rests on a fundamental working assumption, the assumption that if all the necessary components are adequately described and properly interrelated, the model system cannot do other than behave as it should. The converse is not true; an endless variety of invalid components and structure can exist to give the same apparent system behavior, but these incorrect structures usually would not open the way to better system designs.

In the design and justification of a model, we need to call upon the full variety of knowledge that is available about the system. Most of our knowledge is in the experiences and the minds of people who have observed and worked with the system. Much information is in the descriptive literature. Only occasionally will there be numerical and statistical evidence sufficient to settle important model-building questions.

Industrial dynamics models are built on the same information and evidence used for the manager's usual mental model of the management process. The power of industrial dynamics models does not come from access to better information than the manager has. Their power lies in their ability to use more of the same information and to portray more usefully its implications.

Can we establish objective and noncontroversial quantitative tests of whether or not a model is properly constituted? Validity of a model as a description of a specific system should be examined relative to

- System boundaries
- Interacting variables
- Values of parameters

System Boundaries. The first and most important question about detailed model design regards the proper selection of boundaries of the system to be represented. Suitable boundaries are related to objectives, as discussed in Section 13.2. If the system chosen for study does not contain the answers being sought, the model cannot help. If the boundaries range unnecessarily far, the model will be distracting and may lead to such confusion that the project is abandoned.

As with the selection of objectives, the choice of system boundaries is not yet guided by a definitive theory that can be objectively tested. The methods and the conceptual background of a science must, at the forefront, be applied with art and judgment based on successful experience.

Basing decisions on judgment does not mean that the choices lack a foundation of fact or contact with reality. It means rather that a choice is made concerning what part of the available knowledge is to be relied upon. For most model-building questions, a wealth of information exists, but with many contradictions, and these contradictions cannot be resolved by objective means that are generally accepted. One man may resolve a conflict by an "objective" test, but in the eyes of another this begs the question unless the justification and pertinence of the test itself are acceptable.

Interacting Variables. The second most important question of model design is to ask if the model contains an effective choice of variables, properly interconnected. In the final analysis, this will also rest on our working knowledge about the system. The most difficult choices will be in the decision functions of the model. These exist largely within the information networks where policies are informal. For most decision functions, no verifying numerical data are available. Therefore the model must be constructed from the descriptive information on policy.

However, consider the few exceptions where data do seem to be accessible. Even this does not satisfy the search for an objective, quantitative criterion for model design. Data may serve to reject a grossly wrong decision-making hypothesis, but they can scarcely prove a correct one. Two or three inputs to a decision point may account for most of the output, but what sort of fit is good enough? A positive feedback loop whose contribution to a decision stream is well below the noise level in that channel may hold the key to system behavior. The feedback information will be highly correlated with other variables. In a statistical analysis, the feedback signal can easily pass unnoticed in the random noise or as a part of one of the other input variables.

Further, the simple magnitude of an input to a decision stream is not closely related to its importance to the dynamic behavior of the system. Time phasing and transient character may far outweigh sheer magnitude in determining dynamic performance.[2]

The methods of objective analysis offer no guarantee that the proper variables are included and properly interrelated. However, we do have the negative test — a test that is necessary but not sufficient. If the decision stream is demonstrably incompatible with the decision-making hypothesis, the test has failed. On the other hand, passing the test is no guarantee of a correct hypothesis. Even the closeness of fit between data and a hypothesis is not an adequate test for choosing between two hypotheses. A hypothesis based on the wrong selection of variables but with compensating adjustment of parameters may pass a statistical test outside the system context more successfully than a decision function based on a more correct selection of causal variables but with parameter errors that degrade the statistical match but not the dynamic behavior of the model.[3]

Failure to include the proper system variables in a model can destroy the utility of the model as a design tool. As the dynamic behavior of systems becomes better understood, we can expect that useful principles will emerge to guide the selection of model content. For now, it is best to acknowledge that the most useful models will be constructed by those who know the actual system and who at the same time have a background in dynamic system analysis.

Parameter Values. The third and least important aspect of a model to be considered in judging its validity concerns the values for its

[2] This is illustrated in Chapter 18, Figures 18-1 and 18-5, where a feedback signal to customer ordering due to varying producer delivery delay is one of the necessary causes of the large producer employment fluctuation, even though the feedback signal accounts for only a few per cent of the order rate.

[3] An examination of Figures 2-10 and 2-11 will show that with carefully chosen coefficients and delays any variable could be closely correlated with any other, regardless of actual mechanisms that underly the system operation.

parameters (constant coefficients). The system dynamics will be found to be relatively insensitive to many of them. They may be chosen anywhere within a plausible range. The few sensitive parameters will be identified by model tests, and it is not so important to know their past values as it is to control their future values in a system redesign.

The numerical values of parameters of some decision functions can be subjected to a formal statistical test. This is done after

- The objectives of the model have been decided.
- The boundaries of the significant system have been determined.
- The pertinent variables have been chosen.
- A hypothesis of the interaction of variables has been formulated.
- An arbitrary judgment decision has been made about what constitutes passing the statistical test of the parameter value.

At this point, the investigator must make another judgment decision. He must determine whether or not the results of the statistical validity test of a parameter are likely enough to affect the objective of improved system design so that such testing must be ranked among the priority tasks that will compete for the limited manpower.

13.5   Behavior Characteristics of a System

A closed (self-contained) industrial dynamics model should generate time patterns of behavior that do not differ in any significant way (judged within the framework of the objectives of the study) from the real system. This realistic behavior should result from the interaction of model structure and policy, each detail of which has a justification that is independent of the fact that the total system behavior is acceptable.

The more indistinguishable the over-all model performance is from a typical history of the actual system, the better our confidence in the model is apt to be. However, close correspondence of model and real-system behavior is not

by itself a reason to assume that the model is a safe basis on which to predict the effect of system design changes.

Since the similarity of the model to the actual character of the system is necessary (but not sufficient), we should discuss what this similarity means. Lack of proper model behavior, especially in the early stages of model formulation, will often point the way to omissions and conceptual errors in the model. What does one look for? What are the tests of similarity between two systems?

The first test of a model is to demand that its behavior not be obviously implausible. Such a criterion may seem foolish, but the literature of economics and management science contains very few examples of system models that meet this test — models which as completely closed, self-contained systems will generate time histories that show no obvious inconsistency with our knowledge of the actual system.

In the early development of a model the implausible performance is apt to be of a gross nature. Levels of physical items, such as inventories, may become negative. Certain flows that by their particular nature must be unidirectional may run backward under some circumstances. Values of variables may grow beyond the range that is conceivable in the actual system. These things may happen, even when careful thought has been given to the details of model construction. The causes of such implausibilities are usually easy to identify. Examination of the detailed changes that occur in the values of system variables usually reveals inadequacies in the decision policies as the cause. Very often, conditions within the model system are becoming more extreme than was anticipated, and the decision policies (usually the nonlinear character thereof) are found not to be reasonable over a wide enough range of system operation.

Another effective test of a model is to attempt to precipitate additional obvious inadequacies by testing the model over an unusually wide environmental range (but still within the objectives of the investigation). This may well be a range wider than has ever been encountered by the actual system. Much of our knowledge of a system is in the form of knowing what would happen under various crisis conditions. A breakdown of the model policies under "reasonable" crisis tests often reveals defects that affect model performance even in more normal circumstances.

Having eliminated the "obvious" implausibilities in a model, we turn attention to more subtle performance. Attention should be directed to all of the model behavior characteristics that can be compared with the real system.

The first step is usually to investigate the similarity of problem symptoms between model and real behavior. If the model has captured the causes of actual system difficulties, the model will exhibit the same trouble symptoms as the actual system, even though these may be far removed from the point of cause.

If fluctuating behavior is involved, the time intervals between peak values of the variables of the model can be compared with the actual system. When substantial differences exist between model and actual system, they may indicate either wrong values of coefficients or the omission of some significant part of the system. Natural periods in the model are apt to be shorter than those of the actual system because most people tend to underestimate decision-making and action time delays. The same may occur from simplifying a system by leaving out minor elements in decision and action channels rather than by aggregating them (particularly their time delays) into preceding or succeeding parts of the system.

The time-phase relationships between variables[4] often indicate how similar the model results are to the historical data of the actual system. Any actual system will show typical (not necessarily invariant) tendencies for one kind of event to precede another.[5] Since these time relationships will depend on the frequencies of disturbances within the system and on the rate of change of certain of the variables, they are not an indisputable criterion of comparison between model and actual system.

The data and plots obtained from a model run can be examined to see if the decision streams being created would have been considered normal in the actual system. If an important defect exists in the decision policies of a model, a set of conditions will often arise in some model run where decisions are taken that would not have been considered reasonable in actuality. This may lead to discovering an omitted factor in the system operation which cannot be ignored. This is essentially a judgment, by comparison with the system being modeled, of the quality of the decisions being made at each point through time. It is the test one would apply to judging the competence of a manager.

Closely related to time phasing and periodicity is the abruptness with which the values of system variables change. Actual industrial systems differ markedly in the rapidity with which changes occur in prices, production rates, order flows, and other variables. The model of a system should show the same transition characteristics as the system.

Many system characteristics that may not be clearly discernible in the actual system can be observed in a model. However, these model characteristics should not conflict with what is known of the actual system. Systems differ in their tendency to amplify or suppress externally imposed disturbances. This is easily observed in the model,[6] but in the actual system it may be known only by inference from isolated incidents. In a similar way, the model reaction to

---

[4] For example, the peaking of inventories after the sales and production peaks, as in Section 17.1.

[5] The leading and lagging time-series groups of economic business-cycle analysis are an example. See Appendix I.

[6] See Chapter 18 and Appendix K.

nonlinear conditions[7] can be observed, and some of these can be substantiated as existing in the actual system.

Many of the preceding behavior characteristics can be quantitatively measured. For these an objective set of quantitative criteria could be established as a basis for comparing models with data from actual systems. Nevertheless, before being justified in doing so, one would need to decide what significance to attach to differences in the results of applying the criteria. Once the general qualitative nature of a particular phenomenon is present in a model and approximately correct (often within a factor of 2 is quite close enough), it can usually be adjusted to any desired value by changing system parameters without moving these parameters outside the range compatible with our knowledge of their values in the actual system.[8] Furthermore, there are usually several parameters, any one of which will do the trick. Making a highly accurate match of model to real system is not likely to affect conclusions about how to design

a better system. The design changes are not dependent on such accurate knowledge of the system; they are dependent on the relationships in the model that can create a reasonable approximation to the system characteristics of interest.

If adjusting the parameters of a model to get a highly accurate match to the system does not make the model appreciably more useful, and if we cannot choose between two structurally different models on the basis of small differences in meeting the quantitative criteria, then there is little need to establish accurate definitions and measures. This seems to be the present situation. It will no doubt change in the future. The possible gains in system design now appear so great that sensitive tools are not yet needed. As managerial systems become better understood and are improved, the tools will need to be refined.

In summary, serious model defects will usually expose themselves through some failure of the model to perform as would be expected of the actual system. If the model behavior is not close enough to reality (what is close enough depends on the purpose of the model and the nature of the discrepancy), we must then return to the consideration of elementary system structure, boundaries, and details. Causes for the discrepancy must be found which can be explained and defended on grounds other than that their inclusion corrects system behavior. A vast number of things could be done to any model to change a particular behavior characteristic. The ones chosen must be picked on the basis of knowledge about the working details of the actual system.

13.6   Model of a Proposed System

The primary purpose of comparing a model and a real system is to demonstrate that we can match the behavior of an existing system and thereby develop credence in the elementary structure of the model. This belief in the sound-

---

[7] Such as the limited capacity in Figure 2-5.

[8] To illustrate how perfecting the faithfulness of a model output need not increase model usefulness, we might consider Figure 2-4, in which the introduction of noise in a system decision imparted a highly realistic appearance to system over-all behavior. Over a very wide range in that system, the amplitude of factory production rate fluctuation is proportional to the noise amplitude at the retail buying decision. This amplitude must be rather close to a particular value, or else the system amplitude fluctuations will be too large or too small to pass as the life history of a particular real system. Here the appearance of realism is apt to come from merely setting the noise amplitude to yield the desired output. But this does not matter. Our conclusions do not derive from the actual amplitudes in the model output. Instead, for a *given* noise amplitude we wish to see if the system becomes *more* or *less* sensitive to the noise as some system parameter is changed. In that particular figure we can expect that conclusions about the effect of a parameter on system sensitivity to noise would hold over a hundred-to-one ratio of noise amplitude. The noise amplitude used could be grossly untrue without affecting our conclusions about the way to make the system less easily perturbed by random events in the environment.

ness of the basic components is then extended to the components of a redesigned model and then to the inference that the total behavior of the new model is reliable as an indication of how the altered real system will behave.

In many situations the test of matching the present before designing for the future is unnecessary. Many system parameters and policies that are obscure and hard even to guess closely in actual systems would be very easy to control.[9] We are then much more interested in whether the new system can be made to conform to the model than in whether the model conforms to the old system. Here the validity test takes the form of whether or not the actual system is being controlled to agree with the model.

There are other models even more remote from actual systems. These arise in an investigation of what would happen *if* the real system had characteristics like the model. Much can be learned by studying systems that *might* exist. In fact, this is an excellent beginning point in model experimentation. One of the first steps is the model that is fashioned merely to the best of the investigator's immediate ability. The emphasis is on plausibility, not accuracy. Defending the detailed accuracy of assumptions is secondary to emphasizing what the model can teach, so long as the model reflects the *kinds* of things that *might* exist in a real situation. If the model is plausible or possible at the level of elementary actions within the system, it will serve to teach much about the dynamics of large systems.[10] Such studies will help to develop awareness of the kinds of factors that are important in dynamic behavior.

[9] This is true in the model in Chapters 17 and 18, where some of the decision-making time delays in the old system would be impossible to prove from the available data but which can be controlled by making the decisions according to a formal rule.

[10] This is the nature of the model in Chapters 2, 15, and 16. It is a representation of the kinds of things found in typical distribution systems. It shows dynamic behavior common to many industries but does not purport to be an accurate representation of any specific one.

13.7 Comments on Model Testing

The preceding discussion of model validity arises from a philosophical foundation which is different from that which seems to underlie discussions of model validity in parts of the management and economics literature. We must say that it "seems to differ," because there is little trace in most of the literature of any explicit treatment of the assumptions underlying the conventional model validation practices. As a way to make doubly clear what is meant in the earlier sections of this chapter, there follows a discussion of certain views with which this book intentionally differs. None of them may be held by large numbers of people, but all are encountered sooner or later in print or in conversation. Some are not explicitly stated but are only implied by corollary attitudes.

Validity without Goals. Validity and significance are too often discussed outside the context of model purpose. Usefulness can be judged only in relation to a clear statement of purpose. The goals set the frame for deciding what a model must do.

The validity of a model should not be separated from the validity and feasibility of the goals themselves. We sometimes see the implication or outright statement that a model is intended to predict the future state of a system. If the reasonableness and usefulness of such a goal have not been argued, the model aimed at the goal is suspect.

Another, often implicit, model goal is to "explain" actual system behavior. A better word for the objective of many such models would be to "reproduce" system behavior. A model built on historical, statistically derived relationships between variables might reproduce a pattern similar to the actual system. It explains the behavior of the system only if a separate defense is made that the model relationships represent the true causes for system actions. This is because a sufficiently elaborate formal curve-fitting procedure can be devised to fit arbitrarily closely any ensemble of data. The resulting *summary* of the data carries with it no implication of containing proper causal mechanisms. It may give no promise whatever of even qualitative behavior similar to the actual system outside the time period for which the data fitting was done.

Assumptions Underlying Test Procedure. Any "objective" model-validation procedure rests eventually at some lower level on a judgment or faith that either the procedure or its goals are acceptable without objective proof.[11]

In tests of statistical significance it is usually taken as self-evident that the "measure of significance" is significant. Yet there is often no answer to the question, "What is good enough?" because the result of the "objective" test has not been related to a goal.[12]

Many persons will protest at the way the earlier sections of this chapter depend on subjective judgment for model validation. Yet at some point the "objective" methods rest on the same foundation. The objective test is useful if the underlying assumptions for it are sound and if the judgment criteria are not amenable to direct application but must be interpreted through an intermediate quantitative procedure. The danger is that the quantitative procedure

will take on an aura of authenticity in its own right. It becomes a pseudoscientific ritual. The underlying assumptions based on judgment or merely faith may be forgotten. The objective test, which may have been sound for the original goals and assumptions, may now move by itself into new areas where it is useless or actually misleading.

Predicting Future System State. A model is a statement of a "law of behavior." It thereby purports to unify what would otherwise be unrelated and perhaps even apparently contradictory information. Within what the "law of behavior" is supposed to represent, we should expect that it predicts future real-life behavior if that behavior is indeed governed by the "law." At this point we must be most careful in defining and understanding what the "law" represented by a model is intended to do. We judge the validity of a model by comparing its predictions against observations of the system whose behavior it is supposed to represent. But in what sense is a specific model expected to predict, and how is the comparison of a model to the real system to be made?

Take the solar system as an example of what we do *not* have in social systems. It is a group of bodies whose motion is governed by a very few basic causes (inertial forces and gravitation). It is an "open" system; it is not an information-feedback system in which there are forces that would return a planet to its "original" orbit were it to be disturbed by the passage of some foreign heavenly body. It is a system in which the forces not explained by the primary laws (inertia and gravitation) are vanishingly small compared with the primary forces. It is a system whose future *specific state* is of interest to us. It is a system whose future specific state will not be disturbed by our having advance knowledge of what that state will be.

A useful solar system model is one that predicts the *times* of *specific* future *events* like sunrise, winter, and a lunar eclipse. The usefulness of the model rests on its ability to make this

---

[11] See Churchman, Reference 11, for a discussion of the philosophy underlying scientific methodology.

[12] As an illustration, I once attended a seminar conducted by a respected member of the management science community, in which a hypothesis was presented for the way in which a particular kind of decision had been made by the people in an actual system. A decision function near the bottom of the management hierarchy had been chosen because historical data did exist. The hypothesis fitted the data to high accuracy for over 90% of the incidents. The *implication* was that the hypothesis was excellent. But the lecturer had no answer to the question whether the decision function was *really* good enough, or to the question whether too great an effort had been expended to make it as good as that (the research had been extensive). No goal or purpose existed against which to measure or judge or even discuss the pertinence and validity of the validity test. In fact, he offered the response that the research merely proved that the real-life decision procedure was not a random process. I believe that no one could have been found to defend the proposition that it was random.

prediction, and its accuracy is measured by the error between model prediction of time and place and the actual observations.

Often we find the same test being applied to a model of a social system without a critical re-examination of whether the objectives and foundations are still valid. Before applying the test of predicting the system state at specific future times to a model of industrial or economic behavior, we must ask several questions. Is prediction of specific events a proper goal of the model? Is there reason to believe that the objective is possible? Is it the most useful application of the model? Is the system sufficiently independent of the prediction process itself?

Economic systems are conceptually of a different class from the solar system. Economic systems are characterized by their information-feedback-system behavior, while the solar system is an open-ended system. There is no reason to believe that either the nature or purpose of models of the two would be identical. Economic behavior occurs in the context of an information-feedback system in which goal-seeking decisions are constantly responding to the state of the system itself. An external disturbance induces closed-loop reactions within the system toward re-establishing the goals.

The economic system contains large "noise" forces that are not explained by the behavior laws that we feel justified in hypothesizing. These unknown influences are often comparable (depending on the way in which the comparison is made) with the orderly forces. This implies that the system is subject to having its future deflected by the unexplained factors.

Even though stable organizational structure, policies, and human reactions exist and these determine the principal dynamic characteristics of a system, we cannot assume a *perfect* model in which *every* relationship is known *exactly*. Therefore, we are committed to models in which *every* decision function has, at least in principle, a noise or uncertainty component. By definition, the exact time pattern of this noise is unknown,

and we have not discovered its generating causes. (We may have useful estimates of its magnitude and statistical characteristics.) The model acts on the noise components as it acts on all other flows within the system. The structure and characteristics of the model determine the nature of the reaction to the noise.

Not knowing the instantaneous values of the noise, we can still study the *kind* of behavior exhibited by the system, including the sensitivity to the noise inputs. In the model the exact paths of the variables in time depend not only on the model structure and the initial values of system variables but also on the unknown noise. The relative determinancy created by the known policies and structure versus the uncertainty introduced by noise will determine the ability of a model to predict the specific state of a system at a particular future time.

From a consideration of the nature of the components and structure of social systems, it appears that we should not soon expect models that can be useful in predicting specific future system state. This type of prediction of future state is possible only to the extent that the correctly known laws of behavior predominate over the unexplained noise. In the model of the solar system the noise components of force on the bodies are vanishingly small compared with those forces which can be explained. In social systems this is not true.

The relationship between structure, noise, and prediction of future system state can be demonstrated entirely within the realm of model experimentation. Consider that there are two identical models[13] which contain noise at one or more decision points in the system. Let these two models represent, respectively, the "real" system and the "model" of the "real" system. The "model" is then known to be identical in structure and in all its coefficients to the "real" system and even has the same statistical charac-

[13] See Appendix K for this experiment conducted on the production-distribution system with noise input, as already seen in Figure 2-4.

teristics in its noise sources. The instantaneous values of noise are unknown and will differ. As time progresses from the initial conditions, the two models will begin to diverge in the numerical values of the variables they are generating. Soon there will be no evident, instantaneous similarity between the quantitative values of corresponding variables in the two systems.[14] Over long periods of time, however, they will both show the same general type of behavior. Although the "model" may not be useful for predicting the condition of the identical "real" system at a specific future time, it has the same performance nature. It has the same kind and degree of sensitivity to the noise disturbances. Furthermore, this nature, in both the "model" and the "real" systems, is dependent on the structure and coefficients of the system. The "model" can be used to predict how a change in the "real" system would make it a more desirable system, even in the continued presence of noise that excludes prediction of specific future system state.

A dynamic system model should therefore be expected to represent and to predict the behavior characteristics (like profitability, stability of employment and prices, growth tendencies, and typical time-phasing relationships between changes in variables) of the actual system. It should not be expected to predict future system state except to the extent that the system has continuity and momentum characteristics that will cause present conditions and trends to persist for a time in spite of noise disturbances.

Indeed, one of the performance characteristics of a system that we might study in its model is the inherent limitation on prediction ability. Using the model, we should be able to study the counterpart of what the engineer calls "bandwidth" in his linear systems. "Bandwidth"

is an undefined concept in nonlinear systems but implies the degree of persistence of existing trends and cycles and the rapidity with which these might be altered by random events. The narrower the bandwith around any component of system behavior, the longer we can expect that component to persist. Actually, it seems clear that we are dealing with wide-band systems in management and economics, in which the noise components strongly determine the state of the system even in as short a time as a part of one "cycle" of disturbance into the future. Nonlinear systems probably show different bandwidths at different times, so that vulnerability to change will vary with the state of the system.

The determinancy of a system in projecting the past and present into the future will be related to the degree of stability of the system. A system with strong, unstable oscillatory tendencies will be more predictable through the next cycle of disturbance than an inherently more stable system in which the fluctuations are more dependent on noise whose source is unrelated to the behavior of the system itself.

The relationship between noise, policy, observation, system behavior characteristics, and prediction can perhaps be illustrated by an absurdly simple analogy to a coin-flipping experiment. The unknown and uncontrolled forces that determine which side of the coin comes up are usually predominant. We do not expect to predict correctly each separate flipping event. The statistical prediction relates to the nature of the pattern of behavior. The model of the process deals with the type of long-term outcome, not the individual specific event at each point in time which is, in the usual coin experiment, supposed to be controlled by random events. But suppose that we do not like the behavior of this system in which each event is independent of the past and is a result of uncontrolled forces. Are there possible policy changes in the system which might affect the nature of the results?

---

[14] For the system of Chapter 2, the divergence between the "model" and the "real" system is substantial after only a few weeks from the identical starting conditions.

The usual coin-flipping experiment implicitly assumes some very specific structure and policies of the system, of which the coin observations are a manifestation:

- A person or device exists for flipping the coin.
- The coin will be flipped.
- A person or method for observing and record-the state of the coin exists.
- The coin will be observed only on the table, not in the hand of the person doing the flipping.
- The coin will be observed only if a flip has intervened since the last observation.

The behavior of the system would be greatly changed if any of these controlling policies were changed. The bandwidth of the system depends on the frequency with which the coin is flipped. If the observations are made more frequently than the flipping, or if both are at random intervals, a degree of predictability of the specific event enters because there is now some probability that the coin was not touched between observations. The sensitivity of the system to uncontrolled influences can be reduced (here to complete insensitivity) by changing the second policy to one of not flipping the coin. The coin is the point at which the symptoms of system behavior appear. As in a managerial system, these symptoms arise from and depend on system policy and structure that encompass much more than the point where the symptoms can be observed.

In a management system, if observations are very closely spaced, we expect little change between them, just as observing the coin several times between flips discloses no change. As we attempt to predict further and further into the future, the present becomes rapidly less significant, and the intervening random events become more controlling. The rapidity with which the present is degraded in affecting the future depends on the structure and policies of the system and on what is being predicted. For example, we might have little confidence in predicting the time of the bottom of the next economic reces-

sion to within one month but still have a high confidence of predicting average economic growth rate for the next decade within a factor of 2 either way. The precision desired (related to the sensitivity of the system to noise with respect to the particular question) and the expected continuity of past conditions (which we have referred to as bandwidth) determine our confidence.

A simple illustration, as in Figure 13-1, may

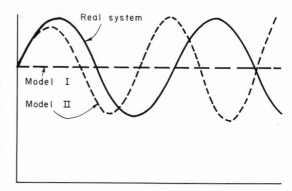

Figure 13-1   Criteria for model validity.

help distinguish between predicting behavior *characteristics* and predicting the state of the system at specific future times. Suppose that a real system has a variable that follows a path shown by the solid line (here a sine curve). Then consider the corresponding variable as given by each of two models that purport to represent the actual system. The first model generates a horizontal straight line. The second generates a slightly growing sinusoid with a period somewhat shorter than that exhibited by the real system. Which is the better model? The answer depends on the test applied and the purpose of the model.

Suppose that we should conceive of the purpose of the model as a means of predicting the value of the real system over the future interval illustrated. Imagine further that we take as a criterion of prediction accuracy the average of the sum of the squares of the differences between model and real-system values at closely spaced points in time. (Such a criterion is common, usually arbitrarily chosen without justification of its significance as a useful measure, and selected because it is mathematically easy to handle.)

By this test, Model I, which gives the straight line, is the better predictor. The above error measure will be smaller and the maximum error will be less than for the second model.

In spite of being a better predictor, according to the above arbitrary criterion, of the specific future state of the system, we should not expect Model I to reveal much about how to redesign the actual system to change its performance. Model I does not exhibit the characteristic fluctuating behavior of the actual system. It does not pass the necessary (but insufficient) test of having dynamic behavior similar to that of the actual system. We should conclude that any defense of its component structure and policies must be inadequate because it seems incapable of creating the sinusoidal behavior which can here be seen and measured in the actual system. Because the components of the model fail to create the nature of the behavior of the actual system, we should not trust modifications in these components to show how the system behavior might be changed.

Consider Model II in Figure 13-1. Suppose that the defense of structure and policy are without serious objection. Does it pass the criterion of having a dynamic behavior similar to that of the real system? There is a predominant sinusoidal behavior pattern. The period of fluctuation in the model is a little shorter than the actual system but only by 25%. Its damping coefficient (indicative of the tendency of the fluctuation to grow or decline) is a small negative value that is not much different from the zero value of the actual system. Because the model, compared with the actual system, has similar amplitude, period, and damping, we should conclude that it may be a close representation of the actual system. This test is meaningful only because we believe *independently* that

the causal relationships of the actual system are represented in the mechanisms of the model. An endless variety of model details having no similarity to the actual system could be assembled that would create the Model II curve. Our confidence in Model II as an experimental tool for studying the effect of structure and policy changes in the actual system is based on our confidence in the model components separately and individually *and* in the fact that in concert they produce the behavior that interests us in the actual system.

One additional point needs to be made regarding the search for models that will predict the state of an actual system at some future point in time. Economic and industrial systems would not, even to a crude first approximation, be independent of a process that would really predict the state of the system far into the future.[15] Predictions are sought as a guide to actions that would take place within the system being modeled. The actions taken as a result of a prediction would directly affect the stream of events whose prediction is being attempted. A successful model for predicting the future state of a business or economic system cannot remain aloof from that system except to the extent that the model fails. Should a model of *proved* ability to predict the specific future course of a social system exist, it would defeat itself if it were put to use.

A "predicting procedure" can be treated only as a decision-making rule, like all other decision rules, in its proper place in the system context. It must be related to the actions that will result from the "prediction." The predicting procedure would not, therefore, exist separately. It would become a component of a larger model of the total system and would contribute to the dynamic behavior character of the system as a whole. Limitations on predicting ability arising from model incompleteness and noise would then become the same as for other models, as discussed above.

[15] See Appendix N.

In a longer-term context, the same could be argued for industrial dynamics models. As they become generally understood, and if they are highly effective as a basis of system design, they become a part of the arsenal of competitive methods whose existence affects the systems they represent. However, a distinction can lie in the immediacy of application. A reliable prediction of an event, like a sales upturn, would be more easily understood and responses to it would be more rapid than the response to a prediction that a specific change in corporate policy would increase the likelihood of higher sales.

Even so, we must never be oblivious to the effect that a system dynamics study may have on the system being studied. It has been my experience that a decision hypothesis, in the context of a system model which reveals the implications of the hypothesis, begins to have an immediate effect on the actual system. The searching questions that must be asked to expose the information flows and the power centers in an organization will cause people to reexamine their action patterns. The investigator who develops a system model (if he wants to compare its behavior, in any of the senses of comparison, against the actual system) must be alert to the fact that he has already himself become a part of the system under study. The more the people in the actual system know of the objectives and possible outcomes of a system investigation, the more they may be affected by the investigation itself. This does not imply that the purposes of a system study should be withheld. It only says that the investigator must be alert to the changes in system structure and policies that may occur simply as a result of exposing those policies. Under some circumstances, the effects might become great enough to affect substantially the dynamic characteristics of the system even before any formal results of a policy design study had been implemented.

Let us summarize the use of dynamic system models for prediction:

- If a model could predict specific events far enough ahead to allow action to be taken as a result of the prediction, there may be sufficient time intervening for random events to alter the predicted values.

- If random disturbances do not exist, if the model is sufficiently near to perfect, and if the momentum of the system is not strong enough to inhibit action in anticipation of the predicted events, then the action taken is itself highly likely to affect the predicted outcome. The prediction procedure would then need to be a system decision function incorporated so that the model recognizes the effect of the prediction on other decisions.

- Any prediction procedure is part of a decision-making process and is properly viewed as a decision function in its system context, where it can be studied to see how it affects the behavior characteristics of the system of which it is a part.

- System models should predict and reproduce the behavior character of a system, not specific events or particular, unique sections of actual system time history.

Nonquantitative Model Validation. We sometimes encounter the attitude that model validity can be treated only in a numerical and quantitative manner. This hardly seems justifiable when such a preponderant amount of human knowledge is in nonquantitative form.

A model will be cast in numerical form in order that our statements will be specific and unambiguous. Such statements, however, often arise from beliefs about relative magnitudes, limiting conditions, and probable consequences. The numbers that appear in such a model often do not derive in any analytical or statistical way from specific numerical data from the operating system.

Quantitative validation of a model should be

done when possible *and* when the anticipated results are expected to justify the cost and effort. However, if most of the content of a model is drawn from nonnumerical sources in the form of individual personal knowledge and verbal and written descriptions, the defense of the model will usually rest on the same kinds of knowledge. The individual expressions in a model should always have meaning in the context of the real system. All variables and parameters should have conceptual meaning that can be individually considered with respect to the actual system. They can then be examined, argued, checked against past incidents and experiences, and considered from the viewpoint of what they imply under both normal and extreme circumstances. Knowledge of all forms can be brought to bear on forming an opinion of whether or not a model is suitable to its particular purpose.

This means that model building and model validation do not stop at the boundary where numerical data fail. It means that both have full access to the vastly richer sources that lie in the nonquantitative areas of business management. By hypothesizing quantitatively about these areas, the day may be hastened when firmer facts and measurements are available.

CHAPTER 14

Summary of Part II

In Part III that follows are examples of dynamic models based on the philosophy and general principles that have just been given in Part II. Before going to Part III, the chapters of Part II will now be briefly summarized.

IN Chapter 4, models are classified. Those that would usefully describe industrial and economic behavior were identified as mathematical, dynamic and nonlinear models. They need to treat both stable and unstable as well as steady-state and transient systems.

Industrial dynamics models, in their purpose and origin, will be more similar to models of engineering systems than to models in the physical, genetic, and agricultural sciences. They are models of information-feedback systems, not models of open-ended systems in which results do not react on causes.

Mathematical models can be constructed from the bottom upward, based on our descriptive knowledge of the elementary parts of the system to be represented (the size of the elements depends on the purpose of the model). A successful dynamic model of nonlinear, noisy, information-feedback systems probably cannot be derived by statistical analysis of the over-all, aggregate, system performance.

Mathematical models make controlled experiments possible and allow us to see the effect of the separate parts of the system. A management laboratory then becomes possible for the design of improved managerial policies. The dynamic model is a tool for the design of policy and organizational form.

The understanding of system dynamics and the tools of electronic computation have now reached the point where systems of hundreds or thousands of linear and nonlinear variables can be readily and economically investigated. The hurdles are not technical nor financial, but intellectual. Success will be paced by the rate of advancement in our descriptive knowledge and insight into the industrial, economic, political, educational, and social interactions that surround us. Models must simultaneously deal with all these factors if they are to serve their purpose.

Dynamic models will be based primarily on our descriptive information already available, not on statistical data alone. Observation of and familiarity with a system will reveal actions, motivations, and information sources that cannot be discovered through historically available quantitative measures. The "intangibles" must appear quantitatively in a model to the best ability of the model builder; to omit them is a more serious failing than to have an error in magnitude. Our real-life systems are not highly sensitive to changes in the components of which they are constructed, neither are models having the proper structure. Severalfold changes in values of parameters and major changes in the shapes of assumed statistical distributions of system responses often have but slight influence. Concentration must be on those factors that determine the characteristics of information-feedback systems — structure, amplification, and delays. A model itself can be used to determine the system sensitivity to a particular parameter or structural change. From the model studies can come an estimate of how much cost is justified in refining our knowledge of the actual system.

Chapter 5 discusses the formulation of a dynamic system model. A model is designed to answer specific classes of questions. The completeness required in a model depends on the questions to be explored. Models of industrial activity usually begin by containing a few hundred variables. Just as verbose writing is easier than the concise, so is the cluttered and complex model easier to formulate than the simple model that still retains the essence of the larger system. Very often it is more expeditious to include a factor than to determine that it can be omitted.

Economic and industrial activities are closed-loop, information-feedback systems. Models of such systems must preserve the closed-loop structure that gives rise to the dynamic behavior of interest. An information-feedback system exists when an environment leads to decisions that in turn lead to actions that are intended to alter the original environment and in turn lead to new decisions.

The essential characteristics of information-feedback systems arise from organization structure (including policy), amplification, and delays. These characteristics must be carefully preserved in going from the real-life system to the model of the system.

A dynamic model can and should be constructed to correspond to identifiable features and policies of the actual system. The terminology should be the same. The units of measure can be identical. The same concepts, objectives, and goals should appear in each.

The dimensional units of measure of variables and of constants should be given meticulous care. The money equivalent of variables having other physical and conceptual meaning should not be substituted for measures in more realistic terms. The conversion points from one kind of flow system to another are the points at which essential policies and conversion factors are to be found.

As a first step, it is strongly recommended that systems be formulated in the continuous form and that discontinuities and periodic review of decisions be postponed until the fundamentals of the continuous system are understood.

We cannot assume that systems will be either stable or linear. Linear analysis over a small range of system operation will often fail to illuminate the characteristics of greatest interest to the manager.

Chapter 6 describes a basic model structure that has been found to correspond to the important dynamic characteristics of industrial and managerial activity. The model consists of an alternating system of levels and flows. The flow channels transport the contents of one level to another, and the flow rates themselves are controlled by the levels. The decision functions are the relationships that describe how the levels control the flow rates. Information channels connect the levels to the decision functions.

This model structure is simple and straightforward. It permits a one-to-one correspondence between the model and the system being represented. It leads to a system of equations that can be individually evaluated one by one at successive time intervals in tracing the time sequence of system operation. No solutions to simultaneous equations are required.

Six interconnected flow systems are used to represent industrial and economic activity. The five elementary flow systems handle orders, materials, money, personnel, and equipment. These five are interconnected by a superimposed information network that links the levels of the five flow systems to the rates in the same and in different flow systems. The information network itself may contain levels and rates.

Chapter 7 describes the types of equations needed for the fundamental model structure. For the symbols to represent variables, English letters and Arabic numerals are used so that they may be readily printed on typewriters and digital-computer output equipment. A time notation indicates the variables as belonging to the present, the immediate past, and the immediately following moments in time. This time notation is for computational purposes and is independent of representation of delays in the system. Level equations describe the contents of the reservoirs or accumulations within the system. Rate equations (decision functions) describe the manner in which the levels control the rates of flow. Auxiliary equations are actually part of the rate equations, but have been algebraically factored out of the rate equations for convenience or because the auxiliary equations define variables of separate interest. The equations constitute a recursive system of first-order difference equations which can be evaluated individually and which yield the numerical values necessary for progressing step by step through time.

Chapter 8 describes a system of flow diagrams that pictorially represent the equations and the relationships within the system. It is strongly recommended that detailed flow diagrams be developed simultaneously with development of the equations for a system. Flow diagrams are often a better means of communication than equations. Also they assist in avoiding errors by ensuring that the equation relationships are as intended.

Chapter 9 discusses the representation of delays in all channels. It suggests that first- and third-order exponential delays are simple and are adequate for most situations that will be encountered.

Chapter 10 discusses decision making and the rate equations that describe the policies controlling the system operation. Decisions are the day-by-day results of applying these policies to the available information. The dynamic model is a tool for studying the influence of policy on system behavior. System nonlinearities appear in the decision-making policies. Much

of the behavior in which we are interested arises from the nonlinear phenomena within the system. Only if these are incorporated in the model can the model show us the behavior characteristics in which we are interested. The rate equations, which represent the policies that control system decisions, can represent only the predominant variables and must necessarily omit a vast amount of secondary detail. In an actual system that part of the omitted detail which arises outside the system and is independent of what happens inside the system can be considered as noise. The estimated statistical characteristics of this noise should be incorporated in many of the decision functions in the system. As with the estimates of other parameters of the system, the exact characteristics of the noise can often be in substantial error without defeating the purposes of the model.

Chapter 11 discusses the aggregation of individual events into continuous flows within the model. This implies that the model takes the viewpoint not of the individual middle manager within the organization, but rather adopts the viewpoint of top corporate officers, wherein the individual event is not of primary significance. Attention is focused on organizational form and on the policies that govern the operation of the system. Aggregate flows will, however, reflect the sequences of actions that would be encountered by any one item in the flow channel.

Chapter 12 discusses exogenous variables. These should be used for establishing the external environment that is to be assumed entirely independent of what happens within the modeled system. Exogenous variables are ordinarily limited to a single variable and are not used to feed real-system data into the model with the hope of achieving greater model accuracy. The exogenous inputs serve as assumed test conditions to allow observation of system reactions. Great care must be exercised to be sure that the externally supplied conditions are indeed independent of what happens within the system insofar as the questions being studied are concerned.

Chapter 13 discusses how to judge the usefulness of a model. Since models of noisy systems cannot be expected to predict future moment-by-moment values of system variables, the model cannot therefore be tested by how well it predicts specific future system events. Confidence in a model arises from a twofold test — the defense of the components and the ac-

ceptability of over-all system behavior. The complete model must be judged on the basis of system behavior — stability, periods of fluctuation, timing relationships between variables, and amplitudes of system fluctuation. These are the variables that describe the general character of a system. The final judgment of industrial dynamics models will rest on the extent to which they are helpful to the manager in designing better industrial systems.

III

EXAMPLES OF

DYNAMIC SYSTEM MODELS

Part III, consisting of the next four chapters, contains two system models to illustrate the model-building discussion of the preceding chapters.

The models themselves are but examples. They are not intended to have universal applicability.

The economics literature has often tended to emphasize the uniqueness, individuality, and the durability of models. One sees many references of the form "so-and-so's model of ————," as if the designer of the model were irrevocably committed to the particular form that may have appeared in a publication. In the system studies that we are discussing here, there is no such thing as "the model." A model will continuously evolve, sometimes a particular form surviving for no longer than a single computer run giving one time history of system operation. Results of studying the model lead to constant change—often extensive change. One phase of a study may center around one model form, while other classes of questions are answered by a different or more extensive model.

● The first model, in Chapters 15 and 16, is the same as the one used in Chapter 2. It is a simple distribution system involving inventories and flows of orders and goods. It is extended to include a simple aspect of the market and sales effort.

● The second model, in Chapters 17 and 18, is similar in complexity but deals with a double-loop system in which the customer-supplier loop interacts with the supplier-labor loop. In that model, money flow has been added.

Neither of these two models incorporates the more subtle factors that are often significant in industrial system behavior. Such factors are beyond the scope of this present volume and also would require the presentation of models that have not yet been sufficiently refined to justify their inclusion. However, to compensate for the incompleteness of the models of Part III, Chapter 19 of Part IV contains discussions of additional facets of the industrial picture that have received partial study in other dynamic system models.

With the same reader classification as in the introduction to Part II preceding Chapter 4, the following suggestions are made:

● *Group A (the manager who is not taking time for study of the more technical material): Sections 15.1 through 15.4 should be read. Section 15.5 develops the equations of the model. It is strongly recommended that the reader work part way into this section. It looks formidable to some people but is actually understandable, with some effort, by anyone interested in the subject of this book. Studying the model equations is the only way to understand fully what assumptions lead to the system results that are illustrated in the time histories generated by the models. Section 15.5 on model equations illustrates the process of going from the manager's descriptive knowledge to a formal model of the industrial enterprise. However, at the point where the reader no longer wishes to study the details of the equations, he can scan the remainder of the section, examine the flow diagrams, and proceed to Sections 15.6 and 15.7. Chapters 16, 17, and 18 should be treated in a similar way by reading the parts of interest. Part IV beginning at Chapter 19 is very different in character from Part III and is especially significant for the manager.*

● *Group B (the reader who is interested in how models of industrial systems can be set up but is not attempting to develop personal skill and competence): Read Part III and finish with Part IV.*

● *Group C (the serious student of industrial dynamics who is developing his personal competence to do system analysis): Study Part III carefully, referring to sections of Part II as necessary. After Part III (or after Part IV), carefully reread all of Part II, which should become more meaningful after studying the examples of Part III.*

CHAPTER 15

Model of the Production-Distribution System of Chapter 2

This chapter illustrates the process of using a statement of the objectives and a verbal description to create a mathematical model of an industrial system. It is a simple example, to the extent that few "intangibles" are involved. A total of seventy-three equations represent retail, distributor, and factory sectors and the initial conditions that describe the starting point of the system. The results of the model runs in Chapter 2 are repeated with details of interpretation and of how the model was altered for each run.

IN this chapter are derived equations for the production-distribution system discussed in Chapter 2. This first simple model will illustrate the application of many of the general principles of Chapters 4 through 13.

This initial example deals with a few of the principal information and material flows in a typical production-distribution system. Although the system is general in nature and can represent many industrial situations, the reader may feel most comfortable if he thinks of the example as applying to the manufacture and distribution of consumer durables, such as household electrical appliances.

We start with a production-distribution system, partly because it is simpler to describe than other corporate functions, partly because plausible behavior patterns and the basis for making the principal decisions are more evident, and partly because production and distribution are the fundamental economic tasks of most corporations.

Because the production-distribution relationships are simpler, this first development of equations can dwell more on method and technique and less on justifying the assumptions about the corporate operations themselves. In later chapters the representation of a company will be extended outward into more difficult corporate functions.

15.1 Objectives

In constructing a useful dynamic model of corporate behavior it is essential to have clearly in mind the purposes of the model. Only by knowing the questions to be answered can we safely judge the pertinence of factors to include in or omit from the system formulation.

Therefore, we shall define our immediate objective as an examination of possible fluctuating or unstable behavior arising from the principal organizational relationships and management policies at the factory, distributor, and retailer. We shall explore the way in which the simple, central structure of the system tends to accentuate or modify the external disturbances that impinge on the system.

Because of the critical importance of time delays in contributing to instability of information-feedback systems, the analysis should include the principal time delays in the flows of orders and material.

Sources of amplification must be perceived and incorporated, since these will be crucial to the dynamic behavior of the whole system. In this simple example such amplification will be found in the factors affecting the generation of purchase orders. Outgoing orders (for replacing stock) from one stage of the system will not necessarily be equal to incoming orders (representing sales). Several factors can amplify changes in the order flow rate.

The first of these amplifications arises from the necessity for filling supply pipelines with orders and goods in proportion to the level of business activity. An increased sales rate requires a corresponding increase in the placing of orders, if inventory is to be maintained. In addition, the higher level of activity requires more orders in transit in the supply pipeline. Building up these orders and goods in the pipeline requires a short-term, transient, additional increase in the outgoing-order rate. If these pipeline-filling orders are not intentionally placed, inventory will fall correspondingly until the in-transit pipeline demands are satisfied.

The second amplification arises from common inventory policies. We observe that an increase in sales level usually creates a desire to carry a higher level of inventory. This inventory can be acquired only by ordering, for a time, in excess of the sales rate.

Such amplifying factors operate as well in the negative direction to cause replacement orders to fall faster than sales.

Forecasting and trend extrapolation methods can also lead to further amplification and further instability of the system. This is a major subject in its own right; it is an "extra" that can easily be added later after we have achieved an understanding of the simple structure of our system. Forecasting will therefore be postponed until later.[1]

For a general approach to this type of study, we must start with a system of sufficient scope to contain interesting dynamic problems. In the first step, however, the investigation of a system should be restricted to a study of its

[1] See Appendix L.

bare framework and major phenomena; otherwise, unimportant detail will obscure the principal lessons that can be learned. Extension of the system boundaries and enrichment of the internal detail can come later as we learn to cope with more complex situations.

15.2 Scope

Industrial activity has been discussed in Chapter 6 in terms of six interacting flow networks:

- Materials
- Orders
- Money
- Personnel
- Capital equipment
- Information

Can some of these be omitted and still retain a system worthy of preliminary study? Production and distribution would seem to require consideration of the flow of materials (physical goods). The materials flow is controlled by orders representing decisions based on information about inventories (material) and about sales. Therefore, we shall include those parts of the information and order networks that directly relate to the material network.[2]

The flow of money represents an accounting for completed transactions. Money flow, however, is not ordinarily the principal determinant of sales and manufacturing decisions. Only in marginal organizations that are operating at a

[2] Some may feel that the omission of economic variables and sales forecasts from the ordering decisions is unrealistic; but these influences on decisions seldom take precedence for long if they run contrary to actual sales and inventory conditions. Forecasting methods can be added once the more basic parts of the system are understood. Furthermore, the model here developed can be interpreted as the response of the system to deviations between actual and forecast sales. Some form of forecast is implicit in any decision making. Here, because nothing more elaborate is included, the implicit forecast is for a continuation of present average sales. The effect of a forecast that is an extrapolation of present trends is shown in Appendix L.

loss, or where cash balances and available credit have dropped too low, would the condition of the money network limit the freedom of decision making otherwise found in the "normal" system. At the other extreme, a high profitability in one sector of the system might be expected to attract competitors, and would lead to increased plant capacities and to pressures affecting pricing. For the moment, if we assume an industry whose character in the short run is not being markedly changed by high profits or losses, it is reasonable to omit money flow and profit calculations from the initial model. This means that we are dealing with an industry in the "mature" phase of its life cycle and only over a time span short enough so that the structural and policy character of the industry is not being changed by profits or available cash. Later we could remove this restriction, to permit study of the financial factors influencing the growth of a new company and to achieve in a single model the evolving structure and policies that represent the entire life cycle of a product.[3]

Also it is reasonable to start without the flows of personnel or capital equipment, since plausible and interesting situations exist in which labor and plant availability are not the factors primarily controlling industrial operations.

Thus we shall concentrate now on the main channel of material flow from factory to consumer and on the principal stream of information flow in the form of orders moving from consumer toward factory. The structure of the system is shown in Figure 15-1. This defines in a general way the problem we are choosing to consider. Only the most pronounced and obvious influences will be included. Refinements of these could be handled later. The purpose here at first is not to achieve a complete representation of all functions but rather to exhibit a method of analysis and to understand the

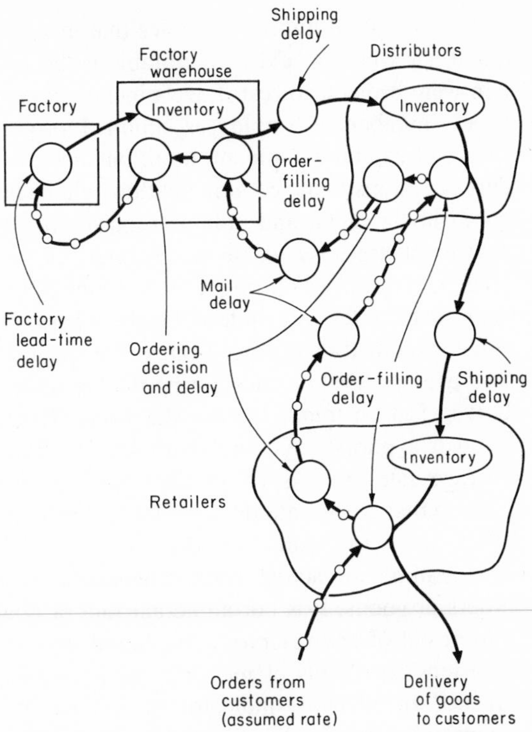

Figure 15-1 Organization of production-distribution system.

contributions to system behavior that can come from organizational form, delays, and management policies.

15.3   Factors to Include

In this analysis the three sectors — retailer, distributor, and factory — will be very similar to one another. We shall begin by formulating equations to represent the relationships that we believe to be of interest in the retail sector.

Listing the principal variables of interest helps direct attention to the proper factors. The types of variables have been defined in Chapter 6 as falling into two major classes — levels and rates. In addition, auxiliary variables (which could be eliminated by substitution into the rate equations) are often defined for convenience or for the representation of component concepts that are significant in describing the corporate system.

[3] See Chapter 19.

Levels represent those variables that measure quantities which would be countable if the system were brought to rest. Levels include inventories, number of employees, unfilled orders, stocks of capital equipment, bank balances, orders in transit in communication channels, goods in shipment, and unfilled requisitions for new employees. As we shall see later, certain other variables, which might at first glance appear to be rates, are indeed levels. These include smoothed or averaged rates. We speak of the "average level of sales during the last year." It is in fact customary to use the word "level" in such a phrase. We shall find later that an average sales rate is of similar mathematical form to that of the equations for levels. Furthermore, an average sales rate does not represent at any time an actual instantaneous flow of orders or goods. It is not an actual rate of flow into or out of any reservoir. In general the "dimensions" or "units of measure" of a variable are not an adequate indicator of whether the quantity is a rate, a level, or an auxiliary variable.

Likewise, we should list all *rates* of flow that we consider significant in the system. By implication this means that we are listing the important decisions, both implicit and overt, which are being made. Decisions control the rates of flow, but in obtaining an initial description of the system it is often more effective to think in terms of the rates than in terms of the corresponding decisions.

In addition to levels and rates, we should initially list the important *delays* that we think may contribute to system behavior. Mathematically, the delays take the form of equations of levels and rates; but in terms of system structure, we can proceed most rapidly by identifying separately the delays that are to be included.

In approaching an actual new situation, many months may be taken in tentative identification of the pertinent variables, in reducing some of these to equation form, and in selecting the factors that are to be included. Within the confines of a book there is not enough space (it would also be unduly confusing) to attempt to trace all of the steps that were actually followed in evolving the present equations. We shall, therefore, take advantage of advance knowledge of the outcome and move directly to a listing of the principal variables.

For the retail sector, the *levels* that are central to the distribution and sales task, and that exist in the orders, information, and materials flow channels, appear to be

- Backlog of orders received from the consumer but not yet filled
- Inventory of goods in stock
- Average sales in the recent past, used in deciding on the desired inventory and supply pipeline levels

Since money flow, personnel, and capital equipment are not yet being included (we assume their adequate availability so as not to restrict actions in the system), the related levels need not be listed.

Correspondingly, the major *rates* of flow that appear pertinent to the objectives cited earlier would be

- Incoming-order flow from customers
- Outgoing-shipping rate to customers
- Outgoing-order rate from retailer to distributor
- Incoming-shipping rate of goods from the distributor

The principal *delays* in these rates of flow will be taken as

- Delay in filling customers' orders at retail
- Delay in deciding on and preparing outgoing orders from retailer to distributor
- Delay in transmission (mail) of orders from retailer to distributor
- Delay in shipping goods from distributor to retailer

15.4 Basis for Developing Equations

We can now develop equations describing activity at the retail level. It should be stressed that these equations are not "right" in any intrinsic or mathematical way. They merely describe what we have chosen as the most significant relationships. The equations are akin to a verbal description of the situation. They are correct if our perception of the system is correct. They are wrong to the extent that we wrongly interpret the organization which is being described.

We are not yet presuming to incorporate *all* of the important factors. We do hope to select the factors that are highest on the priority list of effects within the specified boundaries of the system being studied. Our assumptions about how the selected factors act on one another should be plausible. They should be of interest to the investigator. They may represent his estimate of the behavior of an existing system, or they may represent the characteristics of a proposed or hypothetical system whose behavior is to be explored. The assumptions implied by the system formulation must be defendable in the view of the investigator. In general the usefulness of the model depends on the reasonableness and pertinence of the individual equations. The crucial importance of the validity of the underlying assumptions, in the context of model testing, has already been discussed in Chapter 13.

Where conflicting assumptions about the true nature of the system exist, the model itself can be used to study the influence of alternative assumptions. Often it will be found that the alternatives lead to essentially the same results, and consequently the choice does not matter. Some assumptions will be found to affect the system operation critically. After the high importance of a factor has been demonstrated, then is the time to attempt detailed measurements in a real organization. In this way costly data collection and data analysis can be concentrated where the results will be most useful.

In this chapter the particular choice of factors from among the many that exist in real life will be defended only to such an extent that

the reader can see why they were chosen. Although we believe that these represent the central framework of the ordinary production-distribution relationships, the purpose of this chapter is to show a method of system analysis. If the reader understands the method, he can then substitute for himself his own choice of principal factors affecting the way the system operates.

For this example, we shall let a group of three capital letters represent each variable and parameter under study. In the retail sector the third letter in each group will be an R. The third letter D will identify variables at the distributor level. Likewise, F will identify variables at the factory level.

It seems best to develop the equations and the corresponding flow diagrams simultaneously. In this chapter the first flow diagram will be evolved in steps corresponding to the development of the equations. To conserve space later, flow diagrams will be given in their entirety, anticipating the equations that are to follow.[4]

15.5 System Equations

15.5.1 Equations for the Retail Sector. We shall begin with two simple equations: one representing the level of unfilled orders and the other the inventory of goods. Figure 15-2 shows these two variables in the first step of our evolving flow diagram. Since IAR is an inventory of goods, solid lines representing flows of material come in and out. Since UOR is the level of unfilled orders, circled lines enter and leave the rectangular box.

The following equation for UOR is of the standard type for defining a simple level into which flows a single input rate and from which flows a single output rate:[5]

---

[4] The symbols used in the flow diagrams are as given in Chapter 8.

[5] The nature of level equations has been discussed in Chapter 7.

$$UOR.K = UOR.J + (DT)(RRR.JK - SSR.JK) \qquad \text{15-1, L}$$

UOR Unfilled Orders at Retail (measured in units of goods on order)

RRR Requisitions (orders) Received at Retail (units/week)

SSR Shipments Sent from Retail (units/week)

DT Delta Time (weeks), the time interval between solutions of the equations

On the right-hand margin is the equation serial number; 15 is the chapter number; 1 is the number of the equation within the chapter. The L indicates that the equation defines a level.

The equation calculates the number of unfilled orders at the present time K, based on the unfilled orders last calculated, at time J, and on the rates of inflow and outflow during the time interval JK between calculations of UOR. Inflow and outflow rates are assumed constant during the interval JK (the time interval must be short enough to make this a satisfactory assumption). The length of the time interval DT multiplied by inflow rate RRR.JK gives the new orders received in the interval JK. Likewise (DT)(SSR.JK) gives

the orders filled in the interval JK. The dimensions of the equation that must be identical in all terms are as follows:

$$\text{Units} = \text{units} + (\text{week})\left(\frac{\text{units}}{\text{week}} - \frac{\text{units}}{\text{week}}\right)$$

$$\text{(UOR)} \quad \text{(UOR)} \qquad \text{(DT)} \quad \text{(RRR)} \quad \text{(SSR)}$$

After multiplication of the product on the right, each term of this equation is measured in "units" of product ordered. (DT) multiplied by (RRR) is a length of time multiplied by a flow rate that gives the unfilled orders added to the order file during the preceding time interval. Likewise (DT) times (SSR) represents the orders filled and removed from the unfilled-order file during the preceding time interval. The solution-time interval DT must be short compared to the delays that are to be represented in the system.[6] The solution interval for this example will need to be a small fraction of a week.

By formulating the equation in this way, rates of flow are measured in their conventional

---

[6] As discussed in Section 7.5 and Appendix D.

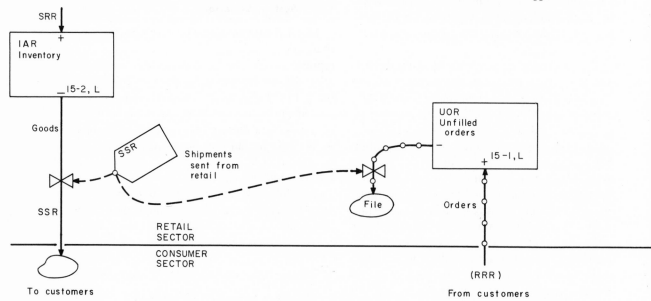

Figure 15-2 Beginning of flow diagram of retail sector.

units, as for example the weekly rate, and the equation is correctly stated independently of the length of solution interval DT. This allows the solution-time interval DT to be chosen arbitrarily, according to criteria to be discussed later, without affecting the validity of the equation.

In Figure 15-2 the meticulous accounting for all the units of flow is indicated by including the discard file for filled orders that are removed from the active system.

The second equation, describing the level of inventory in the retail sector, is entirely similar:

$$\text{IAR.K} = \text{IAR.J} + (\text{DT})(\text{SRR.JK} - \text{SSR.JK}) \qquad \text{15-2, L}$$

IAR Inventory Actual at Retail (units)
SRR Shipments Received at Retail (units/week)
SSR Shipments Sent from Retail (units/week)

This states that the present inventory equals the previous inventory plus the units received minus the units shipped.

The preceding equations for levels are simple and noncontroversial. They represent some of the "basic physics" of the system description. They are simply accounting statements of the fact that the balance on hand is found by a repetitive integration (adding and subtracting) of inflow and outflow rates.

The equations for rates, on the other hand, are not so obvious and straightforward. It is in the rate equations that the decision mechanisms of the system are expressed. The rate equations represent our understanding of the factors that determine actions. These decisions, which control the rates and are the rate equations, must be formulated so that they remain plausible and adequately correct over the extreme ranges that may be encountered in the shifting values of the system variables. Rate equations will often contain nonlinear functional relationships to describe proper behavior under any eventualities.

Contrary to usual expectations, this insistence that rate equations cope with extreme excursions of the variables on which they are based makes the construction of a *useful* model easier rather than more difficult. Very often, by considering all of the extreme conditions that might exist, we can develop a good picture of the boundaries within which an action must take place. With these boundaries established, the intermediate functional relationships between cause and effect frequently become restrained within a small, unimportant range of uncertainty. All relationships satisfying the known boundary conditions may often give about the same results. A rate equation should be developed by thinking through the various circumstances that affect the rate. To illustrate this, we shall begin by considering the rate of shipments sent from retail to the customer SSR.

Here the rate of delivery of goods to customers is assumed to be an implicit rate. This means that it is controlled by the state of the system and not by a free managerial decision. Although it is quite possible to imagine a decision not to deliver available goods, this is infrequent and will be neglected. There are no mathematical differences between an implicit and an overt decision equation. However, identifying which type of decision is being considered usually helps to clarify our thoughts in developing the equation.

The rate of delivery of goods to customers must depend on the existence of an unfilled backlog of orders ready for processing. In the extreme, if there are no orders, there will be no shipping. Likewise, the ability to deliver goods must depend on the existence of an inventory from which to make the delivery. Delivery rate does not depend on any of the other rates in the system at the same instant of time. Present ability to deliver depends on unfilled orders but does not depend on the present *rate* of arrival of new orders, since these are not yet accessible to the shipping location. Only the level of goods on hand, but not the rate of receipt SRR nor the rate of placing new orders to the distributor, affects ability to deliver at the *present* time, even though these levels that determine present ability to deliver have reached their present values because of certain past rates of flow. The present values of other rates affect future but not present ability to deliver. If the reader feels that other present flow rates affect present shipping rate, he is failing to distinguish instantaneous from average rates or is not taking a short enough view of what is meant by the present.

There would be a large number of satisfactory ways to formulate the equation for delivery rate of goods. We shall follow here a procedure whereby the size of the order backlog and the variable delay experienced in filling orders determines the rate of filling. In turn, the delay in the filling of orders will be formulated as a function of existing inventory.

If we start by basing the shipping rate on unfilled orders and on the delay in filling those orders, we could consider a simple relationship like the following:

$$SSR.KL = \frac{UOR.K}{DFR.K}$$

SSR   Shipments Sent from Retail   (units/week)
UOR   Unfilled Orders at Retail   (units)
DFR   Delay (variable) in Filling orders at Retail (weeks)

This equation is of the form of a first-order exponential delay[7] except that the delay may vary with time. It says that the present weekly rate of shipment is a fraction, 1/DFR.K, of the unfilled orders. In one sense this equation defines what is meant by the delay DFR. We can examine how satisfactory this relationship might be under a variety of simple circumstances.

First, consider steady-state conditions of a uniform inflow of orders RRR and an equal and steady shipment rate SSR. The equation says that for a constant rate the longer the delay DFR, the larger must be the unfilled backlog of orders

[7] See Section 9.3.

in process UOR. This is correct. In the steady-state the orders in process must be proportional both to the rate of sales SSR and to the average delay in handling orders.

Consider some transient implications of the above equation. If the order rate RRR should suddenly increase, the corresponding change in the shipping rate SSR would be as shown in Figure 15-3. This is the first-order exponential-delay response. Such a relationship between orders and deliveries seems plausible. If the items being sold (for instance, refrigerators) are not over-the-counter items, then the actual delivery rate will gradually rise after the sales rate increases. The shaded area between the sales curve RRR and the delivery curve SSR represents the additional in-process orders accumulated in the backlog UOR, as just discussed for the steady-state conditions.

We can examine another transient condition. Suppose that there have been no incoming orders RRR, as shown in Figure 15-4. Then, a sudden "batch" of orders are placed, after which the order rate is again zero. This is the "impulse response" condition discussed under the behavior of first-order delays. We see that unfilled orders jump from zero to the number introduced in the "batch." The shipping rate jumps to its maximum value and then declines as the backlog declines. This might be defended on the basis that orders for some items are easily filled and these orders are processed immediately. However, we might properly feel that an initial delay before a delivery rate is established and a more gradual rise of shipping rate would be more realistic under this transient test condition. If so, the order-filling process could be broken into two or more steps. One equation, like the above, could recognize delays due to out-of-stock conditions. Preceding this, a third-order delay in the flow of incoming orders RRR could represent the clerical delays in the system. Such alternatives can be tested to see their effect on the system. We shall depend here on past experience and assume that

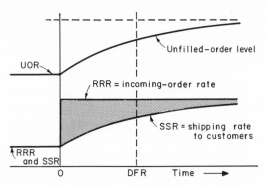

Figure 15-3 Response of shipping rate SSR and unfilled orders UOR to step change in input rate RRR.

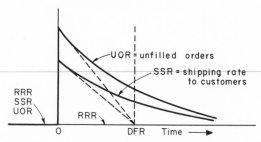

Figure 15-4 Response of shipping rate SSR and unfilled orders UOR to impulse input RRR.

such a refinement will not be necessary for the present purposes.

The equation in the above form gives no positive assurance that the shipping rate SSR existing over the forthcoming time interval KL might not remove more than the remaining inventory. As will be noted later in a discussion of the delay DFR.K, the delay will increase as inventory declines, thus tending to reduce the rate of shipment. However, to provide positive protection against negative inventories, two auxiliary variables will be introduced in the following equations.

The next three equations exhibit a poor model-building practice. They were introduced during the early development of this particular model and are left here as a basis for the discussion in Footnote 8. The previous equation will

be rewritten using, instead of the shipping rate SSR, the auxiliary variable STR:

$$STR.K = \frac{UOR.K}{DFR.K} \qquad \text{15-3, A}$$

STR Shipping rate to be Tried at Retail (units/week)
UOR Unfilled Orders at Retail   (units)
DFR Delay in Filling orders at Retail   (weeks)

This is an auxiliary equation, and STR is an auxiliary variable even though its units of measure are those usually designating a rate. The computation of STR is an auxiliary computation. The value of STR has not yet been accepted as a rate within the system. Its position in the computational sequence will follow the computation of the auxiliary variable DFR. There is often a required sequence for the computation of auxiliary variables, as discussed in Section 7.4.

Before accepting the *proposed rate* of shipment STR as the correct value for the shipping rate SSR, we shall test the value of STR against the rate of shipment that would cause inventory to become negative. This rate is given in the following equation:

$$NIR.K = \frac{IAR.K}{DT} \qquad \text{15-4, A}$$

NIR Negative Inventory limit rate at Retail (units/week)
IAR Inventory Actual at Retail   (units)
DT Solution Time interval   (week)

By carrying the time interval DT to the left-hand side of the equation, we can see that

(NIR.K)(DT) = IAR.K

In words: The rate NIR.K is the limiting rate that if it exists over the solution interval DT will lead to complete depletion of the existing inventory IAR.K.

We are now ready to write our equation for the shipping rate:

$$SSR.KL = \begin{Bmatrix} STR.K & \text{if } NIR.K \geq STR.K \\ NIR.K & \text{if } NIR.K < STR.K \end{Bmatrix} \quad \text{15-5, R}$$

SSR   Shipments Sent from Retail   (units/week)
NIR   Negative Inventory limit rate at Retail
       (units/week)
STR   Shipping rate to be Tried at Retail   (units/week)

Equation 15-5 gives the shipping rate of goods sent from retail which will exist over the forthcoming time interval KL.[8] It says that if the

limiting rate NIR.K, given by Equation 15-4, is greater than or equal to the desired shipping rate STR.K, as given by Equation 15-3, then the rate STR.K will be used. If NIR.K is less than STR.K, then the rate NIR will be used as the shipping rate. Under ordinary circumstances where inventory will not reach zero, the shipping rate is determined by Equation 15-3.

The relationships of Equations 15-3, 15-4, and 15-5 appear in a flow diagram, Figure 15-5.

[8] It should be noted that this "clipping" of the travel of a variable as in Equation 15-5 is a poor practice which is easily abused, to the detriment of sound model formulation. Such abrupt limiting is seldom encountered in actual systems. It was introduced in this model after some absurd negative values of inventory were generated. Should the clipping mechanism as incorporated in Equations 15-3, 15-4, and 15-5 actually be called into play, it probably indicates a defect in the basic equations of the system. Such is the case here. In an aggregate inventory situation of many different catalog items, we should expect the inability to match orders and stock to keep inventory from falling all the way to zero. However, in the ordering equations, to be discussed later, there is no "discouragement" factor. Ordering does not de-

cline when faced by very excessive delivery delays. For example, if the factory capacity is less than the assumed average retail demand, orders accumulate without end in the unfilled-order levels. Under the condition of orders persisting for very long times above production capacity, the model is not valid, and the high levels of unfilled orders can create shipping rates that in one solution interval can exhaust inventory and produce a negative inventory level. Better than using the clipping equation would be a formulation of a more comprehensive equation in place of Equation 15-3 that could not create the conditions against which we wish to guard. Actually, since in the model runs of this chapter and the next, Equations 15-4 and 15-5 are never active, Equation 15-3 determines shipping rate SSR.

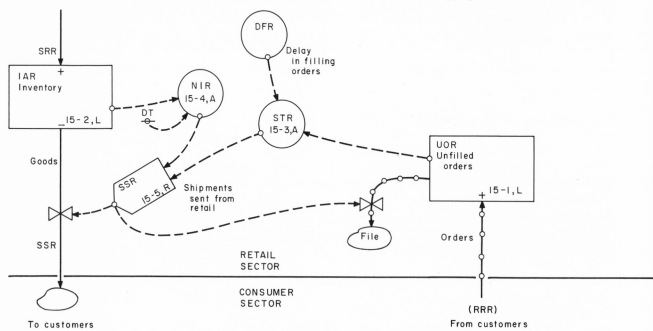

Figure 15-5   Order filling added to retail sector.

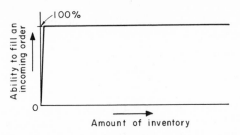

Figure 15-6 One catalog item in one warehouse.

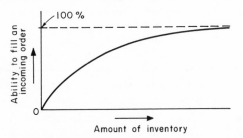

Figure 15-7 Many catalog items in many warehouses.

Before developing an equation for the relationship between inventory and the variable delay **DFR** used to calculate the delivery rate of goods, we must discuss somewhat further the nature of the inventory under consideration. If we are considering a single catalog item in a single warehouse, it is clear that orders can be filled as long as any inventory is left. This is illustrated in Figure 15-6, in which all incoming orders can be filled so long as there is any inventory in the warehouse. The ability to fill an incoming order drops suddenly to zero at the moment inventory is exhausted.

On the other hand, a different situation exists if we consider a single catalog item existing in a number of warehouses, or if we consider many catalog items existing in one warehouse, and of course if we consider a number of catalog items existing in a number of warehouses. Under any of these circumstances we can expect that some items will be depleted before others in some locations and that our over-all ability to fill orders will progressively decrease as total aggregate inventory falls. This is shown in Figure 15-7. We shall take this for our ex-

ample because we are thinking of a product line consisting of several separate catalog items sold by many retailers across the country.

The amount of delay experienced on the average in filling orders will be inversely related to the ability to fill orders. As fewer and fewer orders can be filled from inventory, a larger and larger backlog of orders will be waiting to be filled from incoming shipments. Figure 15-8 shows the general shape of the relationship that *must* exist between the delay in filling orders and the level of inventory. With a sufficiently large inventory, the delay experienced by the average order will approximate the minimum order-handling delay necessary for processing the order and making shipment. As total aggregate inventory decreases, more and more orders in more and more warehouse locations encounter out-of-stock conditions. While these orders wait for incoming shipments of the proper items, they contribute to increasing the average delay experienced by all orders. As the inventory approaches zero, the delay

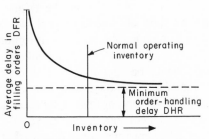

Figure 15-8 Delay versus inventory.

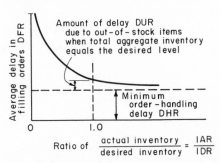

Figure 15-9 Delay versus adequacy of inventory.

must necessarily increase rapidly and approach "infinity." By this we mean that so long as the no-inventory condition persists there is no possibility of making shipment.

We can now discuss a method of computing the variable delay DFR. Referring to Figure 15-8, we see that a curve of delay versus inventory should be of a shape starting from some minimum value at high levels of inventory and climbing more and more steeply as the inventory is depleted. The following equation form is one of many that meet these requirements:

$$\text{Delay} = \text{delay minimum} + \frac{\text{constant}}{\text{inventory}}$$

For very large inventories the right-hand term approaches zero, so that the delay equals the minimum delay in order handling. As the inventory approaches zero, the right-hand term approaches infinity, giving an increasing value of delay.

This formulation has a serious drawback. It is based on the absolute level of inventory which takes no recognition of the average level of business activity that would be an indicator of the amount of inventory that would be "enough." As the measure of inventory, it seems better to use the ratio of actual inventory to what would be "enough" or "proper." We shall therefore introduce the concept of a "desired" inventory against which actual inventory will be compared. The curve of Figure 15-8 then becomes the curve of Figure 15-9.[9] Using the inventory ratio IAR/IDR in the preceding equation and bringing it from the denominator to the numerator of the fraction yield the following equation, which represents the inventory to shipping delay relationship that has been discussed:

$$\text{DFR.K} = \text{DHR} + \text{DUR}\, \frac{\text{IDR.K}}{\text{IAR.K}} \qquad \text{15-6, A}$$

DFR   Delay in Filling orders at Retail (weeks)

DHR   Delay due to minimum Handling time required at Retail (weeks)

DUR   average Delay in Unfilled orders at Retail caused by out-of-stock items when inventory is "normal" (weeks)

IDR   Inventory Desired at Retail (units)

IAR   Inventory Actual at Retail (units)

In Equation 15-6, the total delay in the filling of orders equals the minimum order-handling delay plus a constant multiplied by the ratio of desired over actual inventories. When this equation is added, the flow diagram becomes Figure 15-10. When the constants DHR and DUR are specified, all of the necessary inputs to Equation 15-6 are defined except the value of desired inventory IDR.

The desired inventory, which we can think of as the "ideal" or the "target" level of inventory, is a very important concept. It will be used here in two ways. As discussed above, it is a reference against which to compare the actual inventory for determining ability to fill orders. Also, the "desired level of inventory" will later be one of the inputs to the ordering decision for generating purchase orders for new stock. (These two concepts of "desired" and "necessary" need not be the same except that desired inventory will reflect what is necessary on the average to maintain customer service.) The use of ideal inventory as a component in the reorder of goods makes it an important

---

[9] Note that if this relationship were needed with more precision, a functional relationship between delay and inventory ratio could be generated by simulation of the filling of individual orders. Such a detailed simulation (or perhaps an operations-research type of analysis) would be based on assumed or known probability distributions of incoming orders, inventory reorder rules, and delays in obtaining new stock. Before appreciable effort were so expended, it would be well to determine the sensitivity of this set of equations to the form of the delay versus inventory relationship. For many purposes we can expect that the results will be indifferent to any reasonable changes from the above relation (obtained merely by consideration of the nature of the processes involved) between inventory and order-filling delay.

concept in the dynamic behavior of the system. The relationship of inventory changes to changes in the average level of sales is one of several major sources of amplification causing fluctuation of industrial activity.

The effects on system stability notwithstanding, a very common industrial practice is to build up and decrease inventories as the sales level rises and falls. This is inherent in the concept of carrying inventory equal to a specified number of weeks of sales. Such a policy is practiced in many warehousing operations. The idea of an "annual inventory turnover rate" likewise is tied to the philosophy that inventories are proportional to the level of sales.[10] National figures for the relationship between

[10] Inventory theory may show a square-root relationship, but financial pressures, the ease of thinking in terms of proportional relationships, and speculative forces tend to increase the effect of business activity on inventory.

sales and total inventories tend to show equal excursion amplitudes. Because the practice is followed in many places, we shall begin by introducing a proportional relationship between desired inventory and average sales as follows:

$$IDR.K = (AIR)(RSR.K) \qquad \text{15-7, A}$$

IDR   Inventory Desired at Retail (units)
AIR   proportionality constAnt between Inventory and average sales at Retail (weeks)
RSR   Requisitions Smoothed at Retail, i.e., average sales (units/week)

The constant **AIR** represents the number of weeks that the desired inventory could supply the average sales rate. Divided into 52 weeks, it would give the annual inventory turnover rate. Again the reader should be reminded that these equations are "right" only in the sense that they are the relationships which will exist in the model being constructed. They are not

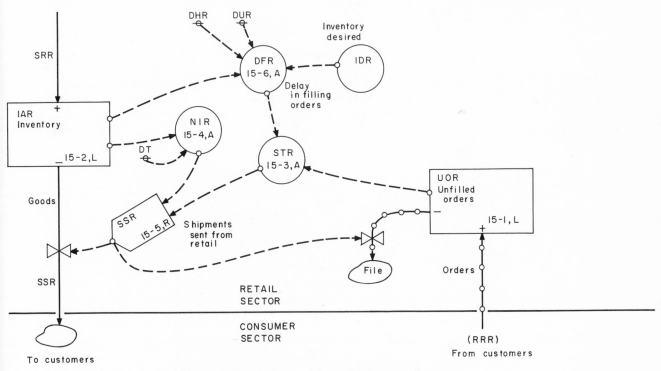

Figure 15-10 Delivery delay added to retail sector.

inherently and necessarily true of every actual industrial situation. In fact, changes in Equation 15-7 may profitably be explored later to find ways of improving stability of the production-distribution system to make it less sensitive to outside disturbances.

In Figure 15-11 has been added the relationship shown in Equation 15-7. The information flow for generating average sales RSR is also shown.

Because the current sales rate RRR will fluctuate from day to day and will not present a smooth flow of information, it is necessary to average the actual sales data to obtain a figure on which to base inventory and ordering plans. The "noisier" the sales data, the longer must be the averaging time. The longer the averaging time, the later the estimate of the average will lag behind true events. Realistic data require smoothing, the smoothing causes a delay, and the delay affects the system be-

havior and stability. The smoothing must be present as a system characteristic even when we test system response to noise-free signals, since the noiseless components are distorted by the same smoothing that must be present to help suppress noise. Various methods of smoothing can be used.[11] Here "first-order exponential smoothing" will be used, giving an equation of the following form:

$$RSR.K = RSR.J + (DT)\left(\frac{1}{DRR}\right)(RRR.JK - RSR.J)$$

$$15\text{-}8, \text{ L}$$

RSR Requisitions Smoothed at Retail   (units/week)

RRR Requisitions Received at Retail, present sales rate (units/week)

DRR Delay in smoothing Requisitions at Retail, the smoothing time constant   (weeks)

Equation 15-8 states that the newly calculated

[11] An elementary discussion of smoothing will be found in Appendix E.

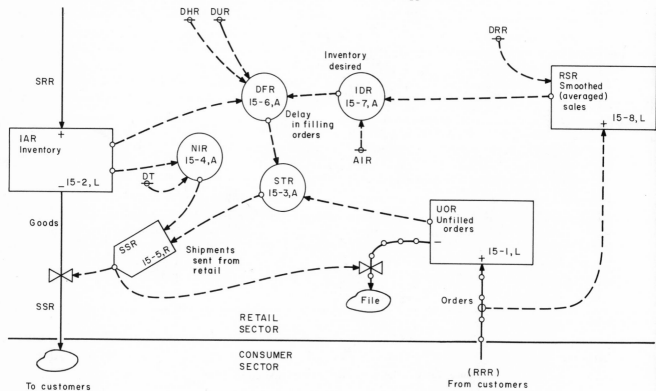

Figure 15-11   Desired inventory and smoothed sales add to retail sector.

level [12] of average sales RSR at the present time K is given by the previous value of average sales, corrected by a fraction of the difference between the rate of sales during the last time interval RRR.JK and the previously computed average sales RSR.J. <u>The constant DRR gives the fraction of the difference that is to be corrected each week;</u> this in turn is multiplied by the time interval DT to get the amount of correction in each computation time interval. Exponential smoothing weights the most recent data most heavily, with progressively decreasing weights on older data.

In the previous equations and in Figure 15-11 have been represented the inflow of orders from the consumer to the retail sector and the factors determining the filling of those orders. The next step is to develop the decision criteria for placing purchase orders from the retailer to the distributor. This will be the equation for the rate of generation of outgoing orders. This is an overt decision because managers at the retail level can order any quantity they wish. Whether or not the goods will be delivered will be determined by implicit decisions at the distributor level that recognize whether or not goods are available for delivery.

At this point there arises a small problem of methodology which should be discussed. Very often, we shall deal with decisions that are reached slowly. They require data collection, proposals, recommendations, and review. The decision is then made and subsequently executed. In placing a purchase order for replacement goods, certain sources of information are used, various subsidiary decisions are reached, requisitions are authorized, and clerical delays are encountered in making up the actual orders. The fine detail of the separate steps in decision making and the separate delays in this process are not of interest to us in studying the broad over-all characteristics of the production-distribution system. In our equation formulation,

[12] See Section 6.1 for a discussion of why an average rate is a level.

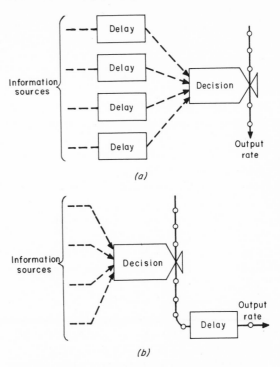

Figure 15-12   Information and decision delays.

delays are handled separately from decisions. We must decide how to group the delays and decisions and relate them to one another. Figure 15-12 illustrates two of many choices in organizing the combination of delays and decision making.

In Figure 15-12a the distributed process of decision-making and information-source delay has been broken into two steps wherein all information sources are separately delayed before coming to the decision point where they are combined to control the output rate. In Figure 15-12b an alternative is shown wherein it is assumed that the decision is made first on the basis of instantaneously available information and then the output of the decision is delayed to represent the total delay encountered in the information sources and the decision process. For greater reality, it might be desirable sometimes to include appropriate delays both before and after the actual decision point. The method shown in Figure 15-12b is simpler than that in Figure 15-12a, because it requires only one set of delay equations rather than several. Since, for the purposes of our present study, there is no obvious preference from the standpoint of reality of system representation, we shall choose the simpler approach in Figure 15-12b. This means that we shall construct an equation which defines the "impending" decision about the purchasing rate and shall then delay this flow of orders by an amount equivalent to the total delay involved in the processing of raw data, in making a purchasing decision, and in the clerical steps of a purchasing office in making up purchase orders.

The first task is to identify the principal sources of information on which the rate of purchasing is to depend. We have already decided to omit any explicit forecasting methods as well as all minor effects. First on the priority list is replacement of goods being sold. Beyond this, orders are necessary for the correction of differences between actual and ideal inventory. Also, in one manner or another it is necessary to recognize the inescapable demands of pipeline filling. If there is a fixed transit time for orders and goods in traveling through the pipeline from the retailer to the distributor and return, it is necessary that the total of orders plus goods in this pipeline be proportional to the level of business activity. If orders are not inserted in the pipeline for this purpose, the deficits will come out of inventories in the system. These factors are recognized in the following equation:

$$PDR.KL = RRR.JK + \left(\frac{1}{DIR}\right)\left[(IDR.K - IAR.K) \right.$$
$$\left. + (LDR.K - LAR.K) + (UOR.K - UNR.K) \right]$$

15-9, R

PDR Purchasing rate Decision at Retail (units/week)
RRR Requisitions Received at Retail (units/week)
DIR Delay in Inventory (and pipeline) adjustment at Retail (weeks)
IDR Inventory Desired at Retail (units)
IAR Inventory Actual at Retail (units)

LDR pipeLine orders Desired (necessary) in transit to supply Retail (units)
LAR pipeLine orders Actual from Retail (units)
UOR Unfilled Orders at Retail (units)
UNR Unfilled orders, Normal, at Retail (units)

The flow diagram with Equation 15-9 added appears in Figure 15-13. A discussion follows of each term of this equation that defines the rate of generating purchase orders during the forecoming time interval KL.

First, the retailer's purchasing rate will depend on the rate of his own sales RRR,[13] during the immediately preceding time interval JK. These are the orders to replace the goods sold. Bear in mind that this is an "impending" decision which is being made on the latest true information, with the delays of the information-gathering process to be put in later.

The delay DIR is a time constant representing the rate at which the retailers, on the average, act on inventory- and pipeline-deficit situations. It is not to be assumed that retailers would respond immediately to the full extent of any theoretical difference between desired and actual inventory. Furthermore, the time lags in observing such differences may sometimes be substantial. The constant DIR allows adjustment of this response time. As an example, a value of four weeks for the constant DIR would give a purchasing rate that corrects any remaining deficit in the term in brackets at the rate of one-quarter of the deficit per week. It will be seen later that this response constant is one of the more critical parameters in determining the system's dynamic performance.

The two inventory terms give the difference

[13] Note that the use of a rate in another rate equation is contrary to the principles of model structure discussed in Chapter 6. More properly, a short-term average rate for the previous day, week, or month would be used here, since in fact the instantaneous rates would not be available in an actual situation. However, as the averaging period becomes sufficiently short, nothing is gained by the extra complexity caused by the introduction of another smoothing equation.

between desired and actual inventory. If the level of desired inventory is above or below actual inventory, a correcting component will be introduced into the ordering rate.

The two pipeline terms act like the inventory terms. The desired level of orders in the pipeline will be defined later as proportional to the

level of average sales, in the same m̲a̲n̲n̲e̲r̲ ̲a̲s̲ desired inventory. A̲c̲t̲u̲a̲l̲l̲y̲,̲ ̲i̲t̲ ̲m̲a̲y̲ ̲b̲e̲ ̲d̲o̲u̲btful that many retail organizations consciously r̲e̲c̲ognize the orders for pipeline filling as p̲a̲r̲t̲ ̲o̲f̲ their ordering procedure. If they do not, t̲h̲e̲ orders necessary to maintain the required d̲e̲livery rate of goods will automatically come ou̲t̲

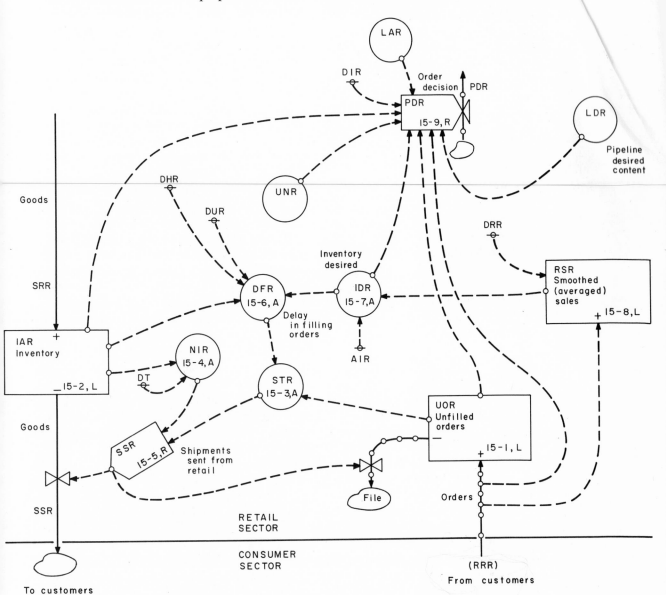

Figure 15-13 Purchasing decision added to retail sector.

of their inventories, and at that time the attempt to readjust inventory will lead to the necessary extra orders. The delay time constant DIR allows for varying degrees of intentional or unintentional delay in making the adjustments.

The term in the ordering equation for unfilled orders at the retail level UOR is inserted as part of the general attempt to make sure that the equations behave reasonably under extreme conditions of operation. In a situation where there is no supply of goods coming from the factory, and where the backlog of unfilled orders at retail is so great as to discourage further incoming customer orders, we find that the retail ordering rate RRR is zero. Under these circumstances the actual inventory would fall to zero. According to Equations 15-7 and 15-8, the desired level of inventory under these circumstances would also fall to zero. This might not appear plausible at first, but under the circumstances of no incoming orders and no supply and a backlog of unfilled orders it is both unlikely and unnecessary to have an inventory in stock. Under these circumstances it will be essential that unfilled orders held at the retail level be reflected in unfilled orders held at the distributor level. Otherwise, there will be no orders on hand at higher points in the system to reinitiate a flow of goods when a supply becomes available. As we shall see later, the actual pipeline orders LAR include the backlog of unfilled orders at the distributor level. However, if the unfilled-order term UOR is not included in Equation 15-9, we see that unfilled orders at the distributor level will be canceled until they cease to exist. This will happen because all other terms in the equation can fall to zero when no supply is available. Then, the negative term for pipeline content LAR (which contains only the backlog of unfilled orders at the distributor under these circumstances) will lead to a negative ordering rate until the backlog is canceled at the distributor. The positive term UOR balances, under the circumstances, the negative

term LAR, so that the backlog of unfilled orders is retained at the distributor.

The term for unfilled orders *normally* held at the retail level UNR is included so that under steady-state conditions the actual inventory at the retailer will run equal to the desired inventory. Under normal steady-state conditions the two pipeline terms will cancel one another and we should expect the two inventory terms to cancel one another. However, if the unfilled-order term UOR were present without the canceling normal level of unfilled orders UNR, the inventory terms would differ from one another by enough to equal the unfilled-order term UOR. While proper values of various inventory constants could be selected to make the actual inventory reach the desired amount, it would be confusing and contrary to our implied meaning of "desired inventory" if actual inventory did not attempt to equal desired inventory under steady-state operating conditions.

Transient interaction between the inventory, pipeline, and unfilled-order terms is important. Suppose that a deficit exists in actual inventory IAR compared with desired inventory IDR. This will cause an ordering rate that places orders in process in the pipeline term LAR. When sufficient orders have been initiated to meet inventory needs, the additional orders in the negative LAR term will cause the total value inside the bracket in Equation 15-9 to fall to zero. As the goods ordered for inventory are delivered, they will transfer from orders in the pipline LAR to inventory IAR, so that the ordering equation continues in balance. The equation thereby avoids duplicate reordering of inventory in succeeding time periods before the goods are delivered.

We can view in another way the terms in the bracket of Equation 15-9 as representing the "stock-commitment gap." By regrouping, (IDR.K + LDR.K + UOR.K) would represent the "desired ownership" by the retailer—desired inventory plus necessary goods in transit in the supply line plus product to fill the orders on hand. Against this will be offset a negative term $(IAR.K + LAR.K + UNR.K)$ representing the commitments that do not need filling—the inventory now present plus the goods now in the pipeline plus the unfilled orders that are considered normal.

Equation 15-9 for purchasing rate depends on values of several variables that have not yet been defined. One of these is the desired or necessary level of orders and goods in transit in the supply pipeline from the retailer to the distributor and return. The necessary orders and goods in transit in this pipeline will depend on the length of the line (total delay) and on the average level of retail sales that are to be supplied. This is expressed by the following equation:

$$LDR.K = (RSR.K)(DCR + DMR + DFD.K + DTR)$$
$$\text{15-10, A}$$

LDR pipeLine orders Desired (or necessary) to supply Retail (units)

RSR Requisitions Smoothed at Retail (average sales) (units/week)

DCR Delay in Clerical order processing at Retail (weeks)

DMR Delay in order Mailing from Retail (weeks)

DFD Delay (variable) in Filling orders at the Distributor (weeks)

DTR Delay in Transportation of goods to Retail (weeks)

The total average orders and goods necessary in transit in the supply line LDR is the product of average retail sales RSR multiplied by the total time required for an order to travel around the supply circuit. The right-hand parentheses give this total delay. Three of the delays are constant—order processing DCR, the mailing delay DMR, and the transportation of goods DTR. It is here assumed these do not vary with the state of the system under study.[14] The delay in

[14] Communications delays probably must be considered a variable in studies of dynamics of the national economy when economic activity reaches a level to cause overloading of existing transportation facilities.

filling orders at the distributor DFD d the availability of inventory at the dis on from which orders can be filled. It is th°r, a variable delay, as it is at retail.

The *actual* content of the supply pip consists of the sum of the orders and the goo in the various separate pieces of the suppl line as follows:

$$LAR.K = CPR.K + PMR.K + UOD.K + MTR.K$$
$$\text{15-11, A}$$

LAR pipeLine orders Actually in transit to Retail (units)

CPR Clerical in-Process orders at Retail (units)

PMR Purchase orders in Mail from Retail (units)

UOD Unfilled Orders at Distributor (units)

MTR Material in Transit to Retail (units)

The actual units in transit in the supply pipeline are defined here as an auxiliary variable that is the sum of the four levels of orders and goods existing in the four sections of the pipeline.

The remaining undefined term from Equation 15-9 is the normal level of unfilled orders at retail UNR. It is equal to the average sales level multiplied by the normal delay in the filling of orders as follows:

$$UNR.K = (RSR.K)(DHR + DUR)\qquad \text{15-12, A}$$

UNR Unfilled Normal level of orders at Retail (units)

RSR Requisitions Smoothed at Retail (average sales) (units/week)

DHR Delay in Handling time at Retail (weeks)

DUR Delay in Unfilled orders at Retail from out-of-stock items at normal inventory (weeks)

The normal delay at retail consists of two components — that represented by the average minimum handling time of orders, plus that contributed by the normal out-of-stock conditions with the associated delay in filling orders. The total delay multiplied by average sales gives the "normal" number of unfilled orders to be expected.

15-1 through 15-12 complete the Equa retail of levels, rates, and the aux-definitbles necessary in the rate equations. iliarmain the equations describing delays. Thseparate delays will be considered — in T'g purchase orders, in mailing orders from ller to distributor, and in shipping goods ɔm distributor to retailer. Figure 15-14 shows Equations 15-10 through 15-18 (including the delays) added to the flow diagram. The delay in filling orders at the distributor will be incorporated later in the description of the distributor sector.

The third-order delay will be used as adequately representing our intuitive "feel" for how order handling, mailing, and shipping respond to various steady-state and transient inputs.[15] Two equations will be written. One equation (a level) will define the quantity in transit in the delay. The other equation, in "shorthand" form, will specify the procedure for computing the output rate. In the computing system being used,[16] this output-rate "functional notation" will be automatically converted into the necessary level and rate equations. The equation used hereafter for defining the output rate of a delay is therefore not the actual difference equation that can be evaluated but is merely an indication sufficient to tell the computing program what computing method is to be used. The actual equations are supplied in the automatic generation of the detailed computer program instructions.

The pair of equations defining the third-order delay in making the purchasing decision and placing purchase orders from retail to distributor is as follows:

$$\text{CPR.K} = \text{CPR.J} + (\text{DT})(\text{PDR.JK} - \text{PSR.JK}) \qquad \text{15-13, L}$$

$$\text{PSR.KL} = \text{DELAY3}(\text{PDR.JK}, \text{DCR}) \qquad \text{15-14, R}$$

CPR   Clerical in-Process orders at Retail (units)

PDR   Purchasing rate Decision at Retail
        (units/week)

PSR   Purchase orders Sent from Retail
        (units/week)

DCR   Delay in Clerical order placing at Retail
        (weeks)

DELAY3   not a variable but a function specifying a
        set of third-order-delay equations[17]

Equation 15-13 is the usual level equation stating that the quantity CPR in transit in the delay equals its former value plus what has gone in minus what has come out. Equation 15-14 states how the output rate is to be obtained. It should be noted that the expression DELAY3 is not a variable like the other letter groups. It is a functional notation (third-order-delay function of the input rate PDR and of the delay DCR). It indicates what is to be done with the quantities. It says that a third-order delay is to be generated, in which the input rate is the variable PDR as defined by its own equation, and the length of the delay is given by the constant DCR.

The output of the clerical processing delay is the input to the mailing delay. The mail will also be represented as a third-order exponential delay:

$$\text{PMR.K} = \text{PMR.J} + (\text{DT})(\text{PSR.JK} - \text{RRD.JK}) \qquad \text{15-15, L}$$

$$\text{RRD.KL} = \text{DELAY3}(\text{PSR.JK}, \text{DMR}) \qquad \text{15-16, R}$$

PMR   Purchase orders in Mail from Retail (units)
PSR   Purchase orders Sent from Retail
        (units/week)
RRD   Requisitions (orders) Received at Distributor
        (units/week)
DMR   Delay in Mail from Retail to distributor
        (weeks)
DELAY3   specifies third-order-delay equations

As before, Equation 15-15 gives the units in transit in the delay. The accompanying special function in Equation 15-16 tells how to gen-

[15] Delays are discussed in Chapter 9.
[16] See Appendix A on the DYNAMO compiler.

[17] Chapter 9 and Appendix H give the actual equations that are evaluated in each time step to generate a third-order exponential delay.

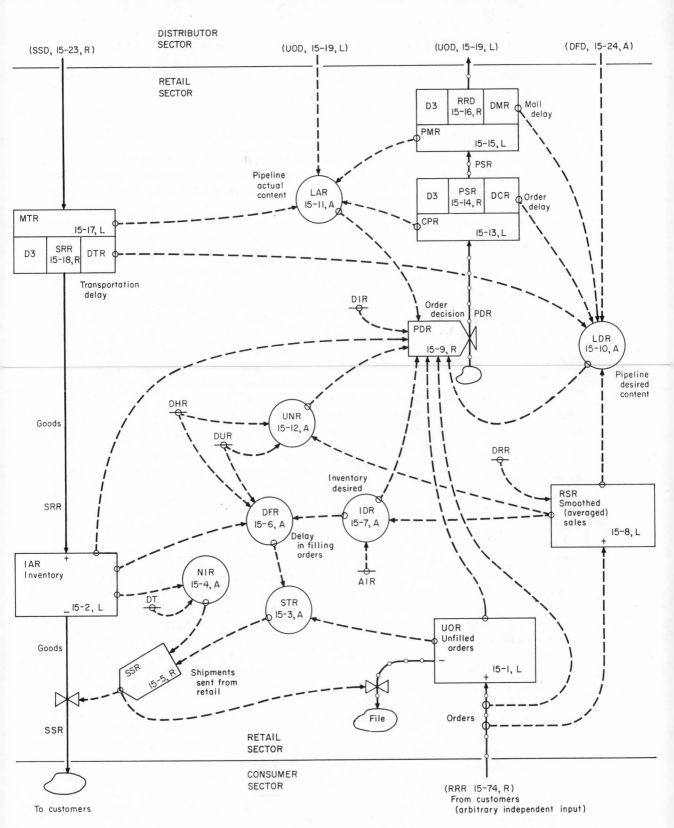

Figure 15-14 Flow diagram of retail sector.

erate the output rate of the mailing delay, which is the rate of actual receipt of orders at the distributor.

We shall define the retail sector to include the transportation of goods from the distributor to the retailer. Therefore we need another pair of equations to define the delay in transportation:

$$MTR.K = MTR.J + (DT)(SSD.JK - SRR.JK) \qquad \text{15-17, L}$$

$$SRR.KL = DELAY3(SSD.JK, DTR) \qquad \text{15-18, R}$$

MTR	Material in Transit to Retail (units)
SSD	Shipments Sent from Distributor inventory (units/week)
SRR	Shipments Received at Retail inventory (units/week)
DTR	Delay in Transportation of goods to Retail (weeks)
DELAY3	specifies third-order-delay equations

The input to the transportation delay is the output rate from the distributor inventory. The output SRR from the shipping delay is the input to the retail inventory as required in Equation 15-2. As before, Equation 15-17 defines the goods in transit, and Equation 15-18 gives the necessary information for calculating the output rate according to the characteristics of a third-order delay.

This completes the set of equations that we propose to use at present for representing operations in the retail sector. The equations are not self-contained, since they depend on knowing the values of certain variables (SSD, UOD, DFD) that exist in the distributor sector.

15.5.2 Equations for the Distributor Sector.

The equations that have already been developed for the retail sector deal with the general characteristics of receiving goods, receiving orders, shipping goods, and ordering replacements. Behavior patterns at the distributor could of course be different. In the absence of wanting to introduce any particular differences between retail and distributor behavior, we can use a similar set of equations to represent the dis-

tributor. We shall therefore set down eighteen equations identical in form with the previous eighteen. The constants defining delays and the other parameters of the system need not of course be the same at the distributor level as at the retail level. If in a particular actual system there were clear differences between retail and distributor decision criteria, there would then be a basis for formulating a different set of equations. Even with the same equation forms, selection of different parameter values can make allowance for different inventory levels, different delay intervals for processing orders and handling goods, and different policies on speed of inventory adjustment.

The equations for the distributor sector will follow with a minimum of discussion, since the explanation of the retail equations applies equally well here. The equation for unfilled orders at the distributor is similar to that given in Equation 15-1 for the retail sector:

$$UOD.K = UOD.J + (DT)(RRD.JK - SSD.JK) \qquad \text{15-19, L}$$

UOD	Unfilled Orders at Distributor (units)
RRD	Requisitions (orders) Received at Distributor (units/week)
SSD	Shipments Sent from Distributor inventory (units/week)
DT	Time interval between solutions of the equations (weeks)

This is an accounting equation giving the new unfilled-order level in terms of previous unfilled orders and the orders which have come in and the orders which have been filled.

Figure 15-15 shows the flow diagram for equations in the distributor sector.

The equation for inventory at the distributor level corresponds to Equation 15-2 at retail:

$$IAD.K = IAD.J + (DT)(SRD.JK - SSD.JK) \qquad \text{15-20, L}$$

IAD	Inventory Actual at Distributor (units)
SRD	Shipments Received at Distributor (units/week)
SSD	Shipments Sent from Distributor (units/week)

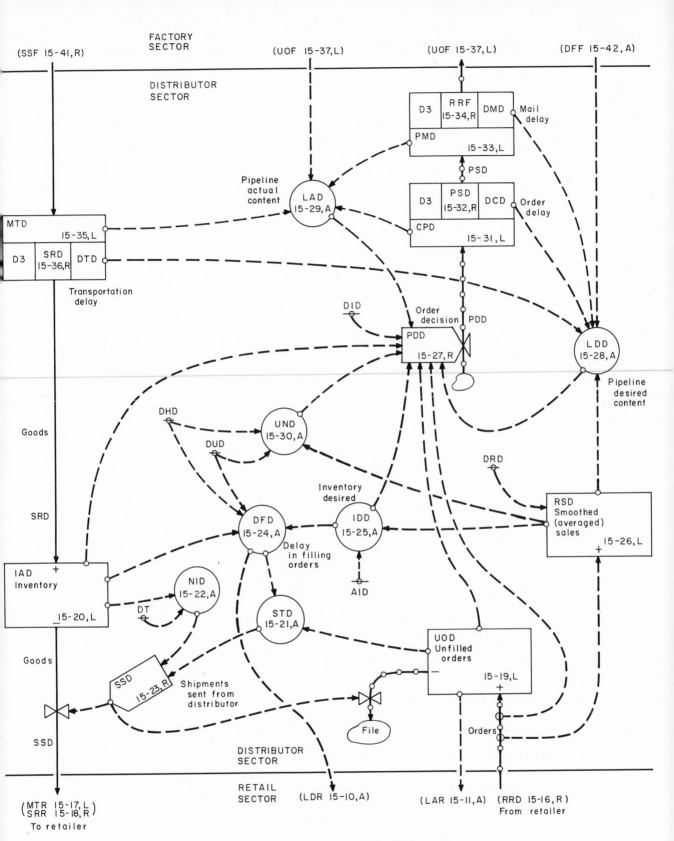

Figure 15-15 Flow diagram of distributor sector.

As in Equations 15-3, 15-4, and 15-5, the shipping rate from the distributor is given by

$$STD.K = \frac{UOD.K}{DFD.K} \qquad \text{15-21, A}$$

$$NID.K = \frac{IAD.K}{DT} \qquad \text{15-22, A}$$

$$SSD.KL = \begin{cases} STD.K & \text{if } NID.K \geq STD.K \\ NID.K & \text{if } NID.K < STD.K \end{cases} \qquad \text{15-23, R}$$

STD Shipping rate to be Tried at Distributor (units/week)
UOD Unfilled Orders at Distributor (units)
DFD Delay (variable) in Filling orders at the Distributor (weeks)
NID Negative Inventory limit rate at Distributor (units/week)
IAD Inventory Actual at Distributor (units)
DT Solution time interval, Delta Time (weeks)
SSD Shipments Sent from Distributor (units/week)

Equation 15-21 gives the tentative shipping rate from the distributor which is to be tested against the rate that would cause a negative inventory (Equation 15-22). The smaller of the two rates is taken as the actual shipping rate (Equation 15-23).

As in Equation 15-6, the variable delay representing the average delay in filling orders at the distributor is

$$DFD.K = DHD + DUD \frac{IDD.K}{IAD.K} \qquad \text{15-24, A}$$

DFD Delay in Filling orders at Distributor (weeks)
DHD Delay due to minimum Handling time required at Distributor (weeks)
DUD Delay, average, in Unfilled orders at Distributor caused by out-of-stock items when inventory is "normal" (weeks)
IDD Inventory Desired at Distributor (units)
IAD Inventory Actual at Distributor (units)

As in Equations 15-7 and 15-8, the desired level of distributor inventory and smoothed incoming requisitions are given by

$$IDD.K = (AID)(RSD.K) \qquad \text{15-25, A}$$

$$RSD.K = RSD.J + (DT)\left(\frac{1}{DRD}\right)(RRD.JK - RSD.J) \qquad \text{15-26, L}$$

IDD Inventory Desired at Distributor (units)
AID proportionality constAnt for Inventory at Distributor (weeks)
RSD Requisitions Smoothed at Distributor (units/week)
DRD Delay time constant in smoothing Requisitions at Distributor (weeks)
RRD Requisitions (orders) Received at Distributor (units/week)

For the purchasing decision equation at the distributor level, we shall use the same form as in Equation 15-9:

$$PDD.KL = RRD.JK + \left(\frac{1}{DID}\right)\Big[(IDD.K - IAD.K) + (LDD.K - LAD.K) + (UOD.K - UND.K)\Big] \qquad \text{15-27, R}$$

PDD Purchasing rate Decision at Distributor (units/week)
RRD Requisitions (orders) Received at Distributor (units/week)
DID Delay in Inventory (and pipeline) adjustment at Distributor (weeks)
IDD Inventory Desired at Distributor (units)
IAD Inventory Actual at Distributor (units)
LDD pipeLine orders Desired (necessary) in transit to Distributor (units)
LAD pipeLine orders Actual in transit to Distributor (units)
UOD Unfilled Orders at the Distributor (units)
UND Unfilled orders, Normal, at Distributor (units)

The equations for the orders and goods necessary in the supply pipeline between the distributor and the factory and the corresponding actual orders and goods in transit in the pipeline are similar to Equations 15-10 and 15-11:

$$LDD.K = (RSD.K)(DCD + DMD + DFF.K + DTD)$$

15-28, A

LDD	pipeLine orders Desired (necessary) in transit to Distributor (units)
RSD	Requisitions Smoothed at Distributor (units/week)
DCD	Delay Clerical at Distributor (weeks)
DMD	Delay in order Mailing from Distributor (weeks)
DFF	Delay (variable) in Filling orders at Factory (weeks)
DTD	Delay of goods in Transit to Distributor (weeks)

$$LAD.K = CPD.K + PMD.K + UOF.K + MTD.K$$

15-29, A

LAD	pipeLine orders Actual in transit to Distributor (units)
CPD	orders in Clerical Processing at Distributor (units)
PMD	Purchase orders in Mail from Distributor (units)
UOF	Unfilled Orders at the Factory (units)
MTD	Material in Transit to the Distributor (units)

As in Equation 15-12, "normal" unfilled orders depend on the normal order-processing delay and the average level of business activity:

$$UND.K = (RSD.K)(DHD + DUD)$$ 15-30, A

UND	Unfilled orders, Normal, at Distributor (units)
RSD	Requisitions (orders) Smoothed at Distributor (units/week)
DHD	Delay due to minimum Handling time required at Distributor (weeks)
DUD	Delay, average, in Unfilled orders at Distributor caused by out-of-stock items when inventory is "normal" (weeks)

The distributor experiences delays in the placing of purchase orders and in mailing these orders to the factory in the same way discussed in Equations 15-13 through 15-16. The four equations for the ordering delay and the mail delay are

$$CPD.K = CPD.J + (DT)(PDD.JK - PSD.JK) \quad \text{15-31, L}$$

$$PSD.KL = DELAY3(PDD.JK, DCD) \quad \text{15-32, R}$$

$$PMD.K = PMD.J + (DT)(PSD.JK - RRF.JK) \quad \text{15-33, L}$$

$$RRF.KL = DELAY3(PSD.JK, DMD) \quad \text{15-34, R}$$

CPD	Clerical in-Process orders at Distributor (units)
PDD	Purchasing-rate Decision at Distributor (units/week)
PSD	Purchase orders Sent from Distributor (units/week)
DCD	Delay in Clerical order processing at Distributor (weeks)
PMD	Purchase orders in Mail from Distributor (units)
RRF	Requisitions (orders) Received at Factory (units/week)
DMD	Delay in Mail from Distributor (weeks)
DELAY3	specifies third-order-delay equations

The remaining two equations for the distributor sector describe the shipping delay for goods coming from the factory. These correspond to Equations 15-17 and 15-18 at retail:

$$MTD.K = MTD.J + (DT)(SSF.JK - SRD.JK) \quad \text{15-35, L}$$

$$SRD.KL = DELAY3(SSF.JK, DTD) \quad \text{15-36, R}$$

MTD	Material in Transit to Distributor (units)
SSF	Shipments Sent from Factory (units/week)
SRD	Shipments Received at Distributor (units/week)
DTD	Delay in Transportation of goods to Distributor (weeks)
DELAY3	specifies third-order-delay equations

This completes the eighteen equations to represent the distributor level.

15.5.3 Equations for the Factory Sector. At the factory many of the order-filling functions are similar to those at retail and the distributor. However, some organizational differences exist at the factory. We shall assume that the factory warehouse and factory are adjacent. Therefore, we shall not provide for mail or transportation delays between the warehouse and the factory; on the other hand, there is a production lead time between a decision to change production rate and the resulting change in factory output.

In developing the equations for the factory, we shall start first with those that are similar to the retail and the distributor sectors. Figure 15-16 gives the flow diagram described by the following equations. The equations for unfilled orders and actual inventory are similar to Equations 15-1 and 15-2:

$$UOF.K = UOF.J + (DT)(RRF.JK - SSF.JK) \qquad 15\text{-}37,\ L$$

$$IAF.K = IAF.J + (DT)(SRF.JK - SSF.JK) \qquad 15\text{-}38,\ L$$

UOF	Unfilled Orders at the Factory (units)
RRF	Requisitions (orders) Received at Factory (units/week)
SSF	Shipments Sent from Factory warehouse (units/week)
IAF	Inventory Actual at Factory warehouse (units)
SRF	Shipments Received at Factory warehouse (manufacturing output) (units/week)

We shall assume that the factory warehouse is stocking and shipping a number of catalog items in the product line. This continues to imply the concept of a delivery delay that increases gradually as the level of inventory decreases. Consequently, our representation of shipping rate will be determined as in Equations 15-3, 15-4, and 15-5:

$$STF.K = \frac{UOF.K}{DFF.K} \qquad 15\text{-}39,\ A$$

$$NIF.K = \frac{IAF.K}{DT} \qquad 15\text{-}40,\ A$$

$$SSF.KL = \begin{cases} STF.K & \text{if } NIF.K \geq STF.K \\ NIF.K & \text{if } NIF.K < STF.K \end{cases} \qquad 15\text{-}41,\ R$$

STF	Shipping rate to be Tried at Factory (units/week)
UOF	Unfilled Orders at the Factory (units)
DFF	Delay (variable) in Filling orders at Factory (weeks)

NIF	Negative Inventory limit rate at Factory (units/week)
IAF	Inventory Actual at Factory (units)
DT	solution time interval, Delta Time (weeks)
SSF	Shipments Sent from Factory warehouse (units/week)

The expressions for the delay in filling orders, the ideal inventory, and the smoothed sales rate can be of the same form as given in Equations 15-6, 15-7, and 15-8:

$$DFF.K = DHF + DUF\,\frac{IDF.K}{IAF.K} \qquad 15\text{-}42,\ A$$

$$IDF.K = (AIF)(RSF.K) \qquad 15\text{-}43,\ A$$

$$RSF.K = RSF.J + (DT)\left(\frac{1}{DRF}\right)(RRF.JK - RSF.J)$$
$$15\text{-}44,\ L$$

DFF	Delay (variable) in Filling orders at Factory (weeks)
DHF	Delay due to minimum Handling time required at Factory (weeks)
DUF	Delay, average, in Unfilled orders at Factory caused by out-of-stock items when inventory is "normal" (weeks)
IDF	Inventory Desired at Factory (units)
IAF	Inventory Actual at Factory (units)
AIF	proportionality constAnt for Inventory at Factory (weeks)
RSF	Requisitions Smoothed at Factory (units/week)
DRF	Delay in smoothing Requisitions at Factory (weeks)
RRF	Requisitions (orders) Received at Factory (units/week)

We now come to the manufacturing rate decision. In a real situation it might be affected by various practical considerations determined by physical manufacturing facilities. Most production lines and equipment can, however, operate over a substantial range of production capacities. Therefore, we assume for this example that manufacturing rate is continuously variable from zero up to some maximum rate.

The desire to produce more than the maximum rate will lead only to maximum factory output. It should be noted that the "overt" de-

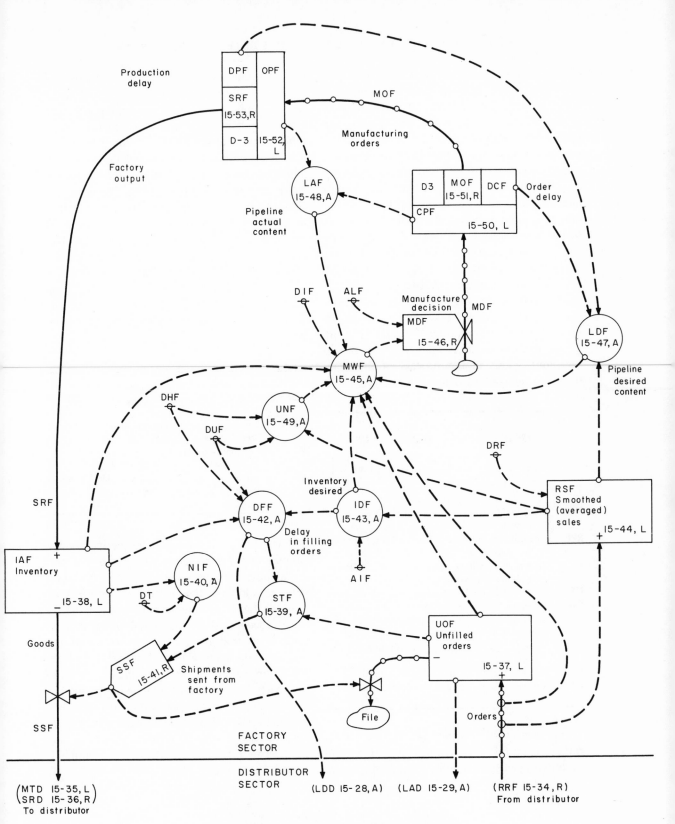

Figure 15-16 Flow diagram of factory sector.

cision to *want* to produce can be made independently of the *ability* to produce. Furthermore, the actual production schedule and flow of manufacturing orders to the factory can exceed the ability to produce; if so, the "implicit" decision, which determines factory output in response to the condition of the schedule, employment, available materials, and equipment, would control the actual output. In order that the equations need not deal with the details of internal factory conditions, the "overt" factory schedule decision is limited to that which the factory can produce. Later chapters will show how to remove this restriction.

The equation for the manufacturing rate desired at the factory will take the form of Equation 15-9 for the ordering of goods at retail. This equation recognizes sales rate, the state of inventory, goods in the process of manufacture, and the state of unfilled orders:

$$MWF.K = RRF.JK + \left(\frac{1}{DIF}\right)\Bigg[(IDF.K - IAF.K) \\ + (LDF.K - LAF.K) + (UOF.K - UNF.K) \Bigg]$$

15-45, A

MWF	Manufacturing rate Wanted at Factory (units/week)
RRF	Requisitions (orders) Received at Factory (units/week)
DIF	Delay in Inventory (and pipeline) adjustment at Factory (weeks)
IDF	Inventory Desired at Factory (units)
IAF	Inventory Actual at Factory (units)
LDF	pipeLine orders Desired (necessary) in transit through Factory (units)
LAF	pipeLine orders Actual in transit through Factory (units)
UOF	Unfilled Orders at the Factory (units)
UNF	Unfilled orders, Normal, at Factory (units)

Equation 15-45 is an auxiliary equation rather than a rate equation because it must still be tested against maximum manufacturing ca-

pacity of the factory. Manufacturing capacity is not an abrupt limit independent of work week, numbers of employees, and production efficiency; however, were we to include these refinements, the model would continue to extend itself outward into facets of the enterprise which we had decided in the initial objectives would be omitted for the present. Therefore, we shall characterize the manufacturing capability as a rate equal to the desired rate so long as this is less than maximum capacity. We shall assume that the factory output is a delayed function of the input schedule.

The following equation will give the actual manufacturing decision at the factory as the lesser of the rate wanted or the factory capacity limit:

$$MDF.KL = \begin{cases} MWF.K & \text{if } ALF \geq MWF.K \\ ALF & \text{if } ALF < MWF.K \end{cases}$$

15-46, R

MDF	Manufacturing rate Decision at Factory (units/week)
MWF	Manufacturing rate Wanted at Factory (units/week)
ALF	constAnt specifying manufacturing capacity Limit at Factory (units/week)

The pipeline terms supplying the factory are simpler than for retail and distributor because we assume there is no mail or transportation delay to and from the factory. Also we assume that the factory manufacturing capacity is already recognized in the initial ordering procedure, so that there is no doubt about the ability of the factory to supply and, therefore, no accumulation of an unfillable-order backlog exists in the factory. Our assumptions are equivalent to saying that availability of labor and materials will not limit production, except as expressed in the maximum production rate limit. The pipeline and the normal unfilled-order equations are

$$LDF.K = (RSF.K)(DCF + DPF) \qquad 15\text{-}47,\ A$$

$$LAF.K = (CPF.K + OPF.K) \qquad 15\text{-}48,\ A$$

$$UNF.K = (RSF.K)(DHF + DUF) \qquad 15\text{-}49,\ A$$

LDF	pipeLine orders Desired (necessary) in transit through Factory (units)
RSF	Requisitions (orders) Smoothed at Factory (units/week)
DCF	Delay in Clerical processing of manufacturing orders at Factory (weeks)
DPF	Delay in Production lead time at Factory (weeks)
LAF	pipeLine orders Actual in transit through Factory (units)
CPF	Clerical in-Process manufacturing orders at Factory (units)
OPF	Orders in Production at the Factory (units)
UNF	Unfilled orders, Normal, at Factory (units)
DHF	Delay due to minimum Handling time required at Factory (weeks)
DUF	Delay, average, in Unfilled orders at Factory caused by out-of-stock items when inventory is "normal" (weeks)

Next is a representation of the delay experienced in converting information into a production-rate decision. This can be represented by an exponential delay similar to Equations 15-13 and 15-14:

$$CPF.K = CPF.J + (DT)(MDF.JK - MOF.JK) \qquad 15\text{-}50,\ L$$

$$MOF.KL = DELAY3(MDF.JK,\ DCF) \qquad 15\text{-}51,\ R$$

CPF	Clerical in-Process manufacturing orders at Factory (units)
MDF	Manufacturing-rate Decision at Factory (units/week)
MOF	Manufacturing Orders into Factory (units/week)
DCF	Delay in Clerical processing of manufacturing orders at Factory (weeks)
DELAY3	specifies third-order-delay equations

Next is the manufacturing process itself. The manufacturing decision has already been limited so that it falls within the range of factory capability. We are assuming that labor and materials impose no other restriction on output.

Manufactured output will therefore be a delayed function of the manufacturing orders received. Depending on the length and nature of the time delay necessary to change production rate, we should choose among the various delay functions available to us. For the present purposes, a third-order delay is satisfactory. If the initial preparations for changing manufacturing required more "lead time" and then the rate were to rise rapidly to its new value, we might select a sixth-order delay.[18] In the absence of wanting to represent any specific production process, we shall take a third-order delay as typical and representative of the normal circumstances we should expect to encounter. This leads to the following pair of equations:

$$OPF.K = OPF.J + (DT)(MOF.JK - SRF.JK) \qquad 15\text{-}52,\ L$$

$$SRF.KL = DELAY3(MOF.JK,\ DPF) \qquad 15\text{-}53,\ R$$

OPF	Orders in Production at Factory (units)
MOF	Manufacturing Order rate into Factory (units/week)
SRF	Shipments Received at Factory inventory (manufacturing output) (units/week)
DPF	Delay in Production lead time at Factory (weeks)
DELAY3	specifies third-order-delay equations

By terminating our system of equations at the factory level and not introducing any new variables, we have now completed the formal mathematical description of the system we propose to study. The only undefined variable in the preceding set is the retail sales rate RRR. The model that we have defined does not presume to represent the characteristics of the consumer market itself. Therefore, retail sales will be taken as various specified time sequences, to see how the production and distribution system itself will respond under various sales conditions.

15.5.4 Initial Conditions. Equations 15-1 through 15-53 are to be evaluated cyclically at

[18] See Chapter 9.

points in time separated by the interval DT. To start this evaluation at the beginning, there must be initial values for certain of the variables. For most explorations using this type of model, it will be easiest and least confusing if the system starts from a steady-state, unperturbed condition. Since retail sales RRR is the only independent input, this means that a past history of continuous unchanging retail sales is assumed. Furthermore, the system will be started in equilibrium, whether or not it turns out that the system is stable. If the equilibrium point is unstable, any disturbance will initiate a growing departure from the initial conditions.

The computing sequence for the main set of equations is first to evaluate the level equations, then the auxiliary equations, and finally the rate equations. To start the computation, constants (or an equation) are required to specify all *levels* at the beginning. Also, if we have taken the liberty of using a rate in another rate equation,[19] we must have a constant or equation for defining all *rates* that appear on the right-hand side of auxiliary or rate equations.

Given these initial values, the levels at the beginning of the problem and certain necessary rates immediately before the beginning are available. From these the auxiliary variables can be calculated. They are derived from the values of the initial levels and from the few specified rates immediately before the initial time. After calculation of the auxiliary variables, the rate variables can then be evaluated for the forthcoming time period immediately following the point of beginning. After that the normal sequence of evaluating levels, auxiliary variables, and rates can proceed in rotation.

In general, it will be best to state initial values in terms of the external inputs and the parameters of the system so that it is possible to change the values of parameters in the equations without making it necessary to respecify

initial-value equations. The equations for initial values will now be developed.

The automatic computer programming system (DYNAMO) which is assumed available for evaluating these equations will place the initial-value equations into the proper order so that one initial value can be defined in terms of initial values already defined (an equation sequence must be possible so that the evaluation of a set of simultaneous initial-condition equations is not required).

The initial and past values of retail requisitions RRR are to be equal to a specified number:

$$RRR = RRI \qquad\qquad \text{15-54, N}$$

RRR initial value, Requisitions (orders) Received at Retail   (units/week)

RRI Retail Requisitions, Initial rate, constant (units/week)

The letter N after the equation number shows that this is an equation defining an initial value. Time-period notation is not used with the variables in initial-value equations.

The first level encountered in the system is the unfilled-order backlog at retail UOR, which is given in Equation 15-1. The normal steady-state level of this variable is indicated by Equation 15-12 defining normal unfilled orders at retail UNR:

$$UOR = (RSR)(DHR + DUR) \qquad \text{15-55, N}$$

UOR initial value, Unfilled Orders at Retail   (units)

RSR initial value, Requisitions Smoothed at Retail (units/week)

DHR Delay due to minimum order-Handling time required at Retail   (weeks)

DUR Delay, average, in Unfilled orders at Retail caused by out-of-stock items when inventory is "normal"   (weeks)

The initial value of actual inventory IAR should equal the desired level as given by Equation 15-7:

---

[19] See Sections 6.1 and 7.1 regarding the expediency of using rates in other rate equations.

$$IAR = (AIR)(RSR) \qquad \text{15-56, N}$$

IAR   initial value, Inventory Actual at Retail   (units)

AIR   proportionality constAnt for Inventory at Retail (ratio of desired inventory to weekly sales) (weeks)

RSR   initial value, Requisitions Smoothed at Retail (units/week)

The next level in the set of equations is that for the smoothed retail sales rate that in the steady state will equal the constant past sales rate:

$$RSR = RRR \qquad \text{15-57, N}$$

RSR   initial value, Requisitions Smoothed at Retail (units/week)

RRR   initial value, Requisitions (orders) Received at Retail   (units/week)

Using the initial values already defined, we can see that the computing program can then evaluate Equations 15-7, 15-6, 15-3, 15-4, and 15-5, yielding values of the auxiliary variables and arriving at a shipping rate from retail SSR that is equal to the specified steady-state condition of retail requisition initial rate RRI.

The initial values of orders and goods in transit in the supply line must also be specified. In the steady state the rates of flow in the supply line between retailer and distributor will equal the retail sales rate. The product of this rate multiplied by the length of a delay will give the quantity stored in the delay. The equations for clerical processing, mail, and transportation are then

$$CPR = (DCR)(RRR)^{20} \qquad \text{15-58, N}$$

[20] In principle, specifying the initial content of a delay is sufficient to specify steady-state initial conditions of the inflow and outflow rates. However, the convention used in the DYNAMO compiler requires that the initial input rate to a third-order delay be given, requiring, in addition to the above, the equations PDR = RRR and SSD = RRR. DYNAMO generates the initial output value of a third-order delay (PSR) so that this will be available to the mail delay.

$$PMR = (DMR)(RRR) \qquad \text{15-59, N}$$

$$MTR = (DTR)(RRR)^{20} \qquad \text{15-60, N}$$

CPR   initial value, Clerical in-Process orders at Retail   (units)

DCR   Delay in Clerical order placing at Retail (weeks)

RRR   initial value, Requisitions (orders) Received at Retail   (units/week)

PMR   initial value, Purchase orders in Mail from Retail   (units)

DMR   Delay in Mail from Retail   (weeks)

MTR   initial value, Material in Transit to Retail (units)

DTR   Delay in Transportation of goods to Retail (weeks)

With the preceding initial values defined and with the initial value of unfilled orders at the distributor UOD, which will be available from the specification of initial values in the distributor sector, it becomes possible to evaluate Equation 15-9 for the rate of generating purchase orders at retail PDR. Under these steady-state initial conditions, all the terms in the bracket of Equation 15-9 total zero, leaving the order-generation rate equal to retail sales.

It is well to test the initial-value equations to make sure that they will generate the anticipated initial values of auxiliary and rate variables. It is sometimes quite easy to formulate a set of equations in which the actual steady-state conditions are not what is desired and not at first glance what they would appear to be. This point was mentioned earlier as one of the reasons for placing the normal unfilled-order term UNR in Equation 15-9 for the purchasing rate.

The additional equations for initial values at the distributor sector follow:

$$RRD = RRR^{\,21} \qquad \text{15-61, N}$$

$$UOD = (RSD)(DHD + DUD) \qquad \text{15-62, N}$$

[21] Since RRD would be generated by DYNAMO as the initial output rate of the mail delay, this equation would not be needed. As before, the ordering and shipping delays would require PDD = RRD and SSF = RRD.

$$IAD = (AID)(RSD) \qquad \text{15-63, N}$$

$$RSD = RRD \qquad \text{15-64, N}$$

$$CPD = (DCD)(RRD) \qquad \text{15-65, N}$$

$$PMD = (DMD)(RRD) \qquad \text{15-66, N}$$

$$MTD = (DTD)(RRD) \qquad \text{15-67, N}$$

RRD initial value, Requisitions (orders) Received at Distributor (units/week)

RRR initial value, Requisitions (orders) Received at Retail (units/week)

UOD initial value, Unfilled Orders at Distributor (units)

RSD initial value, Requisitions Smoothed at Distributor (units/week)

DHD Delay due to minimum Handling time required at Distributor (weeks)

DUD Delay, average, in Unfilled orders at Distributor caused by out-of-stock items when inventory is "normal" (weeks)

IAD initial value, Inventory Actual at Distributor (units)

AID proportionality constAnt for Inventory at Distributor (weeks)

CPD initial value, Clerical in-Process orders at Distributor (units)

DCD Delay in Clerical order processing at Distributor (weeks)

PMD initial value, Purchase orders in Mail from Distributor (units)

DMD Delay in Mail from Distributor (weeks)

MTD initial value, Material in Transit to Distributor (units)

DTD Delay in Transportation of goods to Distributor (weeks)

A similar set of initial-value equations applies to the factory sector as follows:

$$RRF = RRR\,^{22} \qquad \text{15-68, N}$$

$$UOF = (RSF)(DHF + DUF) \qquad \text{15-69, N}$$

$$IAF = (AIF)(RSF) \qquad \text{15-70, N}$$

$$RSF = RRF \qquad \text{15-71, N}$$

$$CPF = (DCF)(RRF) \qquad \text{15-72, N}$$

$$OPF = (DPF)(RRF) \qquad \text{15-73, N}$$

[22] The **DYNAMO** compiler would not require this equation, but for the ordering delay would need MDF = RRF.

RRF initial value, Requisitions Received at Factory (units/week)

RRR initial value, Requisitions Received at Retail (units/week)

UOF initial value, Unfilled Orders at the Factory (units)

RSF initial value, Requisitions Smoothed at Factory (units/week)

DHF Delay due to minimum Handling time required at Factory (weeks)

DUF Delay, average, in Unfilled orders at Factory caused by out-of-stock items when inventory is "normal" (weeks)

IAF initial value, Inventory Actual at Factory (units)

AIF proportionality constAnt for Inventory at Factory (weeks)

CPF initial value, Clerical in-Process manufacturing orders at Factory (units)

DCF Delay in Clerical processing of manufacturing orders at Factory (weeks)

OPF initial value, Orders in Production at Factory (units)

DPF Delay in Production lead time at Factory (weeks)

Equations 15-54 through 15-73 state the initial values necessary for starting the computation of Equations 15-1 through 15-53.[23]

15.5.5 Parameters (Constants) of the System. Having completed the equations describing the system behavior and the equations defining initial values, we now need a set of numerical values of the parameters (constants during any particular solution) of the system.

The first parameter encountered in the equations is really a parameter of the solution procedure rather than of the system as such. It is the solution-time interval DT.[24] The solution interval should be a small fraction (less than one-sixth) of the length of time represented in any third-order delay in the system. Since we shall represent delays as short as a half-week in the system, we take the solution-time interval as

DT = 0.05 week, solution interval

[23] All of the equations are tabulated together in Appendix B.

[24] The basis for choosing a solution-time interval is discussed in Appendix D.

Since we are discussing in this chapter a typical or plausible system and not one representing a particular company, we shall not dwell at length on the selection of numerical values for the parameters. Plausible values will be taken, and later we shall see the effect on system performance if some of these values are changed.

Consider first the delays in the filling of orders at retail, the distributor, and the factory. The first parameter is that arising from the minimum order-handling delay when the item is in stock in inventory. We might expect these delays to be about 1 week at each of the three levels; therefore:

DHR = 1.0 week, Delay in Handling time at Retail
DHD = 1.0 week, Delay in Handling time at Distributor
DHF = 1.0 week, Delay in Handling time at Factory

We must also select values for the delays in handling unfilled orders which are caused by out-of-stock items. These are the constants DUR, DUD, and DUF. In discussing Equation 15-6, we arrived at a functional relationship on the basis of intuitive reasoning, which set the delay due to out-of-stock items as proportional to the ratio of desired inventory divided by actual inventory. We could explore in the model the effect on the system of using different functional relationships and various values of the out-of-stock delay constant.

Figure 15-17 gives the reciprocal function that we have chosen to use. On the vertical scale is plotted the out-of-stock contribution to total average delay as a multiple of the minimum order-handling delay DHR. The horizontal scale is also nondimensionalized to show actual inventory in terms of a multiple of desired inventory. The several curves show different ratios of the delay DUR (the delay due to out-of-stock items in inventory when aggregate inventory IAR is at the desired level IDR) to the delay DHR (the minimum clerical order-processing time).

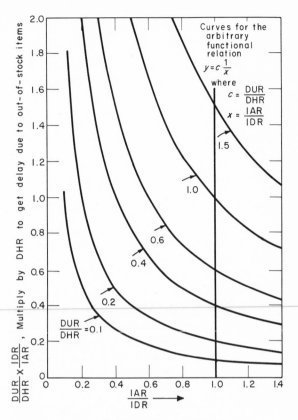

Figure 15-17 Delay versus inventory ratio.

In Figure 15-17 the heavy vertical line is drawn where actual inventory equals desired inventory. The points where the curves cross this line show the multiples of the minimum order-handling time which out-of-stock items will add to the average order-filling delay under conditions of "normal" inventory. Following a particular curve shows us how rapidly the average order-filling delay will vary with changes in inventory. So long as we stay with the functional relationship given in Equations 15-6, 15-24, and 15-42, we cannot choose independently the delay added at *normal* inventory and the rate at which delay will increase with *reductions* of inventory. These two effects could be separately selected by using different functional relationships between inventory and de-

169

lay.[25] Let us assume that the curves marked 0.4, 0.6, and 1.0 agree adequately with our estimate of the out-of-stock order-filling delays at retail, distributor, and factory levels. Since these are to be multiplied by the minimum order-handling delay, which in each case has already been selected as 1 week, the values of the constants are also

DUR = 0.4 week, Delay due to Unfillable orders at Retail

DUD = 0.6 week, Delay due to Unfillable orders at Distributor

DUF = 1.0 week, Delay due to Unfillable orders at Factory

These values in their respective equations state, for example, that as retail inventory drops to half the desired amount, the average order-filling delay at retail will rise from 1.4 to 1.8 of the minimum order-handling time. At the distributor, the delay would rise from 1.6 to 2.2 times the minimum; at the factory from 2.0 to 3.0 times the minimium. Estimates of correct values, in a particular situation, could be reached by examining a sample of orders to determine typical order-filling time and the frequency of out-of-stock delays.

The next parameters encountered in the equations are the ones that relate the level of desired inventory to the average sales rate. This constant is defined as the number of weeks of average sales which could be supplied out of the "normal" inventory. Let us take these at the retail, distributor, and factory levels as

AIR = 8 weeks, proportionality constAnt for Inventory at Retail

AID = 6 weeks, proportionality constAnt for Inventory at Distributor

AIF = 4 weeks, proportionality constAnt for Inventory at Factory

If these are divided into 52 weeks, they give the inventory turnover rate per year. The above

figures are equivalent to an inventory turnover of 6.5, 8.7, and 13 times per year at the three levels.

The parameters in Equations 15-8, 15-26, and 15-44 give the exponential smoothing time constant used for averaging present sales into smoothed sales. We shall assume, to start our investigation, that an 8-week time constant is being used at each of the three levels. Therefore:

DRR = 8 weeks, Delay in smoothing Requisitions at Retail

DRD = 8 weeks, Delay in smoothing Requisitions at Distributor

DRF = 8 weeks,[26] Delay in smoothing Requisitions at Factory

In Equations 15-9, 15-27, and 15-45 are the rates of inventory and pipeline correction given by the parameters **DIR**, **DID**, and **DIF**. As we shall see later, the system performance is sensitive to the values of these inventory and pipeline correction rates. These parameters are ones about whose values we might lack a basis for estimating closely. They might be indicated by historical ordering data. They are controllable by management. We shall start by selecting values that seem plausible and later see the effect of changing these values. Initially, we shall assume that at each level the ordering rate corrects inventory and pipeline deviations at the rate of one-quarter of the imbalance per week:

DIR = 4 weeks, Delay in Inventory (and pipeline) adjustment at Retail

DID = 4 weeks, Delay in Inventory (and pipeline) adjustment at Distributor

DIF = 4 weeks,[27] Delay in Inventory (and pipeline) adjustment at Factory

Next we must specify the delays in the processing of purchase orders. We shall assume that this takes longer at retail than at the distributor,

---

[25] Or a table of values could be provided with interpolation for the delay corresponding to any inventory ratio.

[26] Under some circumstances this is equivalent to a 16-week moving average. See Appendix E.

[27] These values are probably smaller than would be usual.

a period which is in turn longer than at the factory, giving

DCR = 3   weeks, Delay, Clerical at Retail
DCD = 2   weeks, Delay, Clerical at Distributor
DCF = 1   week, Delay, Clerical at Factory

For the mailing delay from retailer and distributor we shall use

DMR = 0.5   week, Delay in Mailing from Retail
DMD = 0.5   week, Delay in Mailing from Distributor

For the delay in shipping goods we shall use the following from distributor to retailer and from factory to distributor

DTR = 1.0   week, Delay in Transportation to Retail
DTD = 2.0   weeks, Delay in Transportation to Distributor

At the factory a 6-week lead time is assumed between the scheduling of a change in production rate and the time when the new rate of flow of goods is available from the factory:

DPF = 6.0   weeks, Delay in Production lead time at Factory

Equation 15-46 requires a value for the maximum factory production rate. For our initial exploration of the system, the factory limitation is not to be active. Therefore it can be set at a very high multiple of the retail sales level.

ALF = (1,000)(RRI)   units/week, constAnt, manufacturing capacity Limit at Factory

In the above expression ALF is the manufacturing limit at the factory and RRI is the retail requisition initial rate.

The initial production and sales rate must be specified at which to have the system operate. This can be arbitrarily chosen to some convenient scale of operations such as

RRI = 1,000   units/week, Retail Requisitions, Initial rate

We now have a complete set of dynamic equations, initial-value equations, and parameters, with the exception of a specification of the input retail requisition rate RRR that will be applied to the system for test purposes.

The retail sales rate will be specified differently for different explorations of system behavior. The specification of this rate will be given with each of the sets of test conditions to which the system is to be subjected.

15.6   Philosophy of Selecting Reasonable Parameter Values

The reader may at first object to the arbitrary liberties just exhibited in selecting values of parameters. The preceding is inconsistent with much of the statistical estimating effort exhibited in the management science and economics literature. However, I feel that extensive data gathering and analysis should *follow* the *demonstration* of a need for more accuracy in a particular parameter. For many purposes values of parameters anywhere within the plausible range will produce approximately the same results.

It must be true that most industrial and economic systems are not highly sensitive to small changes in parameters; otherwise their whole qualitative dynamic character would be much more changeable than it is. To a first approximation, economic fluctuation continues decade after decade in similar patterns although many details of the system have been greatly modified. In the last two hundred years we have changed our form of government and our banking system; government expenditure has risen to a substantial fraction of our national production; the country has shifted from largely agricultural to largely industrial activity; and transportation and communication speeds have increased by a factor of 100. Yet in spite of these changes our capitalist economic system persists in similar fluctuations and trends in growth and monetary inflation. We shall find that the complexity of the system structure, the existence of delays distributed throughout, the decisions that introduce amplification, and the time constants that arise from human memory and action and life span all combine to produce system behavior that is independent of reasonable changes in most of the parameters.

An information-feedback system is a system of counterbalancing influences. An error in one factor is often balanced within the system by self-induced changes in other factors. The more complete and realistic the system, the less sensitive we should expect it to be to small changes in *most* of the individual parameters.

Tests on the model itself can be used to determine model sensitivity to values of parameters. When a peculiarly sensitive parameter is identified, we are faced with more problems than merely measuring its value. Perhaps we can measure it accurately, but we must have confidence that the value is constant with time. Otherwise it may be an important system variable, and if its source of variation cannot be identified, our model behavior may be misleading. Maybe the parameter is one that can be controlled, once its importance is realized. If the parameter cannot be measured accurately, or is not constant, or cannot be controlled, then perhaps we can redesign the structure of the industrial system so that the system behavior is no longer vulnerable to the value of and changes in the parameter.

15.7 Test Runs of Model

In Chapter 2 were given figures showing the way that a typical distribution organization would react to some simplified retail sales inputs. In the preceding sections of this chapter have been given the equations that describe the organization and the management policies of the system of Chapter 2.

In following sections will be given a more detailed description of each of the figures in Chapter 2, including the equations used to produce the test-input conditions.

It might at first be assumed that the most informative type of system test input would be a time series of actual sales taken from a real-life situation. In general, this is not the most useful beginning point. A historical time series is too complex a pattern for a beginning exploration of system behavior. Simpler test conditions are preferable during attempts to understand the fundamental characteristics of the system itself. At a later time, we can study responses to typical historical data or can combine properly selected pure inputs consisting of growth trends, seasonal and other periodic fluctuations, and noise to create a test input of known composition which generates the statistical characteristics of historical sequences.

Figures 2-2 through 2-8 of Chapter 2 will now be repeated and individually discussed. The equations and parameters of Section 15.5 apply except as indicated.

15.7.1 Step Increase in Sales. A very informative, and one of the simplest, test inputs for the study of system dynamics is the "step function." This is a sudden disturbance caused by changing an external system input to some new value that is then held constant. A step function is a shock containing, in principle, an infinite band of component frequencies. It can serve to "excite" any mode of response that may be inherent in the system model being tested. If the system has oscillatory behavior, the step input gives an immediate indication of the natural period of oscillation and the rapidity of damping or of growth of the oscillation. The step input will usually serve to trigger any cumulative tendencies toward sustained growth or decline.

The equations needed for generating a 10% step in retail sales RRR follow:

$$RRR.KL = RRI + RCR.K \qquad \text{15-74, R}$$

$$RCR.K = \begin{cases} 0 & \text{if time is less than 0} \\ 100 & \text{if time is greater than 0} \end{cases} \text{15-75, A}$$

RRR   Requisitions (orders) Received at Retail (units/week)

RRI   Retail Requisitions, Initial rate, constant (units/week)

RCR   Requisition Change at Retail (units/week)

These two equations define retail sales as being a constant steady-state value RRI before the beginning of the test run (given as 1,000 units per week). After the beginning of the run, the value of RRR is increased by 100 units per week, giving a 10% upward step in retail sales.

In Figure 15-18 is seen the progression toward the factory of the disturbance that is created by the change in retail sales. This is summarized in Table 15-1.

Table 15-1   Times of Peak Order and Production Rates after 10% Sales Increase

Variable	Peak Value of Change (%)	Time of Occurrence after Retail Change (weeks)
Retail Sales	+10	constant
Distributor Orders from Retail	+18	11
Factory Orders from Distributors	+34	14
Manufacturing Orders to Factory	+51	15
Factory Output	+45	21

This progressive increase in the peak ordering rate as the disturbance moves upward in the system is a result of the two sources of amplification in the policies controlling the ordering decisions — the unavoidable necessity of increasing the orders and goods in transit in the supply pipeline to and from the next higher stage, and the practice of increasing the "desired" inventory as the level of average sales increases. In the system of this example, the delay from retailer to distributor and return is 6.1 weeks (3 weeks in the ordering process DCR, 0.5 week in mail DMR, 1.6 weeks order handling at the distributor DHD plus DUD, and 1 week in transportation DTR). A 10% increase in retail sales rate requires a 10% increase in

the orders and goods in transit if the delay is constant (more if the delay increases as the model does and as actual business usually will during the early phase of increasing business volume). This increase in supply-line content in Equation 15-10 must then be

$$(\text{Delay})(\text{sales rate change}) = (6.1 \text{ weeks}) \left(100 \, \frac{\text{units}}{\text{week}} \right)$$
$$= 610 \text{ units}$$

In addition, if the rather common practice of increasing the desired inventory in proportion to sales level is followed, there is, according to Equation 15-7, another nonrecurring component of extra orders that equals

$$\text{Increase for Inventory} = (\text{AIR})(\text{sales rate change})$$
$$= (8 \text{ weeks}) \left(100 \, \frac{\text{units}}{\text{week}} \right)$$
$$= 800 \text{ units}$$

AIR proportionality constAnt relating Inventory to average sales rate at Retail

This total of 1,410 units of orders is partly offset by the increased level of normal unfilled orders at the retail level, in Equation 15-12, which will equal

$$\text{Increase in normal retail Unfilled Orders} = (\text{retail delay})(\text{sales rate change})$$
$$= (\text{DUR} + \text{DHR})(\text{sales rate change})$$
$$= (0.4 + 1.0)(100)$$
$$= 140 \text{ units}$$

These 140 units of orders reside in the retail unfilled-order pool. Total deliveries lag behind total incoming orders by this amount, so that this quantity is not deleted from inventory.

These net extra orders for 1,270 units (1,410 minus 140) of product must be placed by the retail stage over the interval of time between the step increase of retail sales and when the retail level has reached steady-state operation

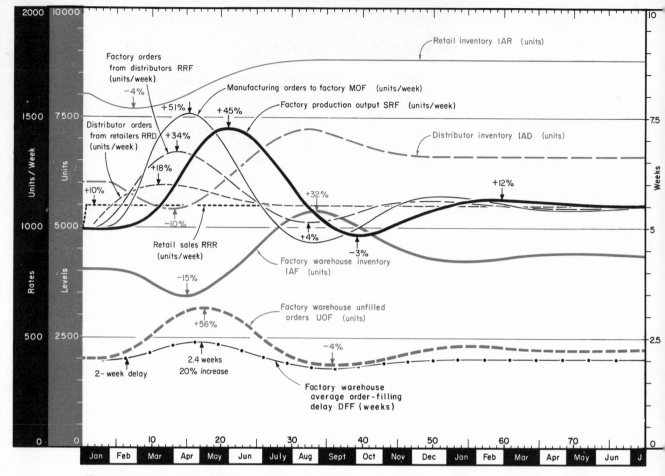

Figure 15-18 (Repeat of Figure 2-2) Response of production-distribution system to a sudden 10% increase in retail sales.

at the new higher sales volume. In Figure 15-18 this occurs over the 30 weeks following the change in retail sales.

For a period of 3 months, the distributor orders from retail are 15% or more above the initial sales level (compared with the 10% increase at retail). This temporary ordering increase in excess of the actual retail sales increase persists for a long enough time to become the new basis for decisions at the distributor level, which in a similar manner causes further multiplication of ordering rate in adjusting distributor inventories and supply lines.

In this example, the retailers have constant sales at the new increased rate. The distributors, however, having built up to a higher level of business caused partly by the retailers' non-continuing orders, find their sales rate declining

between the 11th and the 25th weeks. Their inventory level and supply-line content, which are both now excessive, are then reduced. This reduction is accomplished by reducing outgoing orders to the factory below distributor sales rate. Factory sales dip some 6% below retail sales. Factory production shows still greater swings as attempts are made to adjust factory in-process and finished inventories.

By the time the system reaches its new equilibrium, inventories, goods in transit, and all order levels will be 10% higher than before. This will require that the 1,270 extra orders (above those placed by retail customers) must be initiated by the retail stage; another net of 1,190 orders by the distributors and 900 by the factory scheduler. As a result, there will have been 3,360 more orders generated within the system than were received at the retail level. These internally generated orders equal more than 3 weeks' worth of actual business. Some of the excess orders go to raise the levels of the various pools of orders traversing the system. They also result in an actual increase of 2,000 units of finished product in the system (nearly 2 weeks' production), which is 90% in higher inventories and 10% in more goods in transit. These changes require production rates that average higher than retail sales while the system is adjusting to the new level of business activity. The policies defined for this particular system do not produce the extra quantities in an orderly manner but by a high production peak around the 20th week, which is counterbalanced by a negative production swing near the 35th week. Inventory and ordering policies have a major effect on system stability.

As with the ordering and production rates, the inventory fluctuations are progressively greater as the disturbance is amplified upward in the system. Table 15-2 lists these effects.

Table 15-2   Minimum Inventory Levels after 10% Sales Increase

Level	Inventory Falls from Initial Value (%)	Time after Input Retail Sales Increase (weeks)
Retail	4	7
Distributor	10	13
Factory	15	15

The system is seen to be oscillatory with negative production rebounds that fall below the retail sales rate, and these in turn are followed by slight positive excess rates of factory manufacturing orders and production. A year and a half is required for the disturbance to subside from the system. Successive peaks in factory output of 45% and 12% above the initial values occur at 21 and 59 weeks. The 38-week interval separating the peaks indicates approximately the "natural period" of the manufactur-

ing-distribution system. This indicates that the system would be highly sensitive to any disturbances which contain a periodic component in the vicinity of a 38-week duration. This interval is very nearly a year, and we should expect any annual seasonal changes at retail to be markedly amplified at the factory level.

The system shows enough tendency toward sustained oscillation that it should be highly selective in its reaction to random-noise inputs. Since random noise contains a broad band of component frequencies, the system can select and amplify those frequencies to which it is sensitive.

These conclusions, drawn from the step response, can be tested directly by imposing on the system either a periodic input or a noise input.

15.7.2 One-Year, Periodic Input. The response of a system to sinusoidal inputs of various frequencies is highly informative in showing system characteristics. We shall now examine only the response to a sinusoidal disturbance with a one-year period.[28] Such might represent an unexpected annual seasonal change in sales rate.

The equations and parameters of Section 15.5 apply, along with an input defined as follows:

$$RRR.KL = RRI + RCR.K \qquad \text{15-76, R}$$

$$RCR.K = 100 \text{ sine } 2\pi \frac{TIME.K}{52} \qquad \text{15-77, A}$$

RRR	Requisitions (orders) Received at Retail (units/week)
RRI	Retail Requisitions, Initial rate, constant (units/week)
RCR	Requisition Change at Retail (units/week)
TIME	calendar TIME measured in weeks (automatically generated and available from the DYNAMO compiler)
sine	functional designation for a sinusoidal fluctuation (here of 52-week period)

[28] In Appendix I will be found the reaction of the system to periodic disturbances of other frequencies.

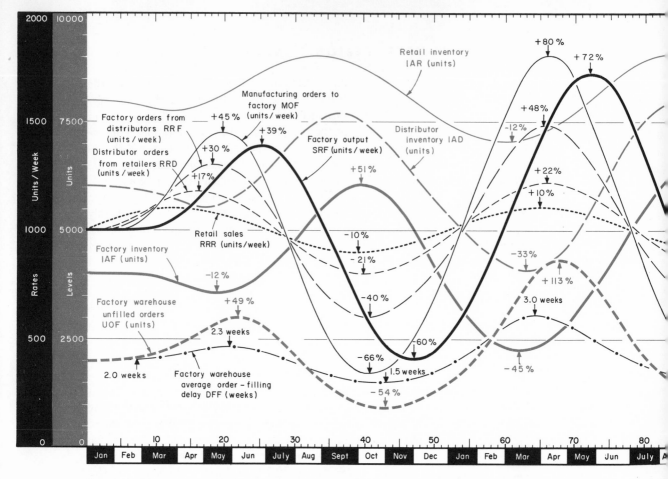

Figure 15-19 (Repeat of Figure 2-3) Response of production-distribution system to a 10% unexpected rise and fall in retail sales over a one-year period.

This gives a system that has been in constant steady-state conditions before the beginning of the computer run. At the beginning of the run a sinusoidal disturbance, with a one-year period and an amplitude of 100 units above and below the average, is generated and added to the input.

It should not be assumed that plans and policies recognizing a seasonal business could suppress the disturbances seen in Figure 15-19. Since there has been no past seasonal history, plans in advance for a seasonal disturbance are not plausible. For a known seasonal business activity, the exact amount of seasonality is not known in advance. The figure might then be interpreted as the response of the system to errors between the predicted and the actual seasonal sales pattern.

The system response in the figure contains two components — the steady-state periodic fluctuation, and the initial transient caused as the system moves from the previous, constant, steady-state conditions into its new periodic mode. The first peaks in the ordering curves occur between 16 and 30 weeks and reflect a combination of transient and periodic conditions. They are different from the peaks between 60 and 75 weeks which repeat annually and no longer show the initial transient.

In Figure 15-18, approximately a year was required for a transient disturbance to subside. In a similar way, approximately a year elapses before the periodic response of the system becomes free of the starting transient conditions. The conditions in the minimuim part of the cycle around the 40th week are very simi-

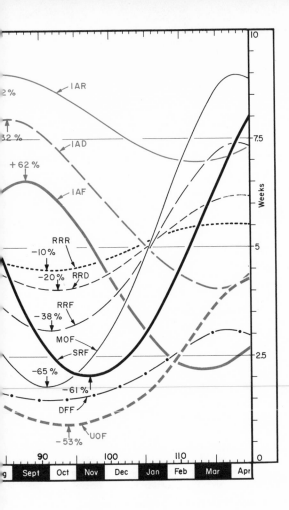

the system and cause the other ordering patterns in stages separated by delays. A system containing amplification can at some disturbing frequencies react in such a way that dependent responses actually appear to occur in advance of the independent driving disturbance.[29]

During the second year the maximum and minimum ordering curves depart from the average sales rate by the percentages given in Table 15-3.

Table 15-3   Maximum and Minimum Order Rates with 10% Retail Sales Seasonal

Location	Maximum (%)	Minimum (%)
Retail	+10	−10
Distributor	+22	−20
Factory Warehouse	+48	−38
Factory Manufacturing	+80	−65

This table shows an amplification of approximately two times at each ordering stage at this annual input period. The nonlinearities of the system (which are slight) cause the upper and lower swings to be unsymmetrical.

It should be observed that the swings in factory inventory are substantial, varying from 45% below to 62% above normal. Inventories are high when orders are low. This is sufficient to affect appreciably the average ability to fill orders at the factory and causes the average order-filling delay to vary from 1.5 to 3.0 weeks. This change in the "length" of the supply pipeline that couples the distributor to the factory introduces another amplifying term which was not readily apparent in the preceding figure. This changing delay creates even more change in factory orders than would be explained by the 22% increase in distributor orders. Factory unfilled orders rise to 113% above the normal and fall to 53% below.

15.7.3 Random Fluctuation in Retail Sales. From the preceding step-input and seasonal responses, it is apparent that this system has in-

[29] See Appendix I.

lar to those around the 90th week. We can therefore assume that from the 40th week onward the curves represent very closely the repeating periodic pattern.

The primary input to the system is the independently generated retail-sales pattern. The initial rises in the ordering curves occur in time succession peaking at the 13th week for retail sales, the 16th week for distributor sales, the 19th week for factory sales, and the 20th week for factory manufacturing orders. This sequence of peaks is similar to that found in Figure 15-18 for a step input.

During the second year, however, it should be noticed that the peaks and the valleys occur almost simultaneously in all of the ordering curves. This may at first seem surprising, since retail sales are clearly the independent input to

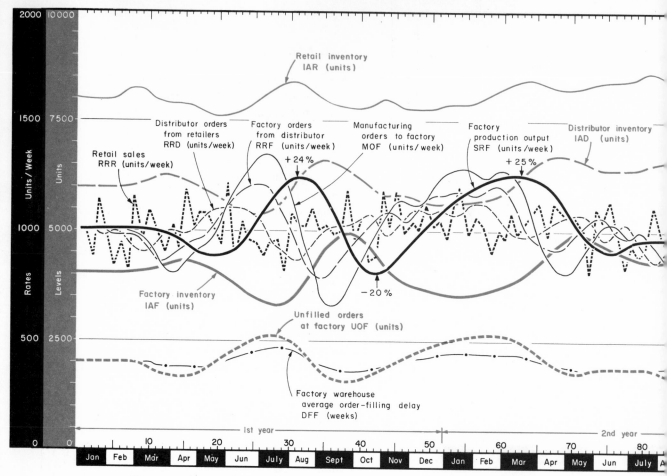

Figure 15-20 (Repeat of Figure 2-4) Effect of random deviations at retail sales.

ternal tendencies toward fluctuation. As shown by the response to an annual sinusoidal sales input, the system tends to amplify input disturbances of certain frequencies.

The preceding "pure" types of system test inputs are not the kinds that are encountered in real situations. Under natural circumstances all decisions in a system will be perturbed by disturbances arising from weather, vacations, length of work week due to holidays, items in the news, and so forth. To ignore these disturbances in studying system behavior would be unrealistic. Yet we usually lack any basis on which to generate these individual, minor, disturbing effects. The multiplicity of disturbing factors can be approximated by introducing a noise (random fluctuating) component. It then becomes possible to see how the system will

react to and modify such sources of disturbance. For the present, it will be sufficient to note how this example system will act if retail sales are constant during each individual week, but successive weeks are selected to have a random variation around the average level of sales.[30] This can be produced by equations as follows:

RRR.KL = RRI + RCR.K	15-78, R
RCR.K = SAMPLE (NSN.K, 1)	15-79, A
NSN.K = NORMRN (0, 100)	15-80, A

RRR	Requisitions (orders) Received at Retail (units/week)
RRI	Retail Requisitions, Initial rate, constant (units/week)

[30] A more complete discussion of generating noise for these purposes will be given in Appendix F.

178

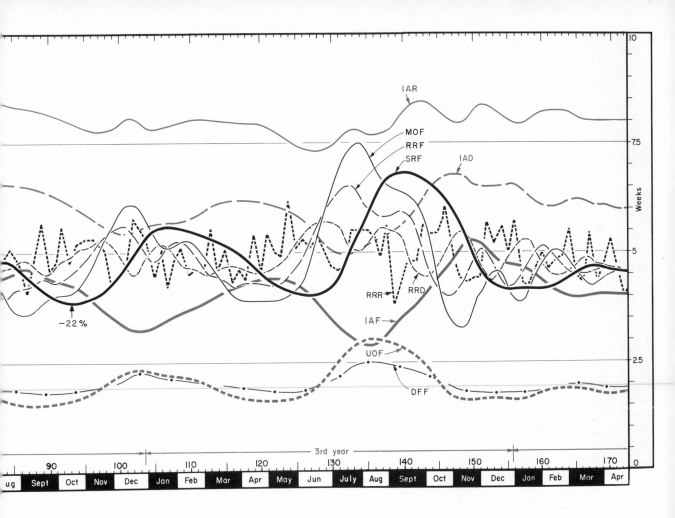

RCR	Requisition Change at Retail (units/week)
SAMPLE	this is a functional notation indicating that the variable NSN is to be sampled at the indicated intervals (1 week) and that these sampled values are to be held and used during the intervening time.
NSN	Normal Source of Noise, a sequence of random numbers having the dimensions of units per week. A new value will be generated at each DT solution interval.
NORMRN	this is a functional notation indicating a pseudorandom (that is, numerically generated) NORMal Random-Noise source measured in units/week. The parentheses indicate the mean value (0) and the normal deviation (100 units per week).

Except for the above, the equations and parameters of Section 15.5 apply.

In Figure 15-20 we can see how the production-distribution system modifies the independent retail input as the retail sales are eventually converted into factory production. The high-frequency week-by-week fluctuation has been suppressed until it is no longer evident in the conditions at the factory. However, a long-period fluctuating condition at the factory has arisen. It clearly is caused by the retail randomness but shows no obvious correspondence to conditions at retail.

As will be discussed in Appendix F, a random-noise sequence contains components of a wide range of different frequencies. A week-by-week random retail sales pattern will therefore necessarily include some monthly, quarterly, annual, and all other intervals of periodicity. If the system to which these are applied is selective and tends to amplify certain

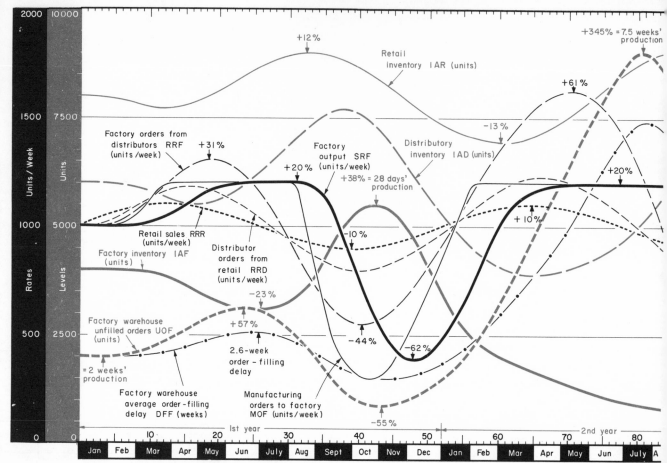

Figure 15-21 (Repeat of Figure 2-5) Effect of fluctuating retail sales on factory with manufacturing capacity limited to 20% above average sales.

frequencies, these particular frequencies will emerge as strongly predominant. This is apparent in Figure 15-20, where the most predominant fluctuations give the general impression of creating peaks that are separated by intervals of 30 weeks to 50 weeks. This is in the range of the 38-week natural frequency that was evident between successive peaks in Figure 15-18.

This tendency of the system to accentuate certain frequencies of disturbance is determined by the nature of the system structure, the delays, and the policies that are being followed. We shall later re-examine this situation to see how a change in management policies might make the system less sensitive to random disturbances.

Orders to adjust inventories and supply-line content are based at each level on average sales.

This average is here obtained by exponential smoothing[31] with an 8-week smoothing time constant. Smoothed sales at retail are not shown in the figure but from the tabular print-out of results were found to lie usually within 2 or 3% of the steady-state initial value, with very infrequent values departing up to 5%. Both the smoothing and the delays of the system tend to remove the high-frequency weekly periodicity from the input but leave the lower-frequency components to which the system is most sensitive.

15.7.4 Limited Factory Production Capacity. The equations for the production-distribution system as given in Section 15.5 contain multiplications and divisions of variables in several

[31] See Appendix E.

180

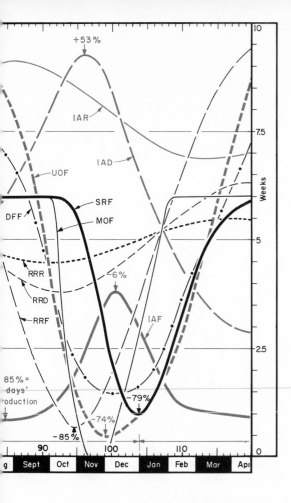

Figure labels: +53%, IAR, IAD, UOF, SRF, DFF, MOF, RRR, RRD, RRF, IAF, -6%, -79%, -74%, -85%, 85% = 7 days' production, Weeks, 10, 7.5, 5, 2.5, 0, 90, 100, 110, g Sept Oct Nov Dec Jan Feb Mar Apr

tion could permit a variable length of work week and number of working shifts, recognize the declining manpower efficiency as plant and equipment are overloaded, and take account of material shortages and other factors affecting actual output.

Changing the following constant in Equation 15-46 will enable us to see how a factory limit that is 20% above average retail consumption might affect conditions at the factory:

ALF = **1,200** units/week, constAnt specifying manufacturing capacity Limit at Factory

For Figure 15-21, a 10% annual periodic fluctuation was used, as defined by Equations 15-76 and 15-77. This input will cause retail sales to rise and fall between 900 units and 1,100 units per week. The upper manufacturing capacity limit at the factory is 1,200 units per week. The factory manufacturing limit is always at least 100 units per week above retail sales. Even so, amplification in the distribution system brings the factory capacity limit into operation.

The factory is unable to meet the demands for adjusting inventories and supply-line contents. A backlog of unfilled orders thereupon develops at the factory. This results in an increasing delay in the filling of orders and in causing the distributors to order still further ahead of their needs because of the growing delay.[33] During the first several months of limited factory production, this effect is regenerative (more orders cause a larger backlog and more delay and more ordering ahead) causing a large backlog of orders and resulting in a sustained full production rate for half of the first year.

Results here look much different from Figure 15-19. A new element of reality in the system has brought into play a number of new effects that were already present but hitherto of minor

places and, therefore, represent a nonlinear system whose degree of nonlinearity is not great.

Actual industrial systems contain many important and decisive nonlinear characteristics. One of these is the upper limit imposed on productive capability by available factory space and capital equipment. We might make a simplified first approximation to the effect of limited capital equipment by simply restricting permissible factory production. This could be done by imposing an upper limit on the rate at which manufacturing orders can be sent to the factory.[32] Eventually a more realistic representa-

[32] Note that this is an arbitrary upper limit that is not a function of the actual variables representing the availability of capital equipment, labor, and materials. No decisions are yet present to account for changing the productive capacity, so that the model in its present form is not suitable for studying growth until the managerial decisions relating to expansion have been incorporated.

[33] This is caused by the supply pipeline term in the ordering equations (for example Equation 15-9). Equation 15-10 contains the term (RSR.K)(DFD.K), where DFD.K is the distributor delay in filling orders.

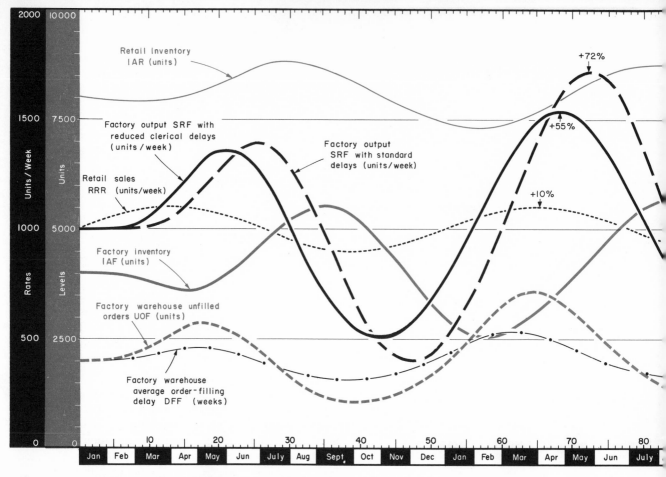

Figure 15-22 (Repeat of Figure 2-6) Effect of reducing clerical delays.

importance in the equations of the system. Already mentioned is the ordering ahead created by lagging delivery (this introduces a third amplification factor, in addition to the inventory build-up and the supply-line content proportional to sales volume). Equations 15-42 and 15-39 define the ability to ship goods in terms of actual inventory and unfilled orders. The increasing delay occurring with falling inventory accounts for the fact that in this illustration the inventory does not reach zero, even in the presence of a large backlog of unfilled orders. This is the practical counterpart of a minimum material handling time in the warehouse, arising because shipments cannot be assembled and sent from an empty stock.

Comparison of Figures 15-19 and 15-21 shows that the factory limit has caused incom-

ing orders at the factory to fluctuate over a wider range than before.[34] Distributor inventories rise sharply when the factory becomes able to meet orders. At that time the factory backlog of unfilled orders is converted to goods and transferred to distributor inventory. This causes a further suppression of orders from the distributor to the factory as the distributor attempts to bring his inventory excess into line with the slight (10%) sales slump that he is experiencing at the same time.

15.7.5 Reduction in Clerical Delays. One of the major objectives of industrial dynamics research is to evaluate changes that might be

[34] It does not follow that the managerial solution to this difficulty is to have excess manufacturing capacity, but rather that we alter policies elsewhere in the system to suppress the undesirable interactions caused by production limitations.

182

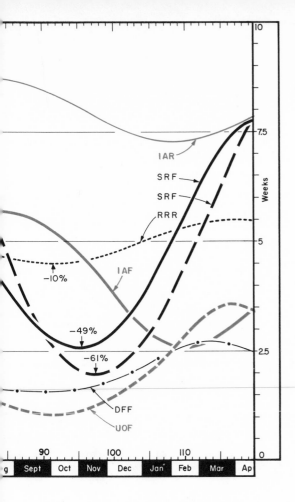

Table 15-4 Clerical Order Delays		
Order-Processing Delay	Reduced Value (weeks)	Previous Value (weeks)
DCR	1.0	3.0
DCD	0.66	2.0
DCF	0.33	1.0

was six times as great as the fluctuation at retail and now is five times after the reduction of clerical delays. At first this small effect might seem surprising, but closer consideration of the system as a whole shows that the clerical processing delays are a very small part of the total system structure. Furthermore, the clerical processing delays are responsible for only a minor part of the total order amplification process. Amplification is being created by inventory accumulation, by changes in supply pipeline length, and by changes in flow rate through the supply pipeline. Clerical delays (if constant) contribute to only a fraction of the latter.

15.7.6 Removal of the Distributor Sector. Some industries have more than the three levels of distribution represented in this model. Others are moving away from the three-level distribution to a two-level system where retail outlets order directly from the factory. Removal of one distribution level removes the amplification caused by inventory accumulation at that point. It also removes one level of pipeline amplification if deliveries can be made as promptly from the factory to the retailer as they were from the distributor.

Some changes are needed in the equations of Section 15.5 to remove the distributor sector. The following equation substitutes a supply pipeline between the retailer and the factory for the one between retailer and distributor and replaces Equation 15-10:

LDR.K = (RSR.K)(DCR + DMR + DFF.K + DTR)

15-81, A

LDR   pipeLine orders Desired (necessary) to supply Retail (units)

---

made in an industrial organization. A change frequently proposed is to substitute machine processes for manual clerical operations in order to speed up the flow of information. To indicate in simplified form how such a step might be evaluated, we could in this elementary distribution system determine the results of reducing the delays in the placing of orders.

The same test input as described by Equations 15-76 and 15-77 will be used to generate an annual 10% periodic retail-sales fluctuation. In Table 15-4 the clerical order-processing delays at retail, distributor, and factory levels are changed from those on page 171.

Figure 15-22 shows that only a small reduction has occurred in the amplification existing between the retail sales and factory production. Previously the factory production fluctuation

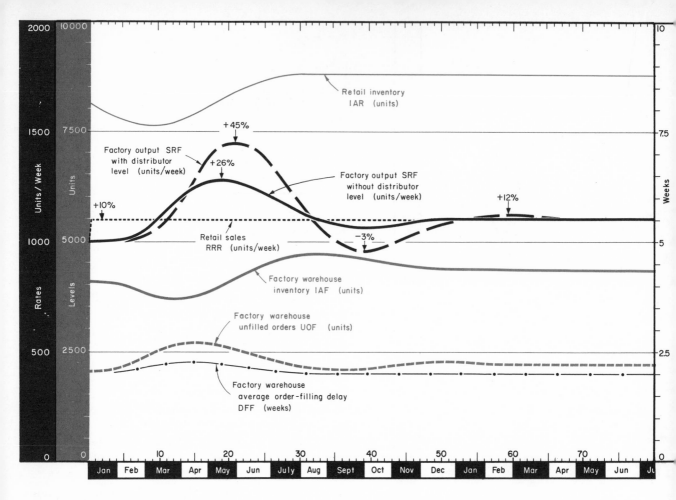

RSR	Requisitions Smoothed at Retail (average sales) (units/week)
DCR	Delay in Clerical order processing at Retail (weeks)
DMR	Delay in order Mailing from Retail (weeks)
DFF	Delay (variable) in Filling orders at Factory (weeks)
DTR	Delay in Transportation of goods to Retail (weeks)

The following equation gives the new content of the supply pipeline and replaces Equation 15-11:

$$\text{LAR.K} = \text{CPR.K} + \text{PMR.K} + \text{UOF.K} + \text{MTR.K}$$

15-82, A

LAR	pipeLine orders Actually in transit to Retail (units)
CPR	Clerical in-Process orders at Retail (units)
PMR	Purchase orders in Mail from Retail (units)

UOF	Unfilled Orders at the Factory (units)
MTR	Material in Transit to Retail (units)

Materials flowing to retail must be coupled to the output of the factory warehouse by replacing Equations 15-17 and 15-18:

$$\text{MTR.K} = \text{MTR.J} + (\text{DT})(\text{SSF.JK} - \text{SRR.JK}) \qquad \text{15-83, L}$$

$$\text{SRR.KL} = \text{DELAY3}(\text{SSF.JK, DTR}) \qquad \text{15-84, R}$$

MTR	Material in Transit to Retail (units)
SSF	Shipments Sent from Factory warehouse (units/week)
SRR	Shipments Received at Retail inventory (units/week)
DELAY3	a functional notation defining the output of a third-order delay in terms of the input rate and the average delay
DTR	Delay in Transportation of goods to Retail (weeks)

← **Figure 15-23** (Repeat of Figure 2-7) Distributor level eliminated.

tor is to be omitted (Equations 15-16, 15-19 through 15-33, 15-35, and 15-36).

These changes in the equations illustrate the process of making a change in the organizational structure of a model system.

The system without a distributor level will be tested with a 10% step in retail sales as described by Equations 15-74 and 15-75.

The results of removing the distributor sector are as expected. Figure 15-23 compares the results with those when the distributor was present and shows that fluctuation at the factory is reduced by omission of the amplification created at the distributor level. The results would not be so significant if the length of the supply pipeline between retailer and factory were longer than taken here, which is the same length as had previously existed between retailer and distributor. The model contains none of the functions of a distributor (such as sales, service, and training assistance to retailers) other than ordering and stocking goods. The results deal with only one facet of the distributor in the system, and the interpretation of the figure must be correspondingly qualified.

The result of a 10% annual sinusoidal input (not illustrated) into the system without a distributor level compares in factory production fluctuation with Figure 15-19 having the distributor level as given in Table 15-5.

The orders into the factory must come from the retail level by changing Equations 15-15 and 15-34 as follows:

$$PMR.K = PMR.J + (DT)(PSR.JK - RRF.JK) \qquad \text{15-85, L}$$

$$RRF.KL = DELAY3(PSR.JK, DMR) \qquad \text{15-86, R}$$

PMR	Purchase orders in Mail from Retail (units)
PSR	Purchase orders Sent from Retail (units/week)
RRF	Requisitions (orders) Received at Factory (units/week)
DELAY3	a functional notation defining the output of a third-order delay in terms of the input rate and the average delay
DMR	Delay in Mail from Retail to factory (weeks)

In addition to these changes in the model equations, the description of the distributor sec-

Table 15-5   Factory Production Fluctuation

	Maximum (%)	Minimum (%)
Without Distributor	+36	−33
With Distributor	+72	−61

These figures are the rise and fall of factory production above and below the average value, as they occur in the second and succeeding years after a sinusoidal input is suddenly applied following steady-state conditions.

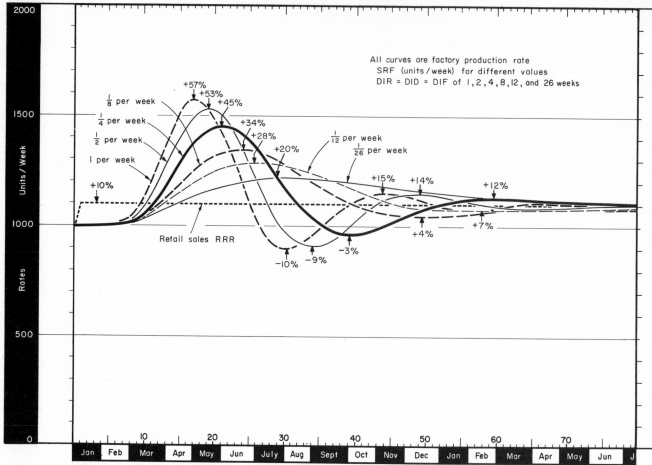

Figure 15-24 (Repeat of Figure 2-8) Changing the time to make inventory and in-process order corrections. (Fractions refer to amount of remaining imbalance corrected in following week.)

15.7.7 Rapidity of Inventory Adjustment.

In the ordering policies as described by Equations 15-9, 15-27, and 15-45, the constants DIR, DID, and DIF determine the rapidity with which orders are placed to adjust inventories and pipeline content. These constants determine the fraction of the inventory and pipeline discrepancies that are introduced into the ordering rate. The larger the constants, the more gradual the adjustment, the lower will be the peak rate of adjustment, and the longer will be the time over which the adjustment will be made.

The effect of different values of these adjustment constants is shown in Figure 15-24. Each curve represents the factory production response from a different computer simulation run. All use the 10% step input at retail sales

as given by Equations 15-74 and 15-75. In any one run the three constants at the three distribution levels have the same value. In the different runs these values are 1 week, 2 weeks, 4 weeks, 8 weeks, 12 weeks, and 26 weeks. A value of 4 weeks has been used in the previous figures.[35]

As the inventory adjustment policy is lengthened, the system changes from a highly oscillatory response with a short period of fluctuation toward a more stable mode of operation and a longer period between peaks. The inventory adjustment rapidity is one of the sensitive parameters in determining system behavior.

---

[35] The 4-week value is probably too short for a realistic representation of the typical distribution system. It is unlikely that changes in average sales are detected and acted upon as quickly as this.

CHAPTER 16

Advertising in the
System Model of Chapter 2

In Chapter 2, the production-distribution system, which has been de-
tailed in Chapter 15, was extended to include one aspect of the many
interactions between advertising and the market. This chapter adds an
advertising-market sector, which is based on consumer deferrability of
purchase and how deferrability might be influenced by advertising. The
implications go much further than the one facet of a market here treated.
Our economy has numerous pools — capital equipment, housing, savings,
management and technical personnel, and research results — where the
"inventory" is very high compared with inflow and outflow rates and
where these flow rates can be raised or lowered for substantial periods of
time without creating apparent serious imbalance in the size of the "in-
ventory." These fluctuating flow rates make possible the long-term cyclic
rise and fall of industrial activity. "Prospective customers" are taken in
this example as one such "inventory," whose variation gives rise to a
long-period system instability.

$\mathbf{A}$N EFFECTIVE approach to a company problem will often require that we first understand the industry of which it is a part. The problems of the company may be primarily the problems of the industry as a whole. The dynamics of the entire industry can often be treated without considering the intercompany relationships that determine market share. In doing so, we must deal, as an industry aggregate characteristic, with those practices arising from competitive pressures that may influence the dynamic character of the entire industry.

This chapter deals with a market-advertising interaction of a total industry such as household appliances. The situation represents advertising practices that arise primarily from intercompany competitive pressures but still have an important effect in creating an undesirable behavior of the total industry structure.

We shall incorporate a pool of "prospective customers," as shown in Figure 16-1. A prospective customer is here defined as one who is going to purchase, who is at least vaguely aware of his need for the product, and whose actual time of purchase will be somewhat affected by sales effort. (Even though the selling effort may be intended to affect market share and not to shift sales from one time to another.) It seems clear that this class of customer represents a significant fraction of the purchasers of consumer durable goods. Here is the housewife who is planning to remodel a kitchen, as well as the builder who knows as soon as he starts planning a house that he will need to purchase kitchen equipment for it.

On the average, the prospective customer will exist as such for a certain period of time before actually buying. This period may be very dif-

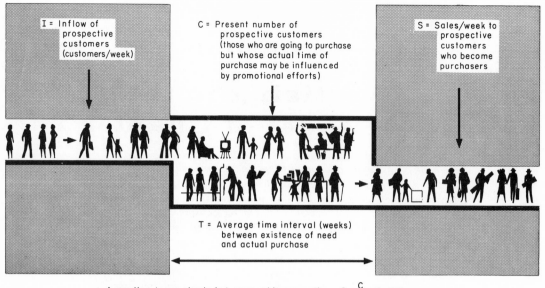

According to one simple but reasonable assumption: $S = \dfrac{C}{T}$ at all times

In the steady state (uniform, stable operation): $S = I$ and $C = I \times T = S \times T$

During changes from one level of sales to another: S does not equal I
C does not equal $I \times T$ or $S \times T$

Figure 16-1 Constant inflow of customers.

ferent for different products — for refrigerators it may be 30 weeks, for raincoats 30 days. By definition, they are customers who will eventually buy, independent of industry selling effort. If advertising has *any* effect on these prospective customers, it must take the form of influencing the average length of time before they make a purchase. Note that we are considering only one possible effect of advertising. In this part of the market, advertising can affect the time of purchase but not total long-term sales. Long-term sales are controlled entirely by the inflow of prospective customers, here assumed to be the independently specified test input. In a more complete model of the dynamic aspects of advertising and marketing, many other factors would need to be included. The purpose now is not to develop a full model of a market but to show the procedure by which a model can be extended.

Before developing a dynamic model we shall examine graphically some of the behavior relationships that might be expected. Figure 16-2 shows the effect of a sudden change in advertising expenditure rate on the prospective customer pool. The curves progress in order from top to the bottom:

- The curve at the top shows the increase in advertising expenditure rate.
- The second curve shows a build-up in what might be called advertising "pressure." By this is simply meant the persuasiveness of the advertising to the prospective customer. It builds up more gradually than the actual increase in advertising because of the time that is required for the advertising campaign to achieve full effect in customer awareness.
- The third curve shows the decrease in the average waiting time, a change that results from increased advertising pressure. The effect of increased advertising is here to induce the average customer to purchase a little sooner than he would have otherwise.

- The fourth curve shows the effect on the number of prospective customers. Reducing the average waiting time before purchase causes the total number of prospective customers in the pool to decrease, because in the steady state the number in the pool is the product of the inflow rate (constant) multiplied by the average waiting interval (decreasing).

- In the last curve, at the bottom of Figure 16-2, is the actual sales rate. It rises as the waiting time is reduced and the pool of prospective customers is partially depleted. Sales then fall again to the initial rate, which in the long run is determined by the constant inflow rate to the prospective customer pool.

The shaded area under the last curve therefore represents sales "borrowed" from the future. The short-term increase in sales comes only from depleting the pool of prospective purchasers.

In Figure 16-3 is an attempt to trace more deeply the interactions of market, advertising, and distribution.

- In the first curve we assume that the inflow of prospective customers to the prospective customer pool has suddenly increased because the need for or popularity of the product has increased.

- As the inflow continues, the second curve shows that the number of buyer prospects in the pool will gradually rise due to the increased inflow of prospective customers.

- Actual sales will correspondingly begin to increase, as shown by the third curve.

- When retail sales increase, it is customary, as discussed in earlier chapters, for higher inven-

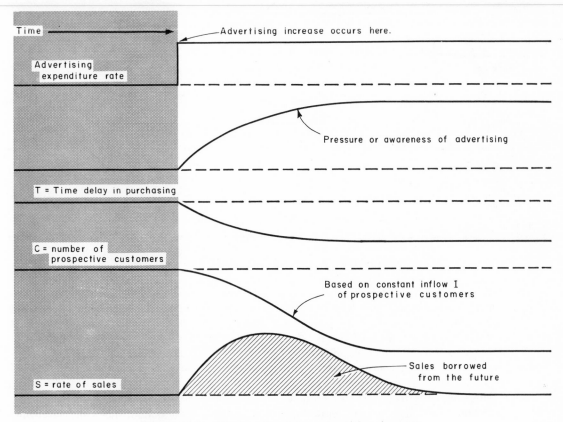

Figure 16-2 Effects of sudden advertising increase.

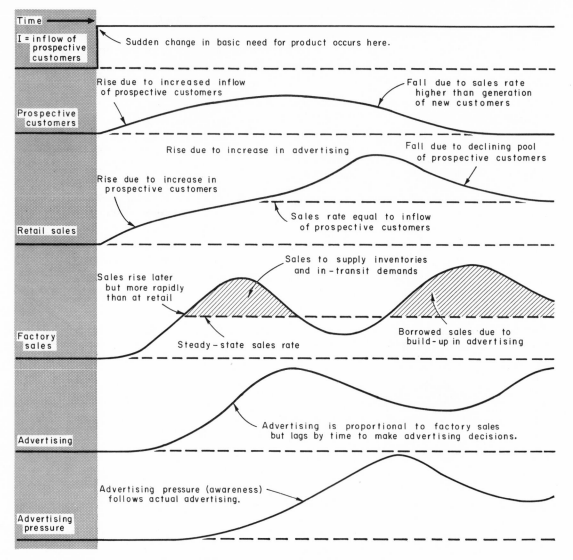

Time

I = inflow of prospective customers

Sudden change in basic need for product occurs here.

Prospective customers

Rise due to increased inflow of prospective customers

Fall due to sales rate higher than generation of new customers

Retail sales

Rise due to increase in advertising

Fall due to declining pool of prospective customers

Rise due to increase in prospective customers

Sales rate equal to inflow of prospective customers

Factory sales

Sales to supply inventories and in-transit demands

Sales rise later but more rapidly than at retail

Steady-state sales rate

Borrowed sales due to build-up in advertising

Advertising

Advertising is proportional to factory sales but lags by time to make advertising decisions.

Advertising pressure

Advertising pressure (awareness) follows actual advertising.

Figure 16-3 Permanent increase in demand.

tory levels to be carried at the different levels of distribution. Also, for a higher level of business there must be more orders and goods in transit in the supply pipelines of the system. Consequently, as in the fourth curve, the factory sales start rising. They are later than retail sales because of the time delays in the placing of orders. However, once they start to rise, they then rise more rapidly and to a higher peak than the actual retail sales because of the inventory and pipeline amplification.

- In many companies the advertising level is effectively determined as a fraction of the sales and production forecast. In the fifth curve is shown the rise in advertising expenditure that might follow the factory sales in such a company. This advertising expenditure is delayed by the length of time necessary to make advertising decisions and to act on them.

- The curve at the bottom shows the advertising pressure or awareness on the part of the prospective consumers.

Note that as advertising pressure begins to rise, it further boosts the rising retail sales curve (third curve). This new retail rise occurs at a time when the factory sales have already satisfied the initial inventory demands of the system and have fallen to their new steady-state level corresponding to the new rate of inflow of prospective customers. The peak in the middle of the retail sales curve then produces another peak at the factory (fourth curve).

This kind of descriptive and pictorial analysis is entirely unsatisfactory except to convey an impression of some of the qualitative interactions that *might* occur. In a system of this kind, with retail sales affecting factory sales, which in turn affect advertising, which in turn affect retail sales, it is necessary to treat the entire system as a closed-loop, information-feedback system to learn its behavior. The overall behavior is not divulged by any analysis of one piece at a time. To make this system study, the additions to the model of Chapter 15 will now be developed to incorporate the kind of behavior that has just been described. Figure 16-4 shows the relationship of advertising and market to the factory and distribution chain.

16.1　Equations for the Advertising Sector

There is substantial evidence that advertising budgets are strongly influenced by the level of sales. As sales increase, advertising expenditures increase, and as sales decline, advertising budgets are cut back. We shall examine here some implications of an advertising policy that is proportional to sales.

Time delays existing in the advertising flow will be included. These will comprise the delay in reaching a decision on advertising expenditure, the delays in advertising agencies and media, and the delays in building up consumer awareness of the existence of an advertising campaign.

At the consumer level, the principal factor will be the influence of advertising on the deferrability of purchase. Only the prospective pur-

chasers are considered whose appearance is independent of advertising. Prospective purchasers are those who eventually do buy and exist for a period of time during which they are susceptible to various influences that can affect the time of purchase. Increased advertising, if it has any effect on this class of prospective purchaser, would almost certainly act to reduce the average waiting time. It will bring information to people earlier so that they can make a decision on purchasing. It may influence some to believe that the time is more propitious than later. It may simply remind others to make a purchase that they have been delaying. If there were no advertising, this class of prospective purchaser would in due time make his purchase. On the other hand, we can assume that

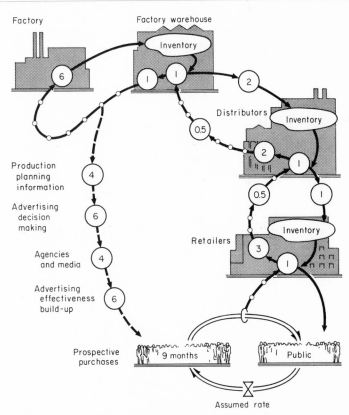

Figure 16-4　(Repeat of Figure 2-9) Advertising and consumer market.

the largest conceivable advertising effort could not reduce the average delay of these prospective purchasers below some minimum time. Therefore the effect of advertising will be to influence the delay between the time that a prospective purchaser becomes susceptible to influence and the time at which purchase is actually made.

The system here under consideration deals with only a part of the market — the purchasers whose *existence* is independent of advertising. The effect of advertising in *creating* purchasers is not being considered here. The deferrability of purchase, as well as the influence on it of many other factors beside advertising, is probably one of the most important determinants of durable goods markets.

A "pool" of prospective purchasers will exist, and the outflow from this into actual buying will be affected by advertising, which in turn is affected by the level of sales at the factory.

This is potentially a "positive feedback" situation in which more sales create more advertising, which in turn creates more sales (but subject to the availability of customers in the "pool"). Our objective is to examine how this phenomenon may interact with the amplification caused by inventory accumulation and pipeline filling that occur in the distribution system.

The first equation to be written will determine the policy for authorizing advertising expenditures. It incorporates advertising expenditure proportional to anticipated gross sales (here taken to equal manufacturing plans). The following equation represents the decision:

$$VDF.KL = (MAF.K)(UPF)(AVS) \qquad 16\text{-}1,R$$

VDF	adVertising Decision at Factory (dollars/week of advertising authorization)
MAF	Manufacturing Average rate at Factory (units/week)
UPF	Unit Price of goods at Factory (dollars/unit)
AVS	constAnt, adVertising as fraction of Sales (dimensionless)

This is an "impending" decision because it does not yet include the delays in actually reaching the decision. The equation gives average manufacturing rate multiplied by the unit price of the product and by the percentage of the gross sales value that is to be devoted to advertising. This gives dollars per week of advertising authorization.

The average manufacturing level MAF will here be given by the simple exponential averaging equation:[1]

$$MAF.K = MAF.J + \frac{DT}{DMS}(MOF.JK - MAF.J) \qquad 16\text{-}2,\ L$$

MAF	Manufacturing Average rate at Factory (units/week)
DT	Delta Time, the equation solution interval
DMS	Delay in Manufacturing rate Smoothing (weeks)
MOF	Manufacturing Orders into Factory (units/week)[2]

This equation will give a short-term average of recent production if the value of the constant DMS is a small number of weeks.

The flow diagram for the advertising sector is shown in Figure 16-5.

After the decision on advertising rate VDF, the delay in reaching the decision will be inserted:

$$VCF.KL = DELAY3(VDF.JK, DVF) \qquad 16\text{-}3,\ R$$

VCF	adVertising Committed at Factory (dollars/week of advertising orders)
DELAY3	functional designation of a third-order exponential delay
VDF	adVertising Decision at Factory (dollars/week of advertising authorization)
DVF	Delay in adVertising decision at Factory (weeks)

Here VCF is the flow of advertising commitments from the factory to the advertising agencies.

[1] Similar to Equation 15-8.
[2] From Equation 15-51.

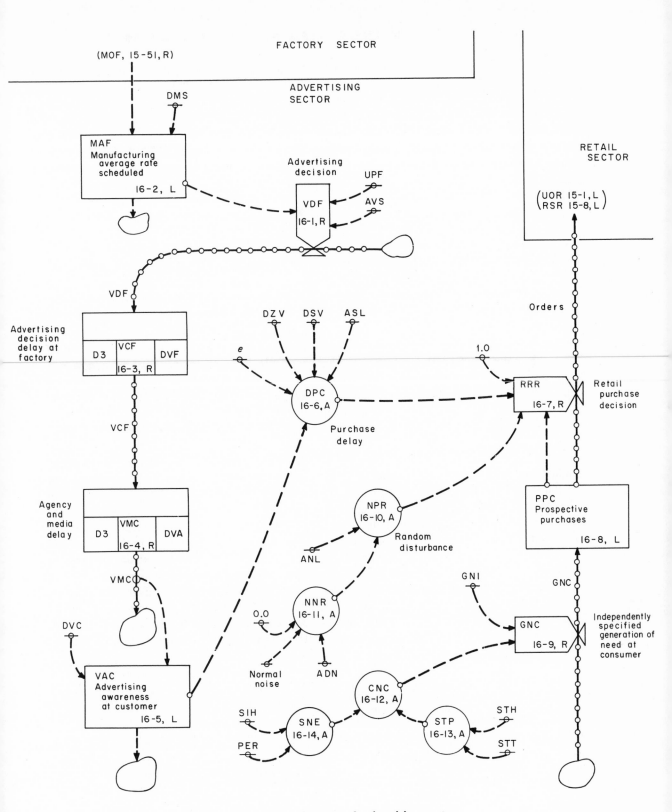

Figure 16-5 Flow diagram of advertising sector.

The flow of advertising orders from the factory should, in turn, be put through another delay representing the time necessary for advertising agencies and the advertising media to carry out their work. The resulting advertising actually appearing before the public is given by variable VMC:

$$VMC.KL = DELAY3(VCF.JK, DVA) \qquad \text{16-4, R}$$

VMC adVertising presented by Media to Consumer (measured in dollars/week of completed advertising)[3]

DELAY3 functional designation of a third-order exponential delay

VCF adVertising Committed at Factory (dollars/week of advertising orders)

DVA Delay at adVertising Agencies and media (weeks)

At this point we face the question of the time response of consumers to an advertising campaign. The first issue of a series of magazine advertisements or the first appearance of a television program certainly does not create as much consumer awareness as a continued series of advertisements. There is, therefore, a gradual build-up of awareness. After a time, we can expect that the full effect of a particular advertising level has been reached in creating awareness in the pool of prospective customers. In this model we are particularly interested in the reactions that may follow *changes* in advertising rate.

There is some evidence,[4] as well as a reasonable intuitive basis, for believing that consumer awareness to an advertising campaign builds up relatively rapidly at first and then begins to level off as the saturation of awareness is ap-

proached. This is the general shape of a first-order delay or smoothing equation, which can be represented as follows:

$$VAC.K = VAC.J + \left(\frac{DT}{DVC}\right)(VMC.JK - VAC.J)$$

$$\text{16-5, L}$$

VAC adVertising Awareness at Consumer (measured in dollars/week of advertising)

DT Delta Time, the equation solution interval

DVC Delay in adVertising awareness buildup at Consumer (weeks)

VMC adVertising presented by Media to Consumer (measured in dollars/week of completed advertising)

Equation 16-5 states the response of consumer awareness to the advertising rate VMC. If, for example, VMC has been zero and should suddenly increase to the level of a new and sustained advertising campaign, the consumer awareness would begin to build up rapidly and then level off.

Next is the problem of estimating the qualitative relationship between advertising and the delay of a prospective purchaser from the time of susceptibility to advertising to the time of purchase. A reasonable relationship is shown in Figure 16-6. The delay in the presence of zero advertising is shown at the level DZV. At the lower limit the average delay in the presence of a saturation intensity of advertising is shown as DSV. It is reasonable to assume that the first small increment in weekly advertising expenditure will create more effect than following increments of the same amount added to the first. In other words, the steepest slope and the greatest dependence of purchasing delay on advertising will occur at low levels of expenditure.[5] As expenditure rises, the curve should flatten out and approach the saturation level DSV. Be-

---

[3] This avoids the fact that dollars are not a suitable measure of advertising rate unless the mix of advertising media is constant. Were a changing emphasis between different advertising media a part of our problem, we should need to convert dollars to hours of television time and inches of newspaper advertisements and to treat separately the proper conversion coefficients and time constants.

[4] See Vidale and Wolfe, Reference 12.

[5] There are those who would wish to use a differently shaped curve that shows little effect from advertising until some threshold is reached, on the basis that a small amount is merely lost in the total barrage of advertising directed at the consumer.

yond this general qualitative behavior we can say little about the probable effects of advertising on our group of prospective purchasers. It seems reasonable, therefore, to take a simple curve that fits the preceding qualitative description. One such curve is an exponential function that describes purchasing delay according to the following equation:

$$DPC.K = DSV + (DZV - DSV)e^{-VAC.K/ASL} \qquad \text{16-6, A}$$

DPC Delay in Purchasing at Consumer (weeks)
DZV Delay for Zero adVertising (weeks)
DSV Delay for Saturated adVertising (weeks)
e natural logarithm base, 2.7183 . . .
VAC adVertising Awareness at Consumer (measured in dollars/week of advertising)
ASL constAnt, Saturation Limit index (dollars/week)

It will be necessary to specify where on the curve of Figure 16-6 we wish to have the system operate. At low levels of advertising, the slope of the curve is relatively steep, while at high levels a corresponding change in weekly expenditure rate will have little effect. An exponential curve possesses an interesting characteristic reference point, which is found by ex-

trapolating the initial slope until it intersects the limit toward which the curve approaches. The intersection of the initial slope with the saturation level DSV has been arbitrarily given the numerical value of 1.0 on the horizontal scale.

The position of "normal" advertising expenditure rates with respect to this arbitrary scale will be established by selecting a numerical value for the constant ASL. A "normal" advertising level would be caused to center around the point A if we select the constant ASL equal to the "normal" advertising rate. (The point A lies 37% of the distance from the level DSV toward DZV.) To investigate operation further to the right in the saturated region, a smaller value should be taken for the constant ASL. Correspondingly, a larger value of ASL would cause a given advertising expenditure rate to fall farther to the left on the curve, as at point B. By the selection of the constant ASL, we can control the point on the saturation curve about which the model is to operate. Here we probably have neither data nor a reliable intuitive feeling about the degree of saturation of a particular level of advertising. We can, however, investigate the way in which our system responds to changes in the operating point.

The delay given by Equation 16-6 will be used to control the fraction of the prospective purchasers who buy each week. This implies an equation as follows:

$$RRR.KL = \frac{PPC.K}{DPC.K}(1 + NPR.K) \qquad \text{16-7, R}$$

RRR Requisitions (orders) Received at Retail (units/week)
PPC Prospective Purchases at Consumer sector (units of orders)
DPC Delay in Purchasing at Consumer sector (weeks)
NPR Noise in Purchasing rate at Retail (dimensionless multiplier with average value of zero)

As described by Equation 16-7, the rate at which retail purchases are made is controlled by the size of the prospective purchase pool

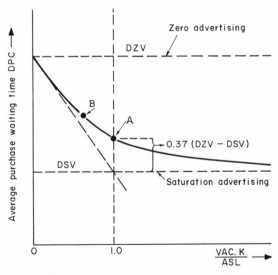

Figure 16-6 Delay in making prospective purchases as dependent on advertising awareness.

PPC and the average delay before purchasing DPC. The noise term NPR is to be used for introducing uncertainty fluctuations into the purchase rate. It will be defined later.

The prospective purchases in the pool PPC is given by a simple level equation based on the inflow and the outflow:

$$PPC.K = PPC.J + (DT)(GNC.JK - RRR.JK) \qquad \text{16-8, L}$$

PPC   Prospective Purchases at Consumer sector (units of orders)

GNC   Generation of Need at Consumer sector (units/week of prospective purchases)

RRR   Requisitions (orders) Received at Retail (units/week)

The inflow to the prospective purchase pool is the rate of generation of need for the product GNC. For the purposes of the present investigation, this will be taken as an arbitrarily specified input. It takes the place of the arbitrarily specified retail sales rate RRR, which was previously used for testing the production-distribution chain itself. By specifying the generation of need rather than arbitrarily specifying retail purchases, we have introduced a further refinement and a higher degree of reality in the system that we are studying. The generation of prospective purchases will therefore be described by the following equation:

$$GNC.KL = GNI + CNC.K \qquad \text{16-9, R}$$

GNC   Generation of Need at Consumer sector (units/week of prospective purchases)

GNI   Generation of Need, Initial rate (units/week)

CNC   Change in Need at Consumer sector (units/week)

The initial rate GNI is the past steady-state level of business activity and is a constant. Any variations on this level which we may wish to use for test purposes can be described by suitable variations in the change term CNC.

There now remain to be defined several variables which are not directly involved in the advertiser and purchase chain and which are independent inputs. One of these is the noise input used in Equation 16-7. Weekly sales ordinarily fluctuate owing to causes that we do not hope to represent in a system model. These will be approximated by a noise input that will vary successive weekly purchasing rates. This noise input will be represented by two equations. The first of these will sample a noise variable and will hold that sample available for the duration of, for example, 1 week. The second describes the noise source that is being sampled.

$$NPR.K = SAMPLE (NNR.K, ANL) \qquad \text{16-10, A}$$

NPR   Noise in Purchasing rate at Retail (dimensionless multiplier with average value of zero)

SAMPLE   a functional notation indicating that the variable is to be sampled and held until the next sampling time

NNR   Normal Noise at Retail (dimensionless)

ANL   constAnt, Noise sample Length (weeks)

The number to be sampled and held by Equation 16-10 is taken from a normal distribution of random noise as defined by the following equation:

$$NNR.K = NORMRN (0.0, ADN) \qquad \text{16-11, A}$$

NNR   Normal Noise at Retail (dimensionless)

NORMRN   functional notation indicating NORMal Random Noise

0.0   mean value of the noise

ADN   constAnt, standard Deviation of Noise (dimensionless)

For the test input CNC in Equation 16-9, we may wish to use one or more of several inputs. One of these might be a step input wherein the rate of generation of need takes on some new value that then continues. A second common type of test input would be a sine wave used to represent seasonal and other fluctuating circumstances. For convenience, these inputs will be combined in the following equation, where each input can be controlled by its own generating equation:

$$CNC.K = STP.K + SNE.K \qquad \text{16-12, A}$$

CNC   Change in Need at Consumer sector
       (units/week)
STP   STeP, test input   (units/week)
SNE   the SiNE variable used as a test input
       (units/week)

The step input to be used in Equation 16-12 can be defined in terms of the amplitude of the step and the time at which it occurs as given in the following equation:

$$STP.K = STEP(STH, STT) \qquad \text{16-13, A}$$

STP   STeP, test input   (units/week)
STEP   a functional notation indicating that a STEP change is to be generated
STH   STep Height, test input   (units/week)
STT   STep Time after beginning of the model run at which the step occurs   (weeks)

A sine-wave input for the test function can be provided by the following equation:

$$SNE.K = (SIH)SIN\left[\frac{2\pi(TIME.K)}{PER}\right] \qquad \text{16-14, A}$$

SNE   the SiNE variable used as a test input
       (units/week)
SIH   SIne Height, test input   (units/week)
SIN   a functional notation indicating a sine fluctuation
TIME   the independent time variable generated by DYNAMO   (weeks)
PER   sine PERiod, test input   (weeks)

This completes the *active* equations needed to add the advertising sector to the production-distribution model of Chapter 15.

16.2   Initial-Condition Equations

Equations are necessary for generating initial conditions for levels and some rates.

Under steady-state conditions the average level of manufacturing MAF will equal the steady-state rate of orders at the factory RRF:[6]

$$MAF = RRF \qquad \text{16-15, N}$$

MAF   Manufacturing Average rate at Factory
       (units/week)
RRF   Requisitions (orders) Received at Factory
       (units/week)

Since the rate of flow VDF is an input to a delay function, its initial rate must be defined by an initial-condition equation. This rate will be determined only by the average manufacturing rate MAF multiplied by unit price and by the fraction of gross receipts going to advertising:

$$VDF = (MAF)(UPF)(AVS) \qquad \text{16-16, N}$$

VDF   adVertising Decision at Factory
       (dollars/week of advertising authorization)
MAF   Manufacturing Average rate at Factory
       (units/week)
UPF   Unit Price of goods at Factory   (dollars/unit)
AVS   constAnt, adVertising as fraction of Sales   (dimensionless)

With the rate VDF given, the initial steady-state outflow rate VCF from the advertising decision delay is necessarily identical.[7] Likewise VCF, as the input to the agency delay, determines the output VMC.

An initial-condition equation is necessary for the level of advertising awareness VAC. In the steady state this will be equal to the constant level of advertising expenditure rate VMC:

$$VAC = VMC \qquad \text{16-17, N}$$

VAC   adVertising Awareness at Consumer   (measured in dollars/week of advertising)
VMC   adVertising presented by Media to Consumer (measured in dollars/week of completed advertising).

---

[6] As determined by Equation 15-68.
[7] The DYNAMO computer compiler, which is here assumed to be available, will automatically generate these subsidiary initial conditions arising from the outflow from delays and will make them available as necessary to other initial-condition calculations.

An operating point in steady state must be chosen on the curve of Figure 16-6. This will be accomplished by selecting a value for ASL; ASL can be defined by an initial-condition equation in terms of the value of the advertising awareness VAC. For initial studies of the system, we shall select a "normal" operating point at A which is obtained when the following equation holds:

$$ASL = VAC \qquad \text{16-18, N}$$

ASL constAnt, Saturation Limit index (dollars/week)
VAC adVertising Awareness at Consumer (measured in dollars/week of advertising)

In order to determine the steady-state level of prospective purchases PPC, it will be necessary to have a value for the steady-state purchasing delay. This value is determined by selecting the steady-state operating delay from Figure 16-6. This is best done by writing the steady-state equivalent of Equation 16-6 as follows:

$$DPC = DSV + (DZV - DSV)e^{-VAC/ASL} \qquad \text{16-19, N}$$

DPC Delay in Purchasing at Consumer level (weeks)
DZV Delay for Zero adVertising (weeks)
DSV Delay for Saturated adVertising (weeks)
e natural logarithm base, 2.7183 . . .
VAC adVertising Awareness at Consumer (measured in dollars/week of advertising)
ASL constAnt Saturation Limit index (dollars/week)

The steady-state level of prospective purchases PPC will be equal to the uniform inflow or outflow multiplied by the average delay experienced in the "pool":

$$PPC = (RRR)(DPC) \qquad \text{16-20, N}$$

PPC Prospective Purchases at Consumer level (units of orders)
RRR Requisitions (orders) Received at Retail (units/week)
DPC Delay in Purchasing at Consumer level (weeks)

Finally, it will be necessary to specify a steady-state initial condition for retail sales RRR for use in the retail sector:[8]

$$RRR = GNI \qquad \text{16-21, N}$$

RRR Requisitions (orders) Received at Retail (units/week)
GNI Generation of Need, Initial rate (units/week)

This completes the initial-condition equations necessary for starting the combined production-distribution-advertising system in steady-state operation.

16.3 Values of Constants

Numerical values of the constants must be chosen before the model is complete.

It should be noted that values must be specified for the unit price UPF and for the fraction of gross revenue being spent for advertising AVS. But neither of these constants has a direct effect on the behavior of the system. The selection of the constant ASL in Equation 16-6 overrides the numbers selected for UPF and AVS and completely determines the relationship between steady-state advertising and the point on the advertising saturation curve in Figure 16-6. The preceding initial-condition equations that set ASL in terms of VAC will cause ASL to compensate for any changes in UPF and AVS. The following arbitrary selection will therefore suffice:

UPF = 100 dollars, Unit Price of goods at Factory
AVS = 0.06 dimensionless constAnt, adVertising as fraction of Sales

Delays in the advertising system must now be specified. The following plausible values will be taken without detailed justification:

DMS = 4 weeks, Delay in Manufacturing rate Smoothing
DVF = 6 weeks, Delay in adVertising decision at Factory
DVA = 4 weeks, Delay at adVertising Agencies and media

[8] This replaces Equation 15-54.

The "time constant" for the rate at which advertising awareness will respond to a change in advertising might lie somewhere between 4 weeks and 12 weeks. We shall here take

DVC = **6** weeks, Delay in adVertising awareness build-up at Consumer level

For consumer durable goods, the length of time taken for a prospective purchase to become an actual retail order may vary over a wide time range from almost immediate purchase up to several years. Our estimates of a normal average time will not be very reliable, nor will our estimates of the maximum value DZV in Figure 16-6 or the minimum value DSV. We can, however, pick some plausible values and then see what effect changes in these values would have on system behavior. Therefore we shall here take the following estimated values for the constants shown in Figure 16-6:

DZV = **70** weeks, Delay for Zero adVertising

DSV = **15** weeks, Delay for Saturated adVertising

As in the study of the production-distribution system by itself, we shall take the steady-state retail sales rate as

GNI = **1,000** units/week, Generation of Need, Initial rate

The following constants determine test conditions for various runs on the model and must be given values for each study to be made:

ADN   constAnt, standard Deviation of Noise (dimensionless)
ANL   constAnt, Noise sample Length (weeks)
PER   sine PERiod, test input (weeks)
SIH   SIne Height, test input (units/week)
STH   STep Height, test input (units/week)
STT   STep Time after start of model run (weeks)

Sections 16.1 through 16.3 have developed the model description of a very simplified advertising and market sector to be added to the production-distribution model of Chapter 15. The equations and constants of Section 15.5 apply except for the following:

- Equation 16-21 replaces 15-54.

- The pipeline and inventory adjustment times of 4 weeks (DIR, DID, and DIF) in Section 15.5.5 have been changed to 8 weeks as a more likely reaction time:

DIR = **DID** = **DIF** = **8** weeks

- The half-week mail delays DMR and DMD in Chapter 15 are the limiting factors in determining how long the solution interval DT may be. Incorporating these delays separately in the model is easily shown (by observing system performance before and after a change) to make a negligible difference in system behavior. In exploring the new modes of behavior with the advertising sector included, we encounter manufacturing fluctuations with periods as long as two years or more. Model runs of several times the natural period of the system are of interest. For run lengths of five to ten years, a solution interval of 0.05 week begins to use appreciable computing time.[9] It is therefore worth while to combine the mailing delay with the clerical order-processing delays at the retail and distributor levels. Such does not change the length of the supply pipelines but only the number of pieces into which they are divided. The retail purchasing delay DCR then becomes 3.5 weeks; and at the distributor, DCD becomes 2.5 weeks. Equations 15-15, 15-16, 15-33, and 15-34 are omitted. Equation 15-14 is altered to define orders flowing to the distributor RRD, and Equation 15-32 to define orders to the factory RRF.

- With the preceding changes, the shortest third-order delays in the system are 1 week, and the solution interval DT can be extended to 0.1 week. Further increase in the solution interval could be obtained (again with no important change in dynamic behavior) by changing the shorter third-order delays in the system to first-order delays, but such has not been done for the following model runs.[10]

16.4 Behavior of System with Advertising

The preceding sections have developed the advertising and market additions to the distri-

[9] Some 15 minutes per run on an IBM 704 computer.

[10] In Appendix B is a tabulation of the equations and constants of this extended model.

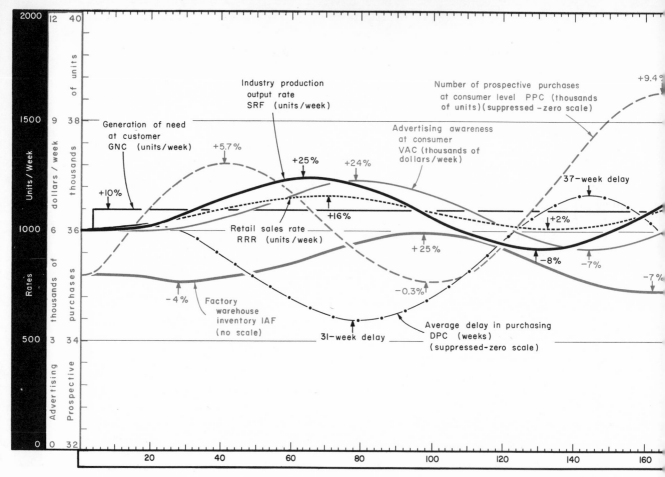

Figure 16-7 (Repeat of Figure 2-10) Effect of advertising that changes with sales.

bution system of Chapter 15. This extended system was used to obtain the results in Figures 2-10 and 2-11. The figures are repeated here and show how extending a system can bring in new kinds of behavior. The extended system is shown in broad outline in Figure 16-4. Previously the distribution system consisted of three cascaded order-supply loops connecting the levels of distribution. Now an outer loop has been added that couples the "top" to the "bottom" of the distribution chain.

16.4.1 Response to Step Input. As with the preceding simpler system, our first investigations should be designed to divulge the inherent internal characteristics of the system itself. One way is to start from a condition of steady-state balance, then to provide an initial disturbance, and to observe the ensuing interactions within the system. Such characteristics as stability, rate of growth or decline of a disturbance, and natural period of oscillation are in a linear system dependent only on the characteristics of the system and not on the nature of any externally applied disturbance.[11] Likewise, a nonlinear system in its mid-operating range will tend to show similar basic system characteristics that are fundamental to the system itself. However, we can expect in nonlinear systems that wide amplitudes of disturbance may carry the system into operating regions where the system characteristics differ substantially.

[11] The *external* inputs will control the periods of *forced* oscillations. For example, once any initial transients have subsided in a stable and linear system, all internal fluctuations will occur at the frequency that is impressed from the outside. By contrast, a nonlinear system may internally generate harmonics of an externally applied frequency.

200

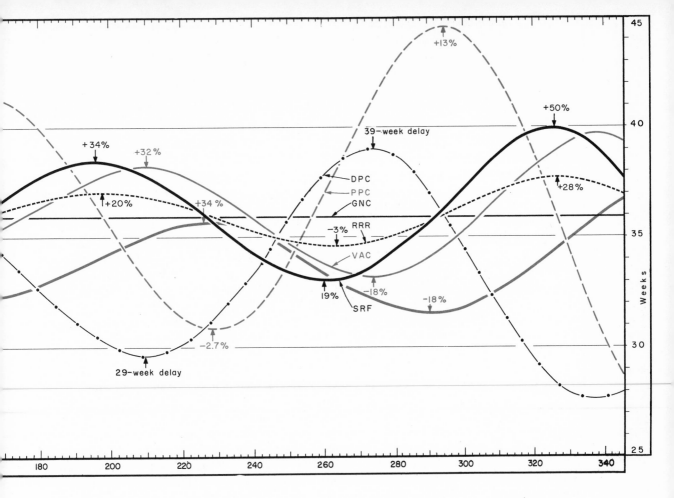

A 10% change in the generation of need at the customer level **GNC** should serve to show the transient path that the system follows in trying to re-establish a new operating equilibrium. The test input conditions will then be

Noise Deviation	**ADN**	= 0	
Noise Sample Length	**ANL**	= **1** week	(value does not matter since amplitude is zero)
Sine Input Period	**PER**	= **52** weeks	(value does not matter since amplitude is zero)
Sine Height	**SIH**	= 0	
Step Height	**STH**	= **100**	(which is 10% of the initial system operating rate)
Step Time	**STT**	= **4** weeks	(gives the time of the step after the start of the system simulation run)

The major aspects of system behavior have been discussed in Section 2.5, which should be reread at this point. The system behavior is shown in Figure 16-7. The increased flow of consumer need **GNC** causes a swelling of the prospective purchase pool. The long delay in this pool (about 35 weeks) implies an equally long time constant in the build-up of prospective purchases; the level of prospective purchases responds gradually to a sudden change in the inflow rate. This in effect acts as a filter that prevents the sudden step in need **GNC** from applying high-frequency (short-period) components at the retail sales rate **RRR**. Therefore we do not observe the short-period (38 weeks) inventory cycle that was predominant in the simpler system.

A new mode of system fluctuation has appeared. It involves the major loop that feeds up through the distribution system and back to the market through the advertising decision chain. It should be realized that the loop called advertising in this example can be considered more broadly as any of the factors that cause a rising factory production rate to increase the pressure to purchase at the retail level. In varying degrees the fluctuating availability of consumer credit, price discounts precipitated by low sales and high inventories, and the flow of rising factory wages into the hands of prospective purchasers can have similar effects.

The system shows a period of about 2.5 years, which furthermore is growing in amplitude of oscillation. We should not attach undue significance to the exact values of either of these observations.

The period of oscillation will be determined primarily by the time delays around the outer feedback loop. Changing the lengths of the delays will affect the period. Here the longest and the most significant delay is the one in the prospective purchase pool. Our economic system abounds in these levels that are large compared with the average inflow and outflow rates. Such levels constitute, in effect, delays with long time constants and are found in money balances (especially in savings), capital equipment pools, inventories of durables (and even textiles) in the hands of consumers, deferrable housing construction, allocation of land utilization, farm commodity storage, the equipment of transportation systems, and the various slowly changing categories of labor and population. All of these and numerous others interact to create economic fluctuation, and should be incorporated in future models of economic systems.

To exhibit oscillation, a feedback system must contain amplification. Here amplification is created by the ordering policies in the distribution system. A variable delay, such as incorporated at the point of prospective purchase DPC, could also create amplification if it had the proper functional relationship to its cause of variation.

The amplification is frequency sensitive; that is, it amplifies different periods of oscillation by different amounts.[12]

The phase relationships between different parts of the system also depend on the disturbance frequency. It may be surprising to note that factory production, although it clearly is a result of retail sales, actually is reaching its peak value earlier than the peak of retail sales. Of course it is equally correct to say that in this example retail sales fluctuate entirely as a result of changes in factory production. However, retail sales do not lag factory production by anywhere near the 20 weeks of delay that exists down through the advertising decision chain. Such characteristics, which may at first be unexpected, are typical of the closed-loop types of systems with which we are dealing. Several factors in the system can cause a phase lead at appropriate disturbance frequencies. This is true in the inventory and pipeline terms of the ordering equation, where these components are partially responsive to the *rate* at which sales are *changing*.[13] Also, the peak of retail sales occurs substantially in advance of the minimum value in purchase delay (because pool depletion as well as delay controls the outflow rate).

Very often a sharp distinction is drawn between stable and unstable oscillatory systems, that is, between ones in which a disturbance diminishes in amplitude versus one in which a disturbance increases. We should not attach any great significance to the point of neutral stability separating these two regions. Better, we should look upon stability as a continuous matter of degree. The rate of decay of a disturbance is important, but it is far from sufficient that a disturbance merely decline. A system showing a natural period that decays slowly will be a

---

[12] See Appendix G.

[13] The derivative (rate of change) of sales rate leads the sales curves by a quarter-period. (The maximum rate of rise in a sinusoidal fluctuation occurs as the curve passes through zero.)

strong amplifier of any externally applied disturbances that contain periods near the natural period of the system. Conversely, an unstable system will contain nonlinearities that will limit the possible amplitude of oscillation. That is, the system is unstable for small amplitudes of disturbance in certain regions of operation but stable for sufficiently large amplitudes. Systems of this latter type are of great importance because probably a majority of our industrial and economic systems fall into the class of being small-amplitude unstable and large-amplitude stable.[14]

Another reason for attaching little importance to the unique condition where stability gives way to instability is that a number of parameters of the system, by changing slightly, can move the system across the point of neutral stability. To ensure even marginal stability, if that were a sufficient objective, would still require that we design a system for sufficiently rapid attenuation of disturbances so that uncertainties in the actual values of parameters would not produce unexpected results.

In the system illustrated in Figure 16-7, the slightly growing oscillation results from the particular parameters that were selected. The important result is not the particular rate of oscillation growth, but the possible occurrence of this kind of 2- to 3-year cycle as a result of a combination of ordinary industrial circumstances. Furthermore, even though a stable system[15] could be obtained by making small changes in some of the assumptions of the model, a really "good" system with a high *degree* of stability may require substantial changes in system structure or in the fundamental system assumptions (like the one of advertising proportional to sales).

In this particular system, the two purchasing delay times DZV and DSV that describe the customer reaction to advertising are among the most sensitive parameters in changing system stability. The system becomes less stable as DZV is increased or as DSV is decreased. This occurs because a wider range between DZV and DSV (see Figure 16-6) increases the limits between which the prospective purchase pool may swing and has the effect of increasing the amplification around the outside loop of the system.

Stability is also affected by the relationship between the natural frequency of the distribution chain by itself (the 38 weeks of Figure 15-18) and the natural period of the larger system as in Figure 16-7. The amplification of the distribution chain depends on the period of disturbance.[16] In the larger system, if the natural period is too much longer than that of the distribution chain, the distribution chain will no longer have enough amplification to sustain oscillation at the lower frequency in the larger system that contains the advertising and marketing segments. For example, this decoupling of the two systems could be expected if a very long term averaging of sales (like two years or more) were made the basis of advertising level. This would lengthen the natural period of the system and would reduce the total amplification around the loop at the natural periods under consideration. It would still allow adjustment of advertising with long-term changes in business volume.[17] Moving in the other direction, if the period of the outer loop is decreased by suitable changes in the system, the two natural frequencies are brought closer

---

[14] Given the realities of human behavior, the unavoidable time delays, and the amplification inherent in such systems, instability is not surprising but can often be expected. Great skill may be required to find those changes that will convert some of our present unstable systems into ones showing strong tendencies to suppress disturbances.

[15] That is, one that tends toward equilibrium in the absence of external disturbance.

[16] See Appendix I.

[17] These observations are not conclusions or recommendations of general applicability but apply here and to only those actual systems that might by chance be adequately represented by the particular model. However, these are the *types* of results that we look for in a practical study of an actual system.

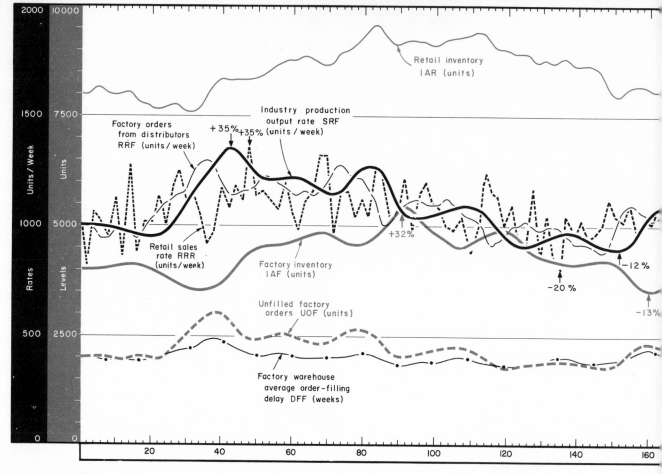

Figure 16-8 (Repeat of Figure 2-11) A composite model with advertising, a factory limit 40% above average sales, and random variation added at retail sales.

together. Under the latter circumstances the advertising loop can feed back enough disturbance at the short period of the distribution chain to convert the distribution chain from a stable system[18] to one in which the distribution chain itself generates growing oscillation at its own natural frequency.

The preceding characteristics of the system, which have been determined by model runs that do not appear in the figures, have been discussed to give some feeling for the ramifications encountered in larger systems. A system model for actual enterprise design will be more complex than these examples. There will be extensive interlocking of functions and many modes of possible behavior, although only some few

[18] As in Figures 15-18 or 15-24.

modes will be of practical interest in any one situation.

One reasonably safe generality appears to be emerging from this type of study—system stability is more apt to be achieved by altering managerial policies to reduce amplification than by adopting forceful countercyclical corrective policies. The two-year mode of fluctuation in this example is, of course, dependent on the advertising policy, among other interacting causes. If the proportionality is reduced, that is, if the advertising level is less responsive to changes in sales, the fluctuation in retail sales will be suppressed.[19] It might appear to follow that an

[19] As stated earlier, keep in mind that many factors other than simple advertising are also coupling the production level to market demand, to combine to create the effects discussed here.

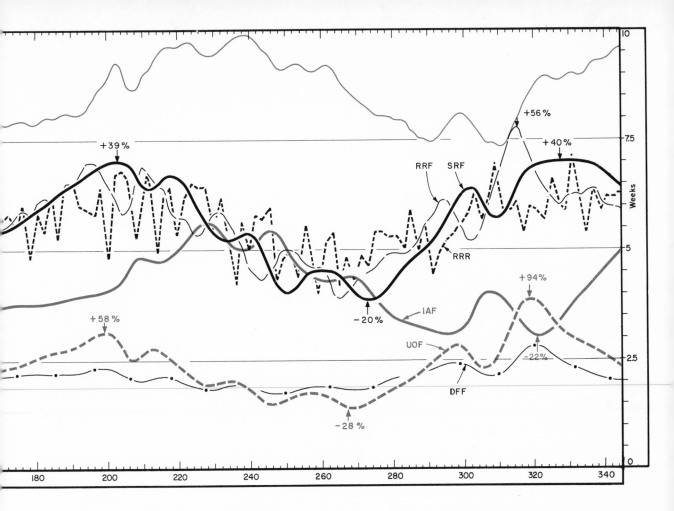

inverse policy of reducing advertising with increasing sales would be countercyclical and produce still greater stability. But it may not.[20] Forceful application of the inverse policy also causes amplification at periods different from those encountered earlier and can create new kinds of difficulty.

Returning to Figure 16-7, we see a typical relationship between factory production and factory inventory. In many industries that exhibit a persistent fluctuation, inventories reach a peak shortly following the peak incoming-order rate. This is a natural consequence of the usual inventory and production policies. Under these typical circumstances, inventories,

[20] See Appendix M for a discussion of this inverse advertising policy and the way it can cause a new mode of system instability.

far from assisting in stabilizing production, are contributing to production fluctuation. Observe, for example, the conditions at the 330th week. Factory production is 1,500 units per week, while average retail sales are 1,100 units per week, and current retail sales are 1,272 units per week. At this point, inventories are increasing at the rates shown in Table 16-1.

Table 16-1 Inventory Increase

Location	Inventory Increase Rate (units/week)
Factory	64
Distributor	82
Retail	70
Total	216

Of the difference between average demand and current production, 55% is going into inventory. When the production facilities are under greatest stress, a major part of the excess production is going into stocks, to heighten the production rate above actual retail demand. Conversely, at the point of *minimum* retail demand, rather than by using inventory to accumulate some of the excess production, sales are being partially filled from inventories, and production has been cut back to a point *lower* than the reduced retail demand. It is not coincidence, nor the result of random bad luck, that inventories reach their lowest point just at the time they are needed to meet a rapidly rising demand; the usual system organization and policies combine to ensure that inventories are lowest just ahead of peak demand and are highest just as demand approaches a minimum.

16.4.2 Random Fluctuations at Retail Sales. Figure 16-7 was examined without random components in any of the system decisions. In a noise-free system we were able to see best some of the inherent characteristics. However, we know that smooth, unperturbed flows are not usually found in actual industrial systems. In principle, we should introduce uncertainty at all decision points in the system, but actually we expect little change in the effect on the system after random inputs have been added at a few of the most sensitive points.

Figure 16-8 shows the effect of a random component in the retail sales rate. The figure results from the equations of Chapter 15, as modified and extended in Sections 16.1 through 16.3, with the test conditions for Equations 16-10 through 16-14 as follows:

Noise Deviation **ADN = 0.1**
Noise Sample Length **ANL = 1.0**
Sine Input Period **PER = 52** weeks (value does not matter since amplitude is zero)
Sine Height **SIH = 0**
Step Height **STH = 100** (which is 10% of the initial system operating rate)

Step Time **STT = 4** weeks (gives the time of the step after the start of the system simulation run)

In addition, for this run the factory limit has been brought down to a point where it will affect system operation when production rates above 140% of the initial rates are called for:

ALF = 1,400 units/week, constAnt specifying manufacturing capacity Limit at Factory

Figure 16-8 retains the underlying characteristics of Figure 16-7. The long-period business fluctuation is still predominant. In addition, the noise at the retail sales rate has reintroduced short-term fluctuation in the inventories and pipelines of the distribution chain, as we saw in Figure 15-20.[21]

Factory production still peaks at about the same time as or before retail sales, and inventory is rising at the time of maximum production and peaks just ahead of the retail sales minimum. Although not shown in the figure, purchasing delay at the prospective purchase pool varies only about 5 weeks from its initial value of 35 weeks as it did in Figure 16-7, yet this is sufficient to cause the fluctuating outflow of retail orders.

In Figure 16-8 the long-period fluctuation of something over two years' duration is still very pronounced and regular. The strength and uniformity of such a mode of operation will depend on both the degree of stability of the system and on the amplitude of those external disturbances that have components with periods near that of the natural period of the system. Here the model system is slightly unstable and will therefore be strongly persistent in its natural mode of operation. Also the noise input that is randomly sampled each week will contain very little energy in the 2-year to 4-year period octave[22] to cause any deviation from the natural internal inclinations of the system. With a more stable system, and with more external disturb-

[21] See Sections 2.4 and 15.7.3.
[22] See Appendix F.

ing forces having periods near that of the system, we should expect a much less coherent character in the natural frequency of the system.[23]

The production-distribution-market system of Figure 16-8 shows behavior somewhat like the business cycle fluctuations of our national economy. So does the system that follows in Chapters 17 and 18. Numerous other industries with similar behavior exist in parallel with these, and still others are cascaded back into capital equipment and commodities. Strong coupling forces existing in the market and in financial circles draw separate industries into a synchronous rise and fall of business activity. The national economy can be viewed as a partially interlocked composite of these separate systems, each of which is organized and managed so that it contains inherent tendencies which contribute to national economic instability.

[23] More like the character of the wandering, broadband fluctuation of Figure 15-20 but showing longer-term components; or like the system in Chapter 18.

CHAPTER 17

A Customer-Producer-Employment Case Study

In this chapter will be developed a dynamic model that resulted from a case study of an actual company. It arose from a search for causes of a fluctuating employment level that varied over a ratio of about 2 to 1, with peak-to-peak intervals of about 2 years. There seemed to be no likelihood that the explanation lay in a similar large variation in the final usage rate of the product. The model shows that the observed employment instability can result from interactions between the purchasing practices of the customers and the inventory, production, and employment practices of the company. The example is of interest here because it is typical of a substantial fraction of American industry. This chapter describes the industrial situation, gives the reasons for choosing the particular factors that went into the model, and develops the mathematical model. Chapter 18 contains the results of the model studies and methods of improving employment stability.

THE problem and system model discussed in this chapter arose from a case study of a company in the electronic components industry. The essential features of the actual study are retained here, although the description and the model have been simplified for clarity of presentation. In some places the model has been generalized to incorporate typical features of other industrial situations that did not appear important in the particular study. Some of the numerical values have been changed to make the model representative of a wider class of industrial situations where the particular case was atypical. None of these changes has had any major effect on the nature of the dynamic behavior nor on the kinds of conclusions that can be drawn.

17.1 Description

The situation relates to a supplier of components that are used by other manufacturers in their products, and these in turn are sold to industrial and governmental customers. The customers of the company under consideration are informed buyers purchasing a product in a manner that is geared to their own production demands.

The product is a high-quality electronic component that is manufactured by a number of suppliers and is used in military and industrial equipment. It had been generally considered, and appears to be true, that the average customer is interested first in quality, second in delivery, and third in price. Price can be relegated to third position partly because it is indeed less important to the customer than quality and availability, and partly because prices are competitive and differ but little between suppliers and appear not to be the principal basis for selection of a supplier.

For the particular company, quality seemed to be high and adequately controlled. Customer service in the form of prompt delivery was therefore left as the principal competitive factor that might affect the company's relationship to its customers. Furthermore, delivery was both variable and longer than desirable.

A characteristic of the system is that week-by-week incoming orders fluctuate over a very wide range. Orders in successive weeks can vary by a factor of 2 or more. It is therefore impossible to tell from current orders in the most recent week whether or not a change has occurred in the average level of sales. Sales must be averaged over a period of 10 weeks or longer (thereby delaying the information) before they give any significant indication of changes in the level of demand.

The industrial customers of the company release specifications from their engineering and manufacturing departments to their purchasing departments as far in advance as necessary to obtain components in time for their planned production. This usually means that orders are placed for various components and materials at different times in advance of production, depending upon the time delays normally experienced in obtaining each separate type of item.

The company had been experiencing fluctuating production and employment in the particular product line. The sales and production fluctuation had been assumed to come exclusively from varying usage rate by the company's customers. However, a review of the nature of the product and of the industry failed to show any plausible likelihood that actual product usage was fluctuating as widely as sales and production had been.

In the product history there had appeared the typical time relationship between production rate and inventory which is found in many industries. In very simplified form, production and inventories are often related as shown in Figure 17-1. Here the significant characteristic

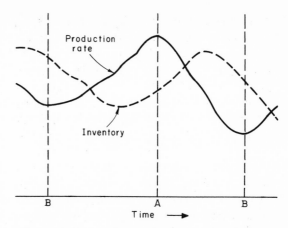

Figure 17-1 Production-inventory relationship.

is that inventory peaks after the production rate peaks and before the production-rate minimum. In the figure, inventory, far from helping to smooth production rate, is contributing to the peaks and valleys in production. At time A, inventory is rising during the production peak. This means that when the system is under stress and is producing at its maximum rate, part of this production is going into inventory. Conversely, at the times marked B, inventory is falling during minimum production. This means that part of the customer demand is being filled from inventory. Therefore, the production rate at the times marked B is lower than the rate of sales to customers, while the production rate at A is higher than the rate of sales to customers. Production rate is therefore fluctuating more than sales. To have a neutral effect on production rate, inventory should not be changing at the peaks and valleys of production. To have a favorable effect on suppressing fluctuations in production, the inventory should be falling at the time of the sales peak, and rising at the time of sales valleys. This is almost never found in the usual industrial situation (except where geared to an obvious annual, seasonal demand), since the normal policies followed by most companies tend to create the kind of relationship shown in Figure 17-1.

These observations raised the possibility that

the fluctuation and instability might arise from the interactions between the practices of the various participants within the industry. If the organizational structure and policies were not the primary cause of fluctuating sales and employment, they might at least be an amplifier that would create larger variations in the product sales rate than in its ultimate usage. The study and the resulting model were aimed at seeing if the symptoms arose from the internal structure and policies of the system. The conclusion, in Chapter 18, is that they do and that changes are possible which should yield a substantially improved system.

17.2   What Constitutes a System?

One of the first steps in exploring the dynamic behavior of an industrial situation is to form a preliminary opinion of what factors are significant to the behavior of the system under study. This is probably the most important and at the same time the most difficult step in the entire procedure of model building. The omission of a part of a system will eliminate from the model those modes of behavior which are dependent on that part. Yet we know that we must limit our area of consideration lest the study of a small company expand to include all of our national and international economic system.

Successful selection of the factors to include in a system model usually implies a preliminary anticipation of the types of system behavior that could be important. There are too many factors that might be included merely to assemble in a model all the considerations that we shall meet in a company. Many factors receiving attention in normal company operations may have little effect, while important interactions may depend on factors that have been given little consideration.[1] A careful and perceptive exploration of the opinions, influences on decisions, historical data, attitudes, practices, beliefs, doubts, and questions that are to be found within an operating organization is necessary to bring to light the phenomena which must be treated. These must be examined against an understanding of how the dynamic interactions within the system may be created. The selection of pertinent factors will be subjective, and one on which opinions may often differ. Where they differ, separate models representing the differences can be developed and will often help to resolve the differences.

Figure 17-2 shows the major sectors that might need to be included in the system boundaries. Starting first with our company supplying components at C, we might ask whether any of the other sectors shown in the figure are necessary in the model. If the orders arriving at the company at point A were independent of any actions taken by the company, then we could consider this order stream as an independent input to a model of the internal company operations. However, this is not true. Incoming orders do depend on how well the company satisfies the demands of its customers. If then the policies and performance of the company affect the attitude of equipment manufacturers who are the customers, and this in turn affects the incoming orders at point A, it follows that we must represent the essential nature of these interactions between the company and its customers.

The same question might now be asked about the flow of orders at point B from the pur-

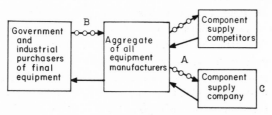

Figure 17-2 Possible major system sectors.

---

[1] For example, in this case study there was no awareness within the company of the dynamic consequences to customer ordering rate that could result from a *varying* delivery delay, even though there was high sensitivity to the competitive importance of prompt delivery.

chasers of final equipment to the equipment manufacturers. Will the policies and practices of the component supplier at C affect the flow of orders at point B? The answer is, probably not. The placing of orders by the purchasers of final equipment will seldom recognize the conditions that surround the supply of ordinary and minor components to the equipment manufacturers. The flow of orders at point B could, in the situation from which this model was derived, be considered independent of the actions of the component manufacturer at C. The purchasers of final equipment could therefore be excluded from the model structure that deals with the component supplier because we feel it is not necessary to include feedback channels from C to the equipment user.

Consider next the competitors of the component supplier. Must the competitors be included in a first model? The answer depends upon the questions that are to be asked. If we wish to study the character of the industry as a whole, which is a reasonable beginning point, then we might include only a single aggregate component supplier sector. This is especially justifiable if there is no indication that the policies of the company and its competitors are appreciably different. A company and its competitors having the same structure and following the same policies would in a model perform in the same way. They can therefore be combined. Once the characteristics of the industry are known, we may discover that some alternative policies appear to be preferable. It might then become necessary to consider a competitor sector, because we should then be proposing that the company would operate according to a set of policies designed differently from those being followed by the competitors. Until that time the initial model can deal with a single supplier sector.

Figure 17-3 shows the two major sectors that were selected for inclusion. The customer sector represents the composite aggregate of all equipment manufacturers who use the type of

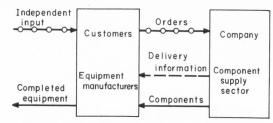

Figure 17-3   Selected major system sectors.

components under study. The company represents the aggregate of all suppliers of the components. Orders flow from customers to company. Components flow from company to customers. In addition, delivery information is available to the customers from the company sector, indicating the time required from the placing of orders until delivery. Orders flowing into the customers are from an independent input, since these are assumed not to be influenced by conditions at the component supplier. This assumption might not be true if a shortage of components actually limited the ability of equipment manufacturers to deliver their product. Such might in turn affect the placing of orders for equipment. However, we here believe that the equipment manufacturer will not permit this eventuality to occur, and that if if it were threatened, he would manufacture the components himself or develop alternative sources of supply.

17.3   Factors to Be Included

Before model formulation was begun, it was tentatively decided that an important interaction between customers and company might arise from the company delivery delay. Figure 17-4 shows the details from which this might arise. Suppose that a constant flow of orders persists at the input to the customers at B, but that for some reason the customers begin to order slightly further in advance of their requirements. This would lead to a short-term increase in the rate of order flow at A from customers to company. Such an increased order flow rate would lead to some depletion of in-

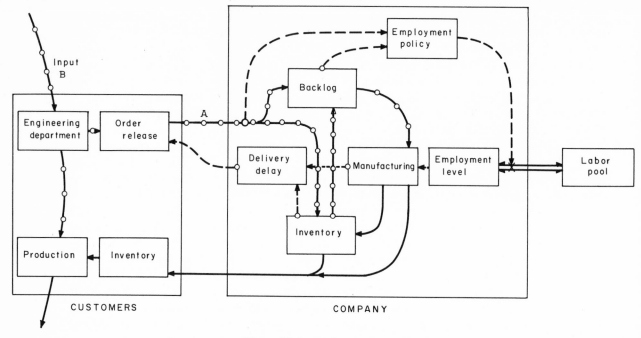

Figure 17-4 Major system details.

ventories at the company and to an increase in the backlog of items manufactured to order. Because of the large ordinary fluctuation in week-by-week orders, it would not be possible to detect this increase in order flow immediately. As a result, the declining inventories and increasing backlog would lead to increasing further the delivery delay. This in turn would induce the customers to order still further in advance, a practice that would continue to deplete inventories and increase the unfilled-order backlog. This cumulative increase in delivery delay, with its accompanying increase in the advance lead time for ordering by the customers, would continue until production rates at the company had been adjusted to the new higher ordering rate. To do this, a higher employment level will then need to consist of two parts.

One part of the increased employment will *seem* to be required to equal the system-induced excess ordering rate so as to stop the increase in backlog and the reduction in inventory.

By this time, inventories will usually be low and backlogs will be high, and the company will feel compelled to employ a second component of additional labor to rebuild inventory and to bring the backlog down to satisfactory proportions. This implies an improving delivery situation. As soon as deliveries begin to improve, the customers need not order so far in advance of their production requirements because they find that they can meet part of their current production from material flowing in from the declining backlog at the company. The apparent ordering rate from the customers to the company then declines even below the customers' current usage rate while they use up the excess materials that are emptying out of the supply pipeline. The company then finds itself with rising inventories, falling backlogs, and rising pressure to reduce production and employment. This again leads to an increasing delivery delay.

Increased ordering increases delivery delay, further increases ordering, and vice versa. Thus, the unstable fluctuation between customer advance ordering and company employment adjustment might persist. This type of interaction appeared to be indicated from the history of sales and production, and the model was constructed to explore the plausibility of the assumption and the conditions under which it might exist.

Should it develop that the managerial practices followed by the customers and the company could lead to fluctuating operations, the next step would be to examine alternative company policies that might lead to greater stability.

The preceding tentative anatomy of the problem was established by two or three men after studying the situation for a period of a year and a half. The same persons in a related situation would progress more rapidly a second time. However, on the first attempt, time is required to develop the ability of the investigators to perceive the central skeleton of the dynamic system through the clutter of necessary but peripheral company activity.

One must always bear in mind that the success of a model depends on what one perceives as important in the system that is being represented. Model details will now be developed based on belief in the preceding analysis of the industrial situation.

At the customers, the dynamic behavior to be explored was assumed to depend on how the customers responded to changes in delivery delays. The important policies and practices represented in the model would show the manner and rapidity with which the customer might respond to changes in delivery delay by the company.

Those factors within the company and in its adjacent environment which affect its ability to deliver its product must be included. Among these would be the level of inventories, the fraction of orders which can be filled from inventory and the fraction which must be manufactured to order, the production rate and factors that affect it, and the production delay.

At the company, some orders were filled from inventory, and some were made specifically to customer order. The state of inventories would determine the fraction of orders that could be filled from stock. Manufacturing to customer's special order would need to be included as well as manufacturing to rebuild inventory.

The product has a high labor content, and production rate is largely determined by the level of employment. The practices and policies regarding changes in employment level appear closely related to changes in delivery capability. Employment practices became a focal point for much of the model attention. The important factors are those on which employment changes are based, and the time delays inherent in detecting the need for adjustment of employment level. The part of the model representing employment policy should take cognizance of important changes in labor productivity, the time to train new employees, the time set by labor agreements or company policy regarding the length of advance notice of dismissal, and the hiring of labor to match the current level of incoming orders as well as to adjust the state of inventories and backlogs.

Several factors that are often important could be omitted. In the particular example, the long-term demand for the product was declining slightly, so that an adequate level of manufacturing space and capital equipment existed. Capital equipment and its purchase could therefore be omitted as factors affecting production capability. Materials were readily available on short notice and seemed to have no major influence on production capability. Therefore procurement and inventories of material could be omitted.

Cash position, profitability, and related considerations seemed not to influence production and employment decisions in this product line. The latter decisions seemed controlled entirely by factors related to the demand for the product. Therefore the financial aspects were not necessary in the model as part of the decision-making structure. However, a simple profit cal-

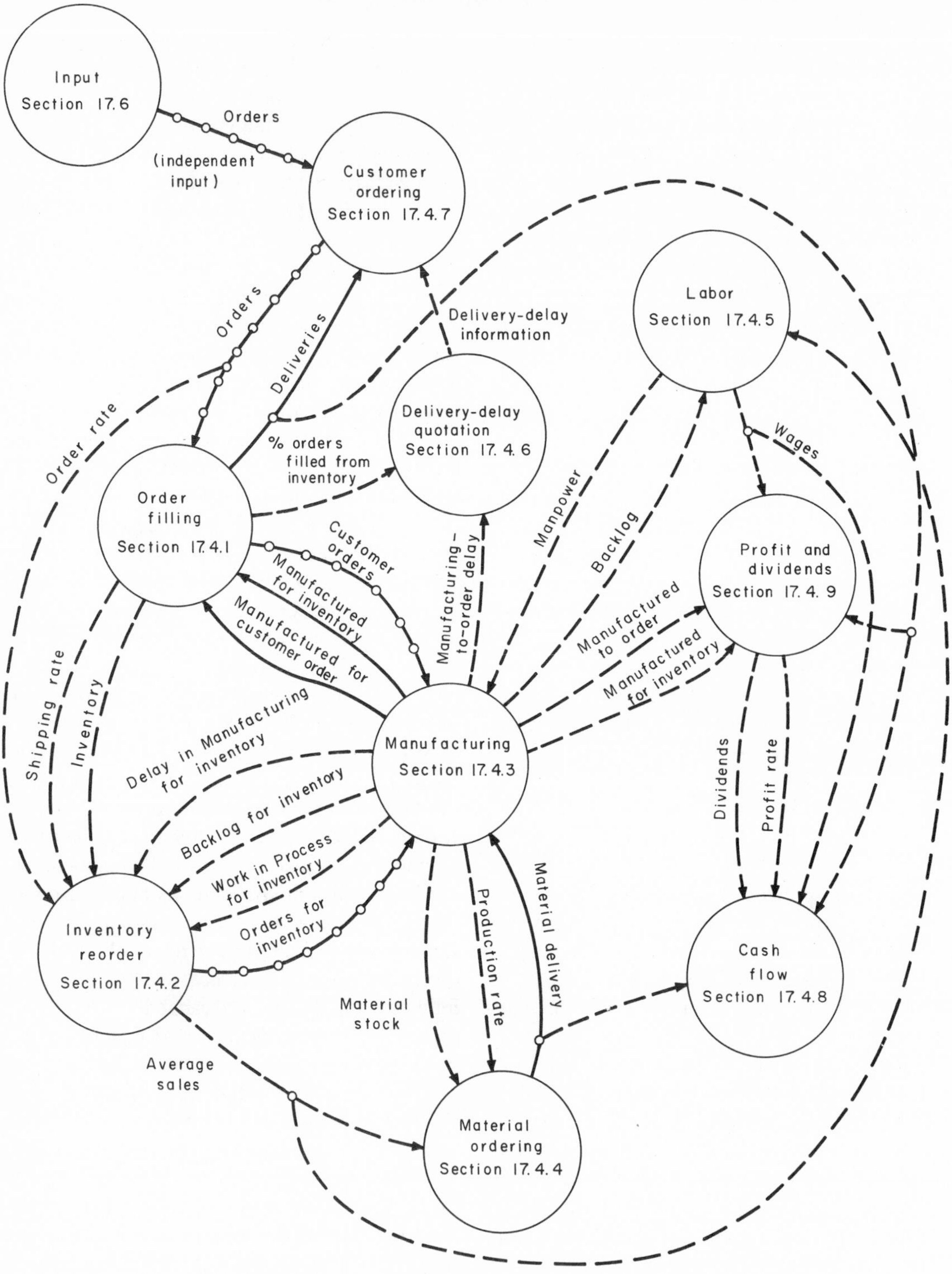

Figure 17-5 Subdivisions of the model, interconnections, and sections where described.

culation and cash flow are included here to provide a rough indicator of system performance to help in judging differences between model runs.

17.4 Equations Describing System

A formal mathematical model will now be developed that fits the preceding verbal description of the company, its labor level and production rate, and its customers.[2] This will be done in nine sections:

- Order Filling
- Inventory Reordering
- Manufacturing
- Material Ordering (as a factor in money flow and cash position)
- Labor
- Delivery-Delay Quotation
- Customer Ordering
- Cash Flow (for model evaluation but not affecting internal system decisions)
- Profit and Dividends (for model evaluation)

As a preview of the subsections to follow, Figure 17-5 shows the interrelationships between these nine sectors of the system and also the independent test input. The full model will involve about 90 active variables plus some 40 initial-condition equations. Approximately 40 constants define the particular system.

17.4.1 Order Filling. The first equations to be developed will be for that part of the company which receives incoming orders and routes some to be filled from inventory and some to be manufactured to customer order. Figure 17-6 shows the functions to be represented. This section of the model includes the inventory of finished goods and the criteria that determine whether or not an incoming order can be filled from inventory.

Equation 17-1 gives the level of requisitions[3] (orders) in clerical processing at the factory RCF, into which flow the incoming requisitions and out of which flow those to be filled from inventory and those to be manufactured especially to customer specification:

$$RCF.K = RCF.J + (DT)(RRF.JK - RFIF.JK - RMOF.JK)$$
$$17\text{-}1, \text{ L}$$

$$RCF = (RRF)(DCPF) \qquad 17\text{-}2, \text{ N}$$

RCF	Requisitions in Clerical processing at Factory (units)
DT	Delta Time, the time interval between solutions of the equations, here taken as 0.25 week
RRF	Requisition rate Received at Factory (units/week)
RFIF	Requisition rate Filled from Inventory at Factory (units/week)
RMOF	Requisition rate Manufactured to Order at Factory (units/week)
DCPF	Delay in Clerical Processing at Factory (weeks)

Equation 17-1 is the standard form of level equation[4] with one inflow rate and two outflow rates. It generates the number of requisitions that have been received at the factory but not yet processed. Flowing in are the requisitions (orders) from customers,[5] and flowing out are two streams: RFIF to be filled from inventory, and RMOF to be manufactured.

---

[2] Most of the work on this model was done by Mr. Willard R. Fey assisted by Mr. Wendyl A. Reis, Jr., and Mr. Carl V. Swanson. An S.M. thesis (Reference 13) by Fey develops these equations in a somewhat different form and interprets them using the methods of servomechanisms theory by linearizing the equations and representing them in Laplace transforms. This thesis provides a bridge between the usual formal methods of servomechanisms theory and experimental simulation in the time domain.

[3] Incoming orders will here be referred to as requisitions, as in Chapter 15, so that they may be more readily distinguished from outgoing orders for materials.

[4] Hereafter, level equations will be given little interpretation because they all take the same form that has been extensively covered in Chapters 6, 7, 8, and 15.

[5] Notice that the flow diagrams can be used to trace the numbers of the equations which generate the input variables to each equation.

The initial-condition equation,[6] 17-2, gives the initial level of RCF as equal to the steady-state incoming flow of orders RRF multiplied by the average clerical processing delay DCPF.

The shipping orders to inventory which have not yet been filled are defined by the level SOF, Equation 17-3:

[6] In Chapters 15 and 16, initial-condition equations were developed in a separate section following the development of the basic model. A different sequence will be followed in this chapter by including an initial-condition equation immediately following the normal operating equation for the same variable. In this way the initial-condition equation can be disposed of while the nature of the variable defined by the level equation is clearly in mind.

$$\text{SOF.K} = \text{SOF.J} + (\text{DT})(\text{RFIF.JK} - \text{SIF.JK}) \qquad \text{17-3, L}$$

$$\text{SOF} = (\text{RFIF})(\text{DSF}) \qquad \text{17-4, N}$$

SOF   Shipping Orders at Factory   (units)

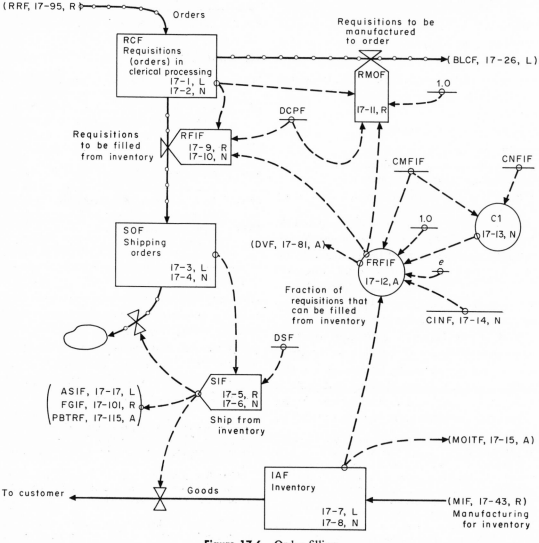

Figure 17-6   Order filling.

RFIF   Requisition rate Filled from Inventory at Factory (units/week)
SIF   Shipments from Inventory at Factory (units/week)
DSF   Delay in Shipping at Factory   (weeks)

In Equation 17-4 the initial steady-state level of shipping orders SOF is equal to the steady-state flow of orders that can be filled from inventory RFIF multiplied by the average delay in making shipments DSF.

The orders shipped per unit of time will be represented as a fixed fraction of the backlog SOF. This is by definition a first-order exponential delay,[7] as given by

$$SIF.KL = \frac{SOF.K}{DSF} \qquad \text{17-5, R}$$

$$SIF = RFIF \qquad \text{17-6, N}$$

SIF   Shipments from Inventory at Factory (units/week)
SOF   Shipping Orders at Factory   (units)
DSF   Delay in Shipping at Factory   (weeks)
RFIF   Requisition rate Filled from Inventory at Factory (units/week)

An average shipping time is here used as

DSF = 1 week, Delay in Shipping from Factory

Equation 17-6 gives the initial value[8] of the delivery rate of goods SIF, which in the steady-state must equal the flow of orders RFIF that are to be filled from inventory.

The inventory of finished components at the factory is given by the usual level equation:

$$IAF.K = IAF.J + (DT)(MIF.JK - SIF.JK) \qquad \text{17-7, L}$$

$$IAF = (CIRF)(RRF) \qquad \text{17-8, N}$$

IAF   Inventory Actual at Factory   (units)
MIF   Manufacturing rate for Inventory at Factory (units/week)
SIF   Shipments from Inventory at Factory (units/week)
CIRF   Coefficient for Inventory Ratio at Factory (weeks)
RRF   Requisition rate Received at Factory (units/week)

In this situation, the target inventory was authorized in terms of a certain number of weeks of sales; therefore the initial steady-state level of inventory in Equation 17-8 is proportional to the constant initial flow of incoming orders multiplied by the target ratio CIRF between sales rate and inventory level.[9] The target ratio CIRF is the number of weeks during which the desired level of inventory would be able to meet the demands of the steady-state order flow rate (assuming all orders could be filled from inventory).

Returning now to the flow rate RFIF, which is that part of the incoming requisition stream that can be filled from inventory, we know that it must depend primarily on three factors — the level of unrouted requisitions RCF, the average delay in routing DCPF, and the fraction of the orders that can be filled from inventory FRFIF (this fraction will be discussed below). These would reasonably be related as in the following equations:

---

[7] See Chapter 9 and Appendix H.

[8] Some initial values for rates are necessary because they are convenient or necessary for defining still other initial conditions. In a few places they are needed in the first time step because other rates or auxiliary equations depend on them (for example, as inputs to third-order-delay functions, see Appendix H). In this chapter we anticipate where we shall want initial conditions for rates (usually discovered as the model formulation is nearing completion) and insert the initial-condition equation while the rate equation is being discussed.

[9] This does not imply that the inventory *has* to be proportional to sales, especially in the short-term business fluctuation. In the long-term, many factors do tend to force inventories to rise with increased business volume. Inventory may be required by wider geographical coverage that requires more inventory locations, by in-process and in-transit inventories that are proportional to sales rate, by larger individual incoming orders that require larger stocks to maintain ability to fill from inventory, and by increased sales that come from additions to the product line.

$$RFIF.KL = (FRFIF.K)\frac{RCF.K}{DCPF} \qquad \text{17-9, R}$$

$$RFIF = (CNFIF)(RRF) \qquad \text{17-10, N}$$

RFIF	Requisition rate Filled from Inventory at Factory (units/week)
FRFIF	Fraction of Requisitions Filled from Inventory at Factory (dimensionless)
RCF	Requisitions in Clerical processing at Factory (units)
DCPF	Delay in Clerical Processing at Factory (weeks)
CNFIF	Constant, Normal Fraction of requisitions filled from Inventory at Factory (dimensionless)
RRF	Requisition rate Received at Factory (units/week)

The total outflow rate from the level RCF would be given by RCF.K divided by the average delay DCPF. To get the flow of orders that can be filled from inventory, this total flow is multiplied by the fraction that can be shipped from inventory FRFIF. The initial value of the rate as given in Equation 17-10 is the normal fraction CNFIF that can be shipped from inventory multiplied by the initial steady-state order flow rate RRF.

The average time necessary to route orders, including the checking of engineering specifications and customer credit standing is here taken as

DCPF = 1 week, Delay in Clerical Processing at Factory

The flow rate of orders sent to the manufacturing department RMOF is simply the fraction that cannot be filled from inventory:

$$RMOF.KL = (1 - FRFIF.K)\left(\frac{RCF.K}{DCPF}\right) \qquad \text{17-11, R}$$

RMOF	Requisition rate Manufactured to Order at Factory (units/week)
FRFIF	Fraction of Requisitions Filled from Inventory at Factory (dimensionless)
RCF	Requisitions in Clerical processing at Factory (units)
DCPF	Delay in Clerical Processing at Factory (weeks)

To complete the order-filling sector, we need to establish the fraction of orders FRFIF that can be filled from inventory. This is intended to be a variable associated with the aggregate flow of orders for all items in the product line. (In the case study, the product line included several thousand separate catalog items.) In general, we can expect that the smaller the inventory, the more inadequate it will be. As the inventory drops, a larger and larger fraction of the incoming-order flow will be thrown into the factory rather than being filled from inventory.

Figure 17-7 shows the kind of relationship that is probable between the fraction of requisitions that can be filled from inventory FRFIF and the level of inventory. Several considerations form the basis for this curve. A certain number of items in the product line are so specialized that they are never carried in finished inventory. Therefore, there is an upper maximum fraction CMFIF beyond which the fraction of orders to be filled from inventory cannot rise. Another significant characteristic of the curve will be the fraction of orders CNFIF that can normally be filled from inventory when inventory is at some particular level. We can think of CNFIF as the fraction of requisitions that we should normally want to be able to fill

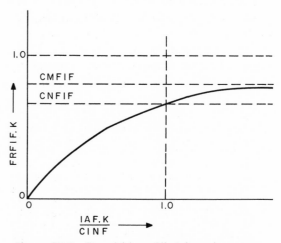

Figure 17-7 Requisitions filled from inventory.

from inventory. It will always be less than the maximum fraction CMFIF. Otherwise, inventory would be unnecessarily high. The horizontal axis is nondimensionalized in terms of the ratio of actual inventory to the amount of inventory CINF that will yield the desired "normal" split fraction.

Figure 17-7 implies that as total aggregate inventory rises, the split fraction rises slowly to approach the limit CMFIF, which represents the items that might conceivably be found in inventory. As inventory declines, the split fraction falls slowly at first and then more and more rapidly as zero inventory is approached and no requisitions can be filled from inventory. The system is set up so that those requisitions that cannot be filled from inventory will be sent to the factory to be made to customer order.

It should be noted that Figure 17-7 asserts that the split fraction depends only on the level of inventory. It does not for example depend on the rate RRF at which requisitions are arriving. This is reasonable under certain plausible conditions. If we assume a reasonably constant inventory proportioning between the various catalog items and a certain inventory level, the probability of filling the next order will be but little affected by requisitions previously filled. This assumes that the average *size* of individual requisitions does not change; that is, business increases because there are more orders, not larger orders.[10]

The relationship as shown in Figure 17-7, giving the split fraction when inventory is known, could be provided in either of two ways in the model. A table of values for FRFIF could be stored for several levels of inventory, and then intermediate values could be found by interpolation. Alternatively, we can choose some mathematical relationship that has the form of the curve in Figure 17-7. This latter procedure

will be followed, whereby the split fraction is determined by the following equation:

$$FRFIF.K = (CMFIF)\left(1 - e^{-(C1)\left(\frac{IAF.K}{CINF}\right)}\right) \qquad \text{17-12, A}$$

FRFIF	Fraction of Requisitions Filled from Inventory at Factory (dimensionless)
CMFIF	Constant, Maximum Fraction from Inventory at Factory (dimensionless)
e	base of natural logarithms, equals 2.7183 . . .
C1	Defined constant selected to fit requisition split-fraction curve to specified initial point (dimensionless)
IAF	Inventory Actual at Factory (units)
CINF	Constant, Inventory Necessary at Factory to make FRFIF equal CNFIF (units)

In Equation 17-12, as IAF approaches zero, the last term becomes e to the zero power, which equals 1. Therefore, the right-hand term becomes $(1 - 1)$, and the split fraction approaches zero. As IAF becomes very large, the last term approaches zero and FRFIF becomes equal to CMFIF, as it should. An exponential function of this kind has one degree of freedom represented by the constant C1; C1 should be chosen so that the curve will go through the value CNFIF in Figure 17-7 when inventory equals the value CINF.[11]

$$C1 = \log_e \frac{CMFIF}{CMFIF - CNFIF} \qquad \text{17-13, N}$$

C1	Defined constant selected to fit requisition split-fraction curve to specified initial point (dimensionless)

[11] When IAF equals CINF, Equation 17-12 becomes

FRFIF = CNFIF = (CMFIF)(1 − e^−C1)

By rearrangement,

$$e^{-C1} = \frac{CMFIF - CNFIF}{CMFIF}$$

By inversion to clear the negative sign in the exponent,

$$e^{C1} = \frac{CMFIF}{CMFIF - CNFIF}$$

By taking the natural logarithm of each side we get Equation 17-13.

[10] Alternative assumptions, such as that the split fraction depends on the ratio of actual to ideal inventory as in Chapter 15, can be readily substituted to see if the system is sensitive to such changes.

$\log_e$ a notation indicating the natural logarithm of the following number

CMFIF   Constant, Maximum Fraction from Inventory at Factory (dimensionless)

CNFIF   Constant, Normal Fraction of requisitions filled from Inventory at Factory (dimensionless)

Equation 17-13 is an initial-condition equation that defines a constant which will cause the curve of Figure 17-7 to pass through the desired point at the specified value of CNFIF; CNFIF must have a value less than CMFIF.

From a consideration of the plans for inventory stocking and the fraction of orders that might conceivably be filled from inventory, we obtain

CMFIF = 0.8, dimensionless Constant, Maximum Fraction from Inventory at Factory

From the average past inventory practice and estimates of the requirements for adequate service for customers, we take

CNFIF = 0.7, dimensionless Constant, Normal Fraction of requisitions filled from Inventory at Factory

Figure 17-7 has been established on the assumption that in steady-state system operation we expect to operate with IAF equal to CINF and therefore with a split fraction equal to CNFIF. In fact this is the definition of CNFIF. Accordingly, CINF is a constant whose value should be that of the initial steady-state inventory; therefore

$$CINF = IAF \qquad\qquad 17\text{-}14, N$$

CINF   Constant, Inventory Necessary at Factory, to make FRFIF equal CNFIF (units)

IAF   Inventory Actual at Factory (units)

This completes that part of the model that relates incoming orders, shipment from inventory, and customer orders sent to the factory.

17.4.2 Inventory Reordering. The inventory reordering decision is shown in Figure 17-8.

In this situation, inventory-reorder procedure did not exist as a formal written policy. However, examination of records and discussion with those who were concerned showed three principal factors as influences on inventory replenishment. One was the average rate at which inventory was being depleted by shipment to customers. A second was the adjustment to bring actual inventory to its desired level. A third was a recognition of the orders for inventory in process in the factory and the factory lead time for filling inventory orders.

This inventory-reorder decision is set up as the typical, isolated inventory-management decision that does not take into account the effect which it may have on total system dynamics. In the next chapter we shall see the implications of this policy and some possible changes in it. As formulated here, target inventory rises with a rise in average sales rate. The inventory-ordering decision is further given the inventory manager's and salesmen's viewpoint by ordering ahead in response to increases in the factory delivery delay.

The following equation combines the pertinent factors in the same way as was used for the placement of orders in Equation 15-9:

$$MOITF.K = ASIF.K + \frac{1}{TIAF}(IDF.K - IAF.K + OINF.K - OIAF.K) \quad 17\text{-}15, A$$

MOITF   Manufacturing Orders for Inventory to be Tried at Factory (units/week)

ASIF   Average Shipments from Inventory at the Factory (units/week)

TIAF   Time for Inventory Adjustment at Factory (weeks)

IDF   Inventory Desired at Factory (units)

IAF   Inventory Actual at Factory (units)

OINF   Orders for Inventory Necessary at Factory (units)

OIAF   Orders for Inventory Actual at Factory (units)

The first term in the equation is the average shipment rate from inventory ASIF. Using the short-term average is the more proper procedure and the more realistic, rather than the expediency used in Equation 15-9, where the current, actual, instantaneous shipment rate was used. Really there is no reason to believe that one representation will give results significantly different from the other. The parentheses on the right-hand side of Equation 17-15 give the difference between desired and actual inventory plus the difference between necessary and actual inventory orders in process at the factory. This inventory and pipeline discrepancy is adjusted by ordering a fraction of the discrepancy per week. The fraction is given by the adjustment time constant TIAF. The numerical value of TIAF will determine the rapidity with which inventory and pipeline adjustments are made, and is here taken as

TIAF = 6 weeks, Time for Inventory Adjustment at Factory

Equation 17-15 is an auxiliary rather than a rate equation because it is tentative. Its purpose is to prevent a negative flow of orders to the factory. Some circumstances arise where excess inventory will lead Equation 17-15 to generate a negative order rate; this would occur at a time when there are no orders in the factory backlog to cancel, and is therefore impossible. The actual rate is taken as

$$\text{MOIF.KL} = \begin{cases} \text{MOITF.K} & \text{if } \text{MOITF.K} \geq 0 \\ 0 & \text{if } \text{MOITF.K} < 0 \end{cases}$$

$$17\text{-}16, \text{ R}$$

MOIF   Manufacturing Orders for Inventory at the Factory (units/week)

MOITF   Manufacturing Orders for Inventory to be Tried at Factory (units/week)

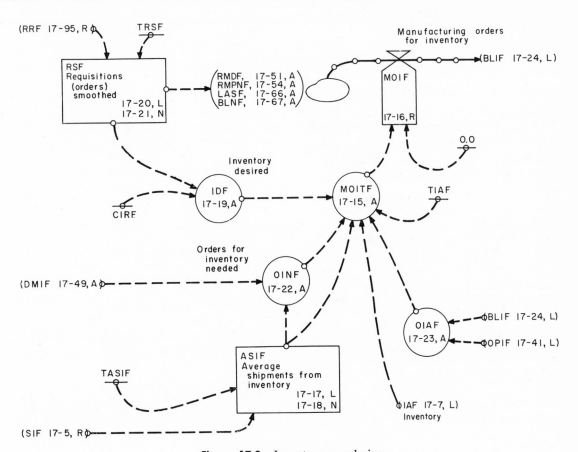

Figure 17-8   Inventory reordering.

The standard first-order exponential averaging equation will be used to convert the shipment rate from inventory into the average shipment rate that was used in Equation 17-15:

$$ASIF.K = ASIF.J + \frac{DT}{TASIF}(SIF.JK - ASIF.J) \quad \text{17-17, L}$$

$$ASIF = RFIF \quad \text{17-18, N}$$

ASIF   Average Shipments from Inventory at the Factory (units/week)

TASIF   Time to Average Shipments from Inventory at the Factory (weeks)

SIF   Shipments from Inventory at Factory (units/week)

RFIF   Requisition rate Filled from Inventory at Factory (units/week)

The average inventory shipping rate is intended to reflect very recent shipments from inventory. A short averaging time is therefore appropriate, such as

TASIF = 2 weeks, Time to Average Shipments From inventory at Factory

As described in Equation 15-7 and as expressed in the preceding discussion of policies, the target inventory will be taken as proportional to average sales:[12]

$$IDF.K = (CIRF)(RSF.K) \quad \text{17-19, A}$$

IDF   Inventory Desired at Factory (units)

CIRF   Coefficient for Inventory Ratio at Factory (weeks)

RSF   Requisition rate Smoothed at Factory (units/week)

---

[12] Again, this proportionality is taken because such was the policy in the case study. In a system redesign, the "desired inventory" concept could give way to a "necessary inventory" concept based on a consideration of the relationships between customer service delay and customer order size, production lead time, manufacturing costs, changing distribution patterns, and changes in product mix that are a function of changing sales levels. The necessary inventory would probably then lie between the square-root relationship (of simple inventory theory) and a linear proportionality to sales.

The target inventory is here taken as equal to 4 weeks of sales, although in many situations a higher level of inventory would be carried.

CIRF = 4 weeks, Coefficient for Inventory Ratio at Factory

Equation 17-19 requires an average level of sales. This is obtained in the same way as discussed in Equation 15-8:

$$RSF.K = RSF.J + \frac{DT}{TRSF}(RRF.JK - RSF.J) \quad \text{17-20, L}$$

$$RSF = RRF \quad \text{17-21, N}$$

RSF   Requisition rate Smoothed at Factory (units/week)

TRSF   Time for Requisition Smoothing at Factory (weeks)

RRF   Requisition rate Received at Factory (units/week)

This averaging of the incoming-order rate was necessary because week-by-week order flow fluctuated widely. Most of such flow rates must be averaged before making management decisions. This averaging may be a formal numerical process or may be an intuitive or psychological averaging of the flow of available information. Equation 17-20 shows that each week a fraction TRSF of the difference between the current and average sales is used to correct the level of average sales. In the steady-state average, sales will equal the incoming sales rate RRF, as given in Equation 17-21.

The averaging time for incoming orders is here used as

TRSF = 15 weeks[13]

The normal number of items on order for inventory to be expected in process at the factory would be the average inventory shipping rate multiplied by the manufacturing delay for inventory in the factory:

---

[13] This is an exponential smoothing time and under certain conditions is equivalent to a 30-week moving-average interval. See Appendix E.

$$OINF.K = (ASIF.K)(DMIF.K) \qquad \text{17-22, A}$$

OINF Orders for Inventory Necessary at Factory (units)

ASIF Average Shipments from Inventory at the Factory (units/week)

DMIF Delay (variable) in Manufacturing for Inventory at Factory (weeks)

The actual orders in process in the factory for inventory will be the sum of two components, the backlog of orders not yet started and the orders in the process of manufacture:

$$OIAF.K = BLIF.K + OPIF.K \qquad \text{17-23, A}$$

OIAF Orders for Inventory Actual at Factory (units)

BLIF BackLog for Inventory at Factory (units)

OPIF Orders in Process for Inventory at Factory (units)

This completes the description of inventory reordering. Next the manufacturing sector of the company will be described.

17.4.3 Manufacturing. The manufacturing operation is here represented by two flows, one for goods that go into inventory and the other for goods made to customer order. In actual practice these were intermixed in the same production lines, but in the model they will be treated separately to provide the necessary variables representing each of the two flows. Consequently, Figure 17-9 shows two backlogs of orders and two production delays. Since in the process being represented the inventory orders and the customer backlog orders were intermixed, there will be no priority given to either category in this first formulation. Chapter 18 discusses some alternative management policies including a priority for customer orders.

The first equation gives the backlog of orders that are to be manufactured for inventory:

$$BLIF.K = BLIF.J + (DT)(MOIF.JK - BLIRF.JK)$$
$$\text{17-24, L}$$

$$BLIF = (RFIF)(DNBLF) \qquad \text{17-25, N}$$

BLIF BackLog for Inventory at Factory (units)

MOIF Manufacturing Orders for Inventory at the Factory (units/week)

BLIRF BackLog Inventory Reduction rate at Factory (units/week)

RFIF Requisition rate Filled from Inventory at Factory (units/week)

DNBLF Delay in Normal BackLog at Factory (weeks)

In Equation 17-25 the initial value for the backlog of orders for inventory is given by the steady-state flow of orders that are filled from inventory RFIF multiplied by the normal length of the backlog DNBLF (value given after Equation 17-67).

The backlog of work in progress for goods being manufactured to customer order is similar:

$$BLCF.K = BLCF.J + (DT)(RMOF.JK - PCOF.JK)$$
$$\text{17-26, L}$$

$$BLCF = (RRF - RFIF)(DNBLF) \qquad \text{17-27, N}$$

BLCF BackLog for Customer at Factory (units)

RMOF Requisition rate Manufactured to Order at Factory (units/week)

PCOF Production to Customer Order at Factory (units/week)

RRF Requisition rate Received at Factory (units/week)

RFIF Requisition rate Filled from Inventory at Factory (units/week)

DNBLF Delay in Normal BackLog at Factory (weeks)

We must now represent the allocation of manufacturing manpower to each of the order flows. In normal operation, the two backlogs were intermixed, giving them approximately the same priority. This implies that manpower was allocated in proportion to each of the backlogs, so this should be the normal mode of operation. However, there may be times when the available labor can produce more items than are in the backlogs during a period in which work force is being reduced. At such times the backlogs should not become negative, but instead the excess labor is devoted to making inventory.

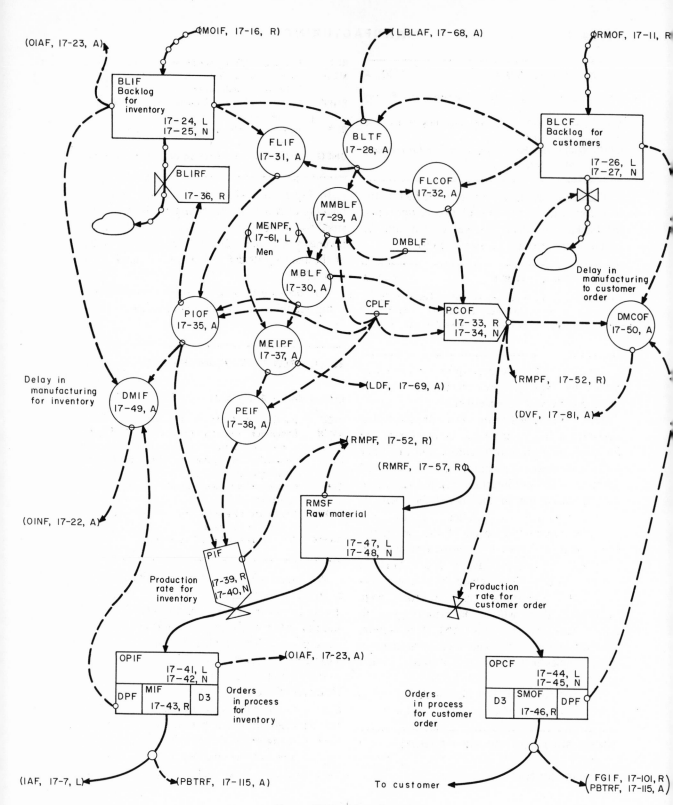

Figure 17-9 Manufacturing.

The following equations describe this allocation of manpower.

The first equation makes available the total backlog of orders:

$$BLTF.K = BLIF.K + BLCF.K \qquad 17\text{-}28, A$$

BLTF   BackLog Total at Factory   (units)
BLIF   BackLog for Inventory at Factory   (units)
BLCF   BackLog for Customer at Factory   (units)

The backlogs must not drop below a length equivalent to some minimum production scheduling time. This minimum delay in the backlogs defines the maximum permissible outflow rate, which in turn tells us the maximum manpower that can be effectively assigned to backlog reduction. Any excess labor will make unordered items for inventory.

$$MMBLF.K = \frac{BLTF.K}{(DMBLF)(CPLF)} \qquad 17\text{-}29, A$$

MMBLF   Men, Maximum, for BackLog production at Factory   (men)
BLTF   BackLog Total at Factory   (units)
DMBLF   Delay, Minimum, in Backlog at Factory   (weeks)
CPLF   Constant, Productivity of Labor at Factory   (units/man-week)

Here the total backlog is divided by the minimum delay in backlog necessary for scheduling the work to give the maximum rate at which orders can enter production. This rate is then divided by the labor productivity to give the maximum work force that can be assigned to making goods to fill orders.

The minimum scheduling time is here

DMBLF = 1 week, Delay, Minimum in Backlog at Factory

The productivity is here taken as

CPLF = 2⅔ units/man-week,[14] Constant Productivity of Labor at Factory

[14] The productivity figure is an arbitrary value chosen to match up a simple set of ratios of costs and production rates for this example. Its value has no effect on system dynamics so long as it is compatible with various other system parameters.

The men to work on backlog reduction should then be the lesser of those available or those permissible (as given by Equation 17-29):

$$MBLF.K = \begin{cases} MENPF.K & \text{if } MMBLF.K \geq MENPF.K \\ MMBLF.K & \text{if } MMBLF.K < MENPF.K \end{cases}$$
$$17\text{-}30, A$$

MBLF   Men for work on BackLog orders at Factory   (men)
MENPF   MEN Producing at Factory   (men)
MMBLF   Men, Maximum, for BackLog Production at Factory   (men)

The men to be used for backlog reduction are to be in proportion to the backlog sizes. The fraction in each backlog is given by

$$FLIF.K = \frac{BLIF.K}{BLTF.K} \qquad 17\text{-}31, A$$

$$FLCOF.K = \frac{BLCF.K}{BLTF.K} \qquad 17\text{-}32, A$$

FLIF   Fraction of Labor on Inventory at Factory   (dimensionless)
BLIF   BackLog for Inventory at Factory   (units)
BLTF   BackLog Total at Factory   (units)
FLCOF   Fraction of Labor on Customer Orders at Factory   (dimensionless)
BLCF   BackLog for Customer at Factory   (units)

The men to work on customer orders is the fraction given by Equation 17-32 of the total working on all orders. This number of men multiplied by labor productivity gives the following production rate for items for customer order:

$$PCOF.KL = (FLCOF.K)(MBLF.K)(CPLF) \qquad 17\text{-}33, R$$

$$PCOF = RRF - RFIF \qquad 17\text{-}34, N$$

PCOF   Production to Customer Order at Factory   (units/week)
FLCOF   Fraction of Labor on Customer Orders at Factory   (dimensionless)
MBLF   Men for work on BackLog orders at Factory   (men)

CPLF   Constant, Productivity of Labor at Factory (units/man-week)

RRF   Requisition rate Received at Factory (units/week)

RFIF   Requisition rate Filled from Inventory at Factory   (units/week)

A similar relationship gives the production rate on inventory backlog orders, but this will first be calculated as an auxiliary equation because it is needed in two different rate calculations:

$$\text{PIOF.K} = (\text{FLIF.K})(\text{MBLF.K})(\text{CPLF}) \qquad \text{17-35, A}$$

$$\text{BLIRF.KL} = \text{PIOF.K} \qquad \text{17-36, R}$$

PIOF   Production of Inventory Orders at Factory (units/week)

FLIF   Fraction of Labor on Inventory at Factory (dimensionless)

MBLF   Men for work on BackLog orders at Factory (men)

CPLF   Constant, Productivity of Labor at Factory (units/man-week)

BLIRF   BackLog Inventory Reduction rate at Factory (units/week)

The actual items produced for inventory will be the sum of those to fill inventory orders plus those to occupy excess employees above the number who can work in response to backlog orders. The excess workers and the rate at which they can produce will first be calculated:

$$\text{MEIPF.K} = \text{MENPF.K} - \text{MBLF.K} \qquad \text{17-37, A}$$

$$\text{PEIF.K} = (\text{MEIPF.K})(\text{CPLF}) \qquad \text{17-38, A}$$

MEIPF   Men for Excess Inventory Production at Factory (men)

MENPF   MEN Producing at Factory   (men)

MBLF   Men for work on BackLog orders at Factory (men)

PEIF   Production Excess for Inventory at Factory (units/week)

CPLF   Constant, Productivity of Labor at Factory (units/man-week)

Equation 17-37 tells us if there is a man-power excess beyond that working on backlog orders. Equation 17-38 multiplies this by productivity to give the excess inventory production.[15]

Total production rate for inventory is given by the sum of the two components already calculated:

$$\text{PIF.KL} = \text{PIOF.K} + \text{PEIF.K} \qquad \text{17-39, R}$$

$$\text{PIF} = \text{RFIF} \qquad \text{17-40, N}$$

PIF   Production rate starts for Inventory at Factory (units/week)

PIOF   Production of Inventory Orders at Factory (units/week)

PEIF   Production Excess for Inventory at Factory (units/week)

RFIF   Requisition rate Filled from Inventory at Factory   (units/week)

The production process is here approximated by two steps. The first is a point at which labor is applied, and thereby controls the rate at which items are started into production. Second, the production start rate is followed by the production process delay before the finished goods become available.[16] If there were justification for further refinement, production could be separated into several stages with allocation of workers and production delay to each. The production delay for inventory items follows as an equation for the work in process and one for the output rate:

[15] Productivity is here being used as a constant. This would not be justified in situations where productivity varies with the state of the demand and the production cycle. During a period of low inventories and factory overload, there may be a special expediting of certain customer orders. Very much of this can increase confusion until productivity actually declines at a time when high productivity is most needed. As demand begins to slacken, the pressure to produce reduces, and lower productivity may persist through the declining market. With an upturn in demand, more output is for a time achievable by the recovery in productivity rather than by actual hiring.

[16] For a discussion of separating rate control and delay, see page 151, before Equation 15-9.

$$OPIF.K = OPIF.J + (DT)(PIF.JK - MIF.JK) \qquad \text{17-41, L}$$

$$OPIF = (DPF)(RFIF) \qquad \text{17-42, N}$$

$$MIF.KL = DELAY3(PIF.JK, \ DPF) \qquad \text{17-43, R}$$

OPIF	Orders in Process for Inventory at Factory (units)
PIF	Production rate starts for Inventory at Factory (units/week)
MIF	Manufactured rate for Inventory at Factory (units/week)
DPF	Delay in Production at Factory (weeks)
RFIF	Requisition rate Filled from Inventory at Factory (units/week)
DELAY3	Specifies third-order-delay equations

Production delay is here represented as a third-order exponential delay of average length

$$DPF = 6 \text{ weeks}$$

The delay in production to customer order is similar:

$$OPCF.K = OPCF.J + (DT)(PCOF.JK - SMOF.JK) \qquad \text{17-44, L}$$

$$OPCF = (DPF)(RRF - RFIF) \qquad \text{17-45, N}$$

$$SMOF.KL = DELAY3(PCOF.JK, \ DPF) \qquad \text{17-46 R}$$

OPCF	Orders in Process for Customers at Factory (units)
PCOF	Production to Customer Order at Factory (units/week)
SMOF	Shipment rate Manufactured to Order at Factory (units/week)
DPF	Delay in Production at Factory (weeks)
RRF	Requisition rate Received at Factory (units/week)
RFIF	Requisition rate Filled from Inventory at Factory (units/week)
DELAY3	Specifies third-order-delay equations

The steady-state flow of orders made to customer specification in Equation 17-45 is the total incoming flow RRF less the flow that can be filled from inventory RFIF.

The raw-material level is produced by the raw-material receiving rate and the two depletion rates into the two production streams:

$$RMSF.K = RMSF.J + (DT)(RMRF.JK - PIF.JK - PCOF.JK) \qquad \text{17-47, L}$$

$$RMSF = (RRF)(CRMSF) \qquad \text{17-48, N}$$

RMSF	Raw-Material Stock at Factory (equivalent product units)
RMRF	Raw Material Received at Factory (equivalent product units/week)
PIF	Production rate starts for Inventory at Factory (units/week)
PCOF	Production to Customer Order at Factory (units/week)
RRF	Requisition rate Received at Factory (units/week)
CRMSF	Coefficient, Raw-Material Supply held at Factory (weeks)

The total delays in the manufacturing department are needed in decisions elsewhere within the system. The delay consists of two parts, that which an order experiences waiting in the backlog and that in the actual production. Equation 17-49 gives the delay expected for inventory production as the actual manufacturing delay plus the size of the backlog divided by the rate at which the backlog is being depleted:

$$DMIF.K = DPF + \frac{BLIF.K}{PIOF.K} \qquad \text{17-49, A}$$

DMIF	Delay (variable) in Manufacturing for Inventory at Factory (weeks)
DPF	Delay in Production at Factory (weeks)
BLIF	BackLog for Inventory at Factory (units)
PIOF	Production of Inventory Orders at Factory (units/week)

In a similar way, the delay for manufacturing to customer order is made up of the constant manufacturing delay plus the variable delay that the order experiences in the backlog:

$$DMCOF.K = DPF + \frac{BLCF.K}{PCOF.JK} \qquad \text{17-50, A}$$

DMCOF	Delay (variable) in Manufacturing to Customer Order at Factory (weeks)
DPF	Delay in Production at Factory (weeks)

BLCF BackLog for Customer at Factory (units)
PCOF Production to Customer Order at Factory (units/week)

Since the production rates are proportional to the backlogs when neither channel has priority, the delays in the two channels will tend to be equal.

17.4.4 Material Ordering. As stated in the initial description, the availability of materials did not appear to control other active decisions within the manufacturing operation. Nevertheless, the ordering equation for materials and the materials delivery delay are included in this model so that they can properly generate the flow of cash for the purchase of materials.

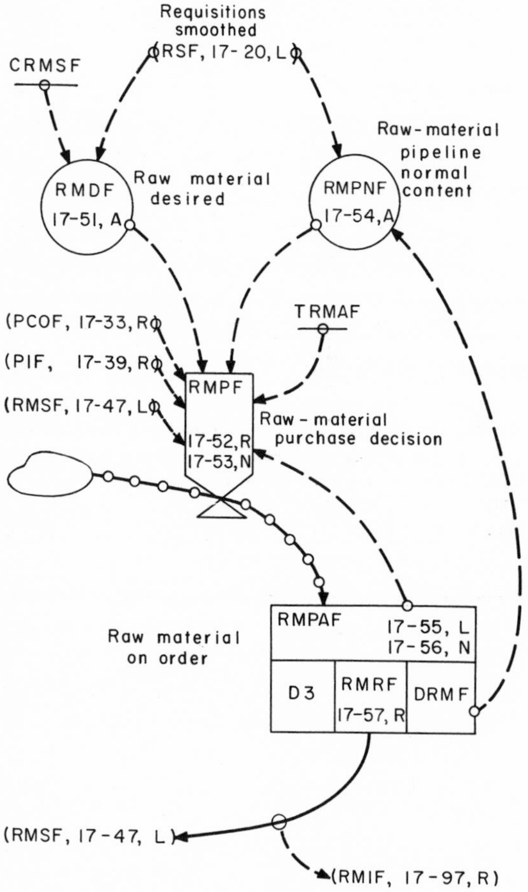

Figure 17-10 Material ordering.

The material-ordering equations that follow are similar to the corresponding equations in Chapter 15. The flow diagram is in Figure 17-10.

First is the concept of raw materials desired at the factory RMDF against which the adequacy of actual stocks can be compared.[17]

$$RMDF.K = (RSF.K)(CRMSF) \qquad \text{17-51, A}$$

RMDF Raw Material Desired at Factory (equivalent product units)
RSF Requisition rate Smoothed at Factory (units/week)
CRMSF Coefficient, Raw-Material Supply held at Factory (weeks)

The desired stock of raw materials will be taken as six times the average weekly usage rate:

CRMSF = 6 weeks

The raw-material purchasing rate will be made to depend upon the rate of usage of materials in production and will also contain a term for adjusting the inventory stock and the raw-material pipeline content. The following equation is similar to Equation 15-9:

$$RMPF.KL = PCOF.JK + PIF.JK + \frac{1}{TRMAF}(RMDF.K - RMSF.K + RMPNF.K - RMPAF.K)$$
$$\text{17-52, R}$$

$$RMPF = RRF \qquad \text{17-53, N}$$

RMPF Raw-Material Purchases at Factory (equivalent product units/week)
PCOF Production to Customer Order at Factory (units/week)
PIF Production rate starts for Inventory at Factory (units/week)
TRMAF Time for Raw-Material Adjustment at Factory (weeks)
RMDF Raw Material Desired at Factory (equivalent product units)
RMSF Raw-Material Stock at Factory (equivalent product units)

[17] In the future, this might be made less dependent on changes in sales level.

RMPNF   Raw-Material Pipeline Normal content at Factory   (units)
RMPAF   Raw-Material Pipeline Actual content at Factory   (units)
RRF   Requistion rate Received at Factory
(units/week)

In Equation 17-52 we again follow the expediency of letting the present raw-material purchasing rate depend upon the present raw-material usage rate even though in principle and in practice the usage rate should be smoothed and delayed slightly to create levels as was done in Equation 17-17. The parentheses in Equation 17-52 give the difference between desired and actual raw-material stock plus the difference between necessary and actual raw-material pipeline content. The time constant TRMAF controls the rate at which inventory and pipeline discrepancies will be corrected. The following value of 8 weeks means that the correction rate is one-eighth of the remaining discrepancy per week:

TRMAF = 8 weeks, Time for Raw-Material Adjustment at Factory

The necessary raw-material orders and material in transit in the material supply pipeline are proportional to the average level of business activity and the length of the pipeline:

$$\text{RMPNF.K} = (\text{RSF.K})(\text{DRMF}) \qquad \text{17-54, A}$$

RMPNF   Raw-Material Pipeline Normal content at Factory   (units)
RSF   Requisition rate Smoothed at Factory
(units/week)
DRMF   Delay in Raw Materials at Factory   (weeks)

Since we are not interested here in the effects of a variable delay in the supply of materials, it will suffice to pass the orders through a third-order delay from which the materials return to factory raw-material stock. This requires a level equation, its initial-value equation, and the designation of the delay function:

$$\text{RMPAF.K} = \text{RMPAF.J} + (\text{DT})(\text{RMPF.JK} - \text{RMRF.JK})$$
$$\text{17-55, L}$$

$$\text{RMPAF} = (\text{RRF})(\text{DRMF}) \qquad \text{17-56, N}$$

$$\text{RMRF.KL} = \text{DELAY3}(\text{RMPF.JK}, \text{DRMF}) \qquad \text{17-57, R}$$

RMPAF   Raw-Material Pipeline Actual content at Factory   (units)
RMPF   Raw-Material Purchases at Factory   (equivalent product   units/week)
RMRF   Raw Material Received at Factory   (equivalent product   units/week)
RRF   Requisition rate Received at Factory
(units/week)
DRMF   Delay in Raw Materials at Factory   (weeks)
DELAY3   Specifies third-order-delay equations

The average delay in procuring raw materials is here

DRMF = 3 weeks

17.4.5 Labor. The labor supply and the policies governing the changes in labor force are important parts of the system under study. The effect under study is the interaction between the varying flow of incoming orders from the customers and the resulting manufacturing rate, which in turn is controlled by the labor force and adjustment.

The labor sector and its governing policies are diagrammed in Figure 17-11. The labor loop itself consists of a labor pool, a hiring decision, an initial training period, the level of manpower available for production, a dismissal decision, and employees who have received termination notices but have not yet left the payroll.

Several concepts are developed in a series of auxiliary equations dealing with the desired number of employees for the average level of business and the number of employees necessary for adjusting undesirable levels of order backlogs.

We shall start with the main labor flow loop by considering the training delay. This is described by the following three equations:

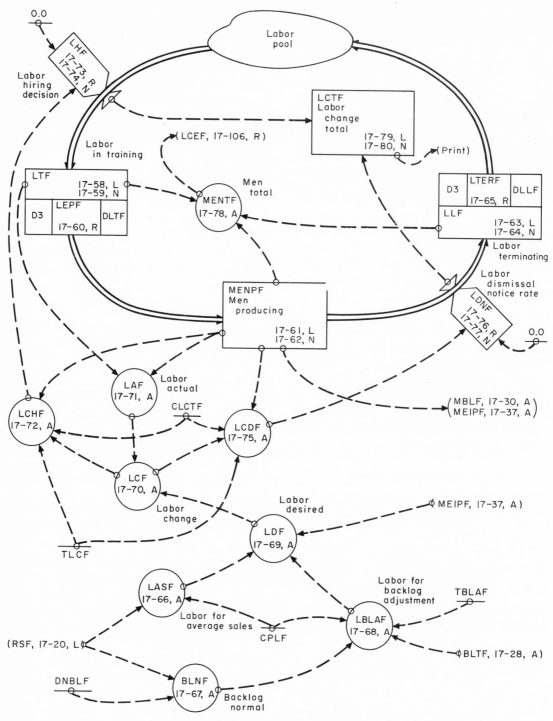

Figure 17-11 Labor flow.

$$LTF.K = LTF.J + (DT)(LHF.JK - LEPF.JK) \qquad \text{17-58, L}$$

$$LTF = 0 \qquad \text{17-59, N}$$

$$LEPF.KL = DELAY3(LHF.JK, \ DLTF) \qquad \text{17-60, R}$$

LTF	Labor in Training at Factory (men)
LHF	Labor Hiring rate at Factory (men/week)
LEPF	Labor Entering Production at Factory (men/week)
DELAY3	Specifies third-order-delay equations
DLTF	Delay in Labor Training at Factory (weeks)

Equation 17-58 is the level of personnel in training. In the steady-state initial conditions when there are no changes in levels of activity within the system, the number of trainees will be zero, as given in Equation 17-59. Equation 17-60 specifies the training-interval delay. The nonproductive training interval is used here both as training and as a method of representing the inefficiencies attendant on an increasing production rate. The greater the rate of addition to the work force, the larger will be the level of trainees LTF. This makes a clean break between a period of no productivity and the entrance into the work force where productivity is considered to be at its full value. In practice, the trainees might start to work soon after being employed, but their addition to the work force would cause rearrangement of production processes, causing some lowering of the effectiveness of workers already in the producing category.

The training delay used here is

$$DLTF = 3 \ \text{weeks}$$

The number of active production workers is given by the level equation:

$$MENPF.K = MENPF.J + (DT)(LEPF.JK - LDNF.JK) \\ \text{17-61, L}$$

$$MENPF = \frac{RRF}{CPLF} \qquad \text{17-62, N}$$

MENPF	MEN Producing at Factory (men)
LEPF	Labor Entering Production at Factory (men/week)

LDNF	Labor Dismissal Notice rate at Factory (men/week)
RRF	Requisition rate Received at Factory (units/week)
CPLF	Constant, Productivity of Labor at Factory (units/man-week)

The initial level of production manpower as given by Equation 17-62 is equal to the steady-state constant level of business activity divided by the productivity per worker.

The labor-termination flow is similar to the labor-training flow as given by the following equations:

$$LLF.K = LLF.J + (DT)(LDNF.JK - LTERF.JK) \qquad \text{17-63, L}$$

$$LLF = 0 \qquad \text{17-64, N}$$

$$LTERF.KL = DELAY3(LDNF.JK, \ DLLF) \qquad \text{17-65, R}$$

LLF	Labor Leaving Factory (men)
LDNF	Labor Dismissal Notice rate at Factory (men/week)
LTERF	Labor TERminating at Factory (men/week)
DELAY3	Specifies third-order-delay equations
DLLF	Delay in Labor Leaving Factory (weeks)

Equation 17-63 gives the number of persons in termination status. Equation 17-64 indicates that this number is zero in steady-state nonvarying production conditions. Equation 17-65 defines the outflow rate LTERF, which is the rate at which labor is leaving the payroll. The termination delay is set up here as a period during which salaries will be paid and the personnel are nonproductive. This could be created by one or both of the following: an actual period of terminal pay following discharge, or a period of lowered productivity caused by the slacking off of pressure and the dislocations created by a reduction and rearrangement in the production rate.

The termination period is here taken as

$$DLLF = 4 \ \text{weeks}$$

We now turn to the several concepts governing the desired labor force and the hiring and layoff rates. First there is a labor level that will produce at the average level of the incoming-order rate:

$$LASF.K = \frac{RSF.K}{CPLF} \qquad 17\text{-}66, A$$

LASF Labor for Average Sales at Factory (men)
RSF Requisition rate Smoothed at Factory (units/week)
CPLF Constant, Productivity of Labor at Factory (units/man-week)

Equation 17-66 divides the average level of incoming orders by the labor productivity to give the number of workers who could produce at the incoming average order rate.

Next is a consideration of the backlog conditions at the factory. A normal backlog of unfilled orders must be defined. Too large a backlog is undesirable for reasons of competition and customer service. Too small a backlog means that the production process is running dangerously close to having no manufacturing authorization on which to work. In many situations, as here, there is a concept of a proper level of backlog in terms of how long it is relative to the rate of production.

$$BLNF.K = (RSF.K)(DNBLF) \qquad 17\text{-}67, A$$

BLNF BackLog Normal at Factory (units)
RSF Requisition rate Smoothed at Factory (units/week)
DNBLF Delay in Normal BackLog at Factory (weeks)

The normal backlog here is

DNBLF = 4 weeks, Delay in Normal BackLog at Factory

Since the actual backlog will often be different from the normal backlog, we now generate the amount of labor that would be necessary to adjust the backlog to the desired level at some specified correction rate:

$$LBLAF.K = \frac{BLTF.K - BLNF.K}{(CPLF)(TBLAF)} \qquad 17\text{-}68, A$$

LBLAF Labor for BackLog Adjustment at Factory (men)
BLTF BackLog Total at Factory (units)
BLNF BackLog Normal at Factory (units)
CPLF Constant, Productivity of Labor at Factory (units/man-week)
TBLAF Time for BackLog Adjustment at Factory (weeks)

In Equation 17-68 the numerator gives the difference between the total backlog at the factory and the backlog that would be considered normal. When this is divided by the productivity CPLF, it gives the man-weeks of work necessary to correct the backlog to its normal value. This must be divided by the period of time over which the correction is to be accomplished to give the number of men who are to be employed for the purpose of adjusting the backlog. The adjustment period [18] is here taken as

TBLAF = 20 weeks, Time for BackLog Adjustment at Factory

This adjustment time may seem long, but is in fact a rather rapid adjustment rate. For example, suppose that the backlog were twice normal, or 8 weeks' worth of work. This would not be an unusual condition in many industries. The excess backlog would then be 4 weeks' worth of production. If this is to be adjusted in 20 weeks, it means devoting to the reduction of backlog a production force equal to one-fifth of that necessary for producing at the rate of current sales. In other words, a backlog of twice the normal amount implies here a 20% increase in the production rate above that necessary for present sales. This is a rapid but not an implausible adjustment rate.

It is now possible to determine the desired level of labor at the factory. This will consist of three parts — the labor necessary to produce at

[18] More properly called the adjustment time constant. This relationship gives the rate at which the *remaining* backlog deficit is to be corrected.

the average sales rate, that necessary to adjust the backlog level, which may be either positive or negative, and a reduction by the number of workers who are producing for inventory beyond the rate covered by inventory production orders:

$$LDF.K = LASF.K + LBLAF.K - MEIPF.K \qquad 17\text{-}69, \text{ A}$$

LDF   Labor Desired at Factory   (men)
LASF   Labor for Average Sales at Factory (men)
LBLAF   Labor for BackLog Adjustment at Factory (men)
MEIPF   Men for Excess Inventory Production at Factory (men)

The desired labor level minus the actual labor level will give the excess or deficit in the present level of production manpower:

$$LCF.K = LDF.K - LAF.K \qquad 17\text{-}70, \text{ A}$$

LCF   Labor Change indicated at Factory (men)
LDF   Labor Desired at Factory (men)
LAF   Labor to be Available at Factory (men).

In Equation 17-70 the present actual labor force is required. Since the producing force in the near future will include both those actually now producing and those in training, the sum of the two is used as follows:

$$LAF.K = LTF.K + MENPF.K \qquad 17\text{-}71, \text{ A}$$

LAF   Labor to be Available at Factory (men)
LTF   Labor in Training at Factory (men)
MENPF   MEN Producing at Factory (men)

Equation 17-70 gives the indicated labor change as a positive number if men are to be hired and a negative number if laid off. One other consideration must be included — how rapidly the indicated change is to be accomplished. In addition, we wish to insert here another possible factor in the hire-layoff decision. Certain managers may follow the practice of making no employment change until the labor discrepancy has reached some percentage

of the work force. For example, as long as the desired employee level is within 2% (or 5%, or 10%) of the value thought best, no change is made in employment. When employment differs by more than the threshold, hiring or layoff takes place until employment again falls within the tolerance. Since we shall later want to study the effect of such an employment-change threshold, the necessary terms will now be incorporated. For the hiring decision, the threshold and the rate at which discrepancies in work force are corrected are represented in the following equation:

$$LCHF.K = \frac{1}{TLCF} [LCF.K - (CLCTF)(MENPF.K)]$$
$$17\text{-}72, \text{ A}$$

LCHF   Labor Change rate for Hiring at Factory (men/week)
TLCF   Time for Labor Change at Factory (weeks)
LCF   Labor Change indicated at Factory (men)
CLCTF   Constant, Labor Change Threshold at Factory, fraction of production labor (dimensionless)
MENPF   MEN Producing at Factory (men)

In Equation 17-72 the threshold for hiring and layoff is given by the percentage **CLCTF** multiplied by the present labor force **MENPF**. The hiring rate is to be effective (see next equation) only for positive values of labor change **LCHF**, so that the threshold is subtracted from the indicated labor change **LCF**. Only the excess of **LCF** over the threshold is effective in causing hiring. The desired labor change in Equation 17-72 is divided by the time constant **TLCF**, which represents the rapidity in making the labor change. Here

TLCF = 10 weeks, Time for Labor Change at Factory

Because the hiring threshold is not to be ordinarily used, its normal value will be

CLCTF = 0.0, dimensionless Constant, Labor-Change Threshold at Factroy

This will be altered only when the effect of a

threshold is being investigated. The reciprocal of the time constant TLCF in Equation 17-72 gives the fraction of the labor discrepancy that will be corrected per week. This factor is used here to represent a number of practical considerations. A period of time is necessary before an actual labor discrepancy will come to the notice of management. There will probably be some delay in a decision, in the hope that no change will be necessary. If the decision is for hiring, time will be required for finding the workers. If the decision is for releasing, various arrangements must be made, and some organizations follow the policy of trying to let normal resignations take care of the reduction. The 10 weeks just mentioned is probably on the short side of the usual delays that will be found at this point in a manufacturing process.

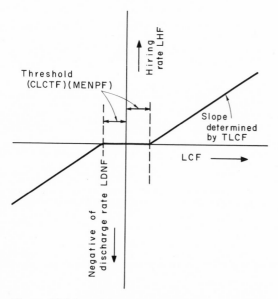

Figure 17-12 Hiring and layoff rates versus indicated labor change.

Figure 17-12 shows the way in which the hiring and layoff rates depend on the labor excess or deficit LCF. If there is no inactive threshold, the horizontal section of the curve disappears, and the sloping sections connect.

Since hiring is to take place only if the result from Equation 17-72 is positive,

$$\text{LHF.KL} = \begin{cases} \text{LCHF.K} & \text{if LCHF.K} \geq 0.0 \\ 0.0 & \text{if LCHF.K} < 0.0 \end{cases} \quad \text{17-73, R}$$

$$\text{LHF} = 0 \qquad \text{17-74, N}$$

LHF   Labor Hiring rate at Factory   (men/week)
LCHF   Labor Change rate for Hiring at Factory (men/week)

The layoff rate is established in a similar way except that the threshold is entered with a positive sign to reduce a negative output from Equation 17-70, and Equation 17-76 reverses the sign to give a positive discharge rate when there is too large a work force:

$$\text{LCDF.K} = \frac{1}{\text{TLCF}}[\text{LCF.K} + (\text{CLCTF})(\text{MENPF.K})]$$

$$\text{17-75, A}$$

$$\text{LDNF.KL} = \begin{cases} 0.0 & \text{if LCDF.K} \geq 0.0 \\ -\text{LCDF.K} & \text{if LCDF.K} < 0.0 \end{cases}$$

$$\text{17-76, R}$$

$$\text{LDNF} = 0 \qquad \text{17-77, N}$$

LCDF   Labor Change rate for Discharge at Factory (men/week)
TLCF   Time for Labor Change at Factory   (weeks)
LCF   Labor Change indicated at Factory   (men)
CLCTF   Constant, Labor Change Threshold at Factory, fraction of production labor (dimensionless)
MENPF   MEN Producing at Factory   (men)
LDNF   Labor Dismissal Notice rate at Factory (men/week)

For use in calculating cash flow for payrolls we shall need the total number of men on the payroll:

$$\text{MENTF.K} = \text{LTF.K} + \text{MENPF.K} + \text{LLF.K} \qquad \text{17-78, A}$$

MENTF   MEN, Total, at Factory   (men)
LTF   Labor in Training at Factory   (men)
MENPF   MEN Producing at Factory   (men)
LLF   Labor Leaving Factory   (men)

As a figure of merit in comparing different model runs with different managerial policies, it will be helpful to know the total employment changes that have occurred. This could be taken as the sum of all hirings and layoffs as given by the following level equation:

$$LCTF.K = LCTF.J + (DT)(LHF.JK + LDNF.JK) \quad \text{17-79, L}$$

$$LCTF = 0 \quad \text{17-80, N}$$

LCTF Labor Change Total at Factory (men)
LHF Labor Hiring rate at Factory (men/week)
LDNF Labor Dismissal Notice rate at Factory (men/week)

Equation 17-79 is not active in the policies governing the system but is a summation from the beginning of a machine run until the end, giving a count of total men hired plus total men laid off. This would be one indication of the degree of labor stability achieved in the system under study.

The added complication of overtime and a variable work week have been omitted from this example. If the work week is normally varied to adjust production rate, the policies leading to adjustment in working hours would be included in a manner similar to the employment changes.

17.4.6 Delivery-Delay Quotation. One of the important coupling channels between the the factory and the customer is the delivery delay that the customer anticipates in placing his orders. The generation and transmission of order-filling delay from the factory to the customer is therefore an essential link in the system under study. Figure 17-13 shows the flow diagram for delivery delay based on variables from Figures 17-6 and 17-9. All incoming orders are delayed in Figure 17-6 at the point where they are being checked for specifications and sorted, either to be filled from inventory or to be made to customer specification. After this sorting, the orders going to inventory experience a constant average delay while those going to the factory experience a variable delay de-

pending on the length of the factory unfilled-order backlog. The delay experienced after the order split at RCF in Figure 17-6 consists of the following two components:

$$DVF.K = (FRFIF.K)(DSF) + (1 - FRFIF.K)(DMCOF.K) \quad \text{17-81, A}$$

DVF Delay, Variable delivery from Factory (weeks)
FRFIF Fraction of Requisitions Filled from Inventory at Factory (dimensionless)
DSF Delay in Shipping at Factory (weeks)
DMCOF Delay (variable) in Manufacturing to Customer Order at Factory (weeks)

In Equation 17-81, the fraction of orders filled from inventory multiplied by the inventory-filling delay is added to the fraction of orders filled to customer specification multiplied

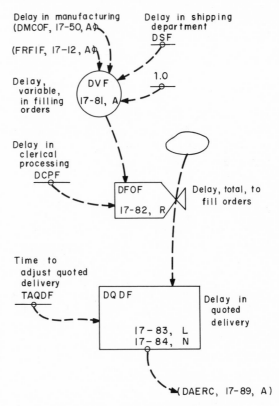

Figure 17-13 Delivery-delay quotation.

by the delay that these orders experience in traversing the factory. The sum of these two components gives the average delay experienced by the entire order flow after it divides into the two channels.

In addition to the average delay for the two flow channels, all orders experience the delay DCPF in Figure 17-6 while they are being separated into the inventory and factory channels. This delay must be added to the variable delay from Equation 17-81 as is done in the following rate equation that controls the primary flow of order-filling-delay information within the factory.

$$DFOF.KL = DCPF + DVF.K \qquad \text{17-82, R}$$

DFOF Delay (variable) to Fill Orders at Factory (weeks)

DCPF Delay in Clerical Processing at Factory (weeks)

DVF Delay, Variable delivery from Factory (weeks)

Equation 17-82 gives the present instantaneous delivery delay. This value is not necessarily known by the sales department for quotation to customers. Awareness of an increasing factory backlog does not permeate the entire organization at once. Quoted delivery delays in many industries are based more on the delays which *have been* experienced by the orders currently being delivered rather than the delays which can be anticipated in the *future,* based on considerations of backlog and present production rate. The delay information given by Equation 17-82 should then be delayed, and the values transmitted to the customer will lag behind the values that may be indicated by present instantaneous operating facts within the organization. The following equation[19] is an averaging equation (producing a first-order infor-

[19] As noted in earlier chapters, the dimensional units of measure are not an adequate indication of whether a variable should be written as a rate or level equation. We have seen numerous examples of level

mation delay) into which flows the delivery-delay information:

$$DQDF.K = DQDF.J + \frac{DT}{TAQDF}(DFOF.JK - DQDF.J)$$

$$\text{17-83, L}$$

$$DQDF = DCPF + (CNFIF)(DSF) + (1 - CNFIF)(DPF + DNBLF) \quad \text{17-84, N}$$

DQDF Delay (variable) in Quoted Delivery at Factory (weeks)

TAQDF Time to Adjust Quoted Delivery at Factory (weeks)

DFOF Delay (variable) to Fill Orders at Factory (weeks)

DCPF Delay in Clerical Processing at Factory (weeks)

CNFIF Constant, Normal Fraction of requisitions filled from Inventory at Factory (dimensionless)

DSF Delay in Shipping at Factory (weeks)

DPF Delay in Production at Factory (weeks)

DNBLF Delay in Normal BackLog at Factory (weeks)

The time delay TAQDF indicates the rapidity with which quoted time delays DQDF approach the value DFOF that is implicit in the state of the manufacturing operation. Here we shall use a 1-month lag; therefore

TAQDF = 4 weeks, Time to Adjust Quoted Delivery at Factory

The initial-value equation, 17-84, is derived by combining Equations 17-81 and 17-82, in which the split fraction FRFIF has been replaced by its normal value CNFIF from Figure 17-7, and the manufacturing delay DMCOF is replaced by the steady-state values.

17.4.7 Customer Ordering. The customer ordering section appears in the flow diagram of Figure 17-14. The independent input INPUT represents the flow of orders to the customer for equipment in which will be used the components made by the component supplier. As discussed in Section 17.2, this will be considered as an independent input to the system under study. It

equations dealing with average rates and having the same dimensions of measure as a rate of flow. In information channels rates will often have the dimensions of the information being delayed and averaged.

may be assumed either constant or variable, depending on what characteristics of the system are being studied. The other outside input to this sector is the delivery delay coming from Figure 17-13. The customer ordering sector includes a representation of the customer's engineering department, from which the pur-

chasing specifications are released, as well as a representation of the customer's purchasing department.

The engineering designs in process at the customer EDPC are given by the level equation connecting the inflow of incoming orders and the outflow of specifications for purchase:

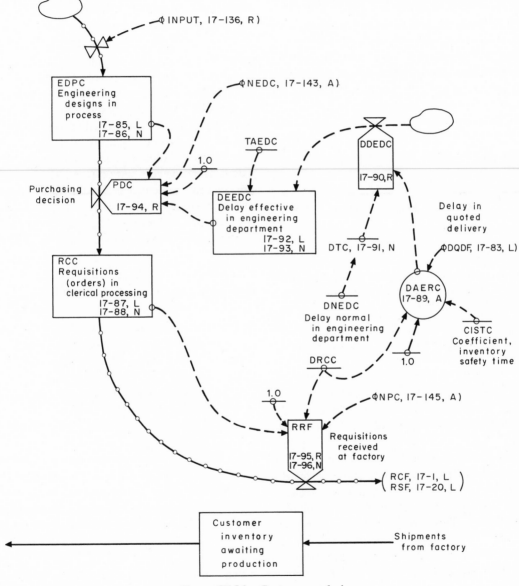

Figure 17-14 Customer ordering.

$$EDPC.K = EDPC.J + (DT)(INPUT.JK - PDC.JK)$$
$$17\text{-}85, L$$

$$EDPC = (INPUT)(DEEDC) \qquad 17\text{-}86, N$$

EDPC   Engineering Designs in Process at Customer (units)

INPUT   INPUT of orders to customer from outside of system (equivalent units/week)

PDC   Purchasing Decision at Customer (units/week)

DEEDC   Delay Effective (variable) in Engineering Department at Customer (weeks)

The steady-state level of designs in process is given by Equation 17-86 as the product of a steady-state input flow and the initial value of the delay in the engineering department.

The next level equation defines the requisitions in process in the purchasing department RCC:

$$RCC.K = RCC.J + (DT)(PDC.JK - RRF.JK) \qquad 17\text{-}87, L$$

$$RCC = (INPUT)(DRCC) \qquad 17\text{-}88, N$$

RCC   Requisitions in Clerical processing at Customer (units)

PDC   Purchasing Decision at Customer (units/week)

RRF   Requisition rate Received at Factory (units/week)

INPUT   INPUT of orders to customer from outside of system (equivalent units/week)

DRCC   Delay, average, for Requisitions in Clerical processing at Customer (weeks)

The steady-state level in Equation 17-88 is given by the initial flow rate multiplied by the average delay in the processing of purchase orders DRCC.

The customer engineering department will be responsive to the component procurement delay as one of the factors controlling the release date of design specifications to the purchasing department. This delay between engineering release and the time that components are assembled into the equipment produced by the customer consists of three parts. The first is the delay in placing orders in the customer's purchasing department, here assumed to be constant, on the average.

The second part of the delay is the time necessary on the average for filling orders at the component manufacturer; this is the variable quoted delivery time DQDF. The third part of the delay is the length of time that components lie in inventory awaiting production at the customer plant. We are especially interested in any variable component in the inventory holding period at the customer, since this can create a variable reservoir that can affect the customer's ordering rate. It is reasonable to assume that the customer will protect himself by carrying an inventory stock that is related to the length of time that he finds is needed to procure components. As the delivery delay increases, so will the inventory holding period, at least to some extent. Here the constant CISTC relates the component procurement delay to the variable part of the inventory holding period. The total time from engineering release date until the components are assembled into equipment is then given by

$$DAERC.K = DRCC + (1 + CISTC)(DQDF.K)$$
$$17\text{-}89, A$$

DAERC   Delay (variable) in procurement After Engineering Release at Customer (weeks)

DRCC   Delay, average, for Requisitions in Clerical processing at Customer (weeks)

CISTC   Coefficient for Inventory Safety Time at Customer, fraction of procurement delay that goods lie in customer inventory (dimensionless)

DQDF   Delay (variable) in Quoted Delivery at Factory (weeks)

We shall here use

CISTC = 0.5, dimensionless Coefficient for Inventory Safety Time at Customer, fraction of procurement delay that goods lie in customer inventory

which means that the customer's inventory holding period goes up and down by half the amount of the actual change in procurement delay.

As stated in the general description of the problem, the customer tends to order critical components as far ahead as necessary. This means that especially critical components are given special consideration in the design process so that they can be released earlier when circumstances make this necessary. If everything required by the customer should be on a long delivery schedule, his own equipment delivery dates may have to be postponed. However, his first attempt will be to maintain his own delivery schedule by accelerating the release date of special items out of his engineering department so that they may be ordered as early as necessary. In other words, normal engineering delay plus average procurement delay controls the customer in promising his own deliveries. The particular delay experienced in each component tends to determine when he places his orders. The permissible time that an order may take in the customer's engineering department will now be defined as the total time he has from incoming equipment order to equipment delivery, less the time that component procurement requires:

$$DDEDC.KL = DTC - DAERC.K \qquad \text{17-90, R}$$

DDEDC   Delay (variable) Desired in Engineering Department of Customer (weeks)
DTC   Delay Total at Customer (weeks)
DAERC   Delay (variable) in procurement After Engineering Release at Customer (weeks)

Equation 17-90 gives the flow of desired release-date information to the engineering department. The normal total time taken by the customer to fill equipment orders DTC is here approximated (omitting details of production and equipment test) by the following:

$$DTC = DNEDC + DRCC + (1 + CISTC)(DQDF)$$
$$\text{17-91, N}$$

DTC   Delay Total at Customer (weeks)

DNEDC   Delay Normal in Engineering Department at Customer (weeks)
DRCC   Delay, average, for Requisitions in Clerical processing at Customer (weeks)
CISTC   Coefficient for Inventory Safety Time at Customer, fraction of procurement delay that goods lie in customer inventory (dimensionless)
DQDF   Delay (variable) Quoted Delivery at Factory (weeks)

The normal engineering department design time used here is

DNEDC = 30 weeks, Delay Normal in Engineering Department at Customer

Equation 17-90 gives the desired time that the engineering department has available for design before purchase specification release. However, the change in engineering design time cannot be instantaneously responsive to a change in component-delivery quotations. To accelerate the design release of a particular component, it may be necessary to start well ahead of the release date to begin making this information available ahead of schedule. The following equation represents the lag in the actual delay effective in the engineering department as it lags behind the changing delivery quotations:

$$DEEDC.K = DEEDC.J + \frac{DT}{TAEDC}(DDEDC.JK - DEEDC.J)$$
$$\text{17-92, L}$$

$$DEEDC = DTC - (1 + CISTC)(DQDF) - DRCC$$
$$\text{17-93, N}$$

DEEDC   Delay Effective (variable) in Engineering Department at Customer (weeks)
TAEDC   Time to Adjust Engineering Delay at Customer (weeks)
DDEDC   Delay (variable) Desired in Engineering Department of Customer (weeks)
DTC   Delay Total at Customer (weeks)
CISTC   Coefficient for Inventory Safety Time at Customer, fraction of procurement delay that goods lie in customer inventory (dimensionless)

DQDF   Delay (variable) Quoted Delivery at Factory (weeks)

DRCC   Delay, average, for Requisition in Clerical processing at Customer (weeks)

The time constant of the engineering response delay is here assumed to be

TAEDC = 15 weeks, Time to Adjust Engineering Delay at Customer

Equation 17-92 is an equation of the first-order exponential smoothing type used here to delay the information **DDEDC** before it becomes actually effective in controlling engineering department releases in the form of **DEEDC**. The reciprocal of the adjustment time **TAEDC** gives the fraction of the difference between the quoted and the effective delay that is adjusted per week within the engineering department scheduling. For example, if **DEEDC** had been 30 weeks and **DDEDC** is now 25 weeks, the difference of 5 weeks divided by the 15-week time constant **TAEDC** would lead to a shortening of the engineering department design schedule by one-third of a week per week.

The variable delay developed in Equation 17-92 is now available to control the fraction per week of the orders in the engineering department which are released for purchasing:

$$PDC.KL = \frac{EDPC.K}{DEEDC.K}(1 + NEDC.K) \qquad \text{17-94, R}$$

PDC   Purchasing Decision at Customer (units/week)

EDPC   Engineering Designs in Process at Customer (units)

DEEDC   Delay Effective (variable) in Engineering Department at Customer (weeks)

NEDC   Noise at Engineering Department output of Customer (dimensionless)

The essential part of the purchasing decision at the customer is the ratio of engineering work in process to the engineering department delay. This gives the outflow rate in terms of the work in progress and the time necessary for its com-

pletion. The term in parentheses adds a noise term that can be used to test the sensitivity of the system to a fluctuating and erratic output from the engineering department pool. In an actual situation this noise can come from the intermittent release of large projects and from changes in economic conditions, changes in military budgeting and contracting, and other factors that may cause customers to accelerate or delay their purchasing decisions. The noise term **NEDC** will be specified later in the input test functions.

In a similar way, the release rate of orders from the purchasing department will be taken as a fixed fraction of the orders being processed. As in the engineering release rate, this can be modified by a noise variable to represent the bunching and irregular placing of orders as they would come out of customer purchasing departments:

$$RRF.KL = \frac{RCC.K}{DRCC}(1 + NPC.K) \qquad \text{17-95, R}$$

RRF = INPUT   17-96, N

RRF   Requisition rate Received at Factory (units/week)

RCC   Requisitions in Clerical processing at Customer (units)

DRCC   Delay, average, for Requisitions in Clerical processing at Customer (weeks)

NPC   Noise at output of Purchasing at Customer (dimensionless)

INPUT   INPUT of orders to customer from outside of system (equivalent units/week)

The average delay in placing purchase orders is here

DRCC = 3 weeks

17.4.8 Cash Flow. The cash flow sector is shown in Figure 17-15. As set up here it is for the monitoring of operation in the remainder of the model. Cash position and the cash flow rates do not enter into any of the essential decisions of the system. However, it is to be ex-

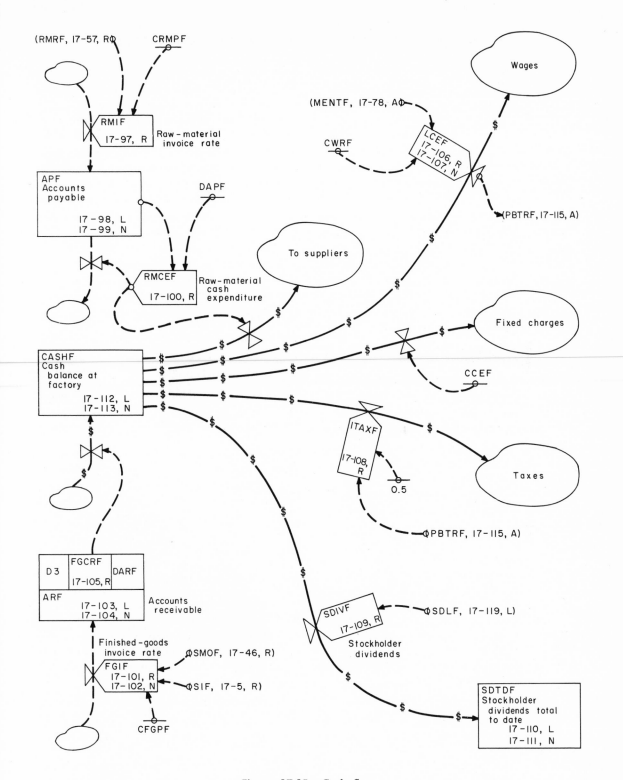

Figure 17-15 Cash flow.

241

pected that different policies will lead to different profitability of operation and to different degrees of fluctuation in cash position. The cash flow sector will be kept simple to serve this monitoring purpose and to show how the financial flows can be added to a dynamic model.

The following equations give raw-material invoice rate, the level of accounts payable, and the rate of cash expenditure for raw materials:

$$RMIF.KL = (RMRF.JK)(CRMPF) \qquad \text{17-97, R}$$

$$APF.K = APF.J + (DT)(RMIF.JK - RMCEF.JK) \qquad \text{17-98, L}$$

$$APF = (RRF)(CRMPF)(DAPF) \qquad \text{17-99, N}$$

$$RMCEF.KL = \frac{APF.K}{DAPF} \qquad \text{17-100, R}$$

RMIF	Raw-Material Invoice rate at Factory (dollars/week)
RMRF	Raw Material Received at Factory (equivalent product units/week)
CRMPF	Constant, Raw-Material Price at Factory (dollars/unit)
APF	Accounts Payable at Factory (dollars)
RMCEF	Raw-Material Cash Expenditure at Factory (dollars/week)
RRF	Requisition rate Received at Factory (units/week)
DAPF	Delay in Accounts Payable at Factory (weeks)

Equation 17-97 gives the raw-material invoice rate as the product of raw-material receiving rate multiplied by the raw-material price per unit of product. Prices here are chosen for this example as a self-consistent set related by reasonable ratios.[20] The raw-material cost is

CRMPF = $20.00 per unit

Equation 17-98 is the standard level equation accumulating the difference between invoices received and paid. Equation 17-99 gives the initial steady-state value of accounts payable as the product of the rate of sales multiplied by the material cost per unit multiplied by the

[20] They are not the values from the case study.

average delay in paying invoices. Equation 17-100 specifies the rate of payment for material as a fraction of the accounts payable. An average invoice payment interval has been taken as

DAPF = 3 weeks, Delay in Accounts Payable at Factory

Money received for goods delivered will be handled in a slightly different but equivalent way to serve as an alternative example. Here invoices for finished goods will traverse a third-order delay (at the customer) before being converted to a flow of cash receipts:

$$FGIF.KL = (SIF.JK + SMOF.JK)(CFGPF) \qquad \text{17-101, R}$$

$$FGIF = (RRF)(CFGPF) \qquad \text{17-102, N}$$

$$ARF.K = ARF.J + (DT)(FGIF.K - FGCRF.JK) \qquad \text{17-103, L}$$

$$ARF = (RRF)(CFGPF)(DARF) \qquad \text{17-104, N}$$

$$FGCRF.KL = DELAY3(FGIF.JK, DARF) \qquad \text{17-105, R}$$

FGIF	Finished-Goods Invoice rate at Factory (dollars/week)
SIF	Shipments from Inventory at Factory (units/week)
SMOF	Shipment rate Manufactured to Order at Factory (units/week)
CFGPF	Constant, Finished-Goods Price at Factory (dollars/unit)
RRF	Requisition rate Received at Factory (units/week)
ARF	Accounts Receivable at Factory (dollars)
FGCRF	Finished-Goods Cash-Receipt rate at Factory (dollars/week)
DARF	Delay in Accounts Receivable at Factory (weeks)
DELAY3	Specifies third-order-delay equations

Equation 17-101 multiplies the shipments from inventory plus the shipments to customer specifications by the price per unit of product.[21]

[21] Here again is the expediency of using flow rates on the right-hand side to determine a flow rate on the left. It would, however, avail us nothing to create a level of invoices in the process of being mailed, since the delay can be included in the delay in Equation 17-105, intervening between the generation of invoices and receipt of payment.

The unit price used in this example is

CFGPF = **$100.00** per unit, Constant, Finished-Goods Price at Factory

Equation 17-102 gives the initial value of the invoice rate as the total sales rate multiplied by the unit price. Equation 17-103 gives the level of accounts receivable; actually this is a variable not used elsewhere in the model, and the equation could be omitted unless its value is wanted in the output information from the model. Equation 17-104 is the initial level of the accounts receivable given by the product of sales rate, unit price, and average accounts-receivable delay. Equation 17-105 represents the sum of all the delays in the accounts-receivable circuit. It includes billing, mailing of invoices, the time necessary for the customer to initiate payment, and the time to receive a check and deposit it. This total all-inclusive delay is taken as

DARF = **5** weeks, Delay in Accounts Receivable at Factory

The next equations give cash flow for the payment of wages:

LCEF.KL = (MENTF.K)(CWRF)	17-106, R
LCEF = (MENPF)(CWRF)	17-107, N
LCEF	Labor Cash Expenditures at Factory (dollars/week)	
MENTF	MEN, Total, at Factory (men)	
CWRF	Constant, Wage Rate at Factory (dollars/man-week)	
MENPF	MEN Producing at Factory (men)	

Equation 17-106 computes the rate of cash flow for wages as the number of total employees multiplied by the weekly wage rate. Overtime and other forms of variable productivity have not been incorporated in this example. An arbitrary wage rate selected to go with the other prices in this example is

CWRF = **$80.00** per man-week

The initial value of the wage rate given by

Equation 17-107 is needed because the wage-rate flow appears on the right-hand side of Equation 17-115.

The basis for taxes and dividends is developed in the next section. The corresponding cash flow is taken here as follows:

ITAXF.KL = (0.5)(PBTRF.K)	17-108, R
SDIVF.KL = SDLF.K	17-109, R
ITAXF	Income TAX at Factory (dollars/week)	
PBTRF	Profit Before Tax Rate at Factory (dollars/week)	
SDIVF	Stockholder DIVidends at Factory (dollars/week)	
SDLF	Stockholder Dividend Level at Factory (dollars/week)	

Equation 17-108 gives taxes as a simple 50% of the profit-before-tax rate. Equation 17-109 gives the rate of payment of dividends as being equal to the level of dividend payments calculated in the next section.

To provide a basis for comparison between computer runs, the stockholder dividend total to date can be accumulated as follows:

SDTDF.K = SDTDF.J + (DT)(SDIVF.JK)	17-110, L
SDTDF = 0	17-111, N
SDTDF	Stockholder Dividends To Date at Factory (dollars)	
SDIVF	Stockholder DIVidends at Factory (dollars/week)	

All the cash flow rates have already been defined, so that the cash balance is given by the level equation:

CASHF.K = CASHF.J + (DT)(FGCRF.JK − RMCEF.JK − LCEF.JK − CCEF − ITAXF.JK − SDIVF.JK)	17-112, L
CASHF = (CNCSF)(CFGPF)(RRF)	17-113, N
CASHF	CASH balance at Factory (dollars)	
FGCRF	Finished-Goods Cash-Receipt rate at Factory (dollars/week)	

RMCEF Raw-Material Cash Expenditure at Factory (dollars/week)

LCEF Labor Cash Expenditures at Factory (dollars/week)

CCEF Constant Cash-Expenditure rate (fixed charges) at Factory (dollars/week)

ITAXF Income TAX at Factory (dollars/week)

SDIVF Stockholder DIVidends at Factory (dollars/week)

CNCSF Constant, Normal Cash Supply at Factory (weeks of cash-receipts rate)

CFGPF Constant, Finished Goods Price at Factory (dollars/unit)

RRF Requisition rate Received at Factory (units/week)

Equation 17-112 adds the inflowing cash rate and subtracts the five outflowing rates. Beside the variable rates, a fixed-charge rate is included as

CCEF = $30,000 per week

Equation 17-113 establishes an initial value for cash in terms of sales rate, price per unit, and the number of weeks of cash receipts that are to be kept in the cash balance, which is here used as

CNCSF = 1 week

It should be noted in Figure 17-15 that symbols for money flow have been used only for actual cash flow. Invoices, accounts receivable,

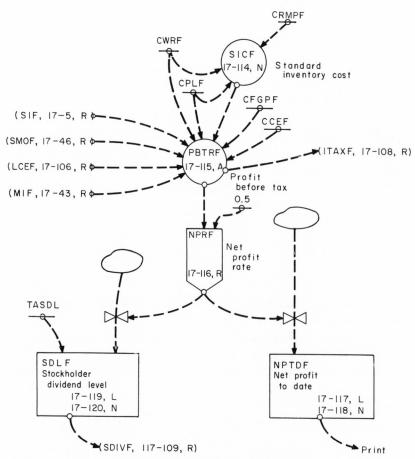

Figure 17-16 Profit and dividends.

and such are part of the information flow system.

17.4.9 Profit and Dividends. In this model a simple computation of current profit rate will be generated to be used as an indicator of system performance. It will be based on a most elementary view of accounting relationships. The profit calculation is shown in the flow diagram of Figure 17-16.

A standard inventory cost will first be generated as an initial-condition equation based on the constants already incorporated:

$$SICF = CRMPF + \frac{CWRF}{CPLF} \qquad \text{17-114, N}$$

SICF   Standard Inventory Cost per item at Factory (dollars/unit)
CRMPF   Constant, Raw-Material Price at Factory (dollars/unit)
CWRF   Constant, Wage Rate at Factory (dollars/man-week)
CPLF   Constant, Productivity of Labor at Factory (units/man-week)

Items in inventory are here being valued at material cost plus the direct labor per unit.

The profit-before-tax rate will be used as the following:

$$PBTRF.K = (SIF.JK + SMOF.JK)(CFGPF - SICF)$$
$$- \left[LCEF.JK - (MIF.JK \right.$$
$$\left. + SMOF.JK) \left(\frac{CWRF}{CPLF} \right) \right] - CCEF$$
$$\text{17-115, A}$$

PBTRF   Profit-Before-Tax Rate at Factory (dollars/week)
SIF   Shipments from Inventory at Factory (units/week)
SMOF   Shipment rate Manufactured to Order at Factory (units/week)
CFGPF   Constant, Finished-Goods Price at Factory (dollars/unit)
SICF   Standard Inventory Cost per item at Factory (dollars/unit)
LCEF   Labor Cash Expenditures at Factory (dollars/week)

MIF   Manufactured rate for Inventory at Factory (units/week)
CWRF   Constant, Wage Rate at Factory (dollars/man-week)
CPLF   Constant, Productivity of Labor at Factory (units/man-week)
CCEF   Constant Cash-Expenditure rate (fixed charges) at Factory (dollars/week)

The first two terms in parentheses give the sum of the production for inventory plus the production to customer specification. This sum is multiplied by the cost of goods sold less the standard inventory costs. This would yield gross profit if the labor were being employed at full efficiency, and the last term subtracts the fixed costs. The middle term of the equation subtracts the total wage payment minus the standard labor value in the goods produced. The wage level **LCEF** will be equal to the labor value in the goods produced if there are no training or termination payments being made. This term subtracts the excess labor costs represented by the losses incurred by training and layoffs.

Net profit rate will be taken simply as half of the profit before tax rate:

$$NPRF.KL = (0.5)(PBTRF.K) \qquad \text{17-116, R}$$
$$NPTDF.K = NPTDF.J + (DT)(NPRF.JK) \qquad \text{17-117, L}$$
$$NPTDF = 0 \qquad \text{17-118, N}$$

NPRF   Net Profit Rate at Factory (dollars/week)
PBTRF   Profit Before Tax Rate at Factory (dollars/week)
NPTDF   Net Profit To Date at Factory (dollars)

Equation 17-117 is a level equation that accumulates net profit to date as one of the indicators of system performance. This level is set initially to zero by Equation 17-118.

The net profit rate will be averaged over a period of time as the basis for the payment of stockholder dividends so that dividends will not fluctuate with short-run changes in net profit rate:

$$SDLF.K = SDLF.J + \frac{DT}{TASDL}(NPRF.JK - SDLF.J)$$

$$17\text{-}119, \ L$$

$$SDLF = 0.5\left[RRF\left(CFGPF - CRMPF - \frac{CWRF}{CPLF} \right) - CCEF \right] \quad 17\text{-}120, \ N$$

SDLF	Stockholder Dividend Level at Factory (dollars/week)
TASDL	Time to Adjust Stockholder Dividend Level (weeks)
NPRF	Net Profit Rate at Factory (dollars/week)
RRF	Requisition rate Received at Factory (units/week)
CFGPF	Constant, Finished-Goods Price at Factory (dollars/unit)
CRMPF	Constant, Raw-Material Price at Factory (dollars/unit)
CWRF	Constant, Wage Rate at Factory (dollars/man-week)
CPLF	Constant, Productivity of Labor at Factory (units/man-week)
CCEF	Constant Cash Expenditure rate (fixed charges) at Factory (dollars/week)

The exponential averaging time of

TASDL = 52 weeks, Time to Adjust Stockholder Dividend Level

used here is equivalent, under many conditions, to a longer moving-average period.[22]

The initial-condition Equation 17-120 is obtained by substituting the initial steady-state values of Equation 17-114 into 17-115, from which the middle term is zero and can be dropped, and including the 0.5 factor from Equation 17-116.

This formulation pays out all profits as dividends, since no other use of profits has been included in the model.

The preceding equations complete the description of the system under study. In addition, it will be desirable to have certain values recorded for study which are not active in the

[22] See Appendix E.

model itself. Also a variety of test input functions will be needed to determine the system behavior. These are covered in the next two sections.

17.5 Supplementary Output Information

In the output tabular and plotted information it is often desirable to present variables in forms or combinations that have not been a part of the model structure itself. These we here call supplementary variables. They are calculated only so that their values may be recorded for study.

For an example system model as presented here, the actual numerical values of the variables are not of great significance since the scale of operations could be increased or decreased to represent any size of business. We are more interested in comparing the *percentage* changes in most of the variables; therefore for the variables of principal importance, we shall calculate them as percentages of their initial steady-state values:

$$BLTPC.K = \frac{(BLTF.K)(100)}{(CINPI)(DNBLF)} \qquad 17\text{-}121, \ S$$

$$CASPC.K = \frac{(CASHF.K)(100)}{(CNCSF)(CFGPF)(CINPI)} \qquad 17\text{-}122, \ S$$

BLTPC	BackLog Total as Per Cent of initial value (per cent)
BLTF	BackLog Total at Factory (units)
CINPI	Constant, INPut, Initial value (units/week)
DNBLF	Delay in Normal BackLog at Factory (weeks)
CASPC	CASh balance in Per Cent of initial value (per cent)
CASHF	CASH balance at Factory (dollars)
CNCSF	Constant, Normal Cash Supply at Factory (weeks of cash-receipt rate)
CFGPF	Constant, Finished-Goods Price at Factory (dollars/unit)

In Equation 17-121, the denominator gives the initial value of backlog as the initial sales rate multiplied by the normal length of the backlog.

The fraction of requisitions filled from inventory needs only be multiplied by 100 to plot on a per cent scale:

$$FRFPC.K = (FRFIF.K)(100) \qquad 17\text{-}123,\ S$$

FRFPC   Fraction of Requisitions Filled from inventory in Per Cent (per cent)

FRFIF   Fraction of Requisitions Filled from Inventory at Factory (dimensionless)

The following equations are self-explanatory for percentage changes in actual and desired inventory, the input sales rate, the employment level, net profit, and orders received at factory:

$$IAFPC.K = \frac{(IAF.K)(100)}{(CIRF)(CINPI)} \qquad 17\text{-}124,\ S$$

$$IDFPC.K = \frac{(IDF.K)(100)}{(CIRF)(CINPI)} \qquad 17\text{-}125,\ S$$

$$INPPC.K = \frac{INPUT.JK}{CINPI}(100) \qquad 17\text{-}126,\ S$$

$$MENPC.K = (MENPF.K)\left(\frac{CPLF}{CINPI}\right)(100) \qquad 17\text{-}127,\ S$$

$$NPRPC.K = \frac{(NPRF.JK)(100)}{(0.5)[(CINPI)(CFGPF - SICF) - CCEF]} \qquad 17\text{-}128,\ S$$

$$RRFPC.K = \frac{RRF.JK}{CINPI}(100) \qquad 17\text{-}129,\ S$$

IAFPC   Inventory Actual at Factory in Per Cent of initial value (per cent)

IAF   Inventory Actual at Factory (units)

CIRF   Coefficient for Inventory Ratio at Factory (weeks)

CINPI   Constant, INPut, Initial value (units/week)

IDFPC   Inventory Desired at Factory in Per Cent of initial value (per cent)

IDF   Inventory Desired at Factory (units)

INPPC   INPut Per Cent of initial value (per cent)

INPUT   INPUT of orders to customer from outside of system (equivalent units/week)

MENPC   MEN producing at factory, Per Cent of initial value (per cent)

MENPF   MEN Producing at Factory (men)

CPLF   Constant, Productivity of Labor at Factory (units/man-week)

NPRPC   Net Profit Rate in Per Cent of initial value (per cent)

NPRF   Net Profit Rate at Factory (dollars/week)

CFGPF   Constant, Finished-Goods Price at Factory (dollars/unit)

SICF   Standard Inventory Cost per item at Factory (dollars/unit)

CCEF   Constant Cash-Expenditure rate (fixed charges) at Factory (dollars/week)

RRFPC   Requisitions Received at Factory in Per Cent of initial value (per cent)

RRF   Requisition rate Received at Factory (units/week)

We may wish to know the total shipment rate to the customer, a value that has not thus far been calculated:

$$SCF.K = SIF.JK + SMOF.JK \qquad 17\text{-}130,\ S$$

SCF   Shipments to Customer from Factory (units/week)

SIF   Shipments from Inventory at Factory (units/week)

SMOF   Shipment rate Manufactured to Order at Factory (units/week)

In evaluating the performance of the model system, we shall sometimes want the kind of financial information ordinarily provided in cash-flow statements and balance sheets. The cash-flow variables for the simple financial considerations of this model have been defined in Section 17.4.8. In addition, we shall want the balance-sheet information giving inventory valuations and the difference between current assets (here cash, accounts receivable, and inventories of finished goods, work in process and material) and current liabilities (here only accounts payable):

$$VIAF.K = (IAF.K)(SICF) \qquad 17\text{-}131,\ S$$

$$VIPIF.K = (OPIF.K + OPCF.K)\left[CRMPF + (0.5)\left(\frac{CWRF}{CPLF}\right)\right] \qquad 17\text{-}132,\ S$$

$$VRMSF.K = (RMSF.K)(CRMPF) \qquad 17\text{-}133,\ S$$

VTIF.K = VIAF.K + VIPIF.K + VRMSF.K   17-134, S

SURPL.K = VTIF.K + ARF.K + CASHF.K − APF.K
$\qquad$ 17-135, S

VIAF	Value of Inventory Actual at Factory (dollars)
IAF	Inventory Actual at Factory (units)
SICF	Standard Inventory Cost per item at Factory (dollars/unit)
VIPIF	Value of In-Process Inventory at Factory (dollars)
OPIF	Orders in Process for Inventory at Factory (units)
OPCF	Orders in Process for Customers at Factory (units)
CRMPF	Constant, Raw-Material Price at Factory (dollars/unit)
CWRF	Constant, Wage Rate at Factory (dollars/man-week)
CPLF	Constant, Productivity of Labor at Factory (units/man-week)
VRMSF	Value of Raw-Material Stock at Factory (dollars)
RMSF	Raw-Material Stock at Factory (equivalent product units)
VTIF	Value of Total Inventory at Factory (dollars)
SURPL	SURPLus, current assets minus accounts receivable (dollars)
ARF	Accounts Receivable at Factory (dollars)
CASHF	CASH balance at Factory (dollars)
APF	Accounts Payable at Factory (dollars)

17.6   Input Test Functions

There remains now only to define the test variables that will be used as the equipment-order inputs to the customer in Figure 17-14 and the noise variation to be used at the output of customer engineering and customer purchasing. These are shown in Figure 17-17.

The input will be defined in terms of its steady-state initial value plus a change variable that includes a step, two ramps (the second to be used to terminate the first), and a sinusoidal fluctuation:

INPUT.KL = CINPI + INPCH.K   17-136, R

INPUT = CINPI   17-137, N

INPCH.K = STP1.K + GTH1.K + GTH2.K + SNE.K
$\qquad$ 17-138, A

INPUT	INPUT of orders to customer from outside of system (equivalent units/week)
CINPI	Constant, INPut, Initial value (units/week)
INPCH	INPut CHange (units/week)
STP1	STeP input no. 1 (units/week)
GTH1	GrowTH ramp no. 1 (units/week)
GTH2	GrowTH ramp no. 2 (units/week)
SNE	SiNE input (units/week)

The average level of system activity is set by the initial sales rate:

CINPI = 1,000 units/week

The following step function, given in the notation for the DYNAMO[23] compiler, specifies the time and height of a step change:

STP1.K = STEP(STH1, STT1)   17-139, A

STP1	STeP input no. 1 (units/week)
STEP	functional notation indicating a STEP generator
STH1	STep Height no. 1 (units/week)
STT1	STep Time no. 1 (weeks)

In the previous and the following test inputs, the defining constants will be initially set to zero so that they will normally be inactive. Those to be active in a particular model run can then be changed to the desired values:

STH1 = 0 units/week, STep Height no. 1
STT1 = 0 units/week, STep Time no. 1

The DYNAMO compiler will generate a straight-line growth ramp with specified starting time and slope. A rising ramp can be terminated and caused to hold its final value by adding to the first continuing ramp a second with negative slope that starts when the input rise is to cease:

GTH1.K = RAMP(GTHR1, GTHT1)   17-140, A

GTH2.K = RAMP(GTHR2, GTHT2)   17-141, A

GTH1	GrowTH ramp no. 1 (units/week)
RAMP	functional notation indicating a RAMP uniform slope generator

[23] See Appendix A.

GTHR1 GrowTH Rate of ramp no. 1
          (units/week/week)
GTHT1 GrowTH Time for start of ramp no. 1   (weeks)
GTH2   GrowTH ramp no. 2   (units/week)
GTHR2 GrowTH Rate of ramp no. 2
          (units/week/week)
GTHT2 GrowTH Time for start of ramp no. 2   (weeks)

As before, the values will initially be set to be inactive:

GTHR1 = 0 unit/week/week, GrowTH Rate of ramp
no. 1

GTHR2 = 0 unit/week/week, GrowTH Rate of ramp
no. 2

GTHT1 = 0 week, GrowTH Time for start of ramp
no. 1

GTHT2 = 0 week, GrowTH Time for start of ramp
no. 2

The sinusoidal input function is:

$$SNE.K = (SIH) \text{ sine } \frac{(2\pi)(TIME.K)}{PER} \qquad \text{17-142, A}$$

SNE   SiNE input   (units/week)

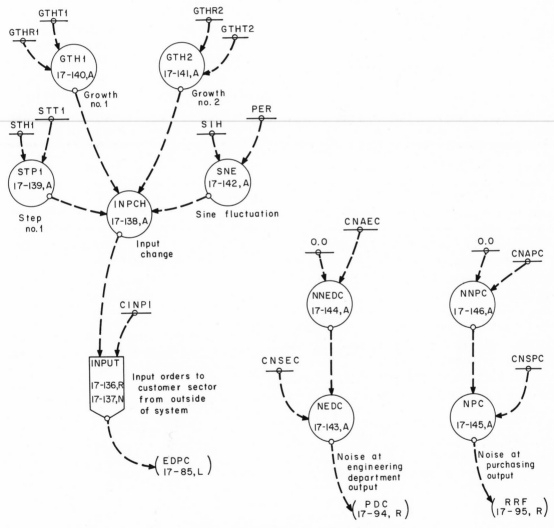

Figure 17-17 Input test functions.

SIH	Sine Input Height (units/week)
sine	trigonometric sine generator
TIME	calendar TIME (supplied by DYNAMO) (weeks)
PER	PERiod of sine test input (weeks)

In the preceding equation, the period cannot be set initially to zero because it is in the denominator of the fraction, but the sine generator is inactive with a zero amplitude variation.

SIH = 0 unit/week, Sine Input Height
PER = 52 weeks, PERiod of sine test input

In the customer ordering section, illustrated in Figure 17-14, a noise input was introduced in the engineering release rate and another in the purchasing department ordering rate, Equations 17-94 and 17-95, respectively. In generating a noise input, care must be given to the disturbance frequencies that are to be created. The most direct methods are apt to introduce a disproportionately large amount of high-frequency noise that has little effect on the system. At the same time they may fail to introduce as much as anticipated or as would be realistic at the lower frequencies. Rather than attempting to specify and control the full noise spectrum for this example, we shall sample a noise source and hold the sample for a period long enough that it will cause a realistic, month-to-month fluctuation in ordering rate.[24]

A sampling function will be used that will select values of a noise source at specified intervals and hold and use each value until the next sample is taken:

NEDC.K = SAMPLE(NNEDC.K, CNSEC)　　　17-143, A

NEDC	Noise at Engineering Department output of Customer (dimensionless)
SAMPLE	a functional notation indicating that the variable in the parentheses is to be sampled periodically and the samples held for the interval given by the constant

---

[24] As discussed in Appendix F, this still concentrates most of the noise power in a frequency region near the noise-sampling frequency.

NNEDC	Noise Normal at Engineering Department output of Customer (dimensionless)
CNSEC	Constant Noise-Sample holding time at Engineering at Customer (weeks)

The sample period will be set

CNSEC = 0 week

until specified for a particular run. A sample period of about a month would be reasonable to consider for this situation where we could expect substantial month-to-month fluctuation in the specifications released by the aggregate of customer engineering departments.

The noise sample will be taken from a normally distributed source by the following:

NNEDC.K = NORMRN(0.0, CNAEC)　　　17-144, A

NNEDC	Noise Normal at Engineering Department output of Customer (dimensionless)
NORMRN	a functional notation calling for an output from the NORMal Random-Noise generator, mean given by first constant and standard deviation by the second
CNAEC	Constant, Noise Amplitude at Engineering output at Customer (dimensionless fraction of total flow)

Here the standard deviation of the noise is set

CNAEC = 0

until needed.

An independent but similar noise source is specified for the purchasing department output by the following:

NPC.K = SAMPLE(NNPC.K, CNSPC)　　　17-145, A

NNPC.K = NORMRN(0.0, CNAPC)　　　17-146, A

NPC	Noise at output of Purchasing at Customer (dimensionless)
SAMPLE	a functional notation indicating that the variable in the parentheses is to be sampled periodically and the samples held for the interval given by the constant
NNPC	Noise, Normal, for Purchasing output at Customer (dimensionless)

CNSPC   Constant, Noise-Sample holding time at Purchasing at Customer   (weeks)

NORMRN   a functional notation calling for an output from the NORMal Random-Noise generator, mean given by first constant and standard deviation by the second

CNAPC   Constant, Noise Amplitude at Purchasing output at Customer   (dimensionless fraction of total flow)

CNSPC = 0 week
CNAPC = 0

Where the reservoir of orders in process in the purchasing department is short (the average delay DRCC equals 3 weeks), a shorter sampling period CNSPC of about 1 week would be reasonable.

This completes the design of the model that will be used in the next chapter.[25]

---

[25] Appendix A contains a tabulation of the equations that have been developed in this chapter.

CHAPTER 18

Dynamic Characteristics of a Customer-Producer-Employment System

Section 18.1 of this chapter examines the dynamic behavior of the system described in Chapter 17. Variations in system parameters are tested in Section 18.2 to determine the policies to which the system is most sensitive. In Sections 18.3 through 18.7 changes in system organization and policies are made to improve the stability of employment, cash position, backlog variation, and delivery delay without requiring larger inventories or inventory fluctuations.

CHAPTER 17 described a case study and developed a model of the practices that had seemed to exist previously within an industrial system. In this chapter we shall first examine the characteristics of the model system that was developed in Chapter 17, to see if its behavior is a reasonable representation of the actual system described in Sections 17.1, 17.2, and 17.3. After that we shall alter certain of the assumptions in the model that was developed in Section 17.4, to see how sensitive the system is to the management policies that were initially included. Then some of the policies of the model will be modified to obtain an improved management system, and the characteristics of the new system will be compared with those of the old.

18.1 The Old System

As in Chapter 15, we shall first test the system under idealized inputs for the insights these will give us into the dynamic character of the system. These will first be a step change in the level of ordering rate entering the customer sector and then a periodic fluctuating change. After that, the effect of random variation at

the engineering department output of the customer will be examined.

18.1.1 Step Change in Demand. To obtain a first indication of the kind of system with which we are dealing, a sudden step input will be used. From this we should be able to see if there are any predominant natural frequencies that die out only slowly. If there are, we can form an estimate of their period and also of rate of attenuation, or if the system is unstable, their rate of expansion in amplitude.

Figure 18-1 shows a 10% step input to the independent ordering rate entering the customer sector of the model.[1] The system shows a periodic fluctuation in response to the step input. The period of this fluctuation is about 100 weeks, with manpower peaks appearing at 68 weeks, 168 weeks, and 266 weeks.[2]

---

[1] This is obtained by setting STT1 = 8 weeks and STH1 = 100 units/week in the input parameters of Section 17.6. This means that at the 8th week the independent-order input will rise 100 units per week, which is a value of 10%. See Figure 17-14 for the point of input, and Figure 17-17 for the input flow diagram.

[2] Many of the numerical values given in the text have been taken from the tabular print-out of results of model runs and may not be readable from the figures included here.

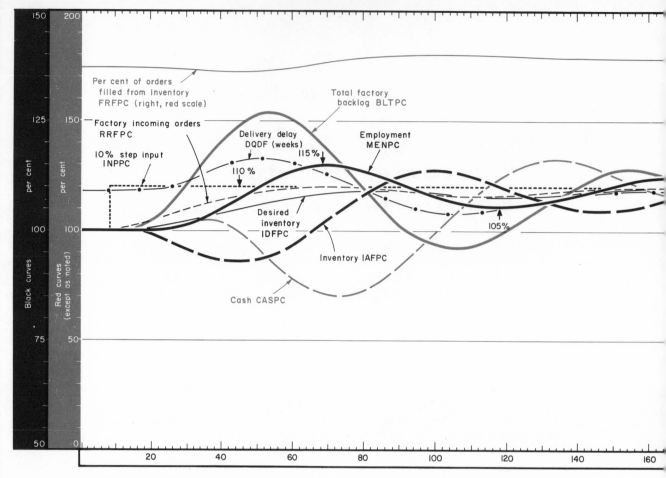

Figure 18-1 Electronic component industry model, old policies, step increase in demand.

The attenuation (tendency for the fluctuation to die out) is seen to be about 50% per cycle. In other words, each employment peak, measured from the steady-state value, is about half of the amplitude of the preceding peak.[3] This should be considered as a very slow attenuation of a disturbance.

Such a small attenuation per cycle indicates a system with rather strong tendencies toward fluctuation at a period somewhat less than two years. As in Chapter 15, this indicates a system that will be easily disturbed by the random-noise components that will exist throughout the system. Also, we expect that it would be a strong amplifier of any disturbances external to or within the system which have a periodicity anywhere near the two-year interval.

Examination of Figure 18-1 shows that the employment peaks and valleys occur at about the same time as those in the incoming-order rate. When the incoming orders and employment are at a peak, the inventory is rising at its most rapid rate, as was illustrated in Figure

[3] Successive peaks are 114.9%, 112.5%, and 111.1% referred to a steady-state level of 110%. This gives amplitude ratios per cycle of $\frac{112.5 - 110}{114.9 - 110} = 0.51$ and $\frac{111.1 - 110}{112.5 - 110} = 0.44$.

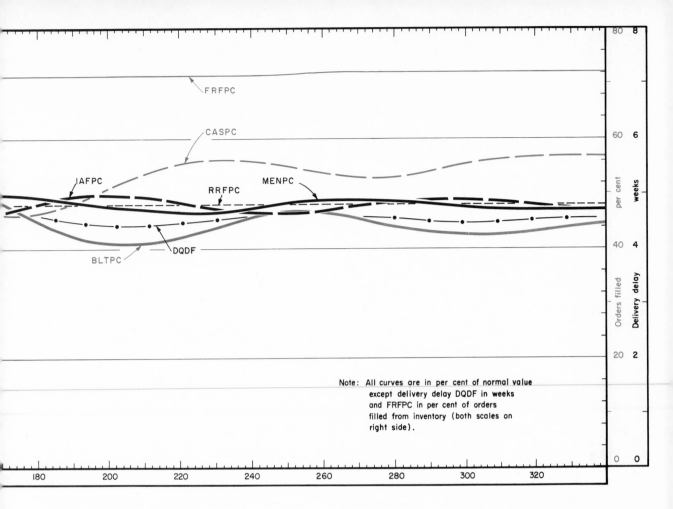

Note: All curves are in per cent of normal value except delivery delay DQDF in weeks and FRFPC in per cent of orders filled from inventory (both scales on right side).

17-1. Conversely, inventory is falling at the time of minimum employment and incoming orders. The fluctuation of employment is more than twice as great as that in incoming orders, with inventory changes making up the difference. As discussed in Section 17.1, inventory is here accentuating the employment fluctuations rather than helping to smooth out employment and production.

In response to a sudden rise in business the cash position rises momentarily while inventory is being depleted. Then cash falls during the period of inventory accumulation before it rises again in response to the higher level of business and profitability.

Figure 18-1 shows two important difficulties in the system:

- For any given fluctuation of incoming orders to the factory, we find that employment, inventories, cash, and backlogs fluctuate a great deal more.
- Conditions feeding back from the factory delivery delay are here causing a fluctuation in the incoming-order rate itself.

Here, as in other oscillatory information-feedback systems, the natural period of fluctuation adjusts itself to near that at which the maximum amplification around the information-feedback loop occurs.

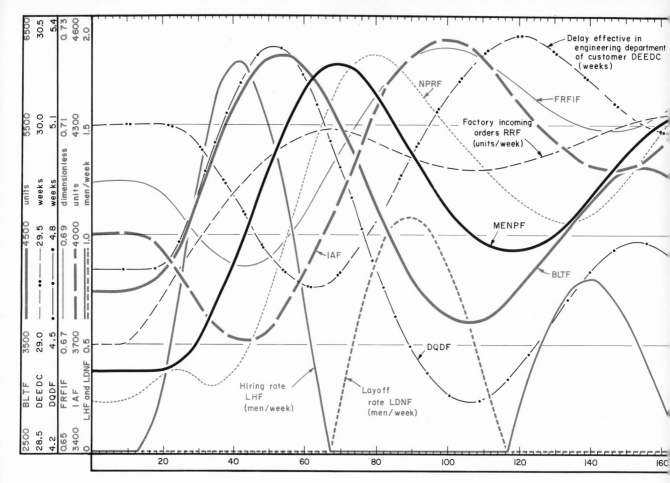

Figure 18-2 Electronic component industry model, old policies, step increase in demand, expanded plotting scales.

Figure 18-2 presents most of the same curves as Figure 18-1 but on a set of expanded scales so that the time relationships between the different variables can be more readily seen.[4] The feedback interactions can be traced in the section of Figure 18-2 lying between 120 and 220 weeks. At 120 weeks, employment is at a mini-

---

[4] Note that in Figure 18-2 each curve has its separate scale, so that care is necessary in comparing the amplitudes of different curves. Also, the variables are plotted in actual quantities rather than in per cent ratios.

mum and inventory is falling. Because inventory is falling, the factory delivery delay is increasing in the interval from 115 weeks to 150 weeks. As the factory delivery delay increases, the customer engineering department, after a time delay to reorganize its work, begins to release component specifications sooner. This is indicated by a falling effective engineering department delay in the interval from 120 weeks to 165 weeks. The rising factory delivery delay and the ensuing acceleration of engineering department specification releases result in a rise

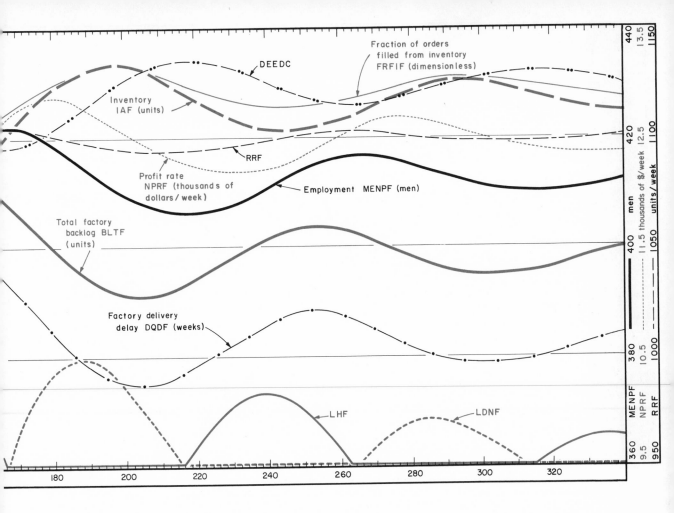

in the orders from customer to component manufacturer over the period from 115 weeks to 160 weeks. The falling inventory and the rising sales lead to increasing employment between 120 and 170 weeks. At 148 weeks the manufacturing rate equals the delivery rate to customers, as indicated by the minimum inventory point where inventory is not changing. Consequently, by 152 weeks factory delivery delay has ceased rising. This causes incoming orders to reach their peak by 162 weeks. Employment, which is now too high, causes a rapidly rising inventory that leads to a decrease in factory delivery delay in the interval from 160 to 200 weeks. This allows the customer to delay somewhat in his order releases, and results in a small cutback in the orders being placed. The excess inventory and the declining orders lead to a curtailment of employment that reaches a minimum at 220 weeks. A complete cycle requiring 100 weeks for completion therefore links manpower, inventory, delivery delay, engineering releases, component ordering, and back to employment.

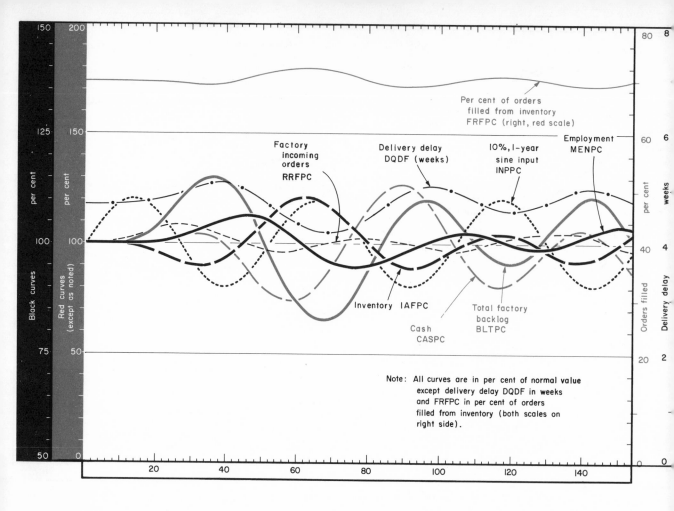

Note: All curves are in per cent of normal value
except delivery delay DQDF in weeks
and FRFPC in per cent of orders
filled from inventory (both scales on
right side).

18.1.2 One-Year Period. Figure 18-3 shows the model response to a one-year period sinusoidal disturbance at the input.[5] All of the flow rates in the system vary by amounts that are small compared with that at the input. This is for two reasons.

First, a delay, such as represented in the customer engineering department, will heavily attenuate any fluctuation whose period approaches (or is shorter than) the length of the delay. The customer engineering delay is here

30 weeks, which is a substantial fraction of a year.[6] The outflow rate from the aggregate of all customer engineering departments will not reflect the full extent of the short-period fluctuations that occur at the input. (Engineering releases may, of course, introduce new fluctuations separate from the input and caused by the way that projects are released for production.)

A second reason for attenuation of the one-year disturbance arises because the time-phase relationships around the feedback loop described in the previous section are now such

[5] Obtained by setting the following in the input test functions of Section 17.6: PER = 52 weeks and SIH = 100 units/week. This provides an input that rises 10% above average, falls to 10% below average, and returns in a year.

[6] From Appendix G we see that a 30-week first-order exponential delay will pass only one-quarter of the amplitude of an input one-year periodic fluctuation.

← **Figure 18-3** Electronic component industry model, old policies, one-year sinusoidal input.

that the variation in delivery delay tends to suppress the system fluctuation. The natural period of the system was seen in Figures 18-1 and 18-2 to be about 100 weeks. At that period of oscillation, the flow of delivery delay information from component supplier to customer tended to reinforce the system fluctuation. We should not expect the same reinforcement for other oscillation periods imposed upon the system. In fact, the reverse is true in the model at a one-year period of disturbance. From a curve for engineering department release delay (not shown), we find that the engineering department is tending to accelerate its design releases at the same time that the work load itself is falling due to the fluctuating input. These two effects tend to cancel one another to some extent and further reduce the amount of the input disturbance that is transmitted from customer to factory.

For the preceding two reasons, incoming orders to the factory vary only 18% as much as the input to the customer. Even so, employment at the factory fluctuates 33% as much as the orders at the customer input. This means that employment is still fluctuating more (185%) than the actual inflow of orders to the factory, so that employment varies almost twice as much as any annual, seasonal component in the factory orders.

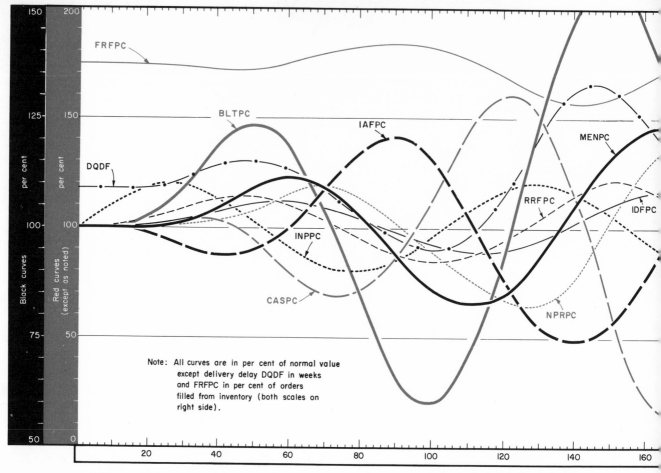

Figure 18-4 Electronic component industry model, old policies, two-year sinusoidal input.

18.1.3 Two-Year Period. Figure 18-4 shows the system response to an input ordering-rate fluctuation to the customer sector of 10% peak amplitude and a two-year period.[7] It is in marked contrast to Figure 18-3. Here some of the weaknesses of the system are coming into prominence. The system is acting as an amplifier of the primary input disturbance. Amplitudes of variation in the system are five to seven times as large as they were for the same amplitude of input having a one-year instead of a two-year period. Employment fluctuation has increased by the largest ratio, now being seven times as great as it was in Figure 18-3. In Figure 18-4 it is 2.4 times as large as the pri-

mary input-ordering variation to the customers.

The orders arriving at the factory from the customers fluctuate by a slightly higher amount than the input orders arriving at the customers. This is true even though there is a 50% attenuation in the amplitude of the input in passing through the customer engineering departments.[8] The customer ordering rate is more than twice what might be expected on the basis of the customer sector by itself because of the amplifying effect of the variable delivery delay fed back from the factory sector. In Figure 18-4 delivery delay from the factory reaches a peak at 250 weeks. This is effective a short time later in producing advance ordering in the customer

[7] The 10% two-year period is obtained with the following constants in Section 17.6: PER = 104 weeks and SIH = 100 units/week.

[8] See Appendix G for the attenuation of a first-order delay of 30-week time constant on which is impressed a sinusoidal variation of 104 weeks.

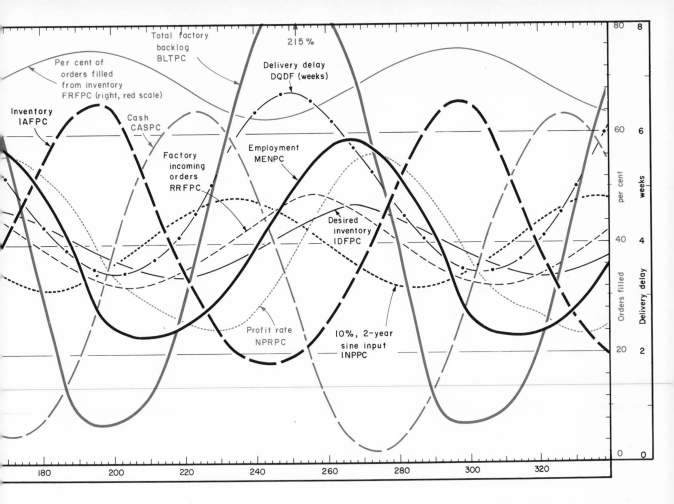

sector and helps create the peak in the factory incoming-order curve, which reaches a maximum at about 255 weeks.

Factory backlog, the "normal" value of which would be equal to 4 weeks of production, is varying in each direction by approximately 4 weeks' worth of sales. Cash position, which is normally equal to 1 week's worth of cash receipts, is falling almost to zero as assets are transferred from the bank balance to inventory and accounts receivable.

As in Figure 17-1, inventory is falling at the time of the sales minimum and rising at the time of the sales maximum. This largely explains the fact that employment is fluctuating 2.2 times as much as the sales rate fluctuation. At the low point in sales, part of customer demand is being filled from inventory; at the high point, production exceeds demand and inventory is accumulating. Some of the nonlinearities of the system (which are not great) are evident in the nonsymmetrical upper and lower loops of the curves.

Inventory is varying less than we might expect, from consideration of the large amplitudes of the other curves. Inventory fluctuation is approximately equal to 1 week of production. (Normal inventory is 4 weeks' worth of production, so a 25% fluctuation is 1 week of production.) Delivery delay varies between 3.5 and 6.7 weeks.

We should judge this as a rather unsatisfactory system structure and policies because of the way they tend to accentuate external disturbances, especially those in the vicinity of a two-year period.

18.1.4 Random Releases from Engineering Department. The preceding figures have been taken with pure, idealized test inputs so that we could look at one component of system behavior at a time. The previous curves have been smooth and have an artificial appearance compared with a real industrial enterprise, because the decisions in the system have all been free of random variations.

We are now ready to examine the response of the system to a more complex environment. For this we introduce a nonuniform design specification release rate from the aggregate customer engineering departments. As set up in Sections 17.4.7 and 17.6, a noise variable is available in the model for insertion in the flow rate between customer engineering department and customer purchasing department. This fluctuating signal does not create or destroy any orders for electronic components. Orders are determined in the long run by those entering customer engineering from the outside world. The random output from the engineering department can affect the time at which orders are released and, therefore, causes a random bunching of the order outflow. Any random signal contains a very broad band of disturbance frequencies and inherently contains excitation for any modes of behavior to which the industrial system is especially sensitive.[9]

We shall now examine the electronic component industry model when the order inflow to the customer engineering department is constant and the order outflow from the customer engi-neering department contains the month-by-month random fluctuation.

The further the random fluctuation penetrates into the system, the more it is smoothed, and the more it takes on the characteristics of the system itself rather than the original noise. Even in the factory incoming requisitions, a certain amount of correlation is evident; that is, orders in consecutive weeks vary smoothly from one to the next because of the intervening customer purchasing department that causes some averaging and smoothing of the order flow.[10]

(The reader should here refer to Figure 18-5, page 264, and the explanation beneath.)

As we had reason to expect from Figures 18-1 and 18-4, the system as shown in Figure 18-5 responds unfavorably to the presence of the unavoidable random disturbances that will always exist in an industrial system. Because the system is highly sensitive to and selective of frequencies near a two-year period, it chooses frequency components in that periodicity range and amplifies them. Any kind of random-noise disturbance is therefore sufficient to trigger into operation the inherent tendencies of the system to develop unfavorable fluctuations owing to the interactions between customer ordering, and factory inventories, delivery delay, and employment level.

18.1.5 Cash Flow and Continuous Balance Sheet. The variables plotted in the preceding figures are the principal ones that determine system dynamic performance even though the financial variables usually receive more attention. The financial data corresponding to Fig-

---

[9] The random disturbance is here obtained by setting CNSEC = 5 weeks and CNAEC = 0.1 in the parameters of Section 17.6. This means that random numbers having a normal distribution and a mean deviation of 10% are added to the customer engineering department output rate. These random numbers are sampled, and each number is held for 5 weeks before a new sample is selected. The sampling and holding are here done as a simple expedient, instead of other methods that might be chosen, to control somewhat the power-density versus frequency characteristic of the noise. See Appendix F.

[10] Section 17.6, Equation 17-145, made available in the model at Equation 17-95 a second noise variable at the output of the customer purchasing department. Activation of this noise input would have produced a week-by-week random scatter in the orders to the factory, but would have had little effect on the system since most of the added noise signal would have been in the very high frequency spectrum to which the system is only slightly sensitive.

ure 18-5 are shown in Figure 18-6 for cash flow and Figure 18-7 as a continuous balance sheet.

Figure 18-6 presents the six cash-flow variables from the cash-flow diagram of Figure 17-15. The scales are plotted from different zero points to superimpose the curves on the same diagram. However, the vertical height of the figure represents a range of $20,000 per week for each of the curves so that vertical changes are comparable for all curves.[11] It should be noted that all of the expenditure rates begin to rise earlier than the cash-receipts rate and reach peaks ahead of the peak in cash receipts. This is clearly evident, for example, in the range from 40 weeks to 80 weeks, and between 120 weeks and 150 weeks. This means that cash drain peaks ahead of cash replenishment, causing the wide cash fluctuations seen in Figure 18-7.

In the continuous balance sheet of Figure 18-7 can be seen the interplay between the various current assets. In-process inventory rises and falls approximately in synchronism with employment level in Figure 18-5.

The most important point in Figure 18-7 is to observe the way in which the value of total inventories and the accounts receivable rise and fall together. This can be seen most clearly in the interval between 50 and 100 weeks, and between 120 and 180 weeks. The result is that both inventory and accounts receivable draw on cash and return into cash at the same times. Cash position must be able to supply the simultaneous demands of both inventory and accounts receivable. This accentuates the cash-position fluctuation. After making changes in system management policies, we shall see in Section 18.6 that the short-term fluctuations of inventories and accounts receivable can be

made to move in opposite directions, so that current assets flow back and forth between accounts receivable and inventory and place less fluctuating demands on cash.

18.1.6 Adequacy of the Model. The results of Sections 18.1.1 through 18.1.5 show a model whose behavior is in accord with the available knowledge about the actual system. The confidence in the model rests primarily on the knowledge we have of the separate parts of the system. Nevertheless, this would not be sufficient if the model exhibited system behavior that was inconsistent with the actual system.

There is usually a paucity of dependable records that can be compared with a system model. The checking consists of examining all aspects of the model behavior about which there is knowledge in the actual system. One examination is to look for any improbable behavior in the decision streams in the model. Phase relationships between variables are important; for example, the model in Figures 18-4 and 18-5 shows the inventories peaking after orders and employment as observed in the actual system, and as shown in Figure 17-1. As appears to be true in the actual system, the model shows production and employment fluctuations that are substantially larger than the changes in order flow to the customer sector from the customers' customers (government and industry orders for electronic equipment). Periods of about two years had existed in the fluctuations of the actual system, and similar behavior is evident in the model. Examination, by the managers familiar with the actual system, of the model assumptions and its over-all performance does not reveal implausible structure or policy assumptions or behavior that is judged to degrade the model for its intended uses.

These are all negative tests. Nothing has been discovered about the model that is believed to invalidate it for the purpose of ex-

[11] Neither Figure 18-5 nor the financial data diagrams of 18-6 and 18-7 give the actual shipping rate of finished goods to the customer. This has approximately the shape of the finished-goods cash-receipts curve if that curve were moved 5 weeks to the left.

Text continues on page 268.

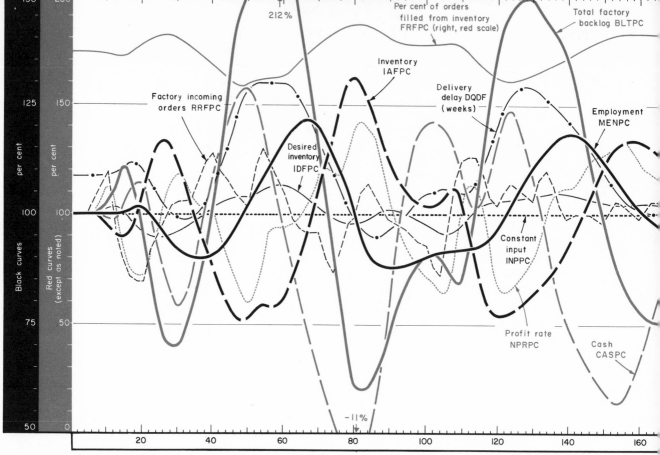

Figure 18-5 Electronic component industry model, old policies, random releases from customer engineering department.

By the time that the random disturbance has propagated to the factory employment level, the predominant variation is not far removed from the 100-week natural period of the system found in Figure 18-1. Employment peaks occur at 66, 141, 250, and 316 weeks. These are separated by intervals of 75, 109, and 66 weeks.[12]

In Figure 18-5 employment covers a range

from 121% of normal down to 87% of normal. Peak employment is therefore 140% of the minimum. The amplitude of the system disturbances is, of course, dependent on the amplitude of the noise variation being inserted in the model. Our main concern here is not the actual amplitude itself but rather the extent to which undesirable system responses for any given input can be suppressed by changes in organization or policy. We are primarily interested in comparative studies between different policies when subjected to the same system input. However, we want such tests to be made in a plausible range of operations, and of course, we must have confidence that the model represents the essential dynamic characteristics of the system under study.

[12] We can expect the apparent natural periods to be pulled down to something shorter than those of the system itself because, as discussed in Appendix F, the higher-frequency (that is, shorter-period) components are accentuated in this kind of noise input. The disturbing signal that is strongly biased in favor of high frequencies will therefore tend to force a somewhat higher-frequency disturbance than the natural frequency of the system itself.

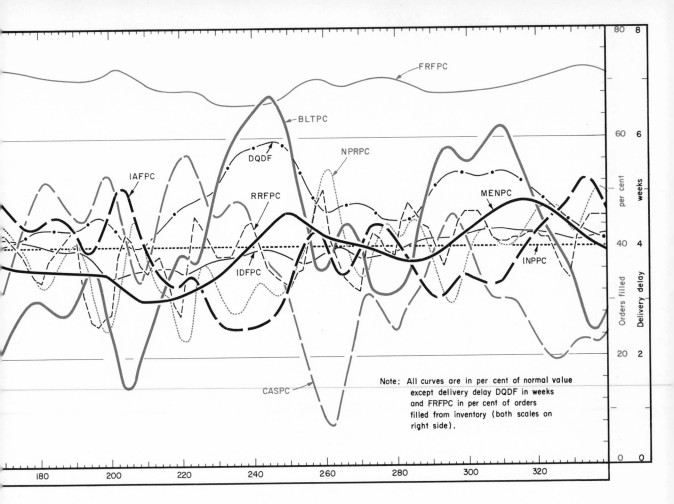

Note: All curves are in per cent of normal value except delivery delay DQDF in weeks and FRFPC in per cent of orders filled from inventory (both scales on right side).

Figure 18-5 contains the major qualitative characteristics that originally led to making the case study. This figure is a seven-year history of model-system interactions in the face of a final demand for the product which is actually constant. Employment fluctuations are large. Periods from one to two years appear between peaks in most of the major variables. Inventories are rising rapidly during the time of the employment peaks, and inventories reach peak value shortly after the peaks of employment. Average delivery delay is fluctuating between 3.6 and 6.4 weeks, although the fraction of orders filled from inventory is varying only between 63% and 75%. Inventories are fluctuating by about 25%, which is one week's worth of production. Unfilled factory backlog, includ-ing both that for the customer and that for inventory, is fluctuating about 4 weeks' worth of production on either side of normal. In general, the time-phase relationships between the different curves are plausible for this kind of industrial situation. Backlog peaks appear ahead of employment peaks that in turn occur ahead of the peaks in inventories. The desired inventory curve can be taken as indicating average sales,[13] and we see that employment fluctuations are wider than average sales and, as in Figure 18-4, inventories rise during high sales and fall during low sales.

Text continues on page 262, right column.

[13] Equations 17-19 and 17-125 combine to make this curve equivalent to percentage change in average sales RSF as well as percentage change in desired inventory.

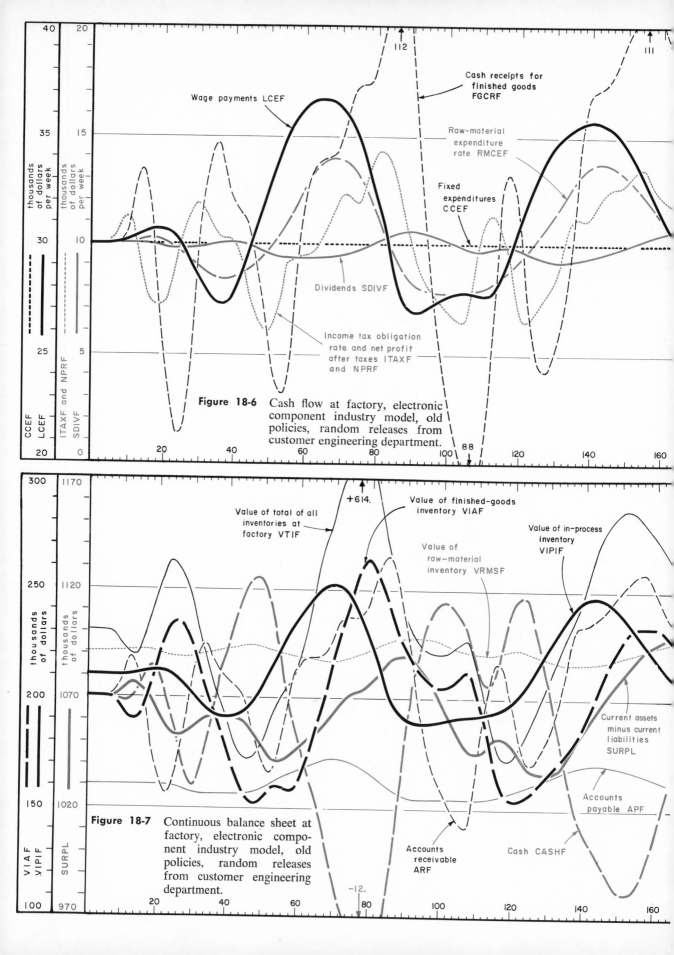

Figure 18-6 Cash flow at factory, electronic component industry model, old policies, random releases from customer engineering department.

Figure 18-7 Continuous balance sheet at factory, electronic component industry model, old policies, random releases from customer engineering department.

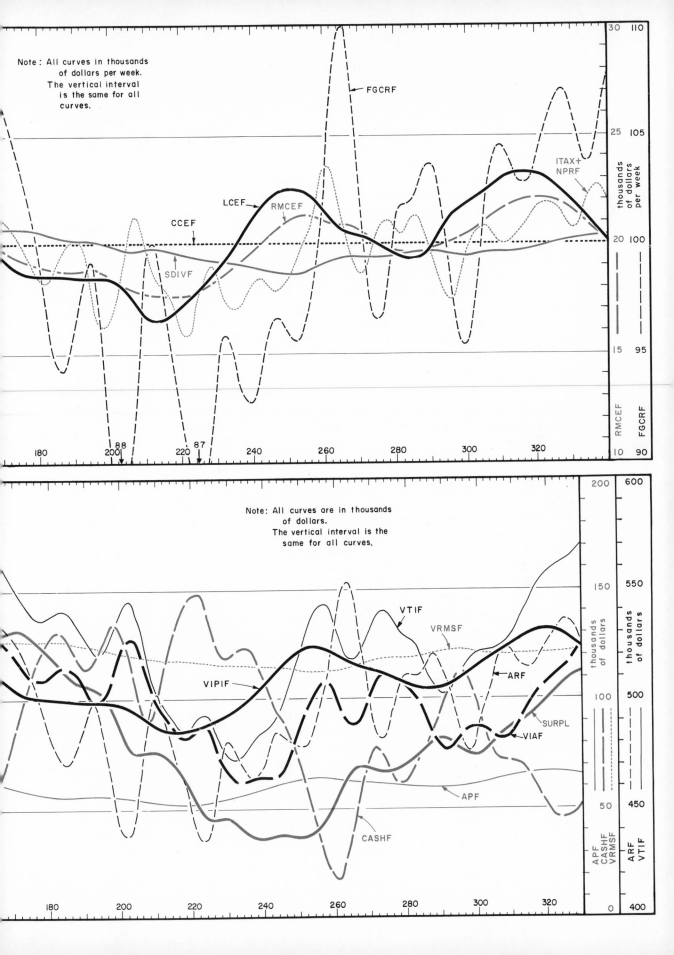

ploring why the system behaves the way it does and then going onward to studies of system redesign. (Validity of each model usage must be considered against the model content. For example, there is no competitive sector, so that intercompany competitive questions are not permissible unless the model is extended.)

These negative tests are very powerful if there are many of them and if they are applied in depth from individual model-component assumptions, through subsystem actions, to total-system behavior. They do not yield definitive proof of model adequacy, and there are no non-controversial ways of doing so. They do yield entirely sufficient confidence to form a basis for continuing into further exploration of the management system. They develop a stronger and more unified confidence than lies behind most management decisions.

18.2 Variations in Parameters of the Old System

Having built, rebuilt, tried, and refined the model until it is an acceptable representation of the major dynamic characteristics of the system under study, we next become interested in the effect of changes in that system. We shall explore changes in two steps. In this section, the effect of changing a few of the parameters of the model in Chapter 17 will be examined. Then in Section 18.3 several of the management policies of the system will be modified toward a more desirable mode of operation, and the effect of those modifications will be examined in Section 18.4.

We are interested for several reasons in trying the model with various changes of parameter values and changes in structure. It is important to know the sensitivity of system behavior to changes in the various components of the system. This acts as another check on model validity, since the changes in model behavior resulting from changes in model assumptions can often be evaluated against the likelihood of similar real-world sensitivity. More important, we should concentrate attention on system design in areas that are critical. Factors to which the system is sensitive must be carefully controlled, or else the system should be redesigned to reduce the sensitivity.

In the following subsection are comments on some of the less important system parameters. In Section 18.2.2 there will be a discussion, with illustrations, of the way the system performance changes when some of the sensitive parameters are altered in the old labor policies.

18.2.1 Sensitivity Analysis. A system like that developed in Chapter 17 is insensitive to changes in most of the equation parameters. A sensitivity analysis will usually be conducted by making changes of a factor of 2 or more in parameters. A change of this size will often make little difference. However, the system may be substantially affected by such a modification of certain factors in the system, and these are highlighted as ones to receive additional attention.

Modifications of the system that produce only slight effects do not justify figures here which would be like those already included. The results can be summarized by identifying some of the system parameters whose values are not critical. We must bear in mind that the conclusions apply only to this particular system. Variables that may superficially appear similar may react quite differently in some other system. In fact, the system changes to be developed later in this chapter will substantially affect the influence that some parameters have on system behavior.

The smoothing time constant TRSF in Equation 17-20 can be changed from 15 weeks to 25 weeks without making any appreciable change in the system step response as shown in Figure 18-1. Likewise, the amount of inventory

CIRF in Equation 17-19 produces almost no difference in step-input response when changed from 4 weeks of inventory to 10 weeks. We must bear in mind here that in neither case is inventory being substantially depleted. Under extreme circumstances when the system is beginning to react to a shortage of available inventory, the response could be expected to differ, depending on the amount of normal inventory carried. The fraction of orders CNFIF in Figure 17-7 that can be filled from inventory can be changed from 0.7 to 0.5 with only slight effect. This nearly doubles the fraction that must be made to customer order.

Random sequences of noise different from the sequence used in Figure 18-5 produce different sets of curves but ones that have the same character. The introduction of a noise variable at the output of the customer purchasing department in addition to a noise variation at the output of the customer engineering department produces almost no additional effect upon the system.[14]

At other points the system is more sensitive to changes. There is a delay in the response of the customer to delivery-delay quotations for the filling of orders. This delay appears in two places. Delivery delay quoted by the factory lags behind the actual condition of production and backlogs as defined by TAQDF in Equation 17-83. The customer responds gradually to delivery quotations as determined by TAEDC in Equation 17-92. If this delay in the response of the customer to delivery conditions is reduced

by a factor of about 3,[15] the system becomes substantially less stable. In response to a 10% step input a *sustained* oscillation results, with about 8% amplitude in employment level and a 90-week period. Here is an example where reducing a time delay unstabilizes system behavior. The reduced stability occurs because the customer responds more quickly to increasing order backlogs at the factory and quickly orders further ahead, a procedure that results in still larger backlogs.

18.2.2 Rapidity of Labor Adjustment. The labor policies in this system are crucial to the dynamic behavior of the industry because they are the mechanism through which production rate is adjusted to meet demand for the product. All plausible policies will in the long run adjust total production to equal total demand, or else inventories would disappear or continue to grow without limit. Differences between various labor policies will exist primarily in the sources of information used and in the rapidity with which indicated labor changes are executed.

We shall now examine some of the effects of decisiveness and speed in executing changes in employment. This is an area where most of the psychological factors tend to delay action. If production rate appears to be running ahead of sales, compassion for the welfare of employees and reluctance to upset carefully laid procurement and production plans may result in delaying employment changes. It is easy to hope that the conditions calling for reduced production rate will disappear and that one can "ride through" the presently developing crisis. There can be situations where this reluctance to act is actually one of the key factors in creating system instability. The reluctance to make small present cutbacks in employment can be one of the causes for still greater employment reductions in the future.

[14] For this particular run a 10% noise deviation sampled once per week was used. This was obtained by setting CNSEC = 5, CNAEC = 0.1, CNSPC = 1 week, and CNAPC = 0.1 in the input test functions of Section 17.6. Likewise, a run without customer-purchasing-department noise but with a larger amplitude and shorter holding period at the customer engineering department produces approximately the same result, which is as would be anticipated on the basis of the discussion in Appendix F. This result was obtained by setting CNSEC = 2 weeks and CNAEC = 0.25.

[15] For example, by changing TAEDC from 15 weeks to 5 weeks and TAQDF from 4 weeks to 1 week.

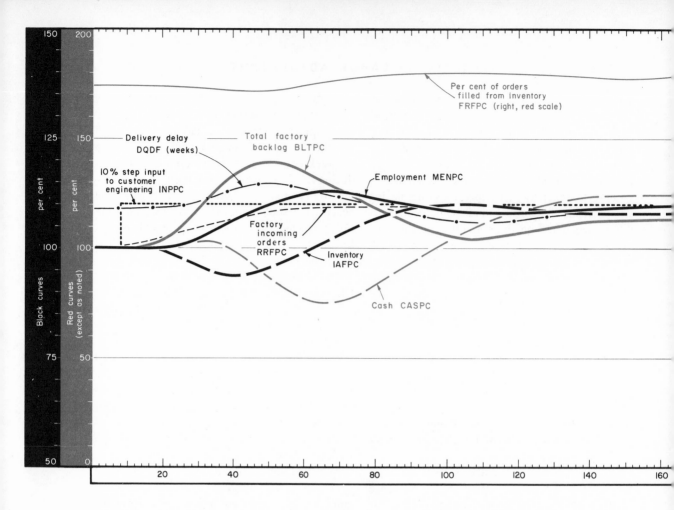

Figure 18-8, when contrasted with Figure 18-1, shows the effect of making the indicated labor changes with twice the previous rapidity.[16] This means, with any given condition of employment, average sales, and factory backlog, that the indicated labor excess or deficit will be

[16] This is a reduction in the time constant for labor change, TLCF, in Equations 17-72 and 17-75. The position of these in the hiring policy structure is shown in Figure 17-11. The normal value is TLCF = 10 weeks. Figure 18-8 is the same as Figure 18-1 except that TLCF = 5 weeks.

adjusted more quickly than before. Comparison of the two figures shows that the more rapid adjustment policy in Figure 18-8 leads to a more stable system than in Figure 18-1. In response to the 10% step increase into the customer engineering department, the disturbances at the component supplier are less severe. Initial amplitudes in factory backlogs, employment, inventory, and cash position changes are all smaller. Furthermore, successive peaks in the response, which still fluctuates slightly, are each

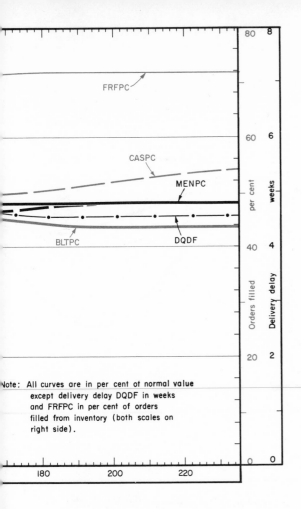

Figure 18-8 Electronic component industry model, old policies, more rapid employment change, step increase in demand.

15% of the previous peak rather than 50% of the previous peak, as in Figure 18-1.

The increased stability of the system can be explained by realizing that employment is brought in line with demand more quickly. Backlogs and inventories do not get so far out of control that they then call for more drastic employment action at a later date to recover the deteriorated situation. We must bear in mind here that this is the effect of a change in the rapidity of employment action in the presence of the same inventory and backlog policies that previously existed. In a later section, those policies regarding inventory and backlog will be changed, and after doing so the system will no longer be so sensitive to the rate of employment adjustment.

There are several other parameters in the system whose predominant effect is on the rapidity of adjustment of the labor force. One is the time for making adjustments in a factory order backlog that may be too large or too small.

271

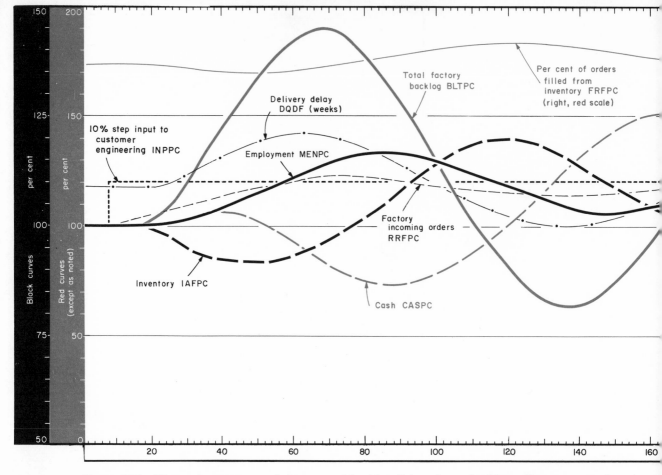

Figure 18-9 Electronic component industry model, old policies, slower backlog adjustment, step increase in demand.

Figure 18-9 shows the effect of doubling the length of time for adjusting factory backlog.[17] The system has become considerably less stable as a result of the increased backlog adjustment time. Manpower peaks occur at 84, 216, and 350 weeks. We see that the natural period of the system has extended from a previous value of about 100 weeks to an interval that is now approximately 134 weeks. The attenuation of disturbances as represented by the amount of fluctuation decrease per cycle is now much less. Manpower peaks are here 80% of preceding

peaks compared with 50% in Figure 18-1 and 15% in Figure 18-8. In Figure 18-9 the longer backlog adjustment time leads to a delivery delay that rises to higher values. This causes more customer ordering reaction and a larger swing in the factory incoming orders.[18] This sys-

[17] The coefficient TBLAF appeared in Equation 17-68 in Section 17.4 and is shown in the flow diagram of Figure 17-11. As introduced in Chapter 17, this was given a value of 20 weeks. In Figure 18-9 TBLAF has been changed to 40 weeks. This means that the factory backlog will contribute to the desired labor force by an amount which would reduce the backlog ¼₀th per week.

[18] I must repeat here warnings given in earlier chapters against generalizing results from a particular system as being applicable in other situations. For example, combinations of parameters can be set up in this model which would be plausible for other industrial situations and in which the above conclusion is exactly reversed. That is, a longer backlog adjustment time can lead to greater system stability. Results from a particular model as given here are presented as examples of how to study a system and as an example of the kinds of effects that can occur as a result of policy and structure changes. The results imply that other systems will contain comparable areas of improvement. We must not, however, conclude that the specific results will be generally applicable.

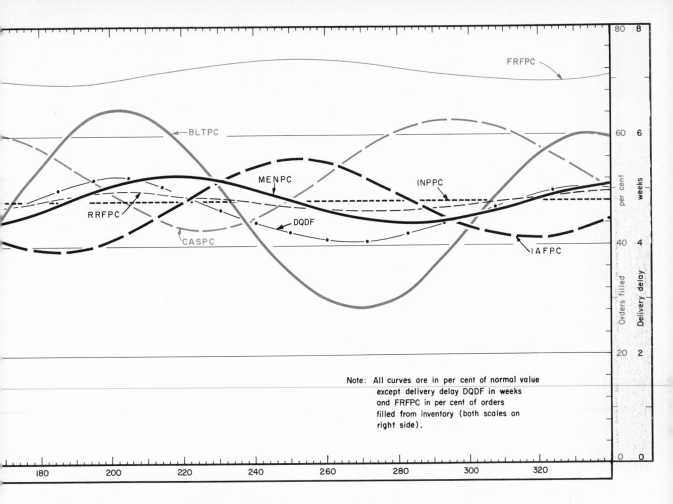

Note: All curves are in per cent of normal value
except delivery delay DQDF in weeks
and FRFPC in per cent of orders
filled from inventory (both scales on
right side).

tem will be more sensitive than the preceding ones to long-period disturbances having periods of two or three years, such as might be imposed by fluctuations in the general economic activity of the country.

If we look back to Figure 18-2, in which the hiring and layoff rates are plotted, we see that the system is continuous in its movement from a condition where employees are being hired to a condition where labor is excess and employees are being discharged. There is no intermediate zone between hiring and layoff during which the employment level is held constant. When a discussion of this system is presented in class to either graduate students or corporate executives, it is very common for someone to comment on the unreality of this kind of continuous transition from hiring to layoff. Further-

more, the suggestion is often made that a dead zone of hiring inactivity would be a stabilizing influence on the system and would suppress the fluctuations seen in Figures 18-1 and 18-2.

Section 17.4.5 allowed for the possibility of a dead zone in employment change, as illustrated in Figure 17-12. Referring to Figure 17-12, we shall now try a hiring and layoff threshold of 2% of the labor force. In other words, no change in the number of employees will be made unless the labor change indicated by Equation 17-70 exceeds 2% of the work force. This is a rather small threshold, and some similar inactivity zone is almost certain to exist in most actual situations.[19]

[19] The hiring threshold is obtained by setting CLCTF = 0.02 in Equations 17-72 and 17-75. See the flow diagram, Figure 17-11.

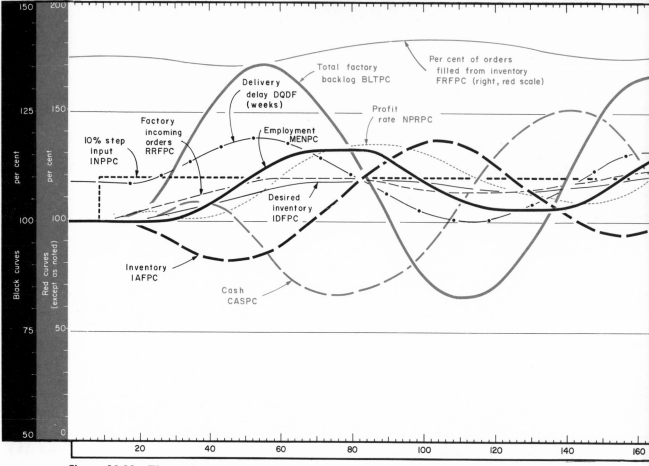

Figure 18-10 Electronic component industry model, old policies, 2% hire-layoff threshold, step increase in demand.

Figure 18-10 shows the effect of the hire-layoff threshold on the system which is otherwise like that in Figure 18-1. Again, a step input of orders to the customer engineering department is used to set up an initial state of disequilibrium from which to watch the system recovery. This time there is no recovery. Here we have a type of system that I believe will be found frequently in industrial and economic behavior. It is unstable for small disturbances and stable for large amplitudes. The 2% dead zone is a very large part of any labor change that might be required during normal small variations in product demand. The effect of a 2% hiring threshold becomes less important in large amplitudes of disturbance. The result is that a sustained oscillation occurs whose amplitude

is in some way related to the hiring dead zone.[20] The hire-layoff threshold in Figure 18-10 constitutes another form of delayed reaction in the employment decision-making processes of the system. As with the other kinds of decision-making delays already seen in Figure 18-1 compared with 18-8, and 18-9 compared with 18-1, the results are detrimental to the operation of the system.[21]

[20] The control engineer will recognize this as being analogous to a servomechanism system that has mechanical backlash in the control or error-detection gear trains. Such an inactivity zone in the control system is notoriously hard to deal with and very often leads to system oscillation through a limited amplitude.

[21] Again we must caution against generalizing, since in Sections 18.3 and 18.4 we shall find that certain combinations of slow response times in the system can lead to substantial improvement.

274

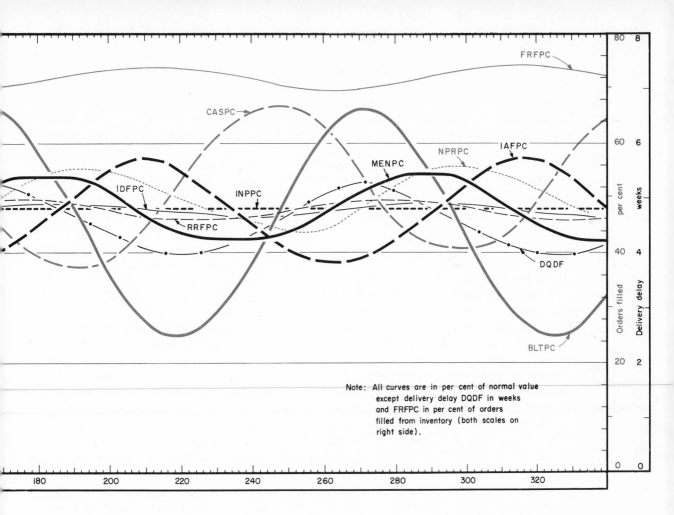

Note: All curves are in per cent of normal value except delivery delay DQDF in weeks and FRFPC in per cent of orders filled from inventory (both scales on right side).

In Figure 18-10 the peaks and valleys of the employment curve are flat because hiring is inactive for a period of 10 or 12 weeks at each peak and valley, and no change of employment occurs. The hiring threshold causes employment to be too high during declining requirements for product and to be too low during the time when demand is increasing. This causes backlogs and inventories to build up further than in Figure 18-1 before being brought under control.

Figures 18-8, 18-9, and 18-10 show that policies regarding employment changes can have a profound effect on the stability of industrial operations. Even more striking is the observation that many of the influences set up by good will and by union pressures and contracts contain the possibility of making matters worse

rather than better. Delays in taking small actions can lead to the necessity for larger changes later. Again we should not generalize from this example to other situations that may seem similar but can be quite different. However, we see here how actions taken in an isolated part of the corporation, perhaps in a union contract bargaining session, can propagate their effects throughout the system. Furthermore, policies and practices that seem reasonable when viewed from the limited environment of the labor relations department may in fact be detrimental to labor stability and the operation of the system as a whole when these conditions have had an opportunity to interact with the other parts of the corporate and market structure.

In the preceding figures we have found that

the old system is sensitive to delays in the employment decisions. These delays are not ordinarily subject to careful consideration or close control. System dynamics are at the mercy of capricious decisions. If the sensitivity to these factors is to remain, the important parameters should be closely controlled. If the system can be redesigned to make it insensitive, that will be even better.

18.3 New Policies

In the preceding sections we have examined the dynamic characteristics of the model that was developed in Chapter 17. The model was designed to represent the operating characteristics of a real case study. The model behavior showed the same problems and difficulties as did the actual system and that originally led to the study. The model passed as a reasonable representation of the system, both in the detailed equations of which it was constructed and also in its over-all performance. The performance of the model was judged by such criteria as natural period and degree of stability, and also by the time-phase relationships that were developed between the internal variables. Changes in the parameters of the model sometimes led to unexpected changes in results. But these changes, when understood, could be traced back to a reasonable likelihood that the effects would be similar in the real system.[22]

At this point, we begin to feel sufficient confidence to believe that the model is an adequate basis for exploring ways to improve the system. Section 18.2.2 has shown that a more responsive hiring policy that acts more quickly produces some improvement. Practical considerations, however, limit the amount that can be

[22] This study of the company and the development and interpretation of the model were made over a two-year period by three or four people.

gained in this direction. Furthermore, we should not expect that the major improvements in system behavior will result merely from changing the values of those parameters which had already been included in the model. More effective changes are often possible. The initial structure of the model was that which had been judged to represent the old mode of operation. The structure is subject to change in many places. Not only can the parameters in decision policies be changed, but those policies can be changed in form as well. Decisions can often be based on different and more effective sources of information.

We shall try to find a redesigned management structure and policies which are achievable in practice and which lead to more favorable operation. The objective is to attain greater labor stability, less tendency for the system to amplify certain critical frequencies of external disturbance, and less tendency for the system to be perturbed by internal or external random variations. This must be done without imposing requirements for unacceptable levels of inventories and without requiring financial resources which are unavailable or which would cost more than the value of the improvements. Since most industrial systems seem to operate so far from a hypothetical ideal, it is reasonable to hope that system improvements can first be obtained without requiring any compromise. Improving one factor may not require paying a penalty elsewhere. That is the situation here. With changes in the model that can easily be made in the real system, we can simultaneously hope to reduce labor fluctuation, reduce peak cash demands, improve the uniformity of delivery delay, reduce the fluctuation in order backlogs, and hold inventory excursions to approximately their present satisfactory value.

In this section, changes for achieving some

of these improvements will be evolved. The improvements to be discussed lie in the management of inventory and of employment changes.

Several points about the old system should be especially noted:

- Inventory changes are not effective in absorbing fluctuations in sales; the reverse holds, so that employment varies more than the usage rate of the product.

- As shown in Figure 17-8, inventory and inventory reordering are managed relatively independently of the remainder of the system.

- Authorized inventory is geared to the average level of sales. If sales rise, more inventory is permissible and considered desirable. If average sales fall, inventories come in for intensive scrutiny and must be reduced. If inventories fall in a period of rising business, there is concern that customer service will be impaired. All of these viewpoints seem reasonable and are common industrial practice. An upturn in sales depletes inventory while at the same time making more inventory permissible and presumably desirable.

- The system is set up as if inventory were a directly controllable variable. An inventory reordering policy exists for the control of inventory levels. Desired inventory changes are discussed as such. The system organization and management discussions tend to contradict what everyone knows — that inventory is not a directly controllable quantity. Inventory is the accumulated difference between what is produced and what is delivered to the customer. Assuming that we sell and deliver what we can, inventory is controllable only by control of production rate.

- Employment changes in Figure 17-11 are made primarily as a result of the average level of sales and the state of factory-order backlogs.

- Before inventory adjustments influence employment changes, the inventory orders must find their way into the factory backlog. If we fol-

low the flow of information from inventory in Figure 17-6 to inventory reordering in Figure 17-8 to manufacturing backlogs in Figure 17-9 and to employment changes in Figure 17-11, we see that three sets of decision-making and decision-executing delays lie between factory inventory and work force. This is not an unusual situation within most companies, in view of all of the steps that couple primary information to production-rate changes.

All of the preceding points suggest revision of the information flows controlling inventory and employment changes. Information about inventories should enter directly into employment-change decisions without going through the intervening step of appearing as factory-manufacturing backlog. Inventory orders need not enter the factory in such a way that they mix with orders being manufactured to customer specification. If customer orders are given expeditious scheduling and priority, they will not be delayed so long in factory backlog and should contribute less to the variability of delivery delay.

Doing these things implies that production for inventory is, as it should be, simply the way to occupy efficiently that fraction of the work force not manufacturing items to customer specification. The factory needs a priority list of the particular catalog items that are most needed in inventory, but this is different from the concept of inventory production orders that are binding on the factory to deliver. The factory produces for inventory the amount that its employment level will permit. Adjusting the flow of goods into inventory is done by adjusting the employment level. This changes the number of employees in excess of those making special orders for customers.

In the following three subsections are the necessary changes to the equations and flow diagrams of Chapter 17.

18.3.1 Changes in Inventory Ordering. Figure 18-11 shows the flow diagram of the new policies that replace those in Figure 17-8.[23] In Chapter 17 one calculation of smoothed sales was made in Equation 17-20, and this was used both for the level of authorized inventory and for determining average factory production rate. There is no reason to believe that the sales averaging time should be the same for these two purposes. The average sales from Equation 17-20 will continue to be used in determining raw-material ordering and the labor needed to supply current sales.

[23] From the inventory reordering in Section 17.4.2, Equations 17-15, 17-16, 17-19, 17-22, and 17-23 are now to be omitted. The flow diagrams give the numbers of the equations and can be used to find the source of each variable in any equation.

A second smoothing equation and averaging time will be introduced for use in determining authorized inventory so that a separate smoothing time constant may be selected:

$$RSF1.K = RSF1.J + \frac{DT}{TRSF1} (RRF.JK - RSF1.J) \quad \text{18-1, L}$$

$$RSF1 = RRF \qquad \text{18-2, N}$$

RSF1   Requisitions (orders) Smoothed at Factory for authorized inventory level (units/week)

TRSF1   Time for Requisition Smoothing at Factory (weeks)

RRF   Requisition rate Received at Factory (units/week)

Equation 18-1 is identical in form to Equation 17-20 except that a numeral is appended to designate the new variable.

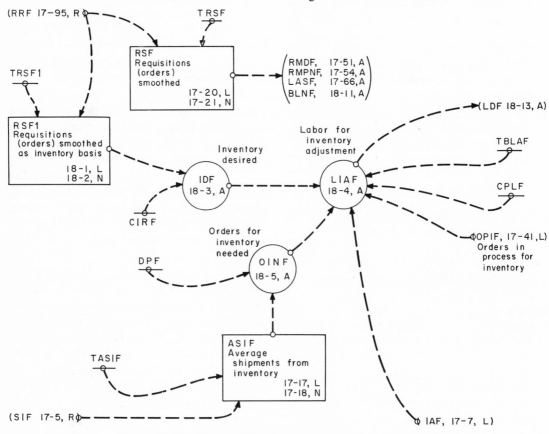

Figure 18-11 New policies for aggregate inventory control.

Desired inventory is determined as in Equation 17-19 but uses the new value of smoothed sales:

$$IDF.K = (CIRF)(RSF1.K) \qquad \text{18-3, A}$$

IDF   Inventory Desired at Factory   (units)
CIRF   Coefficient for Inventory Ratio at Factory (weeks)
RSF1   Requisitions (orders) Smoothed at Factory for authorized inventory level   (units/week)

The important change to be made is that aggregate inventory is no longer to be controlled by the separate independent inventory orders for individual catalog items. A control of total aggregate inventory is to be established by control of employment, and within this framework priorities will indicate to the factory the preferences for various catalog items. The policy therefore becomes a statement of how many employees are necessary for adjusting an improper level of inventory:

$$LIAF.K = \left(\frac{1}{TBLAF}\right)\left(\frac{1}{CPLF}\right)(IDF.K - IAF.K$$
$$+ \; OINF.K - OPIF.K) \quad \text{18-4, A}$$

LIAF   Labor for Inventory Adjustment at Factory (men)
TBLAF   Time for BackLog Adjustment at Factory (weeks)
CPLF   Constant, Productivity of Labor at Factory (units/man-week)
IDF   Inventory Desired at Factory   (units)
IAF   Inventory Actual at Factory   (units)
OINF   Orders for Inventory Necessary at Factory (units)
OPIF   Orders in Process for Inventory at Factory (units)

The last term in parentheses in Equation 18-4 gives the difference between desired and actual finished inventory and the difference between necessary and actual in-process inventory. The sum of these differences gives the inventory adjustment which is desired. This is divided by labor productivity to give the man-weeks of work necessary for inventory adjustment. The result is further divided by a time period over which the adjustment is to be accomplished, yielding the number of men needed to adjust inventory. Since inventory and factory backlog are in many senses the inverse of one another, the same adjustment time constant will be used here as previously for the correction of backlogs. Information from Equation 18-4 flows directly to the labor decision in Equation 18-13.

In the manufacturing sector there will no longer be a formal backlog of orders destined for inventory, so that the former Equation 17-22 must be modified:

$$OINF.K = (ASIF.K)(DPF) \qquad \text{18-5, A}$$

OINF   Orders for Inventory Necessary at Factory (units)
ASIF   Average Shipments from Inventory at the Factory   (units/week)
DPF   Delay in Production at Factory   (weeks)

Equation 18-5 gives the normal work in process in the factory for inventory as the product of the average usage rate from inventory multiplied by the factory-production delay.

Actually the new equations in Figure 18-11 no longer represent inventory ordering in the usual sense. They become one of the considerations entering into the new employment policies. The responsibility for total aggregate inventory would now rest at the point in the organization where central system control is being exercised. At the stock room, the responsibility becomes one of establishing priorities based on stock levels and usage rates. The stock room controls preferences for the sequence of goods to be manufactured but not the total amount to be manufactured for inventory.

18.3.2 Changes in Manufacturing Department.

The essential information and control in the part of the model that represents the manufacturing department are now to be simplified as shown in Figure 18-12.[24] In the revised operation a normal amount of time will be taken for the planning and scheduling of special orders for customers, and then the full resources of the factory are first available for making these orders. The excess labor remaining is then applied to making product for finished-goods inventory. The factory is no longer obligated to deliver the full amount ordered for inventory but rather determines the production for inventory by the labor available after production obligations for special customer orders have been met.

$$LDCOF.K = \frac{BLCF.K}{(DNBLF)(CPLF)} \qquad \text{18-6, A}$$

$$LCOF.K = \begin{cases} LDCOF.K & \text{if } MENPF.K \geq LDCOF.K \\ MENPF.K & \text{if } MENPF.K < LDCOF.K \end{cases}$$
$$\text{18-7, A}$$

LDCOF   Labor Desired for Customer Orders at Factory (men)
BLCF   BackLog for Customer at Factory (units)
DNBLF   Delay in Normal BackLog at Factory (weeks)
CPLF   Constant, Productivity of Labor at Factory (units/man-week)
LCOF   Labor for Customer Order at Factory (men)
MENPF   MEN Producing at Factory (men)

The objective of Equation 18-6 is to apply to the customer backlog of orders a work force in proportion to the size of that backlog so that the backlog delay will be held constant. The backlog is divided by the labor productivity giving the man-hours of labor represented in the backlog. This in turn divided by the normal scheduling interval for backlog orders gives the number of men that should be used. Equation

[24] The following equations from Section 17.4.3 are to be omitted: 17-24, 17-25, 17-28 through 17-33, 17-35 through 17-39, and 17-49.

18-7 checks this against the number of men available, and the lesser of those needed or those available are taken to work on customer orders.

The production rate on customer orders is therefore

$$PCOF.KL = (LCOF.K)(CPLF) \qquad \text{18-8, R}$$

PCOF   Production to Customer Order at Factory (units/week)
LCOF   Labor for Customer Order at Factory (men)
CPLF   Constant, Productivity of Labor at Factory (units/man-week)

The remainder of the labor force not working on customer orders is available to manufacture for inventory:

$$LIF.K = MENPF.K - LCOF.K \qquad \text{18-9, A}$$

$$PIF.KL = (LIF.K)(CPLF) \qquad \text{18-10, R}$$

LIF   Labor for Inventory at Factory (men)
MENPF   MEN Producing at Factory (men)
LCOF   Labor for Customer Order at Factory (men)
PIF   Production rate starts for Inventory at Factory (units/week)
CPLF   Constant, Productivity of Labor at Factory (units/man-week)

Figure 18-12 does not show the flow of instructions for determining what items will be manufactured for inventory with the available labor. The manner of determining the actual items to be made will not affect the dynamic behavior of the system. It is assumed that items will be chosen to keep the material in the aggregate inventory as useful as possible in filling customer orders. The relationship in Figure 17-7 continues to control the fraction of customer orders that can be filled from inventory.

Total employment level is determined at the point where the over-all system dynamics are being controlled. The responsibility of the factory management is for efficient production using the authorized people.

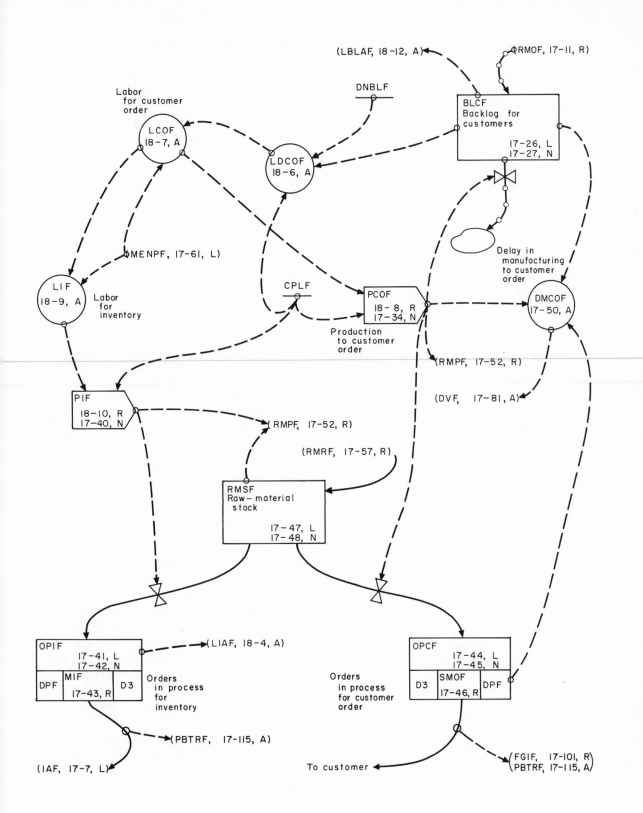

Figure 18-12 New policies for manufacturing.

18.3.3 Changes in Labor Hiring. Labor hiring is now to be based as before on average sales and the labor necessary for backlog adjustment. But in addition, a factor in the labor-adjustment decision is to include the labor necessary for the adjustment of aggregate finished-goods inventory. The new flow diagram is shown in Figure 18-13.[25]

The normal factory backlog now contains only that part of the order flow that is made to special order, for customers, so that Equation 17-67 is to be replaced by the following:

$$\text{BLNF.K} = (\text{RSF.K})(1 - \text{CNFIF})(\text{DNBLF}) \qquad \text{18-11, A}$$

BLNF   BackLog Normal at Factory (units)
RSF   Requisition rate Smoothed at Factory (units/week)
CNFIF Constant, Normal Fraction of requisitions filled from Inventory at Factory (dimensionless)
DNBLF Delay in Normal BackLog at Factory (weeks)

The backlog-adjustment Equation 17-68 must be revised because it referred to total backlog and must now refer only to customer-order backlog:

$$\text{LBLAF.K} = \left(\frac{1}{\text{TBLAF}}\right)\left(\frac{1}{\text{CPLF}}\right)(\text{BLCF.K} - \text{BLNF.K})$$

$$\text{18-12, A}$$

LBLAF Labor for BackLog Adjustment at Factory (men)
TBLAF Time for BackLog Adjustment at Factory (weeks)
CPLF   Constant, Productivity of Labor at Factory (units/man-week)
BLCF   BackLog for Customer at Factory (units)
BLNF   BackLog Normal at Factory (units)

Previously the desired labor given by Equation 17-69 consisted of that necessary to support the average rate of sales, plus that for

backlog adjustment, less the excess manufacturing for inventory. In the revised system, the desired labor consists of that necessary to support average sales, plus that for factory-backlog adjustment, plus that for finished-goods inventory adjustment:

$$\text{LDF.K} = \text{LASF.K} + \text{LBLAF.K} + \text{LIAF.K} \qquad \text{18-13, A}$$

LDF   Labor Desired at Factory (men)
LASF   Labor for Average Sales at Factory (men)
LBLAF Labor for BackLog Adjustment at Factory (men)
LIAF   Labor for Inventory Adjustment at Factory (men)

The decisions for employment levels are now the ones in which the control of the dynamic behavior of the system is concentrated. The employment policies establish the inventory and the rate at which changes in inventory are accomplished. By determining employment directly on the basis of other system variables, the former cascaded delays have been removed at the inventory-ordering and factory-backlog points. No new information sources have been demanded, but the use of information has been reorganized.

This completes the changes that are to be examined in this book in the structure of the management system and in the decision policies to be followed.[26] In the next section the effect of these structural changes will be examined using the same numerical values for parameters as previously.[27] In Section 18.5 some changes in the values of parameters will be made, and finally the total effect of new policies and new parameter values will be shown in Section 18.6.

---

[25] The equations of this section replace Equations 17-67 through 17-69 in Section 17.4.

[26] The equations of the revised system with new policies are tabulated in Appendix B (with the new parameter values from Section 18.5).

[27] The new parameter TRSF1, time for requisition smoothing in Equation 18-1, introduced in this section will initially be given the same value (15 weeks) as its counterpart TRSF. A new value will be chosen in Section 18.5.

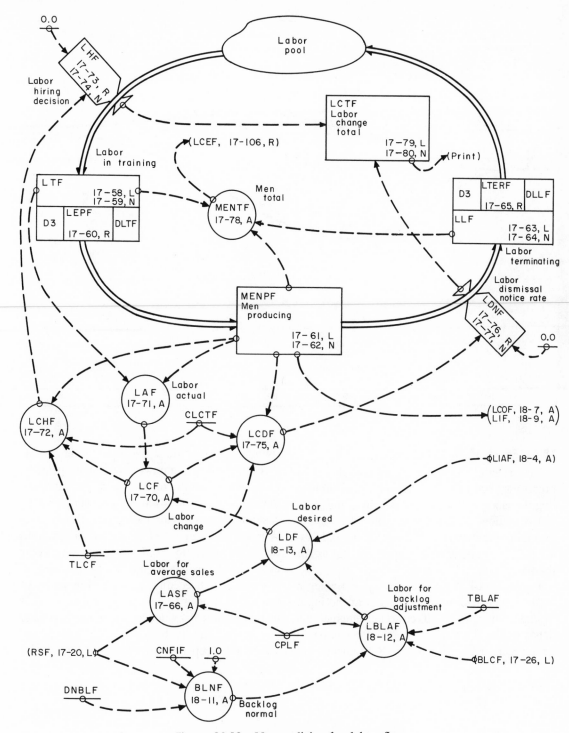

Figure 18-13 New policies for labor flow.

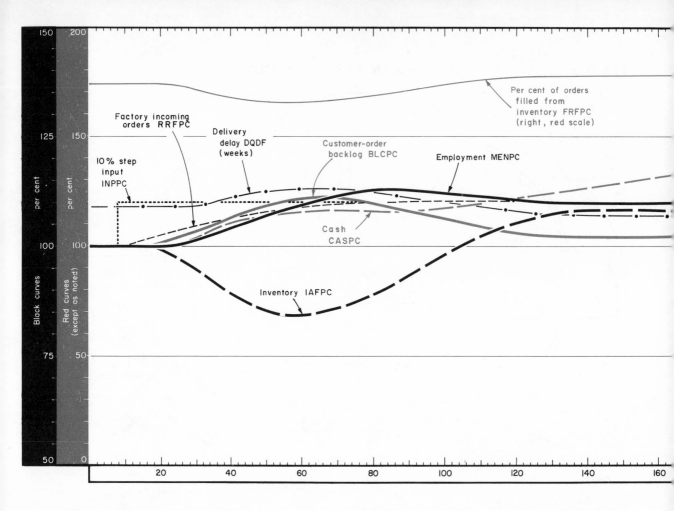

Per cent of orders
filled from
inventory FRFPC
(right, red scale)

Factory incoming
orders RRFPC

Delivery
delay DQDF
(weeks)

Customer-order
backlog BLCPC

Employment MENPC

10% step
input
INPPC

Cash
CASPC

Inventory IAFPC

18.4 Effect of New Policies

The effect of the policy changes made in the preceding section is best divulged by submitting the model of the revised system to the same test conditions as used in Section 18.1 for determining the dynamic behavior of the old system. The parameter values are the same as for the old system.

Figure 18-14 shows the new system when a 10% sudden rise in orders occurs at the input to the aggregate of all customer engineering departments. This figure should be compared with Figure 18-1. In Figure 18-14 manpower rises 25% more than the input orders, while in Figure 18-1 manpower rose at its peak by 50% more than the input orders. The tendency toward a two-year cyclic interaction between customer and supplier has nearly disappeared. Whereas in Figure 18-1 manpower fluctuations decreased only to 50% in each succeeding cycle, they now in Figure 18-14 decrease to 7% in each succeeding cycle. In Figure 18-14 the effect of priority given to customers' orders in the factory is evidenced by a much smaller rise in the factory backlog and a longer and somewhat deeper drop in finished-goods in-

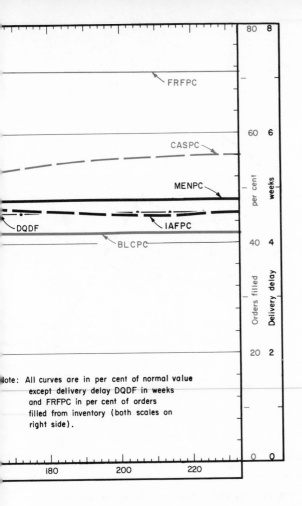

Figure 18-14 Electronic component industry model, new policies, old parameters, step increase in demand.

ventory. Inventory is now being controlled by Equation 18-4, in which the recovery time constant **TBLAF** is 20 weeks. In Figure 18-1 inventory ordering was controlled by Equation 17-15, in which the inventory recovery time constant **TIAF** was 6 weeks, and inventory captured the product that in Figure 18-14 fills customer orders, and helps hold down customer-order backlog.

This slower rebuilding of the inventory initially depleted by the upsurge in orders has several effects. It leads to the smaller peak in manpower requirements, which in both figures are devoted primarily to inventory rebuilding. In Figure 18-1 the inventory rebuilding is done more quickly through a higher production rate than in Figure 18-14, where the peak employment is lower and inventory recovery is spread over a longer period.

Another conspicuous difference between the two figures is in the area of cash balance. In Figure 18-1 the cash balance rose briefly and then dipped as inventory accumulation was superimposed on an increasing level of accounts receivable. In Figure 18-14 the peak dollar flows for employment and materials are some-

285

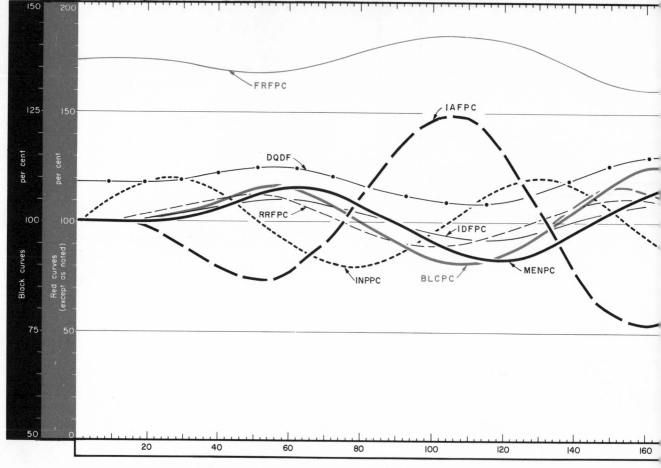

Figure 18-15 Electronic component industry model, new policies, old parameters, two-year sinusoidal input.

what less. The result is that the usual decrease in cash position accompanying the early stages of a sales rise does not occur in Figure 18-14.

In the scales used for these plots, the one-year sinusoidal input to the new system appears substantially like Figure 18-3 for the old system. However, some of the numerical values are appreciably different and worthy of note. As in Figure 18-3, but not illustrated here, only a small fraction of the annual cyclic input to customer engineering penetrates to the component supplier. For a 10% variation in primary input, orders to the factory in the old system varied 1.6%, and in the new they vary 2.4%. In spite of the larger fluctuation in orders, manpower variation is smaller in the new system, being 2.1% compared with 3.2% in the old. The ratios of these order and employment fig-

ures yield the significant observation that in the old system manpower fluctuates 185% as much as the order-rate fluctuation into the component supply factory, when the fluctuation is annual and seasonal. With the new policies, the manpower fluctuates only 85% as much as the incoming-order flow.

With the old system, it was the two-year sinusoidal input that showed the strong tendencies of the system to be undesirably sensitive to longer-period disturbances. Figure 18-15 is the corresponding response of the system with new policies when a two-year sinusoidal input is introduced as the ordering rate to the customer engineering department. A substantial improvement has occurred.

The principal effect is a reduction in the amount of ordering fluctuation that feeds

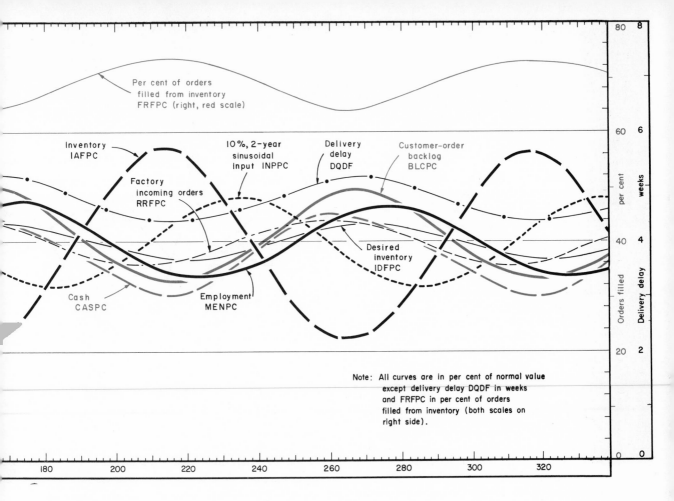

Note: All curves are in per cent of normal value except delivery delay DQDF in weeks and FRFPC in per cent of orders filled from inventory (both scales on right side).

through from the customer to the component-supply factory. In Figure 18-4 ordering rate from customer to component supplier was 107% of the input fluctuation, while in Figure 18-15 it is 53%. The reduction occurs because the delivery delay in Figure 18-15 is varying by only 0.8 week instead of 3.2 weeks, as in Figure 18-4. This reduces the tendency of the customer to order ahead and then to cut back as deliveries change. The ordering rate feeding through from customer to supplier is approximately that 50% fraction of the primary input fluctuation which will penetrate through the 30-week delay of the customer engineering department. In Figure 18-15 manpower is fluctuating 85% as much as the primary input, whereas in Figure 18-4 it fluctuated 230% as much. The improvement comes from two

sources. As just described, the ordering rate from customer to factory is more uniform. Furthermore, in Figure 18-15 manpower is fluctuating 160% as much as incoming orders to the component-supply factory, whereas in Figure 18-4 manpower fluctuated 220% as much as incoming orders to the factory.

We note that the system improvement has occurred in all of the variables plotted in Figure 18-15. Cash varies from 76% to 113% of its initial value, whereas in Figure 18-4 it varied from 5% to 160%. In Figure 18-15 inventories vary between 78% and 120% of normal, while in Figure 18-4 they varied between 73% and 132% of normal. Inventory fluctuation has declined somewhat but not by as much as the other variables. This is satisfactory, since inventory fluctuation is not excessive.

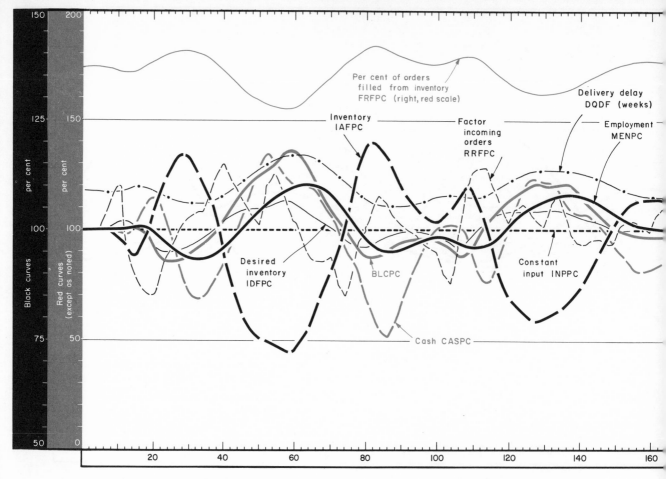

Figure 18-16 Electronic component industry model, new policies, old parameters, random releases from customer engineering department.

The most interesting comparison between the old and new policies occurs when the random[28]

variation is allowed to occur in the specification release rate at the customer. Figure 18-16 should be compared with Figure 18-5. The enhanced stability of the new system is clearly evident. Equation 17-79 accumulates the sum of all hirings and all layoffs to give the total labor change. For Figure 18-5 this is 760 per-

[28] For ease in making comparisons between the old and new policies, the same sequence of noise is taken in Figure 18-16 as was used in Figure 18-5. These noise sequences come from a pseudorandom-number generator which can be started at different specified points at the beginning of a computer run, to make the noise sequences the same or different as desired.

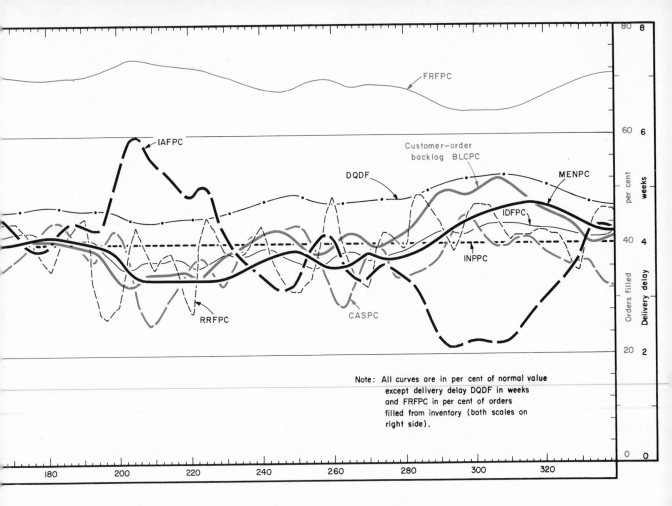

Note: All curves are in per cent of normal value
except delivery delay DQDF in weeks
and FRFPC in per cent of orders
filled from inventory (both scales on
right side).

sons over the 350 weeks, counting both additions to and subtractions from the employment level. In Figure 18-16 the corresponding labor change is 482, or 63% as much. Similar improvement occurs in most of the other variables. In Figure 18-5 cash fluctuates from −11% to 156% of normal, while in Figure 18-16 the range is 53% to 133%. The peak excursions

of inventory are about the same in the two situations. In Figure 18-16 peak employment is 122% of the minimum, while in Figure 18-5 it is 140%.

In Figure 18-5, delivery delay fluctuates from 3.6 to 6.4 weeks, and in Figure 18-16 the amount of variation in delivery delay is only 35% as great.

18.5 Improvements in New Policies

We have seen that the system containing the new inventory and employment policies as represented in Figures 18-14, 18-15, and 18-16 is more stable and less sensitive to disturbing forces than the old policies represented in Figures 18-1, 18-4, and 18-5. However, we have not necessarily arrived at the best solution that we can find.

Figure 17-1 in Section 17.1 discusses the way in which inventory fluctuations so often accentuate employment changes compared with the sales curve. In Figure 18-4, the old system was one in which inventories had begun to rise before the peak of the sales curve. This means that production rate is exceeding sales rate at some time before peak sales, and that the peak of production rate must necessarily be higher than the peak of the sales curve. The policies that lead to an inventory upturn before peak sales are reached account for much of the system tendency to amplify disturbances in the region of a two-year period. Figure 18-15 shows that the new policies have alleviated this situation somewhat. The point of minimum inventory is at the time of the peak incoming factory orders. This means that the manpower curve, representing production capacity, still crosses the top of the incoming-order curve and must necessarily have its own peak later at some higher value. The situation here is better than for Figure 18-4 in that inventories are doing less harm. They are, however, contributing nothing to the reduction of employment fluctuations. The greatest amount of good would be obtained if the maximum rate of fall of inventories were to coincide with the peak of the incoming-order curve.

It is interesting to note that in Figure 18-15 the amount of inventory fluctuation which is present would be entirely sufficient to carry all of the incoming-order fluctuation if the timing of the inventory fluctuation were correct. In other words, a constant level of employment would cause no more inventory fluctuation than is already here present. The incoming-order curve to the factory rises to a peak of about 5% above the average sales rate. This represents a peak of 50 units per week in excess of the average level of sales. The inventory curve shown in Figure 18-15 is falling at the rate of 50 units per week at its steepest point, and if that point coincided with the peak sales, the inventory could supply all of the excess demand at the peak and absorb the excess production at the sales minimum. We can work toward this, but there is no reason to believe that we can fully achieve the ideal inventory relationship over all of the fluctuation frequencies of interest.

One way to make better use of the inventory fluctuation is to adopt policies that will hold a more uniform employment and production rate. This in itself will cause the sales and production excesses to flow in and out of inventory. In both Figure 18-4 and Figure 18-15 the large amount of employment fluctuation is caused by the rapidity with which employment is being adjusted to equal sales rate and also the rapidity with which inventory corrections are being demanded. In the system policies three time constants predominantly influence the responsiveness of employment to changes in sales and inventory.

The smoothing constant TRSF in Equation 17-20 determines how quickly the average sales RSF will follow changes in the incoming-order rate. The average sales RSF determine the work force to be employed for meeting sales demands.

The constant TRSF1 in Equation 18-1 determines the rapidity with which the average sales RSF1 follow the changes in incoming factory orders. This value of average sales determines the authorized level of desired inventory. It is clear that the desired level of inventory need change little if at all during any short-term fluctuations in orders. However, in the very long run, there will necessarily be inventory change in the same direction as sales rate, since the inventory necessary to provide customer service does vary as sales change over very wide ranges and over long periods of time.

The constant TBLAF in Equation 18-4 determines the rate at which inventory excess or deficit conditions will be corrected.

From the viewpoint of Figures 18-4 and 18-15, all three of the preceding time constants have thus far been too short. A longer value of TRSF would cause a slower response in employment changes as a result of changes in incoming orders. This would cause more of the incoming-order fluctuation to be carried by inventory if it were not that the inventory-adjustment Equation 18-4 responds quickly, thereby pushing employment up for inventory stabilization even if it were not pushed up by the average sales rate. It can be seen that a very small value of the inventory and backlog-adjustment time constant TBLAF would cause sufficiently large employment changes to keep a nearly constant inventory.

As an example of the effect that these three parameters may have, we shall explore the system with the new policies of Section 18.3, in which the following new parameter values are established:

TRSF = 26 weeks, instead of 15 weeks
TRSF1 = 52 weeks, instead of 15 weeks
TBLAF = 50 weeks, instead of 20 weeks

The preceding means that the exponential time constant for smoothing sales as a basis for employment level is to be 26 weeks. A longer time constant of 52 weeks will be used as the sales-smoothing basis for authorized inventory. A time constant of 50 weeks will be taken for adjusting inventory toward the desired level. The latter means that sufficient manpower will be employed or discharged to correct any remaining inventory imbalance at the rate of 2% per week.

The preceding are rather long time constants. The hazard in taking averages over long periods of time is that the system may not be sufficiently responsive to long-term changes in the level of business activity. This may or may not have become true here with these changes, depending on the kind of market to be expected. In the next section this new system with new parameter values will be examined for its response to the step, noise, and sinusoidal signals previously used as test inputs. Then the nature of the test inputs will be expanded to see at what point the sluggish response caused by the lengthened time constants would lead to difficulty.

18.6 Characteristics of the New System

In Sections 18.1 and 18.2 the dynamic characteristics of the old system, the model of which was developed in Chapter 17, were explored. In Section 18.3 some changes in system structure and policies were introduced, and the effects of those changes were examined in Section 18.4. In Section 18.5 some changes in the parameter values of the new policies were introduced. Now the dynamic characteristics of the new system will be examined.[29]

The step response of the new system in Figure 18-17 should be compared with Figures 18-1 and 18-14. Compared with Figure 18-14, the effects of the longer smoothing and inventory-adjustment time constants are clearly evident. Employment rises more slowly than factory orders, causing a deeper and longer fall-off

[29] The new system consists of the equations of Chapter 17 modified as in Section 18.3 with the new parameter values from 18.5, which are TRSF = 26 weeks instead of 15 weeks, TRSF1 = 52 weeks instead of 15 weeks, and TBLAF = 50 weeks instead of 20 weeks. A complete tabulation of this new model will be found in Appendix B.

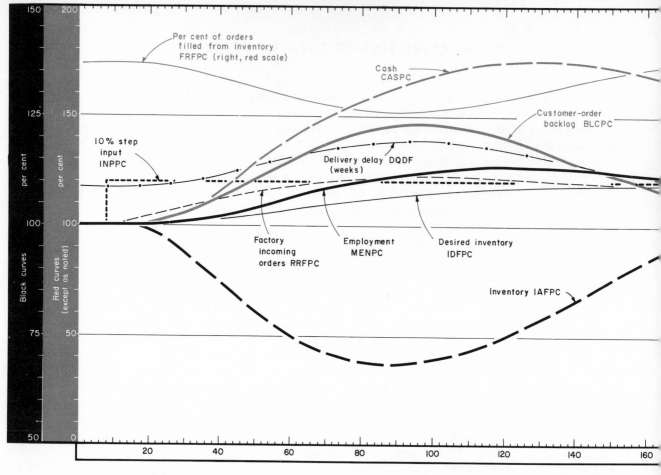

Figure 18-17 Electronic component industry model, new policies, new parameters, step increase in demand.

of inventory. The longer inventory recovery time causes the inventory rebuilding to be more gradual. The inventory dip represents only a little over 1 week of production. In Figure 18-17 employment reaches a peak of 113.4% compared with 112.5% in Figure 18-14. The manpower peak to accomplish inventory recovery in Figure 18-17 is 34% over the amount of the input step increase, whereas in Figure 18-14 it was 25%. Comparison of Figures 18-17 and 18-14 shows that Figure 18-17 represents slightly less desirable behavior. However, the

improved response of the new system to other types of inputs may cause us to judge it the better system. In Figure 18-17 as in Figure 18-14 the rise in level of sales causes a rise in cash with no initial dip as seen in Figure 18-1. Should a 10% drop in business occur, the curves of Figure 18-17 would be inverted; inventory would rise and cash would fall.

The response of the new system to a one-year fluctuation of 10%, comparable with the old system in Figure 18-3, is so small that some of the more significant ratios will serve better

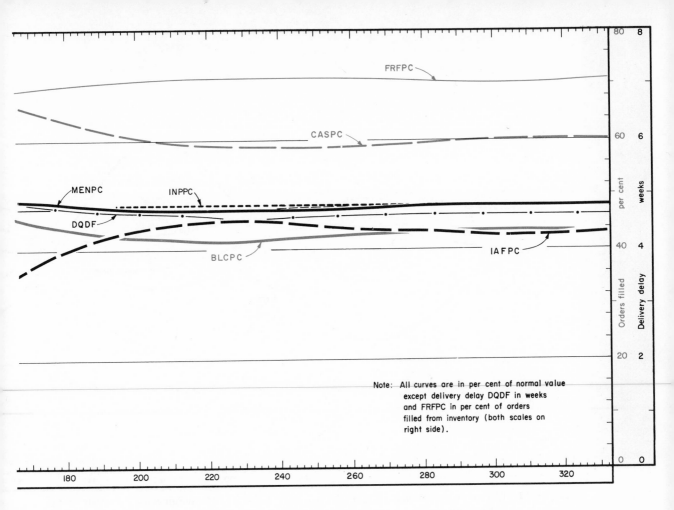

Note: All curves are in per cent of normal value except delivery delay DQDF in weeks and FRFPC in per cent of orders filled from inventory (both scales on right side).

than a graph. In Figure 18-3 employment fluctuated 32% as much as the primary ordering input to the customers' engineering departments. With the new policies and old parameters of Section 18.3 employment fluctuated 24% as much as the input. With the new system and new parameters employment fluctuates 9% of the input. Considering only the interactions within the component supplier company and neglecting the effect of the customer, we can look at the ratio of employment change to change in incoming factory-order rate. For the old system this was 185%, for the new system and old parameters it was 86%, and for the new system and new parameters it is 38%. We see, therefore, a marked tendency for the new system to suppress the annual seasonal component of any fluctuations that may be occurring. This attenuation of seasonal disturbances is obtained while at the same time inventory and cash fluctuations are also being reduced. The new system with new parameters has an inventory fluctuation 63% as great as the old system, and a cash fluctuation 67% as great as the old system.

The most striking change has occurred in

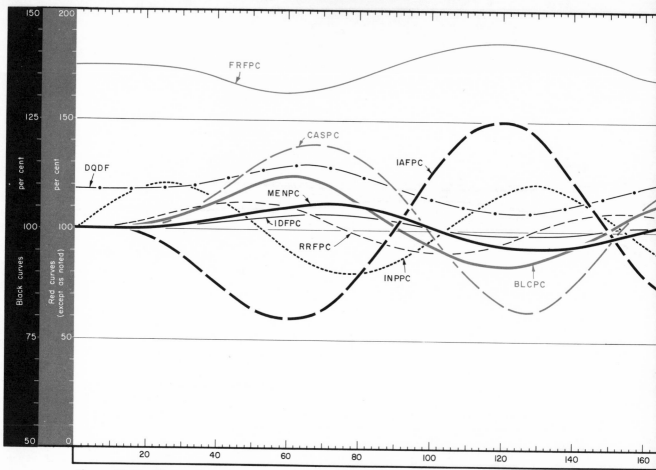

Figure 18-18 Electronic component industry model, new policies, new parameters, two-year sinusoidal input.

the sensitivity of the system to a disturbance having a two-year period of fluctuation. Since this was the particular weakness of the old system, it is the kind of input for which the greatest improvement is needed. Figure 18-18 shows the response of the new system to a 10% two-year fluctuation in orders to the customer engineering departments. In response to the input fluctuation of 10%, the orders from the customer to the component supplier vary 4.6%, and employment varies 3.9%. Cash position fluctuates 25.4% and inventory 15.5%. The orders to the component supplier are varying only 46% as much as orders to the customer engineering department. Employment at the component supplier is varying 85% as much as the orders from customer to component

supplier. All these are improvements over the earlier responses to a two-year input.

Some comparisons between Figure 18-18 and Figure 18-4 show the extent of the improvement in the system sensitivity to the two-year fluctuation of input orders. In Figure 18-18 fluctuation in employment is 17% of that of Figure 18-4, in incoming orders to the component supplier 43%, in cash 42%, in inventory 48%, and in delivery delay 20%. These show substantial improvement in all the variables, with the largest improvement in employment stability.

The improved behavior of the system in Figure 18-18 arises from two factors that are not really independent of one another. The first is the reduced fluctuation in delivery delay,

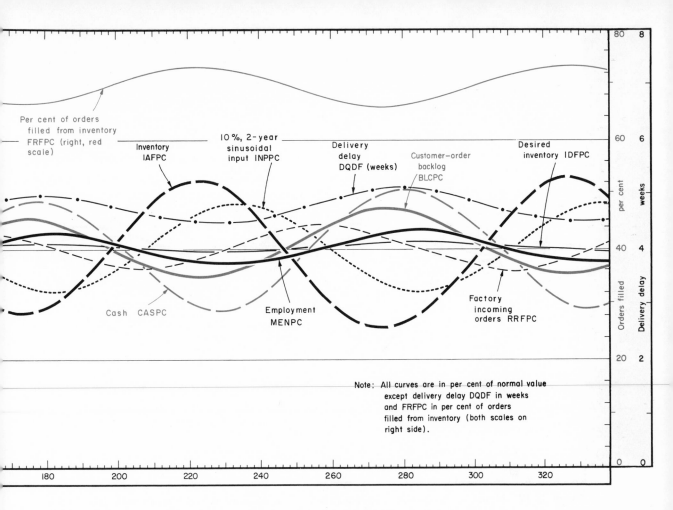

Note: All curves are in per cent of normal value
except delivery delay DQDF in weeks
and FRFPC in per cent of orders
filled from inventory (both scales on
right side).

caused partly by the way in which factory back-logs are handled, and the effect that this reduced fluctuation has in reducing variation in the ordering rate from customer to component supplier. The other factor is the improved use of inventories in Figure 18-18.

Figure 18-18 shows for the first time inventories that are clearly falling at the time of the peak incoming-order rate to the factory. In Figure 18-4 the minimum inventory occurred 22 weeks before the peak of the incoming-factory-order curve. In Figure 18-15 the minimum inventory appeared 4 weeks after the incoming factory orders. In Figure 18-18 minimum inventory occurs 18 weeks after the peak of incoming orders. This means that the production-rate curve crosses the incoming-order-rate

curve 18 weeks after the peak of the incoming-order curve, making it possible for the peak of the production-rate and employment curves to be lower than the peak of the ordering curve. The input variation is here being suppressed in two stages. The natural tendency of the long delay in the customer engineering department is allowed to be effective. Furthermore, employment is now fluctuating less than even the incoming-order rate to the factory. The ratio of fluctuation of employment to incoming-order rate is 85% in Figure 18-18, 160% in Figure 18-15, and 220% in Figure 18-4.

In the system of Figure 18-18 peaks in orders are filled by depleting inventory, and after sufficient time to be sure of the sales the employment level is raised to rebuild inventory.

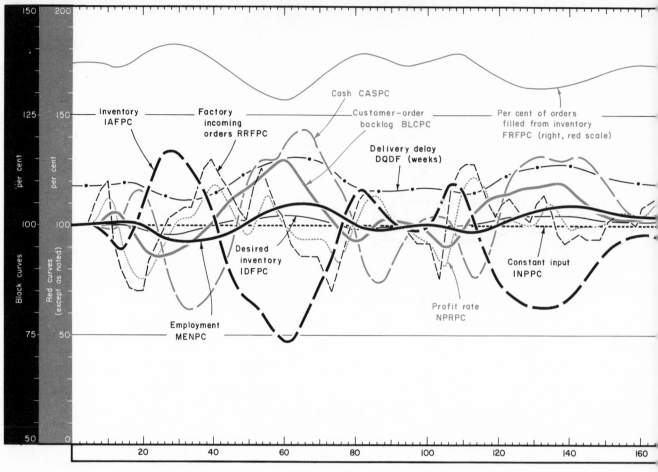

Figure 18-19 Model of electronic component industry, new policies, new parameters, random releases from customer engineering department.

We shall now examine how the new system responds to the same random variation at the output of the customer engineering department as was observed in Figures 18-5 and 18-16. The new system is shown in Figure 18-19. Here the total labor change (the sum of all hirings and layoffs) is 231 men for the 350-week period. In Figure 18-16 it was 482 men, and in Figure 18-5 it was 760 men. The new system has only 30% as much labor change as the old. The peak employment is only 112% of the minimum, compared to 122% in Figure 18-16 and 140% in Figure 18-5. The minimum to which cash position dips in Figure 18-19 is 45%, compared to 53% in Figure 18-16 and minus 11% in Figure 18-5. The peak excur-

sions in inventory are approximately the same in all three situations.[30]

The cash-flow and balance-sheet changes in Figures 18-20 and 18-21 can now be compared with Figures 18-6 and 18-7 for the old system. Cash flow in the new system in Figure 18-20 is much more uniform than in Figure 18-6. This is true both of cash expenditures and cash re-

[30] Model runs to test the sensitivity of the new system to changes in the labor hiring delays are not shown. Unlike the old system, as described in Section 18.2.2, the new system shows little difference when the time for labor change TLCF or the time for inventory and backlog adjustment TBLAF are altered. A hiring dead zone or threshold as in Figure 18-10 will still cause inventory to rise and fall because corrective changes in employment are lost until an appreciable system imbalance has developed.

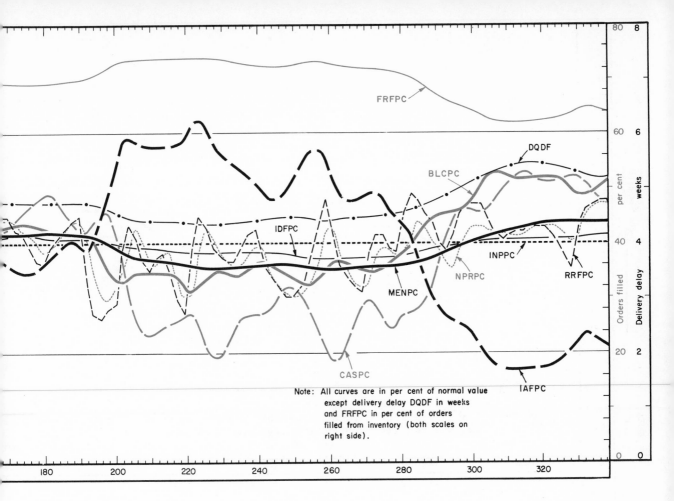

Note: All curves are in per cent of normal value
except delivery delay DQDF in weeks
and FRFPC in per cent of orders
filled from inventory (both scales on
right side).

ceipts. The more uniform production rate produces a more uniform expenditure rate, and the better customer service and more uniform shipping and billing rates produce improved uniformity of the cash-receipts rate.

The result of the stabilized cash flow is shown in the continuous balance sheet of Figure 18-21 for the new system. Compared with Figure 18-7 there is much less variability in the current assets accounts. Of particular interest is the new relationship that now exists between total inventory and accounts receivable. In Figure 18-7 inventories and accounts receivable moved together in the same direction, so that their effect was added in creating fluctuation in the cash account. By contrast, in Figure 18-21

total inventory and accounts receivable tend to move in opposite directions. Current assets therefore tend to exchange between inventories and accounts receivable, which counterbalance each other and result in lower demands on the cash account. The reason for this can be seen in comparing the cash-flow accounts in Figures 18-20 and 18-6. In Figure 18-6 expenditures rose before receipts, whereas in Figure 18-20 receipts rise before expenditures. This happens because of the slower rate of inventory rebuilding in the new system. Current assets are transferred from inventories to accounts receivable, and the incoming cash flow from the sale of goods begins before the cash flow associated with inventory rebuilding.

297

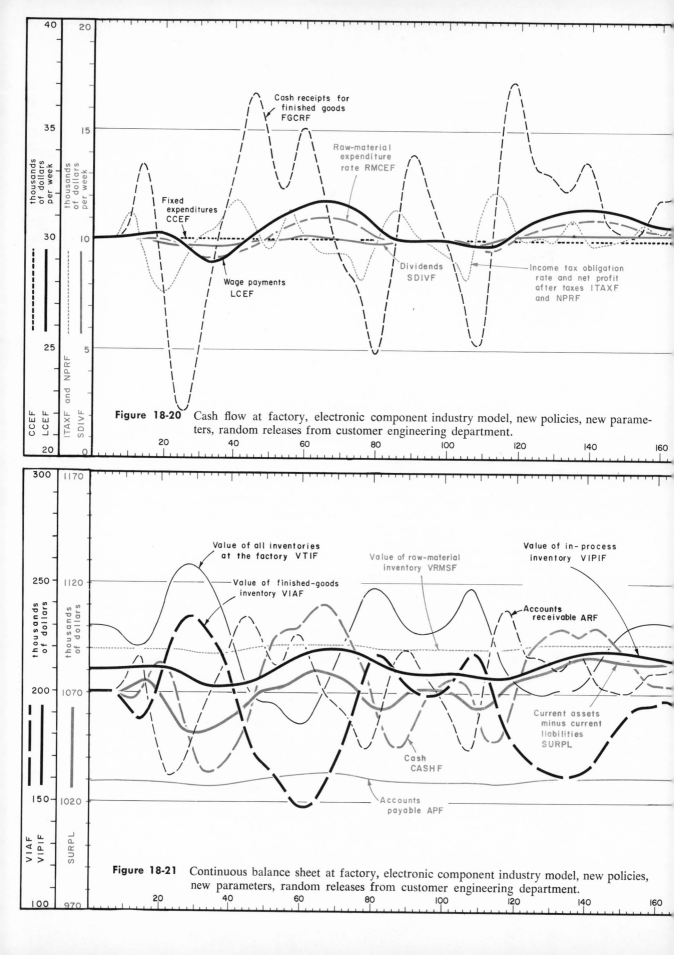

Figure 18-20 Cash flow at factory, electronic component industry model, new policies, new parameters, random releases from customer engineering department.

Figure 18-21 Continuous balance sheet at factory, electronic component industry model, new policies, new parameters, random releases from customer engineering department.

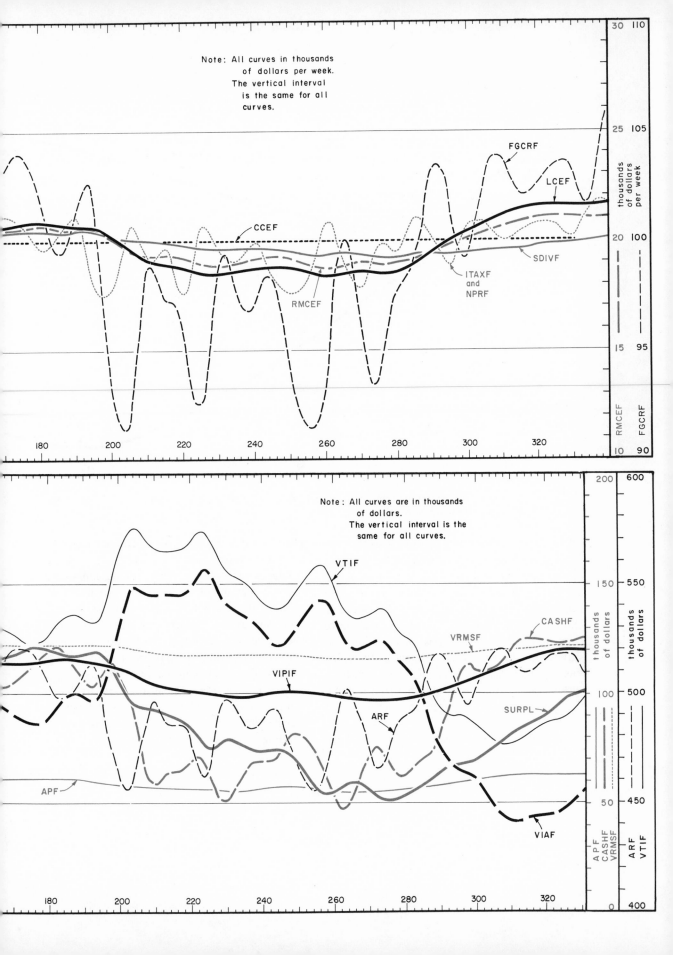

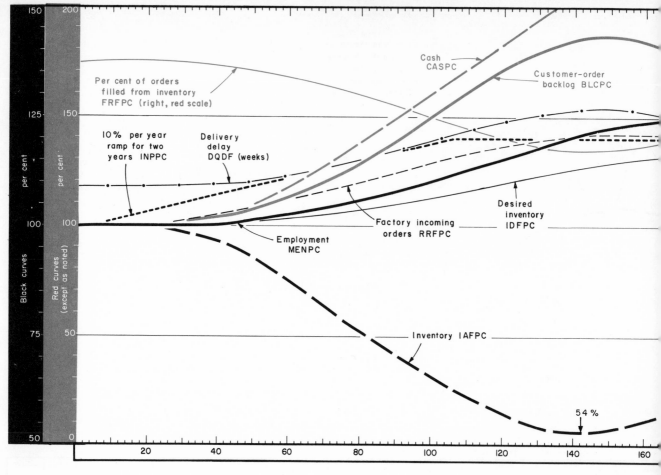

Figure 18-22 Electronic component industry model, new policies, new parameters, 10% per year ramp input for two years.

18.7 New System with Large Disturbances

The emphasis in this chapter and the previous one has been on an industrial situation where the average sales over any two-year period have been about constant, and where the future is expected to be similar. System policies have been designed with emphasis on reducing sensitivity to short-term disturbances. In doing this, we may, as has happened here, create a system that is more sensitive than the original to large long-term changes in average sales level.

Quite different approaches would have been taken had this been a product characterized by rapid growth or decline.

The assumption of fairly stable, long-term average sales might lead one into a vulnerable position. It therefore becomes necessary to explore various kinds of extreme situations to see where the system could break down. These areas of vulnerability may be considered so unlikely that they can be neglected. On the other hand, if the system reaction to large changes in sales level is serious, further design changes may be necessary to ensure proper handling of the extreme possibilities.

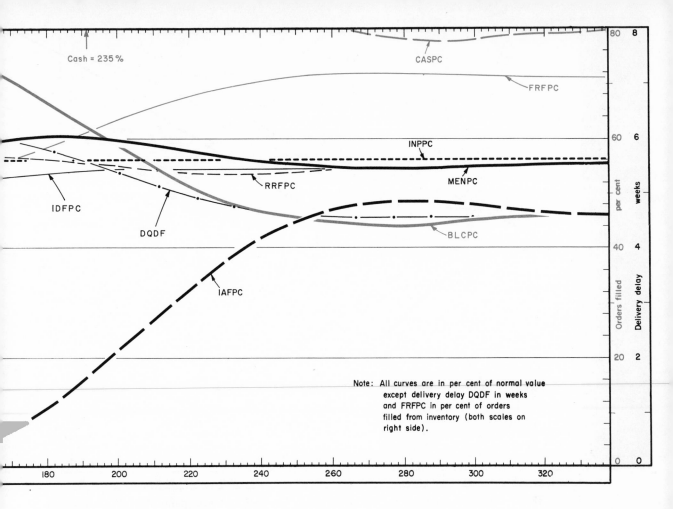

Cash = 235 %

CASPC

FRFPC

INPPC

RRFPC

MENPC

IDFPC

DQDF

BLCPC

IAFPC

Note: All curves are in per cent of normal value
except delivery delay DQDF in weeks
and FRFPC in per cent of orders
filled from inventory (both scales on
right side).

per cent weeks

Orders filled Delivery delay

80 8

60 6

40 4

20 2

0 0

180   200   220   240   260   280   300   320

There follow the responses of the preceding new system to ramp inputs representing sales growth and also to a long-period (four-year) sinusoidal disturbance. The results will be neither ideal nor serious. Their significance depends on long-term judgments that have not been incorporated in the model about the future of the particular product.

Figure 18-22 shows the way the new system will react to a 10% per year rise in sales that persists for two years. At the end of two years, sales level out at a point 20% higher than the beginning. Inventory falls to about 50% of its initial value and recovers very slowly because of the long time constants of the inventory rebuilding that have been made a part of the system policies. Average delivery delay rises from 4.7 weeks to a peak of about 6 weeks. Whether or not the results pictured here would be considered unsatisfactory depends on the nature and practices of the particular industry. There is nothing here that is especially alarming. However, if the sales had fallen for two years instead of having risen, the inverted curves with a high inventory and low cash balance might have become a cause for concern.

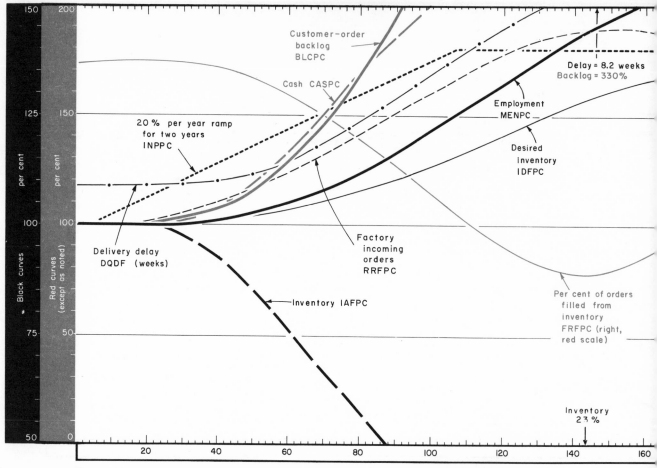

Figure 18-23 Electronic component industry model, new policies, new parameters, 20% per year ramp input for two years.

To explore the ultimate limits of the ability of the system to deal with large inputs, we might select a still steeper ramp increase in sales. Figure 18-23 shows the system reaction to a rise in orders at the customers' engineering department input which increases at 20% per year for two years. Sales reach a level 40% above the initial rate. Inventory falls to 23% of normal, which is about 1 week's worth of production. Delivery delay rises from 4.7 to over 8 weeks. Inventory does not recover to its initial value until more than two years after the end of the sales rise. This would no doubt be considered unsatisfactory in any situation where the large increase in sales might be a reasonable possibility.

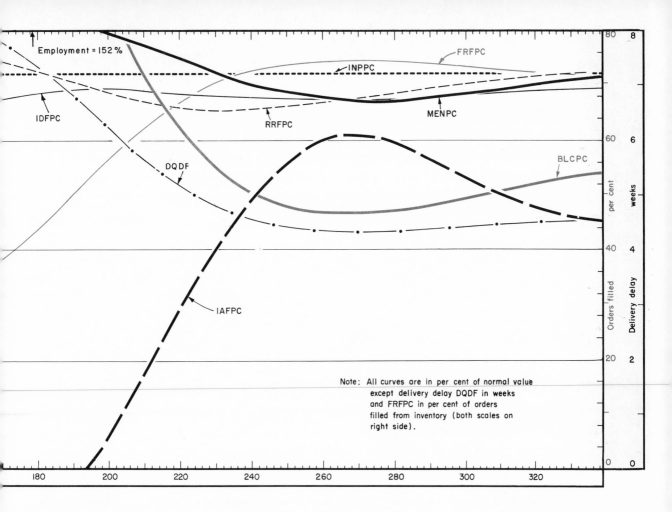

Note: All curves are in per cent of normal value except delivery delay DQDF in weeks and FRFPC in per cent of orders filled from inventory (both scales on right side).

In Figure 18-18 we saw that employment fluctuated with a much smaller amplitude than the input ordering rate for a periodic change having a two-year period. It is obvious that as the length of the period of a sinusoidal disturbance gets long enough, the sales and production must rise and fall by similar amounts, or else inventory fluctuation will become excessive.

We must therefore be interested in how our new system responds to even longer term disturbances than those with a two-year period. Such long-term fluctuations can come from external disturbances representing changes in national economic activity. Economic behavior is here assumed to lie outside of the system that is being modeled.

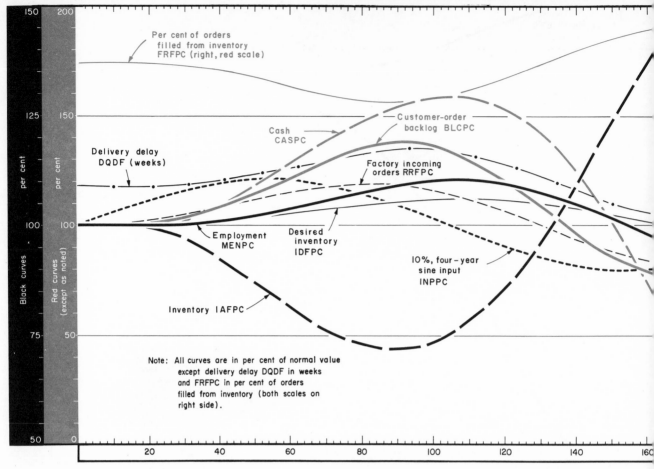

Figure 18-24 Electronic component industry model, new policies, new parameters, four-year sinusoidal input.

Figure 18-24 shows the new system when a fluctuation having a 10% amplitude and a period of 4 years is imposed on the customer engineering department input. At this longer period, inventory fluctuations are creeping back into a position where they cease to provide employment stabilization. Minimum inventory at 300 weeks occurs approximately at the point of maximum incoming orders to the factory. The input to the customer engineering department is fluctuating 10%. Incoming orders to the factory are fluctuating 11% and employment by 13%. Inventories are covering a range about 50% on either side of normal.

The severity of inputs used in Figures 18-22, 18-23, and 18-24 might seem unlikely in many situations. Under other circumstances the disturbances might seem probable, but the responses in the preceding figures would be accepted as tolerable. In still different situations these disturbances or even more severe ones might be plausible, and greater stabilization of cash, delivery delay, and inventory might be desired. Various avenues would then be explored until a satisfactory set of system relationships had been discovered. One possibility is to reduce some of the long time constants introduced in Section 18.5. These can be reduced somewhat without seriously affecting the gains made in stabilizing the system against short-term disturbances. Another approach would be to improve upon the very simple first-order exponential smoothing formulas that have been assumed. A third approach would be to

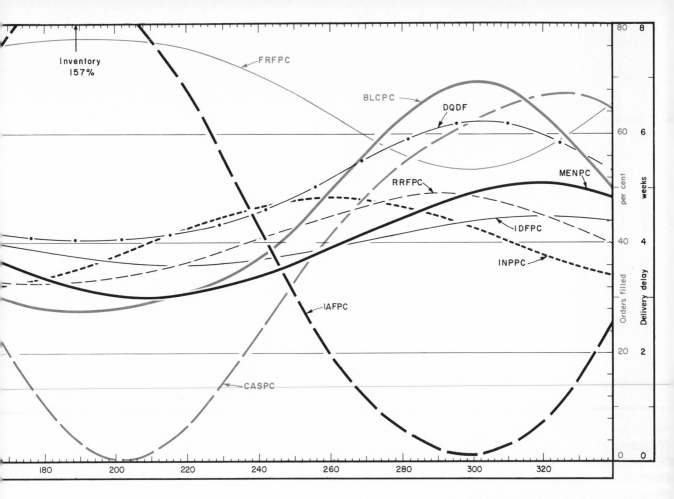

Inventory
157%

FRFPC

BLCPC

DQDF

MENPC

RRFPC

IDFPC

INPPC

IAFPC

CASPC

180   200   220   240   260   280   300   320

80   8

per cent

60   6

weeks

Orders filled

40   4

Delivery delay

20   2

0   0

include various nonlinear relationships in the control policies so that employment would be more than proportionately responsive to large excursions in cash and inventories.[31]

[31] The recovery time TBLAF in Equations 18-4 and 18-12 has been 50 weeks in the new system of Figures 18-17 through 18-24. This is so long that inventory must depart far from normal before it has much effect on production rate. This is good under normal operation but leads to the kind of extreme response seen in Figure 18-23 when input demand changes severely. Figure 18-25 is based on a variable backlog-adjustment time BLATF that equals TBLAF when inventory equals desired inventory but decreases as inventory departs from normal. The following changes create the difference between Figure 18-23 and Figure 18-25:

1. Remove equations 18-4 and 18-12 and replace them by

$$\text{LIAF.K} = \frac{1}{(\text{BLATF.K})(\text{CPLF})} \, (\text{IDF.K} - \text{IAF.K} + \text{OINF.K} - \text{OPIF.K})$$

Just to illustrate that gains are still possible, Figure 18-25 shows a third approach to system improvement — a nonlinear modification to the inventory recovery time. For small

$$\text{LBLAF.K} = \frac{1}{(\text{BLATF.K})(\text{CPLF})} \, (\text{BLCF.K} - \text{BLNF.K})$$

2. Add

$$\text{BLATF.K} = \frac{\text{TBLAF}}{1 + \left[\dfrac{(\text{CIRF})(\text{IDF.K} - \text{IAF.K})}{(\text{CBLAV})(\text{IDF.K})}\right]^2}$$

$$\text{CBLAV} = 1.0 \text{ week}$$

CBLAV is the inventory deviation (in terms of weeks of production) away from normal at which backlog-adjustment time falls to half of TBLAF. This is the point where the value inside the bracket in the denominator is 1. At twice this deviation, adjustment time will be one-fifth of TBLAF. The value of TBLAF remains 50 weeks. CIRF gives the normal inventory relationship to average sales in Equation 18-3; here, as before, it equals 4 weeks.

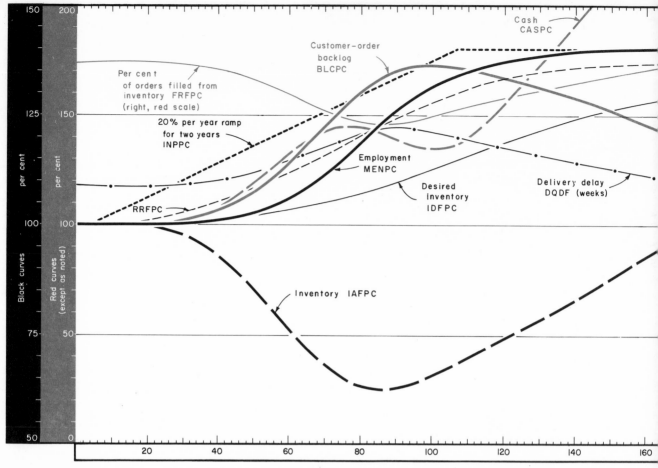

Figure 18-25 Electronic component industry model, new policies, new parameters, variable inventory recovery time, 20% per year ramp input for two years.

inventory excursions the recovery time **TBLAF** remains long. As inventory begins to reach high or low values, the recovery time shortens, so that the inventory deviation has more effect on production rate. Figure 18-25 shows the same 20% per year ramp for two years that led to the extreme conditions of Figure 18-23. The contrast is marked. Inventory dips to only 62% instead of 23%. Manpower rises to the new 140% level of sales with no overshoot, whereas in Figure 18-23 manpower peaked at 152%. Delivery delay rises from 4.7 to only 5.7 weeks instead of 8.2 weeks. Customer-order backlog peaks at 171% instead of 330%. The percentage of orders filled from inventory falls to only 58% instead of 31%. Because of the better control of delivery delay, customer advance or-

dering does not reach such proportions as to cause incoming factory orders to exceed the new-order input rate to the customers as it did in Figure 18-23.

These improvements occur because the large downward excursion in inventory in Figure 18-25 causes a more rapid rise in employment than in Figure 18-23. Production equals shipments in the 86th week instead of the 144th week.

The effect of the system changes in Figure 18-25 is to retain the advantages of the new system with its long inventory-recovery time when the system is encountering the short-term sales fluctuations that cause only small aggregate changes of inventory. When changes in long-term sales occur, they are manifested by

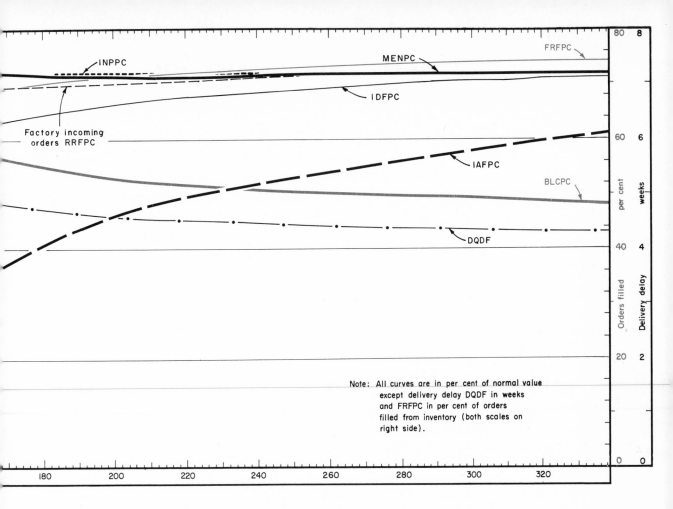

Note: All curves are in per cent of normal value
except delivery delay DQDF in weeks
and FRFPC in per cent of orders
filled from inventory (both scales on
right side).

larger inventory changes, and these in turn create more decisive action in changing employment level.

The nonlinear inventory-recovery time as used in Figure 18-25 produces almost no discernible change in the system response to a noise variable like that seen in Figure 18-19. Likewise, since the inventory excursions in the response to a two-year period of 10% amplitude as in Figure 18-18 are small, the change of inventory-recovery time makes negligible difference for this test input.

By contrast, the new basis for inventory adjustment makes a substantial alteration in the response to an input with four-year period. Figure 18-24 showed the new system before altering the inventory-recovery control. After the change, inventory varies from 64% to 141% instead of 52% to 157%. Inventory peaks are not greatly reduced, but their relationship to other variables has changed so as to create a greater reduction in cash fluctuation. After the change, cash varies from 65% to 123% instead of from 1% to 167% as in Figure 18-24. This improvement in inventory and cash-position extremes has been achieved with only slight loss in employment stability, which now varies from 87% to 115% compared with 88% to 114% previously.

Consequently, it was here possible to find structural and policy changes to alleviate a specific undesirable system characteristic without serious detrimental effect on aspects of system behavior that were already satisfactory.

18.8 Summary

In Chapter 17 a model was designed to represent the organization and policies in a particular case study.

In Section 18-1 the dynamic characteristics of this "old" system were studied. The model showed the kinds of behavior that had been experienced in the actual system. The characteristics of the closed, self-contained model seemed reasonable compared with the actual system. The model was acceptable at the level of component parts and agreed with the extensive descriptive knowledge of the separate facets of the company and of customer motivation. When these acceptable pieces were assembled, they interacted in agreement with past knowledge of the total system. The model was thereupon judged sufficiently reliable to be a basis for system redesign.

Need for redesign was indicated by the strong tendency of the system to amplify disturbances having a periodicity near two years. The undesirable responsiveness to such periods made the system susceptible to being badly perturbed by short-period random effects of even small amplitude.

In Section 18.2 the sensitivity of the old system was explored to determine what parts were most crucial in determining the undesirable behavior. Delays in the hiring policies, which had not been subjected to careful managerial control, were found to be of great importance.

In Section 18.3 the sources of information and the policies for controlling labor level were changed. This brought substantially better system stability but still left room for improvement.

In Sections 18.4 and 18.5, the equations and parameters controlling inventory were changed to make inventory variations assist in providing employment stability instead of contributing to the instability as before.

These changes led to a system, as explored in Section 18.7, that was somewhat unresponsive to large, enduring changes in sales level. Figure 18-25 indicated the way that long-term, large-amplitude inputs could be controlled without giving up the improvements which had been achieved in random and short-period system sensitivity.

We might well assume that improvements over the policies studied in this chapter are still possible. Nevertheless, we should not make the mistake of expecting perfection in a set of system policies. Neither should we delay unduly the implementation of improvements while pursuing some illusive optimum design.

All of the changes made in the system model in this chapter are ones that can be readily made in the actual system. Their implementation requires the formalizing of critical policies of system control to ensure that they are consistently and routinely executed. Certain information flows are required which had not previously been used but which can readily be made available.[32]

---

[32] A report on the experiences in implementing these changes in the actual system is not yet timely and will be left to other authors at a later date.

IV

FUTURE OF INDUSTRIAL DYNAMICS

Part IV completes the book with a discussion of the future of industrial dynamics.

● Chapter 19 summarizes the high points of industrial dynamics models that have been developed for a variety of industrial situations. These are included to illustrate how the methods discussed in this book can go beyond the examples given in Part III. Dynamic models of market, growth, commodities, and research and development have been explored.

● Chapter 20 relates industrial dynamics to management education as providing a central system structure around which to integrate the separate functional subdivisions of management.

● Chapter 21 discusses the organization, staffing, and use of industrial dynamics within the corporation.

Part IV is suggested for all readers.

CHAPTER 19

Broader Applications of Dynamic Models

The purpose of this chapter is to indicate how aspects of the industrial enterprise beyond the functions treated in Part III can be represented in dynamic models. Examples are drawn from thesis and staff research at M.I.T. A model of an automobile market has related design decisions to market penetration. A model of product growth represents a transient dynamic situation that continuously changes in character, as distinguished from the models of Part III, which are "steady-state dynamic" systems in which any five- or ten-year period is similar in general character to any other period. Studies on commodity markets show unexpected ways in which price stability may be improved. The process of military research and development management is being formulated into a system involving characteristics of the product, the development organization, the military contracting office, and the government budget cycle. Long-range-planning models can be used to integrate present facts and assumptions to see the ways they might interact in the future.

THE two models presented in Part III treat very restricted aspects of the total industrial picture. Capital equipment, one of the six flow systems in Section 6.2, was not used. Money flow has appeared in the model of Chapters 17 and 18 but only in an incidental, reporting capacity. Of the major functional subdivisions of management we have given some attention to production and distribution, have barely touched the market area, and have entirely omitted research, capital investment, long-range corporate planning, and many others.

These omissions do not occur because they represent areas of less importance or because they are less amenable to study by dynamic model methods, or because excursions are lacking into these somewhat more elusive aspects of corporate life. Rather, the more advanced models fall beyond the work to be undertaken in this volume, and the ones that have been developed have not been explored and revised to the point where they yet justify formal presentation in their entirety. However, in associated work done to date in other models of industrial activity, many interesting points have arisen and tentative conclusions formulated. This chapter will try to round out the description of the models given in Part III by discussing the high lights arising from studies of other industrial situations that have followed the methods outlined in this book.

19.1 Market Dynamics

There seems to be a common inclination to assume that the mechanisms of the market are obscure, that psychological factors predominate to the exclusion of tangible factors, and that there is little possibility of useful modeling of the market interactions. I hold the opposite view. To be sure, many of the market charac-

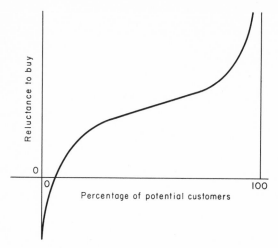

Figure 19-1 Example of functional relationships existing in the market.

A market is composed of numerous interacting concepts, many of them concepts of whose existence we can be confident but which have been ignored in quantitative analysis because of the misleading assumption that we must omit those factors which we are presently unable to measure accurately. By employing only the information available in our own practical experience, we can establish at least the *shape* of many of these conceptual relationships. Often we shall even have the necessary knowledge to establish one or more points on the curve and thus define it closely. This figure shows one possible intrinsic characteristic determining the relationship of potential customers to a product. This characteristic, called "reluctance to buy," is intended to be that factor arising from the inherent situations and psychological natures of the potential customers and not from their state of knowledge of the product or other factors which would be separately but similarly incorporated. This relationship suggests an ordering of customers into a sequence according to their "reluctance to buy." The negative reluctance to buy at the bottom end of the curve represents those experimenters and innovators who will seek out a new product and offer to buy, even before any overt attempt has been made to sell. Such customers are limited. The broad market is represented by a wide middle range where selling persuasiveness rises more slowly to expand the market. Last, the reluctance curve rises steeply as we approach those customers who require greater persuasion, price concessions, and technical and environmental assistance before the product is attractive to them (for example, selling electric clothes dryers to nomadic desert dwellers who have $600 per year income and no

teristics are harder to observe than the factors in, say, the production processes. This makes the understanding which can come from model construction more, rather than less, essential.

Most of the dynamic interactions within a market are noncontroversial as to their existence — price versus utility and substitute products, purchase deferrability dependent on the consumer stocks, rebuy dependent on degree of satisfaction with previous purchases, influence on new customers from attitudes and opinions of past customers, wear-out and replacement rate of past purchases, degree of market saturation, distribution of potential customers from the daring experimenters who readily try new products to the conservatives who wait for proved performance (see Figure 19-1), fad and prestige factors, style,[1] secondhand markets and price, obsolescence, and buying in expectation of price rises and shortages. The uncertainties lie in the relative importance and possible interactions between the factors that are known to exist to some degree. As elsewhere in the industrial system, our intuition about market dynamics is unreliable. We act on a mental model wherein the basic component assumptions may

[1] See Reference 14 by Robinson for a good discussion of the dynamics of style change.

electric power). A careful consideration of even the shape of such a curve for a particular product is rewarding. Some products depend on the innovative customer to launch the market growth. Other products may be in trouble if product attractiveness is marginal in the wide, flat, central region of the curve wherein small shifts in the curves can cause large changes in demand. The rapidity with which such a curve breaks from one region to another will, depending very much on companion curves describing the product, determine the alertness that management must exercise in changing production and marketing plans as the industry changes. The first step in a study is to identify the fundamental market and product characteristics that are thought to be important, to estimate the shapes and magnitudes of these functions no matter how tenuously, and next to determine which assumptions are critical in determining system behavior.

never be able to lead to the assumed results. We never discover the incompatibilities that exist within the assumptions that we accept. Constructing a formal model clarifies our thinking and illuminates the inconsistencies that have intruded.

As with other parts of the industrial system, a market model will consist of levels and rates of flow. The model may be limited to steady-state dynamics like that in Chapter 16; or it may incorporate growth and life-cycle factors that show how demand grows, competition appears, and newer innovations arise.

In a model of the entire life cycle of a product, customer levels would be set up for those who have never heard of the product (the initial condition), those who know of it but have not purchased, those using their first purchase, satisfied users who rebuy, and dissatisfied customers who revert to nonbuyers. Flows between these would be nonlinear functions of these same customer levels, of information and awareness levels existing in the system, of product levels, and of levels of competition and alternate choice.

Such a study would aim initially at an understanding of how the market dynamics *could* function under various circumstances. What conditions make a particular factor predominate? What are the indicators that the dynamic character of a market is changing, as from high-profit initial growth to saturation and profitless competition? What is the relative importance of various dynamic (that is, time-varying) characteristics of the product itself — for example, its showroom appeal versus the owner satisfaction it creates, or the length of time to develop or tool-up for a product versus the time necessary for the customer to learn to use it?

The preceding types of levels, rates, and relationships between a product and its market exist for any product. We can imagine a general-purpose model good for any market. Differences between railroad diesel engines and hula hoops would appear as different times for perfecting the design, different degrees of fad appeal and herd instinct, different lengths of time to educate the customer to the product use, etc. Before we can expect a useful general-purpose model of dynamic market interactions we shall first need to explore numerous limited aspects where only certain of the more predominant factors are included.

As one of these partial representations of a market, Walter formulated an interesting dynamic model of the automobile market and one of the managerial policies affecting a company's share of the market.[2]

Walter set out to explore the possible relationships between the market penetration (percentage share of the market) of a particular company and its policies governing the lead time between the date that a design is frozen and the date on which that design is available for sale. Automobiles in the United States have been restyled annually. The design lead time may vary from eighteen months to four years, so that the lead time is longer than the market life of the product.

Several questions immediately arise. From the viewpoint of this study and this dynamic model, what is an automobile? Also, what is the dilemma faced by the policy maker in deciding on a suitable design lead time? It is characteristic of any significant decision that the action must be neither too much nor too little. The decision is always seeking a satisfactory middle ground. A properly constructed information-feedback-system model should be able to generate the differing consequences of too much or too little action. Executive decisions are characterized by their intention to thread a path between the undesirable extremes. This existence of a dilemma was obscured in the models of Chapters 15 and 17 by the simplicity and obviousness of the systems being considered.

---

[2] See Reference 13 for the S.M. thesis written under the author's direction by Franklin Walter, as a Sloan Fellow at M.I.T. from the Chrysler Corporation, 1958-59.

In Chapter 15 the major decision was that of ordering goods from a supplier. An incorrect ordering rate would result in a level of inventory that was too high or too low, and this would become an indicator of necessary correction in future ordering decisions. In Chapter 17 one of the important decisions was the employment level, and too high a level would lead to excess inventory, and too low a level to a deficient inventory. What then is to be the dilemma representing the consequences of a too long or a too short design lead time in the automobile production-marketing system? Many answers might be given depending on the particular facet of the industry being studied. There is no pretense that the particular study related here deals with all of the important facets of the automobile industry, but only that it deals with one of the top-priority sets of interactions.

As developed by Walter, a shorter design lead time would permit a company to be more cognizant of recent customer preferences and design and styling trends. New innovations and market fads could be recognized and incorporated before they had run their courses and expired as market factors. However, shortening of the design lead time means a shorter interval in which to build experimental models, to find design defects, and to perfect the multitude of small details. Conversely, a longer design lead time should, within limits, lead to greater physical perfection while at the same time introducing the hazard of designs that are out of step with current market trends. Here then is the dilemma faced by the executive who must decide when to freeze designs and proceed with the long sequence of construction, testing, and tooling steps.

In this context, what is the dynamic description of an automobile? The obvious physical attributes such as gasoline mileage, height of tail fins, and colors, do not seem to lend themselves to direct incorporation. How does the result of the lead-time decision enter into the time-dependent behavior of the market? In this study the question of system formulation was resolved by defining two characteristics of an automobile — showroom appeal and ownership satisfaction. These two characteristics can relate design lead time to market reactions.

Showroom appeal combines all those characteristics of the product which affect its sales attractiveness to a potential customer who has had no prior personal experience with the product. This may be thought of as primarily style but also includes advertising, existing prevalence of the product and consequent awareness of it, reputation regarding operating characteristics, and so forth.

On the other hand, ownership satisfaction deals with those characteristics of the product which affect the opinion of the person who has purchased it and which ultimately affect his inclination to rebuy the same brand of product.

It is clear that a particular physical characteristic of the product may not lie exclusively in one or the other of these categories. Yet, given a particular characteristic, it is usually clear whether the characteristic is more persuasive at the point of sale or exerts its strongest influence on the person who has already purchased and is an owner. For example, style will affect primarily the showroom appeal; but door handles that fall off will affect ownership satisfaction.

Figure 19-2 shows reasonable shapes for these two curves as a function of the design lead time. Within a certain short lead-time interval, further shortening of design lead time would not be expected to increase appreciably the showroom appeal, because market trends would not be perceptible and significant over such brief periods of observation. The curve must therefore be horizontal as it approaches the introduction date. On the other hand, showroom appeal may decline rather rapidly as the design lead time is lengthened into the three-year interval that characterizes the life of short-term style fads. Beyond four years of lead time, the product must concentrate on timelessness, adopts a neutral style, relies no longer on current style fluctuation, and again the showroom appeal versus lead time levels out.

By contrast, ownership satisfaction will increase only gradually for design lead times greater than four years, but would certainly deteriorate very rapidly as design lead times are shortened toward the one-year point.

This dynamic model then generates a stream of automobile designs for each of the competitors in the industry. Moving with each design throughout its life in the hands of the consumer is a measure of the ownership-satisfaction coefficient with which it was endowed as a result of design lead time. This, of course, assumes that the same average caliber of skill is applied each year and that the product result depends on the time available for bringing that skill to bear on the product. The showroom-appeal coefficient needs be associated with the product only during the design interval plus the year in which it is on sale.

Certain customers like their previous brand of automobile well enough to rebuy the same make automatically. Others feel that they have had such an unsatisfactory experience that at the time of rebuying they look only to alterna-

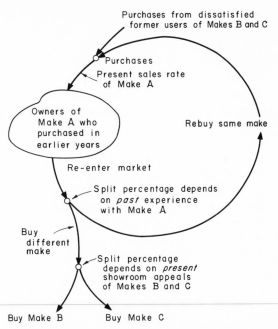

Figure 19-3   Automobile repurchase flows.

tive makes. For this particular study, the substantial gray zone between these two extremes was clarified by assuming a clean, although variable, separation into these two categories. An intermediate zone with rebuy dependent on both showroom appeal and owner satisfaction would be possible but would add almost nothing to the pertinence of an initial examination of the manufacturer-marketing system. Each year after the purchase of a particular model of automobile a certain (and variable) percentage of the owners will re-enter the market. This is shown in the flow diagram of Figure 19-3. A certain variable percentage of former owners of Make A rebuy the make they previously owned, depending on their past experience with that make. Those who do not rebuy turn to either Make B or Make C, depending on the relative showroom-appeal factors of the latter two manufacturers. A similar market sector exists for the other two makes yielding a three-company competitive market. The final loop in the information-feedback system would be to

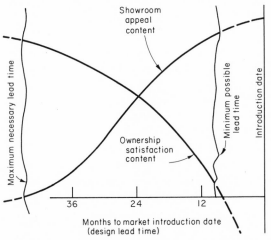

Figure 19-2   Showroom appeal and ownership satisfaction versus design lead time.

define for each manufacturer his policies that determine the way in which market information leads him to choose future product design lead times.[3]

This model serves to show how differently a system can operate in different regions of the parameters. If a high ownership satisfaction is unachievable, that is, if no one can build up loyalty on the part of former owners, then showroom appeal takes on a major significance. On the other hand, it also shows that a manufacturer who can develop ownership satisfaction and customer loyalty can hold his own against almost any conceivable disadvantage in the area of showroom appeal. A very high ownership satisfaction coefficient serves to hold former customers, while even a small showroom appeal persistently captures a small flow of dissatisfied customers from other makes until an equilibrium is reached in which the make with the high ownership satisfaction has a disproportionate share of the market.[4]

Some simple aspects of the situation are shown in the curves of Figure 19-4. Here we take the unlikely initial condition that three competitive automobiles are identical in showroom appeal and owner rebuy ratio and therefore share the market equally. Suppose that showroom appeal is given the arbitrary value

of 1, and that for each of the products the ownership rebuy ratio is 0.8. This means that 0.8 of the former owners of each make rebuy the same make when they replace their automobiles. Now suppose that one of the manufactures discovers a way to double forever after the showroom attractiveness of his product, but that at the same time the ownership rebuy ratio may drop to 0.5. Would this be a wise move? In many kinds of products the emphasis on style, advertising promises, and planned obsolescence would indicate that the managerial policy does indeed stress showroom appeal over the customer-satisfaction coefficient. The figure shows the general nature of the transient that would follow such a change in the characteristics of Make A. The rebuy ratio remains high so long as the product in the hands of the customer is primarily that purchased before the time of the product policy change. Because showroom appeal is twice that of the competitors (meaning that Make A captures twice as much of the conquest market as its competitors), sales begin to rise. This may persist for one, two, or three years. Then as the less satisfactory product becomes predominant in the hands of the

---

[3] Walter did not have time to carry the system formulation into this final feedback loop but looked rather at the way in which externally assumed production lead times would affect market transients.

[4] Conversely, an automobile make that consistently encounters a low rebuy ratio must depend on the stream of dissatisfied customers from other makes. I have many times had the experience of going into the showrooms of one particular make of luxury automobile that has a low market share and low rebuy ratio and finding that the major sales pitch revolves around the large number of sales made to former owners of the sales leader in that price class. This sounds impressive; it sounds like a customer swing from the leading brand to the former underdog. On more careful inspection it merely means that former owners of the underdog make are still not rebuying; low sales are being sustained on that small percentage of owners that have had unfortunate experience with the more successful make.

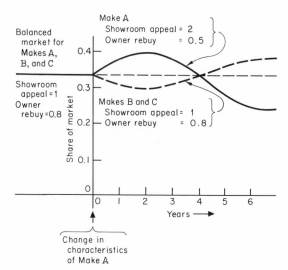

Figure 19-4 Market transient from a single change in product characteristics.

customer, rebuy ratio begins to fall, and eventually the total amount of product in use declines. A new equilibrium develops in which the market penetration of Make A is lower than initially, and the loss in market has transferred to the other makes.

If this situation underlies the character of the company and market relationship, the managerial dangers are evident. An actual situation will never be the pure uncluttered case pictured. Many other factors will also intrude, to which changes in market share can be attributed. If sales should rise for a period of two or three years, this is apt to confirm the wisdom of an initial decision even though that same decision sets the stage for a lower market share after the fourth year. Any fall-off in sales as much as four years after a decision is apt to be attributed to some current unfavorable effect at the later time. Here we see the usual conflict of short-term and long-term actions, the effect of which may be accentuated by the short tenure of men in managerial positions. Executive E entering the picture and making the change of policy at the beginning of the interval in Figure 19-4 might be judged favorably and be promoted before the period of declining sales sets in. Executive F entering the picture after the fourth year is faced with falling sales. If he now reverses the original decision and places emphasis on those factors controlling owner satisfaction, he is faced with the inverse of the preceding curve, and sales will fall still more rapidly during the fourth and fifth years before the long-range recovery begins.

The management decisions on design lead time are apt to be heavily influenced by market information, but will this information be properly interpreted? The feedback loop that can be traced from lead-time decision through the design interval, through the period of ownership and use of the automobile, through the factors affecting rebuying, through the interpretation of the meaning of this market performance, and finally back to a decision on design lead time can be four to eight years long. All these past decisions on lead time are "stored" in the system. More than one full cycle of fluctuation of management attitude toward the question of lead time may be contained in the system at one time. The situation is replete with opportunities to draw the wrong conclusions from the available information unless there is a full understanding of the many dynamic interactions that are present.

19.2 Growth

A most exciting use of industrial dynamics is in the study of corporate and product growth. By developing an explicit statement in dynamic model form of our concepts of the growth process, we can separate the crucial from the unessential assumptions. The dynamics of growth is most important in those areas of our economy where technology is advancing rapidly and product life cycles are short.

What combinations of characteristics of the products, corporate policies, and market conditions favor high profitability and rapid growth? What is the effect of equipment financing, tax policies, and the timing of market development expenditures? Why do many new companies fail, and on the other hand what are the characteristics that sustain 30% and 50% annual growth rates in other organizations? We commonly say this is the difference between poor and good management. But what factors are being introduced by the good manager, and can these be more widely understood?

In a frontier area of analysis such as the dynamics of growth, it seems, contrary to first impressions, that we should start with an abstract and general example rather than with a model of a particular product. Of course, this general-purpose model must be undertaken from a background of extensive knowledge of the processes to be represented; but it should not be exclusively dependent on a single case study.

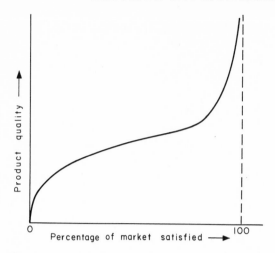

Figure 19-5 Quality versus percentage of market satisfied.

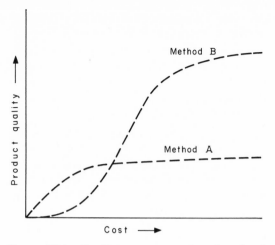

Figure 19-6 Quality versus cost of product development.

If we force ourselves to generalize, we shall necessarily consider fundamental relationships that are important. In the absence of a specific product to use as a crutch, we cannot be misled into a premature concern about detailed physical characteristics of a particular item. We must consider those fundamental relationships having dynamic meaning and that are descriptive of any product. Later there will be time to interpret these into the terms of a unique product.

To illustrate what is meant here by fundamental and general-purpose relationships, consider Figure 19-5. This shows the product quality[5] necessary to satisfy various percentages of the potential customers. Until a certain level of quality is achieved, the product will not meet the needs of any of the potential market. Thereafter, increases in quality rapidly expand the possible market. Finally, large increases in quality are necessary to satisfy the remaining exacting demands.[6]

Some companion curves as in Figure 19-6 describe achievable product quality as a function of man-hours of research-and-development effort. A choice between different technical methods of fulfilling the product function are usually available. These different approaches are apt to present different quality-versus-cost relationships as in the figure. Method A represents a well-known approach that can be rapidly exploited to produce the fastest initial gains with small cost. As shown, it has a lower

[5] "Quality" is here taken to mean a combination of all characteristics such as life, precision, compactness, and performance that are not elsewhere being represented in the model.

[6] Curves of interacting price and value would simultaneously be at work at the various quality levels.

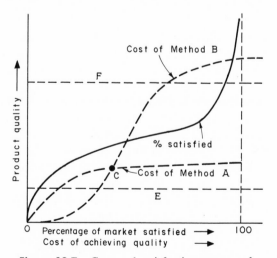

Figure 19-7 Cost and satisfaction compared.

top-quality potential than that illustrated by Method B. Method B is less far along when the choice is to be made and will incur considerable research and development cost before one can hope for an acceptable product. However, Method B has a much higher ultimate quality potential.

The effect of these two curves becomes more apparent when they are superimposed on the market characteristic from the previous figure. Figure 19-7 shows the combination. The horizontal scales are different for market percentage and cost, so that horizontal positioning between the cost and satisfaction curves is not significant. However, a horizontal line as at E or F allows us to compare costs and market satisfactions. Line E intersects both the cost curves of Method A and Method B. For this level of satisfaction, it shows that Method B is more expensive. However, Line E represents a very small penetration of the potential market. At the crossover point C, the costs are the same for the same product quality. At costs higher than C, Method B returns a better product for the expenditure. At the quality represented by Line F, only Method B is technically feasible.

By transferring points graphically we can obtain Figure 19-8, which shows percentage of market satisfaction versus cost to the same horizontal cost scale as before and with the market-satisfaction scale now vertical.[7] Both Method A and Method B are shown. In Figure 19-7 for Method A, the initial curvature is concave downward as is the initial part of the satisfaction curve. This produces a curve for Method A in Figure 19-8 which is very straight in the lower regions indicating market penetration in proportion to development expenditure. However, this curve for Method A in Figure 19-8

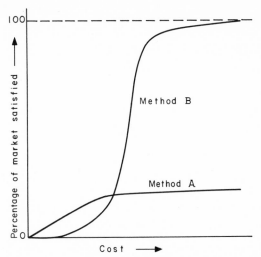

Figure 19-8 Satisfaction versus cost for two different technological approaches to a product.

saturates even more abruptly than the cost curve in Figure 19-7. The satisfaction-versus-development cost for Method B is very different. In the region above point C in Figure 19-7 the quality is rising rapidly with cost while the market penetration is increasing rapidly with quality. This leads to a region of very abrupt rise in market penetration versus development cost, as seen in Figure 19-8 for Method B.

The short-term advantage of Method A is apt to lead some managers to select a technological approach that lacks adequate growth potential. The "easy" method is selected; financial and emotional commitments are made to it. The effort is vulnerable to competition that succeeds in anticipating the advantages of Method B, even if it holds less immediate promise.

The type of analysis in Figure 19-8 is static; it shows the eventual steady-state outcome if development cost, quality, and satisfaction were the only variables. It does not show the time path that would be traversed nor does it show the contribution which would be made by the many other variables that are changing at the same time.

[7] Done by entering Figure 19-7 vertically from a selected point on the horizontal cost scale, rising to the chosen method curve, moving horizontally to the quality-versus-satisfaction curve, reading the percentage of satisfaction from the second horizontal scale, and using this value for the vertical scale of Figure 19-8.

Curves such as in Figure 19-5 and 19-6 are typical of the nonlinear functions that would be used to describe the general character of a market and of a product.[8] Many other characteristics would also be involved to define the various dimensions pertinent to the product, the market, the manufacturer, and the competition.

An application of the methods in this book was made to product and company growth by Kinsley.[9] Kinsley developed a dynamic model of 160 variables illustrating the interrelationships that his experience had shown him to be of importance in the development and marketing of a new product. The model might be looked upon either as the model of one product in a larger company or as the growth factors in the early life of a small company while it is still dependent on an initial product concept.

This mathematical model describing the life cycle of a product includes a preliminary treatment of six sectors internal to the company (manufacturing, personnel, capital equipment, money, marketing, and research and development) and two external sectors (the market, and pricing and competition).

Speaking of the typical but hypothetical company represented by the model, Kinsley says, "It can be broadly defined as a research-based company with no other identifiable characteristics. This definition is purposely vague. . . . The characteristics assigned to the ABC company in the course of building the mathematical model can be typically found in many companies." In discussing the nature of the new product process one is apt to be diverted by the peculiarities of a particular product and trapped into a treatment of its superficial characteristics. In the previous section on the discussion of the automobile market, we saw how the physical characteristics of the product could be converted into a more fundamental viewpoint dealing with product concepts as they affect various phases in the life of the product. Here, in discussing the *typical* life cycle of a new product, Kinsley avoids the dangers of prematurely concentrating on a special case. The model is built around a nonspecific description of the *general* class of situations that he is exploring. To succeed in this approach requires intimate familiarity with the actual problems, processes, and factors that are involved. These are in general accessible only to the person with keen insight into the type of endeavor being represented. However, in drawing on specific personal experience it is necessary to distill out the underlying fundamentals in order that the resulting dynamic model is suitable to *any* member of the entire class of products being represented. In this example,

---

[8] We often see comparative situations like Method A and Method B in new, rapidly evolving technological areas. Method A illustrates the approach which is initially easier and which people attempt to exploit beyond its inherent capabilities. Programs become committed to such improvement and can involve large expenditures. An entirely different approach aimed at much higher goals can often be the least costly in the long run because it requires embarking upon an approach of much greater inherent potential. Each industry can find its examples, especially during the early period of rapid technological change. An example from the last decade is the high-speed internal memory for digital computing machines. Method A in the period around 1950 was represented by the use of beam-deflection electrostatic storage tubes. These represented only a small step beyond the cathode-ray-tube art, and yielded very promising results for small amounts of effort. However, the method was inherently limited by the short life and unreliability of vacuum tubes and secondary-emission surfaces. Method B was at that time represented by the random-access coincident-current magnetic-core storage. This held the promise of indefinite life and superior performance but required much preliminary effort in the development of materials before it could be competitive. At the time of crossover of the two technologies (about 1953), as at point C in Figure 19-7, the rate of rise of performance of the magnetic-core storage was so rapid that within a period of two or three years it had superseded all competitive methods of high-speed internal storage in digital computers.

[9] See Reference 13 for the S.M. thesis written under the author's direction by Edward R. Kinsley, as a Sloan Fellow at M.I.T. from the Texas Instruments Company, 1958-59.

the mathematical model deals with the *nature* of the processes that bring a new product to the market. The constants and parameters that are included are the ones which the formulator believes are the fundamental descriptors that define the relationships of the various sectors of the system to one another.

As a simple example of these fundamental descriptors of the situation, consider the following. A product has as one of its characteristics the length of time necessary for its development from a research idea to a design ready for production. (In fact, this parameter has at least two dimensions, since the development may require a certain number of man-hours and, in addition to this, at least a minimum length of calendar time.) Considering simply the length of time necessary for development, other conditions being momentarily considered constant, we can examine this length of time in relation to other comparable times in the system. The product will be characterized by the length of time necessary to establish production facilities. It will also be characterized by the length of time that the typical customer in the market place requires to develop familiarity with the product and to expand his usage. The relationships between the required lengths of time in development, for production, and for market learning imply different expectations for the life cycle of the product. Two contrasting products will be used for illustration.

For the first product, the development time is very long, the market learning time is very long, and the production preparation time is very short. This defines a product in which a high-profitability period of technological monopoly is improbable. Designs and samples must be available to the market for a long period of time in order that the customers learn how to use the product. Information is therefore available to any potential competitor who can set up for production faster than the learning characteristics of the market will permit market

growth. Such a product is one in which the early developers may not be the major producers after the market has developed.[10]

For the second product, the market learning time is very short, and the production facilities are expensive, technologically complicated, and require a long period of development. In such an example the market can develop almost simultaneously with the production output and well ahead of competitive production. Such a combination of characteristics helps to ensure a period of good profit margins for recovering the development and capital equipment costs.[11]

19.3 Commodities

Chapter 15 dealt with an industrial distribution system from the retail customer back to the factory. If we look at the economic system lying behind the factory, we find various commodity supply systems handling agricultural products and natural resources. In many of these commodities, price is conspicuously unstable.

The question arises, to what extent is the commodity price fluctuation a reflection of the managerial practices within the industry, and to what extent is it imposed by disturbances from outside of the industry? A majority of those in industries such as textiles from fibers to clothing, metals from mine to manufactured products, and beef from the ranch to the table all seem to feel that the fluctuating conditions of the industry are imposed by changes at the final retail sales level and propagate back toward the supply point. Contrary to this feeling there is good reason to expect that many of these unstable manifestations might continue

[10] Transistors would fit this description.
[11] A new photographic color film is an example. Also, we could class here the extremely high-quality, high-performance version of almost any standard product in which the unique characteristic of the product is achieved by unusually high levels of engineering and production and sales skills that either cannot or will not be duplicated by competitors.

even in the face of a constant and unchanging final demand for the commodity. The organizational and policy interactions within the industry may be unstable enough to generate the system fluctuation from nothing more than being perturbed by the continuous flow of noise and uncertainty that will exist at all points in the system.

As we go back into a commodity supply system, the structural character begins to change from that represented within the stages of the distribution system in Chapter 15. In the distribution system for manufactured products, goods are shipped in response to orders. A product is delivered to a customer only if he wants it. Stresses within the system manifest themselves more by a change in the flow rate of goods than by changes in price. To a first approximation, prices tend to be administered in such a way as to reflect costs plus the typical profit margin of the particular industry. To be sure, these profit margins rise and fall with the state of supply and inventories. However, we commonly observe that a factory will adjust production rate to market demand by production-rate changes that are larger and faster than are the price changes.

By contrast, the commodity system tends to be one in which supply rates can be adjusted but slowly. The commodity is not produced to the specific order of the customer. The purchaser of a commodity from the primary source of supply tends to take the full quantity offered. The producer usually is not in a position to stockpile the commodity for long periods of time and delivers his output to the market as it is produced. Price fluctuates more rapidly than supply rate. As the price changes, a relatively slower adjustment in supply rate is made. As prices fall, some producers find themselves unable to operate profitably and eventually curtail production. The producer is usually "frozen in" so that he has little short-term alternative except to continue producing and to hope that prices will recover. In fact, to maintain his former income, he may try to produce more with falling prices. Losses over an extended period of time may be necessary to force business failures and thereby the reduction of excess production.

In a commodity system we therefore expect to find some stage in the chain at which inventories tend to fluctuate according to the difference between an order-filling output rate and a market-clearing input rate. At this point, inventory level and the rate of change of inventory level are two of the principal inputs to the price-changing process. This price is then propagated downward through the order-filling part of the distribution system to a final output point where it determines, eventually, the demand for the commodity. At the same time, the price is propagated upward to influence eventually the primary rate of production. Both demand and production may respond very slowly to price changes. And, for example, a reduced price will proceed to curtail production and to increase demand without any necessarily inherent tendencies to settle at an equilibrium balance.

At any or all levels in this system, speculation and hoarding, which vary with the price and supply situation, may accentuate the system instability.

An interesting study was started by Ballmer dealing with the international copper industry.[12] Ballmer developed a dynamic model of some 160 variables representing his concept of the major policies and organizational relationships in the six sectors of the copper industry from mine to final manufacturer. His six sectors are shown in Figure 19-9, treating mines, concentrators, smelters, refiners, fabricators (finished metal forms such as wire, sheet, brass, and so forth), and the manufacturers of products which use copper. This chain overlaps in the final stage with the factory level of Chapter

---

[12] See Reference 13 for the S.M. thesis written under the author's direction by Ray W. Ballmer, as a Sloan Fellow at M.I.T. from the Kennecott Copper Corporation, 1959-60.

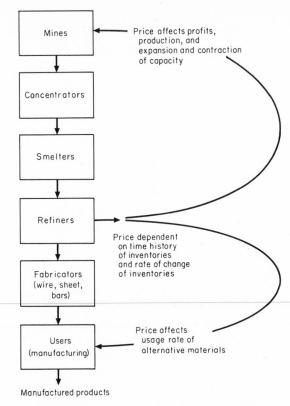

Figure 19-9 Stages in copper production.

15. Ballmer set up two independent mining sectors, with the intention that the study would eventually be extended to explore the question of how large a fraction of the total copper supply must be controlled in order to stabilize the large fluctuations in copper price — if indeed a change in mining policies turns out to be one of the points where stabilizing control can be exercised.

There seems to be reason to expect, both from certain actual experiences and from a consideration of the nature of the dynamics of the systems involved, that with the proper policies even a minor participant in an unstable commodity system may have enough leverage to exert a strongly stabilizing influence. Stabilization here means reducing fluctuation but not the artificial control of long-term average price.

Several interesting questions arise. Can stability and steady growth be achieved rather than wide fluctuation? Where in the system is stabilization easiest and least risky? How large a segment of the industry must be controlled to achieve this stabilization? Can a single company, perhaps not even the major company of the industry, do it, or is stabilization possible only through aggregate action that would first require a modification of antitrust laws? There is strong reason to believe that ways can be found whereby unilateral action by a single organization may be sufficient to provide a strong measure of industry stability.

These commodity systems have pronounced, internal, regenerative, unstabilizing forces, many of which are subject to managerial control. The effect of some unstabilizing policies is even subject to reversal. Changed policies could provide stabilizing rather than unstabilizing action and might thereby improve the dynamic character of the entire industry.

For example, in the preceding work by Ballmer there are two very interesting characteristics of the system to be explored. One is the tendency at all levels toward speculative inventory accumulation. As prices start to rise and as the supply situation seems to be tightening, there is a tendency to lay in extra inventory. This is, of course, well known as a strongly unstabilizing policy that tends further to drive up prices and shorten the available supply. Again, as we have seen in Chapters 16, 17, and 18, inventories tend to rise at a time that accentuates the stress on the system. This effect is self-regenerative and means that the final unfavorable effect is considerably stronger than might seem to follow from the small amount of inventory accumulation that is actually accomplished. An interesting question, not yet explored, is the extent to which a part of an industry at one or several levels might counter this entire effect by exercising aggregate inventory policies that reduce inventories as the supply situation begins to tighten and accumulate inventories in the slack periods. This would not necessarily imply any larger swings in inventory than are current at the present time, and would, if it is possible to do so, actually make inventories available at the time they are most needed.

Another control point for management policy to affect system stability may exist at the mining level. There is some reason to believe that the pressures on the mine manager may lead to actions which further help to accentuate system fluctuation, although this point needs further exploration. The question arises concerning how mine production rate varies with price, both in the short run and the long run. In the long run, and this may represent two years or more, a rising price will no doubt lead to greater production as additional marginal mines are opened for operation. In the short run there are factors tending to suppress an increase in production with an increase in price and may actually create a reduced output as price rises. This could result from the following combination of circumstances:

- The mining, concentrating, and smelting levels tend toward a fixed tonnage capacity of raw ore.

- The available ore at a mine varies in yield per ton, and therefore in the copper output per hour of machine operation, depending on what part of the property is mined.

- The mine manager attempts not to make the maximum possible profit at every moment but rather is under pressure to meet the profit levels expected within the industry and, if he can accomplish this, is further expected to maintain stability of profitable operations and is also expected to use eventually both the high-grade and the low-grade ores available to him.

Under the foregoing circumstances, as the price rises, it becomes possible to maintain the accepted level of profitability while mining lower grade ores and thereby using up those mineral resources which might not be profitably mined under lower-price operations. If the equipment is being operated at maximum tonnage capacity, this means that the same amount of ore, yielding less final copper, would be mined. The result of a rising price can then be a continuation of the same profit levels and a reduction in the mine output. Whether or not this factor is predominant, any tendency in this direction may, if it exists to a significant extent, contribute to reducing the industry stability.

Ballmer found the system stability to be considerably influenced by both of the preceding factors, but the study has not yet been carried far enough to be conclusive, nor has any consideration been given to the risks that might be encountered by anyone attempting such policy changes. The risks would no doubt be strongly influenced by the size of the industry sector whose policies could be changed. Such explorations show great promise of successful managerial results, but a large number of questions would need to be explored before definitive recommendations would emerge. Such explorations would not be unduly lengthy or costly in comparison with the possible importance of the outcome.[13]

Other promising areas of exploration exist in agricultural products. Perhaps the most promising for initial exploration are those in which there are individual suppliers or users controlling at least 10% of the market and in which governmental politically inspired controls are not an overriding market factor.

19.4 Research and Development Management

The management of research and development is a new facet of corporate life that has grown to its present importance during the last two decades. Because of this newness, research

[13] The study begun by Ballmer was continued in 1960-61 by Kenneth J. Schlager, as a Sloan Fellow from the AC Spark Plug Division of General Motors Corporation. He also developed a dynamic model of the aluminum industry to make a comparison with copper and further examined the factors influencing price and supply. See Reference 13.

management is fraught with more uncertainty than most other parts of the management picture. It might seem that any attempt at formal dynamic analysis of the research management process would be foolhardy. Yet because the need is so great, this may indeed be one of the points of major payoff even though it is a frontier for analysis and one filled with intangible factors. The first result of such an attempt is a clarification of one's thinking about the importance of various factors and how they are related to one another. From such a study, the combinations of circumstances leading to success or failure become clearer.

An initial approach to a dynamic model of military research-and-development management was begun by Katz.[14] This work is being carried forward by Roberts.[15]

Research and development is commonly connected in our thinking with invention and innovation, an area of human endeavor surrounded by much controversy. On the one hand we argue the unpredictability of invention, while on the other recognize certain orderly relationships in the expectations of probable success. The size of a group that we assign to a research project is always a function of our need for the result, our estimate of the magnitude of the undertaking, and the uncertainty that we feel in the outcome. In other words, we usually operate on an assumption that a larger effort will either shorten the time or increase the probability of success. Throughout the picture there are similar recognitions of acknowledged orderly relationships between the variables.

In approaching a representation of the research-and-development process, we are well

advised to think first of a typical project rather than a particular one. In Section 19.2 we dealt with the general factors in the growth of a product. We concentrated on the characteristics that had significance in the time sequence of events under study. Likewise, here we should concentrate on those fundamental characteristics of the process that are independent of the particular research project but are representative of all. What are those common and fundamental descriptors which will define any project and which will separate one from the other when suitable numerical values have been assigned to the parameters of the system? As with the dynamic models already discussed, the research-and-development activity will consist of several interacting sectors. Depending on the objectives of the study, these might be

- The end product including size, complexity, novelty, and stringency of performance requirements, all relative to the present standard practice of the art
- The customer for whom the research is done. This may be a military contracting office with which a company has a contract, or in other circumstances it might be the management of the company for whom the research department is engaging in projects.
- The characteristics of the research organization and its management
- The characteristics of the need that the results of the research are to fill

This latter item — the nature of the need being filled — brings into the open a characteristic of the information-feedback model which may have been obscured by the simplicity of the models in Chapters 15 and 17. What is the measure of system success? What is the system attempting to accomplish? What are the criteria by which the system adjusts itself toward an attempted better meeting of the objectives? In both of the models in Chapters 15 and 17 the system was attempting to provide the flow of goods called for by an externally supplied de-

[14] See Reference 13 for the S.M. thesis written under the author's direction by Abraham Katz, as a Sloan Fellow at M.I.T. from the Radio Corporation of America, 1957-58.

[15] As a doctoral thesis and staff research in the Industrial Dynamics Project at M.I.T. by Edward B. Roberts on the staff of the School of Industrial Management.

mand input. The decisions within the system attempted to adjust to this demand. What is the corresponding demand in the context of a research-and-development program? By what is the success of the result to be measured? Put another way, what is the value of the result of the research as a function of time? Figure 19-10 shows the typical shape of such a curve. For any specified product, one can conceive of its being available too early. Prior to point A, the need for the product or the surrounding circumstances necessary for its use have not yet arisen. Should we actually have it before point A, storage, maintenance, and invested money would represent a negative value. After this point, the value rises to some peak, as at B, after which the need declines or the product has become obsolete and superseded by later developments. This curve should be thought of as the *true* value of the product to society. It is not what the customer is willing to pay. It is not what either the customer or the research group *estimates* to be the value. All of these are other and different variables in the system. The project is undertaken against a background of *true* value to society versus time, whether or not this true value is ever to be known.

The characteristics of the research organization and of the customer for whom the research is being done will lead these organizations to form their own *estimates* of the true value of the project. These estimates of the true value will change with time and will depend on the characteristics of the organizations that derive the estimates. In practice, we find the estimates of the true-value curve in error by as much as a factor of 10 or 100 under various circumstances. In the early stages, when time is at a point earlier than A in Figure 19-10, organizations look ahead and underestimate the peak value at point B. On the other hand, after the rapidly rising value has been discerned between points A and C, there may often be a tendency to extrapolate this continued rise of importance to values in the future higher than B.[16] Accuracy of estimates will depend on the degree of foresight exhibited by an organization, on the extent to which it is immersed in current crises or is guided by a sound evaluation of the future implications of forces and factors at work in the present, on the amount of experience which its management has in the field of interest, and on the extent to which it has a proper balance of imagination versus hardheaded evaluation. A concept like Figure 19-10 becomes an independent input to the dynamic model of the research-and-development process.[17] Identifying such fundamental concepts clarifies our thinking. Even though we do not expect to have available accurate numerical values of such a curve as Figure 19-10, it does focus our attention on a basic parameter on which we are depending. A spotlight turned on the fundamental assumptions will usually lead to better estimates than if these assumptions are glossed over into an amorphous, intuitive haze.

In other parts of the picture, there are important intrinsic parameters of the task itself, of the participating organizations, and of the surrounding technological environment and its

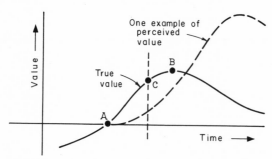

Figure 19-10 True value of results of research project versus time of project completion.

[16] Consider the change in perceived value, in the United States, of space vehicles before and after the first Russian sputnik, yet the basic technology had been advancing steadily and in an orderly manner for twenty-five years.

[17] Discounted values based on the earning rate of money must be introduced. Such has been undertaken by Katz in Reference 15.

rate of change. These combine to generate different project-cost curves for different managerial practices.

For example, those policies and organizational characteristics that lead to different project starting times set the stage for results as in Figure 19-11. Here final completion cost is plotted against project duration for various calendar times for the starting of a project. The horizontal scale is the length of the time interval taken after the beginning of the task. Two curves are shown. The one for an early start shows a longer minimum completion interval. This occurs because the earlier a project begins, the more embryonic will be the development of the surrounding technology. The shortest possible completion interval is increased by the greater effort and uncertainty introduced by the earlier start. The curves reach a minimum value and then rise slowly because, if a project is stretched out over too long a period, the total costs will rise as enthusiasm lags, personnel shift, and momentum is lost. The curve for a later project starting date can have a shorter minimum completion interval and a lower minimum cost level because it can draw more heavily on the advancements that have been made in the meantime in independent ac-

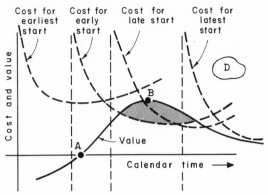

Figure 19-12   Curve of project value versus time, with four superimposed curves of cost versus completion date, each for a different starting time.

tivities. Because the curves are plotted with respect to duration after the separate starting dates, they are not properly related to one another in calendar time. It may be more informative to view them superimposed on the true-value curve.

Figure 19-12 is a repetition of the true-value curve of Figure 19-10 in which are shown four different cost curves to represent four different starting dates. Each curve represents a sequence of possible completion dates for the same beginning date. In the curve for the earliest starting date, we see that the project was premature, regardless of how long a period is taken for project completion.[18] The point of minimum completion cost is high and occurs earlier than the time at which the results have substantial value. Results obtained by such an early project start may also be deficient in performance and reliability. (These are other system variables.) Few large, established, industrial organizations run any risk of project failure from too early a starting date. This project history is more apt to be encountered by the visionary or crackpot

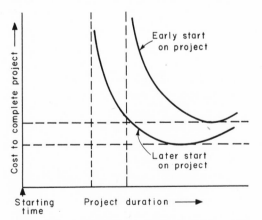

Figure 19-11   Project cost versus time for completion as dependent on various dates of starting project.

[18] This of course assumes that certain other system policies are constant for the different scheduled completion intervals, such as the manpower distribution over the life of a project and the cost of invested money.

who undertakes to reach a sound objective but long before the society has come to the point where it can use and appreciate the results.

The second cost curve for an early start has a substantial area of overlap with the value curve, as shown by the shaded region. Furthermore, we see that the maximum margin of the value curve above the cost curve is not very sensitive to the horizontal position of the cost curve in time nor to the length of project completion interval. In other words, a misjudgment of proper start and completion dates is less serious than it would be if operations are shifted into the region of the third curve, labeled as a late start. The third curve, as shown, has a substantial maximum point for value in excess of cost, but this value diminishes very rapidly if conditions shift the curve to the right. Here the late start increases the risk. The fourth cost curve for the latest start has the lowest minimum cost but again is a curve in which cost always is in excess of project completion value.

The preceding cost curves are not input functions to the model like the true-value curve but are in the nature of the kind of results to be derived from a sufficient number of model runs. Their discussion brings forth the kinds of factors that underlie the research and development processes.

An important factor in research and development management is apt to be the nature of the financial approvals and budget cycle involved. This is especially pertinent in large military projects in which several years may be involved in preparing military estimates, getting these approved, defending and securing the appropriations from Congress, and writing contracts for the work.

In the budget cycle, the managerial characteristics of the contractor and contracting office enter heavily into the probability of success. Competence, foresight, courage, and integrity exert a major influence on the realism of estimates. A low level of courage and integrity

can lead to initial budgeting plans influenced more by the available funds than by a realistic estimate of the task to be undertaken. By the time that the full magnitude of the job is apparent and its cost implications have become unavoidable, it may already be too late to reorganize the effort for completion at a time and cost that fall under the true-value curve. However, this does not necessarily mean that the project will be terminated. The momentum of the organizations involved, the unwillingness to recognize an error in judgment, wishful thinking, and the obscurity of the true-value concept can easily lead to actions based on an assumption of value in the region marked D in Figure 19-12. As the cost rises and as the estimated cost exceeds by greater and greater margins the true value to society, the probability increases that a collapse will occur in the emotionally created value structure. The internal system will be under stress and on an unfirm foundation, making it an easy target for penetrating inquiry and criticism from outsiders who have become unsympathetic.

Here we see the crucial importance of so-called "intangibles" in a model of system dynamics. Anyone intimately familiar with research and development recognizes the crucial importance of factors such as courage, imagination, foresight, integrity, hope, optimism, and emotional defense of past decisions. The relative ranking of several organizations with respect to any of these characteristics will usually be noncontroversial if the selected organizations cover the wide range encountered among actual organizations.

Some organizations are conspicuously and continuously successful in the projects they undertake. To a considerable extent this arises through the initial selection of projects that can succeed, coupled with the foresight to generate an early enough start even if it may mean the risk of spending their own internal funds before the value of the undertaking is recognized by financial sponsors. Such organizations have the courage to develop realistic estimates of job cost and are willing to lose contracts to more optimistic bidders when they are dealing with a source of funds that is unable to appreciate a realistic even though high-priced plan. Such organizations establish a reputation of being successful, and this increasingly gives them their choice of new undertakings that *can* be successful. A regenerative cycle of success breeding success is then set in motion. Conversely, the organization whose bid is influenced largely by the funds available rather than those needed and by the time schedule desired rather than that which is realistic finds itself saddled with projects which have already passed the crisis point. Work is completed too late in the cost-versus-value relationship. These organizations develop a reputation that tends to force them more and more into those jobs which are refused by the successful organizations. Again a negative regenerative cycle tends to reward failure by offering tasks for which the probability of success is lower.

19.5 Top-Management Structure

We quite properly attribute the success or failure of an industrial organization to the quality of its managerial leadership. However, in beginning the development of a dynamic model of industrial activity, we usually start with the more concrete factors of the business. Production, distribution, and inventories come first, followed closely by employment policies, and then by a consideration of the dynamics of the market. After this may come the inclusion of those factors affecting new-product development.

As the basic functions of the business become integrated into a dynamic model, we face more and more the necessity of completing the picture by adding the decision-making characteristics of the higher management levels.

Top-management structures have different forms, different attitudes, and different histories. They differ in courage, conservatism, flexibility,

rapidity of reaching decisions, and in the objectives being sought. These differences are reflected in different dynamic characteristics of the management processes in the companies. Just as the operating functions interact with one another to produce important dynamic behavior characteristics, so will the interaction between the top-management structure and the operating departments favor different growth and stability patterns.

The desired dynamic characteristics of the management structure depend on the kind of markets, rate of technological change, and the other characteristics of the industry. Different organizational forms are seen to favor different classes of products. The management attitudes that work well in one situation falter in another because the life cycle of the product is longer or shorter, the ratios of times needed to develop a product in comparison with the time for putting it into production are different, or the market is more sensitive to certain of its characteristics and less sensitive to others.

The existence of different dynamic characteristics of the management structure can perhaps be illustrated by comparing a company with functional subdivision as in Figure 19-13 with a company using a project organization as in Figure 19-14. In the functional organization, the coordination between the various stages of each product is carried out by top management, often leading to large, central staff groups that participate in the actual decision-making processes that mark the successive stages in the evolution of an idea from research through to marketing. Here the stress is on efficiency within each of the separate functional specialties. It is an organizational form having advantages in a very slowly changing product situation. For example, if a product and its manufacturing processes are relatively stable, then improvements in efficiency can be measured by the year-to-year trends in labor costs, and material usage. Quotas, sales trends, and financial criteria can be generated to measure the effectiveness of the marketing operation.[19]

On the other hand, the functional organization runs into difficulty as the product life cycle shortens. When the product life cycle becomes short enough so that the top-management decision-making delays and the rate of build-up of knowledge and enthusiasm for an idea within a functional department require appreciable fractions of the product life cycle, then we find that the functional decision-making system of Figure 19-13 begins to break down in competition with a more mobile operation, as shown in Figure 19-14.

In the pure form of the project organization, top management exists in a very different dynamic relationship to the active working part of the organization than it does in the extreme functional form. Rather than having continuous top-management participation in the line decisions with respect to each particular product, the entire sweep of operating decisions from research to market rests in the hands of a director

[19] Perhaps such products are characterized by Shredded Wheat, Ivory Soap, and black-and-white film for amateur photography.

and his immediate project staff who have no other conflicting responsibilities. The project director is not a coordinator of work done by independent functional departments but has all the required resources responsible directly to him, and these resources have no other obligations.

In the project organization, top management takes a view that is longer than the individual project. Control is exercised through the initial approval of project objectives and scope, by approval of the budget, and by approval of the person to take charge of the project. Top management exercises continuing control by reviewing changes in the planned budget and scope, by observing progress, and by the passive action of permitting the project director to continue unless matters take such a serious turn that the director must be replaced or the project terminated. Progress and results are continuously observed, but primarily as a source of information for the approval of other new projects and for the identification of those men showing superior managerial skill.

In the project organization of Figure 19-14 a product does not suffer the shock of being separated from one functional group and having its rate of progress slowed until the knowledge and enthusiasms of the next functional group in sequence can be built up. The project organization makes it easier to act on the necessary early preparatory steps that anticipate the problems of the later product stages such as manufacturing and marketing. To be successful, the project organization must of course have project leadership able to exercise the required skills in each of the functional activities. The perpetuation and expansion of the leadership is one of the longer-term processes involving the management structure.

The dynamics of the long-term evolution of management structure are interesting in that most small new companies begin with the project form of Figure 19-14. As they grow, they break into the functional subdivision of Figure

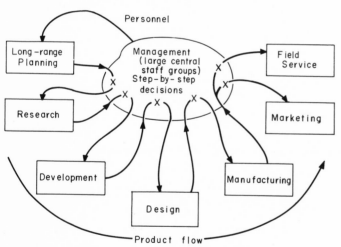

Figure 19-13 Functional subdivision in which successive decisions in progress of a product are made at the company management level, usually by coordinating committee.

19-13, driven by a desire to achieve an apparent gain in effectiveness. This gain may be short-term, lasting but a few years. The functional organization provides a poor training ground for the type of man necessary for project or top-management leadership, so that any transition back from Figure 19-13 toward Figure 19-14 becomes less and less possible as the organization ceases to regenerate the kind of wide-ranging leaders necessary for perceiving the interactions of all facets of an enterprise.[20]

At this point the construction of a dynamic model begins to enter areas of managerial decisiveness, decision-making delays, sources of information used for decisions, factors affecting morale, and the measures of manager and organizational success. The industrial models already discussed contain some of these management implications. The delays and policies in advertising in Chapter 16 arise from conditions in the company management. In Chapters 17 and 18 the attitudes of management are reflected in the amount of emphasis on order-filling delays and in the excitability implied by the changes of employment and desired inventory levels. Such management factors have, however, entered indirectly. They can be specifically incorporated where they are crucial to the problems being studied.

An interesting set of dynamic interactions arose in a study of the effect of allocation of managerial time. The phenomenon is best ex-

[20] It seems to me that executive rotation as practiced in functional organizations from one department to another fails to provide the broad training that the project executive in Figure 19-14 receives. Rotation between departments does not provide experience in taking full personal responsibility for one's decisions. A man enters a department that is already operating and receives men and partially completed products over whose history he has had no control. The results of his decisions will be passed on to someone else and are not made with the expectation that he himself must live with the deficiencies in those decisions. Such an environment tends to focus attention on short-term expediency and encourages long-term irresponsibility.

plained in connection with the early stages of development of a new product, although it can be seen in companies that manage by crisis and unduly shift management concentration and emphasis to those points that currently show the strongest trouble symptoms.

The basis for this particular dynamic behavior came to light after exploring possible causes for a sales and profit fluctuation as shown in Figure 19-15. A new product was developed, which at first sold slowly and then developed a highly encouraging upswing in sales, accompanied by a very favorable rise in profit rate. This was followed by a deep dip in profits and then recovery. The curve for general economic conditions is merely to show that the rise occurred during a declining condition and that recovery of the economy set in ahead of the fall-off in sales and profit. The nature of the product, the market, and the customers would indicate that any effect from economic conditions would be inverse to that actually experienced. The reason for the profit fluctuation

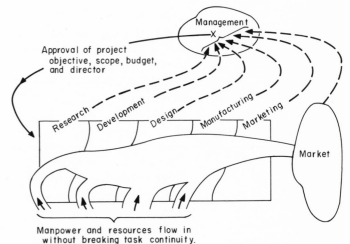

Figure 19-14 Project organization in which the resources are acquired by the project, rather than remaining in separated functional entities, and coordinating decisions are made within an organization having but one single common objectve.

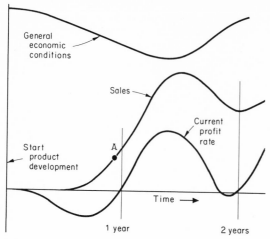

Figure 19-15 Instability of profit rate resulting from pressures for allocation of management attention.

probably lay elsewhere than in the general economy. The rather rapid decline in profits clearly came from the smaller downturn in sales, because the fixed costs of the organization had risen to a point where a sales downturn would cause the typical, accentuated fall in profits.

A consideration of the market could lead to the belief that the fall-off in sales was a seasonal factor reflecting less activity during the summer on the part of potential customers. However, the investigators were led to examine some of the dynamic aspects of the market and the way in which the organization had been allocating the time of its most skilled management personnel. As a result, they began to suspect that the character of the market and the pressures on the management were interacting to cause the observed behavior. A dynamic model of this system is being developed. This dynamic model combines some simplified characteristics of product development, of market growth, and of allocation of management and personnel resources. It is a highly nonlinear system.

Figure 19-16 shows the principal relationships under study. In this case, and as often happens with new companies, the initial conditions consisted of a reservoir of ideas and two men who planned to start a new company. Under these beginning circumstances the top-priority job is to design the ideas into manufacturing drawings and specifications for a salable product. As designs are completed, the emphasis shifts from the development area to the manufacturing facilities and manufacturing processes. This pressure to concentrate on manufacturing is enhanced if the designs are attractive to the market and if a small amount of sales effort produces a backlog of unfilled orders. As the manufacturing processes are established, the backlog of orders is filled, and an inventory begins to develop. The emphasis shifts to the sales area, to ensure an inflow of orders to use the available production.

In this situation, sales efforts seemed at first to produce little effect. Then orders began to flow in, and the sales curve started climbing. Operations reached the vicinity of point A of Figure 19-15. Orders continued to rise. Manufacturing which had been launched quickly had received less management attention during the period of sales crisis. For lack of skilled management concentration, manufacturing was by then suffering from quality rejects and lower than needed production rate. Demand in excess of manufacturing output shifted the high priority for management attention back to the manufacturing area. Orders continued to flow in from the in-process prospective orders stored up in the six-month customer-ordering delay (to be discussed later). As production was brought into line with the rising ordering rate, orders began to fall off because now the delayed effect of reduced sales effort was becoming apparent and the stored prospective orders in the customer's processing procedures were being depleted. Sales continued to fall after emphasis was shifted back to the sales area until the customer-order-processing delay was again filled. At that point another upswing in orders developed.

In this situation the market seemed to have several significant characteristics:

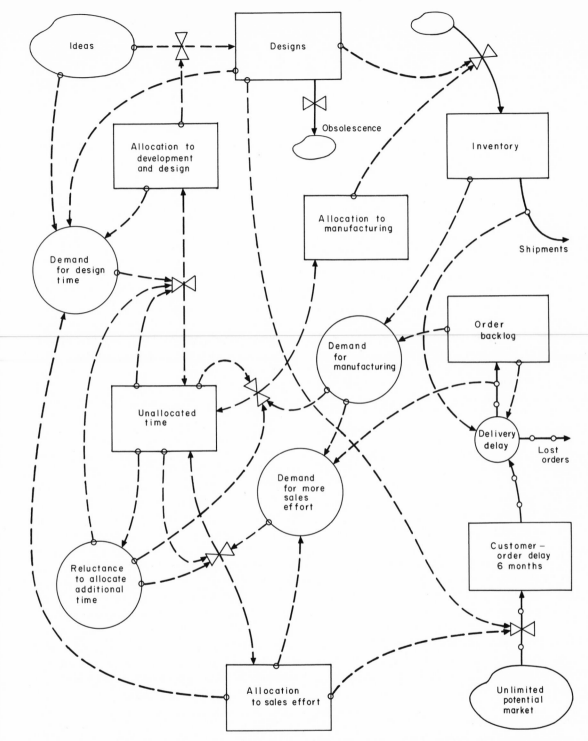

Figure 19-16 Principal interactions in allocation of management to development, manufacturing, and sales.

333

- For all practical purposes, the size of the market was unlimited in comparison with the size of the company and its potential production ability.

- The product was capital equipment in which an individual order might run several thousands or a few tens of thousands of dollars. Such equipment has been traditionally made to order. Skilled technical salesmanship plus effective designs, sample equipment, and demonstrations were sufficient to initiate discussion of a customer order.

- Because it was an item of capital equipment, several levels of management approval within the customer organization were necessary to place an order, and under some circumstances approval of the purchase was required from a military contracting officer if the product was to be used in connection with military projects. These approvals, the answering of inquiries, the demonstration to various superiors of the person actually desiring the equipment, and the preparation by the customer of specifications and an order require about six months on the average. Orders become ready for placement a substantial period of time after the initial demonstrations and selling.

- The field is so competitive that orders, when they are ready for placing, can be lost if deliveries are too slow. The state of the order backlog and the shipping rate are indicative of the delivery delays, and in the model these control the split of order flow between those placed and those lost.

The allocation of managerial and personnel time consist in this model, as shown in Figure 19-16, of a system of needs and priorities that balance themselves by causing flows of time allocation by way of the pool of unallocated time. A pressure, or allocation priority, is generated for each of the four levels, including the unallocated level. Differences in allocation pressure cause a flow of time allocation from one part of the system to the other. Figure 19-17 illustrates one of these functions as it applies to the pool of unallocated time. This shows the resistance

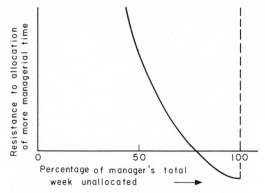

Figure 19-17 Resistance to time allocation versus percentage of unallocated time.

to the allocation of more of the unallocated time as a function of the percentage of total time unallocated. This is presented as a percentage of a full 168-hour week. The resistance to allocation of additional time is approximately neutral in the vicinity of a 35- or 40-hour work week. At a larger amount of unallocated time the coefficient is shown as negative, indicating an actual desire to be more active. As unallocated time falls, the resistance to the allocation of any more time rises steeply as the limit of human endurance is approached.

Quite differently shaped curves will be generated for the "pressure" for allocation to the working areas. These pressures will depend on need for work in those areas. For example, an average order level higher than production rate and a falling and low inventory would create pressures for allocation of more time in the production sector. Pressure to allocate to the sales area would rise as the rate of orders declines, as the backlog of unfilled orders declines, and as inventories become excessive. In addition, relative priorities between sectors would be assigned to represent various managerial policies. A short-sighted organization is apt to have a low-priority coefficient on the conversion of ideas to designs if there is stress in the manufacturing area. Likewise, trouble in the sales area is apt to take precedence over either of the other two. As would be expected, the behavior of the system is highly dependent on the priority coefficients, the allocation functions, and the time constants with which allocation changes are made. The model becomes a basis for studying the results of various management attitudes in different product and market situations.

19.6 Money and Accounting

In view of the central position held by accounting information and money flow in most discussions of business operation, the reader may feel that financial variables have been slighted in this book. At first thought, money may appear to constitute the network which ties together the parts of a business. In this book, the coordinating position has been taken by the information flow network. Furthermore, most information flow has been presented as nonfinancial in form. I believe this change in emphasis is proper, although money flow and the generation of accounting-type information cannot be ignored in the construction of more complete models. A beginning to the inclusion of money and accounting data was introduced in Chapter 17. In that chapter, however, the financial information does not form an integral part of the decision-making functions in the model. Rather it is a reporting system to indicate to the investigator how the system has behaved.

Is not reporting of performance the principal use commonly made of accounting information? It measures past success of the organization. It performs a policing action to ensure honesty. It acts as a basis for rewards — both to managers and to stockholders.

To be sure, accounting information is held in great esteem as data for current operating decisions. But is this justified? Does it matter whether the profit-and-loss statement and the balance sheet for the last month are available on the fifth or twenty-fifth of this month, since both reflect the results of decisions taken six months, a year, or five years ago? Does three more weeks of delay matter, added to a year or more?

To the extent that accounting information actually is used in making management decisions, it must be represented within a model. The peculiar characteristics of accounting information can impart important dynamic characteristics to a business. If this information, which is so long delayed after the primary decisions on which it is based, is an important influence in the managerial decisions, then those delays and the amplification caused by them have serious implications for system stability.

When a business has gotten itself into such a serious condition that its operations are jeopardized by low cash position and unavailable credit, then the financial inputs may be the principal controllers of the operating decisions. Such a state is, however, usually the result of a long earlier history in which the implications of other kinds of information have gone unheeded.

Barring these drastic marginal situations, accounting information does not seem to find its way directly into operating decisions in the same tangible way that sales become one of the ingredients in the orders for replacement goods. Instead, the profit-and-loss statement may have an important effect in setting the general tone of optimism and care surrounding the making of decisions. As current profitability increases, there is apt to be a relaxing of careful, disciplined consideration of new expenditures. Flimsier arguments will suffice for hiring additional personnel into service and staff groups where they become a fixed cost to the organization, which persists well into a later time of financial difficulty. On the other hand, unfavorable indicators from the accounting system may serve to delay seriously and perhaps prevent those expenditures that alone hold the key to financial improvement. In this manner we can anticipate that the psychological attitudes deriving from the financial information have the potential for influencing decisions in such a way as to reinforce other difficulties to which the organization may be subject.

The skeleton framework of primary effects within the organization can often be represented without financial and accounting information. On the other hand, as models become more subtle and begin to deal with the very important aspect of top-management decision making, the accounting system becomes an essential part of internal information loops affecting attitudes and decisions.

However, we must not lose sight of the fact that financial information is but one small part of the total information within the organization. Usually it measures symptoms, not causes. It is dangerous because it is easier to derive than other more important kinds of information. We have a tendency to gather easily available information, whether or not that information is the most efficient and effective.

19.7 Competition

In the conduct of a business, the competitive aspects of product and market seem to take precedence in management thinking. Yet many of the problems which a business faces are characteristic of the entire industry. Companies are often sufficiently alike and the factors interlocking their behavior are sufficiently strong so that the ups and downs in the fortunes of the industry are usually reflected in the welfare of each company within the industry. The coupling forces between companies within an industry serve to impress upon each a similar timing of environmental circumstances. If one company has a better delivery capability than another, sufficient business can shift to provide equal loading during times of production stress. Labor union demands tend to equalize costs. A rising demand will tend to stimulate concurrent expansion of production capability throughout the industry.

If the industry has conspicuous characteristics whose changes with time affect all the companies within the industry, then it seems wise to start a study of dynamic characteristics with the industry as a whole. This was done in Chapter 17. It presumes that the company of immediate interest is similar to others in the industry and that the problems of the industry are the problems of the company. After the dynamic characteristics of the industry are better understood, then company differentiation can be studied. If substantially different policies would be desirable for the industry, there then arises the question of what will happen should one company unilaterally adopt these policies. We can expect that there will be circumstances where unilateral adoption of different policies is feasible, and on the other hand situations wherein the risk will be too great. Once the nature of the industry is adequately understood, the study of different policies between companies becomes important.

The policies of a company can control not only its share of the total market but also that part of the market to which the company caters. We are accustomed to thinking of market differentiation on the basis of product performance, price, quality, and other physical characteristics. There is also another dimension of possible differentiation, and that is in the dynamic characteristics of the market. In some industries we find one company whose policies attract the fluctuating part of the market demand, whereas another company has policies that attract a stable underlying continuity of demand. The company that follows the policy of pursuing every possible sale and having product available to push into the hands of the customer even in peak periods of demand may find it is unknowingly selecting the peaks of demand as its share of the market. This will be especially true if the intrinsic value of the product in the eyes of the customer is less than that of competitors and if the company is taking advantage of sales that come to it because of the unavailability of preferred competitive products. On the other hand, a contrasting company policy could be to establish a preferred position in design, quality, and sales effectiveness so that all production is salable in the periods of lowered market demand. This company might forgo possible higher sales in periods of increased demand in the interest of greater continuity of operations and to prevent dilution of its quality and skill. In this situation the first company has a much higher percentage fluctuation in its operations than has the industry as a whole.

Differences in policy that tend to differentiate a company on the basis of its dynamic characteristics will be an important aspect of competitive models. In Section 19.1 the study of design lead time and market penetration are examples. In Section 19.3 the discussion of a model of the copper industry illustrates the process of starting first with the industry as a whole and then distinguishing policies between competitive companies.

The dynamics of early growth discussed in Section 19.5 shows, on the other hand, a study where the industry is amorphous and poorly defined and has no conspicuous characteristics other than growth and wide opportunity. The challenge there tended to be not the competitive interactions between companies, but rather the matching of company characteristics to the dynamic characteristics of a field of rapid technological advancement.

Other types of questions deal primarily with competitive behavior. The life cycle of a new product such as discussed in Section 19.2 is strongly influenced by the factors which give rise to competition and which control changing profit rates throughout the life cycle of the product.

19.8   The Future in Decision Making

Forecasting and long-range planning are terms arising from the manager's concern about the future effect of his decisions. Forecasting usually implies a search for decision-making help in the near future, and often rests heavily on a statistical analysis of past data in a search for trends and seasonal behavior. Long-range planning implies a time period beyond that of forecasting, and rests more on the manager's interpretation of the way present factors will interact to shape the future. Interpreted in this way, forecasting and long-range planning are related in quite different ways to dynamic model building.

Forecasting. Forecasting, as referred to in business and especially in sales practice, usually depends heavily on the analysis of past data in an attempt to detect patterns of growth, fluctuation, and changes in trends. The results are used as predictions of the future on which to base operating decisions.

It should hardly be necessary to point out that there is no such thing as information available from the future. Yet the aura of mystery surrounding sales forecasting in many companies seems to give it an esoteric character setting it apart from the straightforward concept of converting presently available information from one form into another.

The forecasting process fits into the concept of the decision equations discussed in earlier chapters. These equations take the available sources of present information and combine them to generate present decisions and actions. The possible complexity of the process does not change its fundamental character. If a forecasting procedure is used, we must be sure to represent it in the decision equations of a model. This is especially true if we believe that the forecasting affects the behavior of the industrial system — either for better or worse.

More folklore seems to surround the process of forecasting than in any other part of the corporation. The high esteem that the results enjoy often belie a comparison between forecast and actual happenings. Be the accuracy as it may, the implications of forecasting go still deeper. Nearly all of the statistical and mathematical treatment of forecasting implicitly assumes that the time series that is being forecast is independent of actions based on the forecast.[21] We have seen in the behavior of the systems modeled in Chapters 15 and 17 how easy it is for actions of the company to affect market behavior. In fact, all of marketing and salesmanship is aimed in that direction. If the past pattern of sales is the basis for a forecast and if the forecast leads to actions that can affect future sales, we have a closed-loop information-feedback system. The assumption that the input time series is independent of the actions based on analysis of that time series no longer holds. The forecasting procedure becomes a part of the system in which it is set. The forecasting procedure has then the potential for creating amplification and a shift in the time responses of the system. The results may be far from those expected. In fact, it is quite reasonable to expect that forecasting methods will often be one of the factors creating system instability and that they can accentuate the very problems they are presumed to alleviate.

It is well to note that any decision-making process implicitly contains some type of forecast. Very often, whether recognized or not, the psychological factors include extrapolation of recent business changes. Even the apparent complete ignoring of a forecast as in the purchasing decision of Equation 15-9, implies some simple attitude like the future being a continuation of the present. Major system effects[22] can be created by even so simple a change in forecasting attitude as there is between

- Continuation of the present level of activity and
- Continuation of the present rate of change of the level of activity

[21] For example, as discussed by Robert G. Brown in Reference 16 and by Peter R. Winters in Reference 17.

[22] See Appendix L. There a simple extrapolation of changes in recent sales has been used to guide the inventory and pipeline terms in the ordering equations. This anticipation of a continuation of present direction and rate of change in business activity in the near future markedly reduces system stability.

Most attempts at forecasting turn out to be heavily dependent on extrapolation into the future of the course of events in the rather recent past. Many other factors presumably go into forecasts, but usually these other factors do not carry enough weight to override extrapolation as the major forecasting component. Sales departments see orders increasing and estimate still greater rises to encourage the production department to meet any possible future demands. Delivery delays are increasing, so that we accelerate ordering in anticipation of still tighter delivery. Prices are falling so that orders are cut back in the hope of still lower future prices.

The forecasting process can affect not only the series of events being forecast but also can affect the future forecasting procedures themselves. A forecast which is too high leads to certain production and inventory crises which create greater forecasting caution in the next year. This conservatism can bias future forecasts downward with the result that they undershoot.

Analysis methods that derive seasonal behavior patterns are also potentially dangerous. A sequence of events can be set up in a closed-loop feedback system in which the detection of a seasonal pattern can lead to actions that presumably anticipate the seasonality, but that can feed back to the market system in such a way as to strengthen and accentuate the apparent seasonality.[23]

In summary, forecasting methods that predict from past and present data are a mixed blessing. They may be good or bad, depending on how they interact with the remainder of the company. On the whole, I am inclined to believe that present industrial forecasting attempts do more harm than good. Either way, they will have a powerful effect on the system and must

[23] See Appendix N for an illustration of seasonal forecasting in the production-distribution-market-advertising model of Chapter 16.

be included in any realistic model of industrial behavior.

Long-Range Planning. There is another use of models about the future which is different in principle from the analysis of the statistical components of a past time series. This we shall call the long-range-planning model that represents the interdependencies which are thought to be significant between the factors that affect a future course of events. In a broad sense, all of the models discussed in this book belong in the category of long-range planning. The models are based on the fundamental factors that are thought to be of importance, and from these one develops an understanding of the system and how its future general behavior depends on the component assumptions and management policies. Such a model is used to anticipate how a system will be affected by changing its organizational form or its policies.

A model to aid in planning will deal with the generation of future time sequences that have not existed in the past. The planning model shows how those present factors about which we have information and confidence can combine in shaping the future. This is like the mental model on which good intuitive judgment is based. We have experience in how various types of factors will evolve. Visualizing the development of a field gives us our estimate of its future potential and how rapidly it will grow. A formal mathematical model for the purpose of planning acts as a check on our estimates and to refine the assumptions on which the future plan is based. Like the other dynamic models discussed in this book, it helps to clarify our thinking and will often divulge inconsistencies in the basic assumptions and will also show unexpected degrees of sensitivity to the various input factors. Some assumptions will be less important than initially assumed, while others of apparently minor significance may turn out to be highly sensitive in their effects on the results. We become alerted and

forewarned about the latter crucial assumptions.

An interesting formulation of such a planning model was undertaken by Hurford.[24] The dynamic model developed by Hurford interrelates 140 variables affecting the transition of the electric utility industry from generation from thermal power to generation from nuclear power. It deals in particular with the manufacture of nuclear fuel for both initial installation and replacement.

This model is primarily one of the learning processes involved in the transition of an industry to a new type of technology. One of the primary inputs is the present distribution of utilities from low-cost to high-cost fuel areas. Against this are set those factors that will contribute to the reduction in cost of nuclear power generation. These include the rate at which technical knowledge might be acquired as a function of the number of plants built, the cost of atomic plants as it declines with manufacturing experience, the construction delays versus design and manufacturing experience, and the rate at which the confidence of the customer can be developed as a function of the number of successful plants already established.

Figure 19-18 shows the general shape of the cost curve in terms of the percentage of utilities that could economically use power generation at various costs. A new technology is very dependent on the shape of the tail of the curve on the right-hand side. Initially a high-cost new technology may be economically usable only by a very small number of customers. The question then arises whether the technical knowledge gained from these first installations will accumulate rapidly enough so that costs will fall quickly enough to open the way to still more penetration of the market. A fortuitous relation-

[24] See Reference 13 for the S.M. thesis written under the author's direction by Walter J. Hurford, as a Sloan Fellow at M.I.T. from the Westinghouse Electric Corporation, 1959-60.

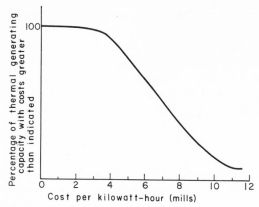

Figure 19-18   Thermal generating capacity versus power cost.

ship between this tail at the high-cost end and the factors governing technological advancement can lead to rapid development of the industry. On the other hand, the absence of a sufficient number of customers over the high-cost region may make the growth of the industry dependent on entirely different factors such as those utilities that might be willing to incur uneconomical costs for the long-run pioneering of a new technology or for the purpose of repelling governmental encroachment into the electric generating business.

The factors studied by Hurford give an idea of the conditions necessary for a take-off of the industry and will indicate the circumstances under which the industry may merely mark time until technological advancement from other areas (here from military and naval propulsion developments) would establish the necessary initial conditions.

19.9   Models of Entire Industries

Very often the individual companies in an industry are similar. The conspicuous manifestations are those of the industry as a whole rather than those uniquely marking one company. Other industries are characterized by the evident differences between the companies, but this is more apt to be in the earlier parts of the life cycle of an industry. In the older, more mature industries we often find similarity of companies, a rather highly competitive environment, and often a marked degree of industry instability. Examples are textiles, automobiles, copper, oil tankers, and electric generating equipment.

Industrial dynamics emphasizes the broad view of factors that influence industrial success. The first step in studying dynamic behavior is to explore in a preliminary way the broad aspects of the industry within the largest set of boundaries that contains interactions that are judged significant. The results of the broad study establish the framework within which narrower and more specific investigations can be formulated. It seems to be a mistake to start in the other direction. If an adequate setting has not already been established, there is no basis for properly formulating the details within a small part of the problem. How extensive the outer boundary must be depends on the problems to be investigated. The decision will be a matter of judgment about which factors must be given consideration. A basis for establishing a boundary was discussed in connection with the model of Chapter 17.

In those industries where the industry characteristics are more conspicuous than the company characteristics, it is especially necessary to begin with establishing the dynamic framework of the industry as a whole. The copper industry, as discussed in Section 19.3, is an example.

In this section two other industries will be briefly mentioned wherein one should develop his understanding of the industry as a whole before embarking on questions of modifying the practices of any one company within the industry — the oil tanker industry and the electric generating equipment industry. For each of these, a preliminary dynamic model has been investigated.

Fundamentally, the tanker and power generation industries are very similar. Their output is to perform a continuing service, in one case ton-miles of oil transportation per year and in the other kilowatt-hours of electric power per year. In both, the cost of the service is a small part of the cost of the products into which it goes. Both involve the construction of capital equipment. They differ markedly in the stability of price of the final service — electric power costs being very stable and tanker charter rates fluctuating widely. Even so, they are similar in showing a wide cyclic fluctuation in the orders placed for new tankers and new electric generating equipment. In both cases, the instability seems clearly to arise out of the interacting characteristics of the participants in the industry rather than from any factors that are imposed by the final demands of society upon the industries.

Oil Tanker Industry. Two recent studies bear on the dynamic nature of the oil tanker charter rates and shipbuilding fluctuations. One of these by Zannetos deals with the economic factors and the market expectations bearing on oil tankship rates.[25] His study develops an extensive descriptive picture of the industry, derives numerical values for descriptive parameters, and interprets industry behavior from the viewpoint of traditional static and dynamic economic models.

A coordinated dynamic model of the tankship industry following the procedures outlined in this book has been done by Raff.[26] Raff uses some 230 variables to interrelate four activities within the oil companies (supply department, chartering department, coordination department, and operating department) and three exterior activities (tankship brokers, independent owners, and shipyards).[27]

[25] See Reference 18.
[26] See Reference 13 for the S.M. thesis written under the author's direction by Alfred I. Raff, graduate student at M.I.T. in the Department of Naval Architecture and Marine Engineering, 1960.
[27] This study was carried to the point of a complete operable dynamic model, the behavior of which was generally reasonable, but not to the point of incorporating revisions resulting from analysis.

Electric Generating Equipment. At first glance the ordering and production of electric generating equipment presents a striking anomaly. The curve of electric power consumption in this country is one of the smoothest and best-behaved time series in our economic life. At the same time the placing of orders for electric generating equipment and the manufacturing of this equipment represent a classic case of one of our most unstable industries. Ordering rates for new generators vary by a ratio as high as 10 to 1, with a typical interval between peaks of five or six years. Manufacturing rates may vary over a ratio of 4 to 1. This fluctuation has existed for several decades.

Such a high degree of fluctuation of industrial activity increases the cost of the product. On the average, factory capacity is far from fully employed. Wage rates must be high enough to compensate at least partially for the periods of unemployment. Expenses are incurred in shifting production rates and finding and training employees.[28] These added costs must appear in the price of the equipment and eventually in the cost of electricity. The social cost to the community and to the companies can be high in poor labor relations and disruption of community well-being.

The dynamic study of such an industry shows how a combination of commonly accepted factors can interact to create a persisting instability.[29]

It is characteristic in such an industry that each participant blames the undesirable phenomena on the practices of others. Here for example, the equipment companies feel that the trouble arises primarily from the manner in which the utilities place orders for equip-

[28] I have received informal estimates from individuals within the industry that equipment costs are increased some 20% by the inefficiencies of fluctuating activity.
[29] A study of the dynamic characteristics of the electric generating equipment industry is being carried out as a research project by Alexander L. Pugh, III, in the Industrial Dynamics Group of the School of Industrial Management, M.I.T.

ment. However, in such a situation it is a reasonable assumption that partial causes of the unhappy symptoms may lie in widely separated places. Here the instability of the industry appears to require the interaction of most of the following factors:

- Deferrability of demand for equipment at the utilities
- Variable advance ordering by the utilities in response to changes in procurement lead time at the manufacturer
- The technological nature of the product which requires a long manufacturing period, and the practice of the industry which stresses equipment that is designed and made to special order
- The practices of the electic utilities in load forecasting
- The existence of excess manufacturing capacity in the equipment supplying industry
- The time delays inherent in increasing and decreasing production rates at the manufacturers

The utilities carry a substantial margin of excess generating capacity as contingency against emergencies. This margin is great enough so that on the average there is flexibility in the time of ordering new equipment. Equipment can be ordered early at the expense of slight increases in the margin, and orders can be deferred for periods of a number of months without serious encroachment on the excess generating margin. This immediately introduces an effect comparable with the prospective customer pool discussed in Chapter 16. It provides a large reservoir out of which a variable demand rate can be drawn without serious consequences to the utilities.

The utilities are sensitive to changes in the length of time necessary to procure generating equipment. If the manufacturing industry is becoming heavily loaded and the order-filling delay is increasing, the utilities tend to increase their ordering rate in anticipation of the longer deliveries being encountered. This is an effect similar to that discussed in Section 15.7 where

the overloading of the factory and the extending delivery delays tend to put upward pressure on the ordering rate. This is also the delivery-delay factor that was essential to the unstable behavior of the model discussed in Chapters 17 and 18.

It has been characteristic in the industry that each generating installation shall be designed to the customer's special order. This has become a practice, and it appears that the utilities expect an individualized product without realizing the price advantages that might come from more standardized design which would permit manufacturers to start some construction ahead of the receipt of orders.

The forecasting procedures for utility load seem to contain a high component of extrapolation, as discussed in Section 19.8. This extrapolation of the growth curve of electric consumption appears in the aggregate to be made with too short a smoothing time constant, with the results illustrated in Figure 19-19. The dashed curve in the diagram is a smooth, long-term, growth line. The actual power generation fluctuates around this line somewhat as the productivity and industrial output of the national economy change. If, as appears to be partially true, the effective past data entering into a forecast are shorter than one period of fluctuation of actual demand, then the forecasts will amplify the actual fluctuation as shown.

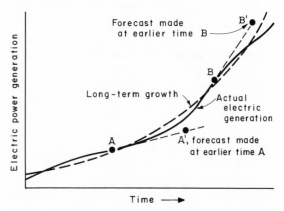

Figure 19-19 Forecasting of electric load.

Extension of the period of slower growth as at point A leads to a forecast of the future at point A'. A forecast made during the period of more rapid increase in demand as at point B leads to a forecast at point B'. Not only does point B' lie above the actual demand curve, but in the interval from A' to B' the deficit represented by the amount that point A' lies below the actual demand curve must be made up in new equipment orders. Formal forecasting procedures will usually recognize a longer base period from which to derive the prediction; but the shorter-range components of forecasting are apt to be introduced by psychological variables applied in the managerial and financial decision-making phases of placing an order.

The existence of excess capacity for the manufacturing of generating equipment is an essential component of the industry instability. If capacity existed only for the long-term growth rate, higher rates of equipment installation would not be possible to create peak periods, and the minimum periods would not be necessary for recovery from the preceding excess rate of installation. Excess capacity develops from the desire of each company to be in a position to capture a maximum possible fraction of the business during the upswing in ordering rates. Yet this motivation leads to a condition that accentuates the upswings that each company is attempting to capture.

Other less tangible factors also appear to be involved and to accentuate the difficulties that would be created by the preceding combination of six industry characteristics. A strong psychological and speculative undercurrent seems to circulate within the industry. This can be exemplified by the following sequence. A few additional orders may be placed by utilities as part of the upswing in cyclic business variation or as a combination of random circumstances. These lead to sales optimism and to increasing expectations within the manufacturing company. The president of a manufacturing company may make a speech on the condition of his industry and cite the favorable near-term prospects and state or imply that order books are beginning to fill up. The presidents of electric utilities take note of these comments and encourage their engineering and operating departments to advance their ordering dates ahead of the rush on manufacturing facilities. This leads to a still more optimistic environment within the manufacturing sales departments. Surprisingly small amounts of this kind of positive feedback can produce substantial amplification in an already oscillatory system.

Another secondary factor affecting this particular industry is a repetitive pattern of price softening as the period of very low manufacturing backlogs develops. This likewise is a self-perpetuating factor in which utilities come to expect a period of price discounting. Some utilities withhold and delay orders in anticipation of a period of more favorable prices. By that time they may have deferred new equipment ordering as long as they can, so that the price discounting arrives along with a mandatory re-entry of the utilities into the equipment market. These combine and cause an accentuated ordering rate.

An even broader outside loop exists here if we should conclude that the manufactures of electric generating equipment represent a large enough piece of the national economy so that the actual variable rate of manufacturing would contribute to the fluctuation of electric power generation. Certainly in the initial study of the behavior of such an industry, we should be likely to take the country's use of electricity as separable from the industry and to assume that the connection is sufficiently slight not to interact from one sector of the industry out through the general economy and back into some other sector of the industry being studied. However, the possibility of this type of interaction must come up for consideration in dealing with the larger industries.

CHAPTER 20

Industrial Dynamics
and Management Education

Industrial dynamics views business as an integrated system. As such, it can provide an academic framework on which to assemble the other management subjects. Mathematical models provide a case method of interrelating as much richness of detail as necessary. Models make this detail specific and show how the system that has been described will evolve through time. Industrial dynamics can be taught as a brief exposure to introduce system concepts or can be extended to a major field with a thread running throughout an undergraduate and graduate academic program. Industrial history (of a kind dealing with the reasons for decisions and the life histories of industries, companies, and products) and also linear information-feedback theory are especially important supporting fields for industrial dynamics. Management games have several points of methodology in common with industrial dynamics but fail to treat structure and policy in the business enterprise; they combine the shortcomings of both complete dynamic model research and real-life experience without the advantages of either. New challenges face academic research in finding new management approaches to modern technology and economic development.

THIS chapter will relate system dynamics to management education. The study of systems can provide a framework to unite subjects in the separate management functions. It can introduce the time dimension into what has previously been too static a treatment of the management process.

Certain principles can be evolved that have general usefulness, but these should not be misunderstood by attempting a close analogy with the isolated laws of open systems as they are known in a field like physics.

An academic program built around system dynamics can be dramatic, challenging, and intellectually demanding. It requires a different emphasis in the selection and goals of people,

both teachers and students, than has been customary in management schools. The impact on academic research in management can be vast, moving research away from mere data collection and explanation and into a position of leadership toward the design of more effective enterprises.

A fundamental view of system dynamics does not require the characterization of a system by the size of the institution — macro- and microeconomics merge, and the growth of the small corporation fits a transient dynamic mold similar to the development of a new national economy. The more successful we are in identifying truly fundamental factors, the more universally they are applicable, whether in research management, production, or market dynamics, in transient or cyclical change, in management or economics.

Emphasis on system behavior removes excessive attention from the particular business decision and places needed emphasis on the controlling policy — policy as affected by both formal rules and informal understandings, by organization, and by tradition, habit, past experience, and accepted practice.

20.1 Industrial Dynamics as an Integrating Structure

One of the conspicuous weaknesses of management education has been that "the subject lacks a well-defined base or well-knit internal structure." [1] This lack of integration appears in the universities[2] as a separation of educational courses into functional areas. These fragments are drawn together only in "business policy" courses or by case-method discussions.

Attempts at integration by discussion of only business cases can easily fail to perceive the consequences of feedback interactions between the separate facets of business. Misconceptions can persist because there is no experimental or objective test of the opinions rendered. Classroom situations are "solved" by using as a guide the commonly accepted practices from the real-life management art. Here is another of our information-feedback loops — the art is

[1] Page ix of Reference 19 by Pierson.
[2] And even the professional societies of management science tend to move away from an integrated concept of the management process. For example, The Institute of Management Sciences has divided its activities into "colleges" (professional subgroups) that disperse the membership into Managerial Economics, Business Computer and Data Systems, Organization Theory, Management Communications, Management Control Systems, Decision Processes, Simulation, Management Games, Planning, Research and Development Management, Management Psychology, and Measurements in Management. This compartmentalization is a caricature of the tendency to subdivide and specialize the subject until the manager's reason for being, that is to integrate and coordinate, has been lost from management education and research.

taught in the colleges, it becomes the basis for future practice of the art, and returns as the voice of experience to support future teaching.

A comprehensive dynamic model can form a bridge between the case approach to management policy and the teaching of principles, analysis, and functions. The case approach, based on discussion and argument, has the advantage of broad perspective and integration; but it lacks precision, it may easily rest on wrong intuitive estimates of the sources of difficulty, it is unreliable in tracing the consequences of policy changes, and it can perpetuate past misunderstandings. On the other hand, the teaching of separate principles and corporate functions is necessarily done in isolation from the interactions with other functions; the principles and analysis methods tend to be static and fail to account for the paramount importance of the time variable; the managerial goals to integrate action and to lead advancement are lost, and for them are substituted description and the mathematical goals of optimum solutions to artificial problems.

In a very real sense, the dynamic model is a business case. It can have as much richness of detail as one needs or has the patience to provide. It can handle any concepts and interrelationships that can be propounded in descriptive language. But it can go further than the language. It can show the consequences of interlinking the components of the system that have been described. It can show how changes in organization and policy will lead to a different system character. It can discipline discussion of the case by showing whether proposed changes in management methods lead to improvement, are of no consequence, or in fact make matters worse.

The objection that a dynamic model cannot incorporate the required subtlety and completeness must be dismissed. It has been amply demonstrated by experience in the use of models in teaching that a dynamic model can encompass more system factors than can be held in proper perspective during a verbal discussion. Whatever the shortcomings, they are inherently less severe than doing without an orderly structure for interrelating the parts of the system that is under consideration.

We can begin to integrate the separate aspects of management when a common language and structure have been found that equally well fit the circumstances of production, marketing, corporate policy, research and development, investment, or human relations. A language and structure are required that can relate hypothesis to implications, and cause to effect. This seems to be provided in the methodology of Part II, Chapters 4 to 14:

- Levels (of inventory, knowledge, need, confidence, money, or equipment) represent the momentary state to which the system has arrived.

- Information about the state of the system is the ingredient from which decisions are generated.

- Policy describes the procedure by which information is converted into decisions. Policy may be formal or informal. It may rest on inescapable physical characteristics of the system or on habit, precedent, and folklore.

- Decisions control the rates of flows to and from the levels. The evolution of the system through time is created. This evolution shows the consequences of the structure and polices that have been used to describe the system.

The dynamic model represents a system as broad as one chooses to describe. It gives quantitative system results that follow from the input assumptions. It allows a study of policy. It therefore becomes a vehicle for integrating the descriptive, the quantitative, and the case-discussion methods of management education.

The distinction between policy making and operations fades away when we realize that all decisions arise from some method (policy) of converting information into action. Policy can be studied at any level from the stock room to the board of directors. The proper scope of a policy can be defined after knowing how broadly it affects the system and what information it should recognize.[3]

The duties of the line manager and the staff advisor will lose their present conflicting objectives as policy is more carefully distinguished from the resulting decisions. Now the line manager is presumably responsible for policy but is too busy with day-by-day decisions to study personally the nature of a wise guiding policy. The staff man is an advisor on decisions, but he is the only one with time to consider policy, yet he does not have the authority and responsibility to put policy into effect. Clarification will come with realization that policy and structure are the principal determinants of success. The architect of the corporation's policies and organization is the most influential person in determining growth, stability, and profitability. It is with him that leadership, authority, and responsibility should reside. He must have time and the ability to design the corporation. He will have the authority and the willingness to take risks that characterize today's line manager but will need the training and a detachment from day-by-day problems that are now commonly associated with staff advisors.

Historical experiences and present activity can be brought into a common frame of reference. This will come as rapidly as we succeed in identifying those physical and human factors which tend to be constant but which cause the superficial and erratic symptoms that are distractingly apparent. For example, much is written on the acceleration of technological obsolescence — how entirely new problems are

---

[3] For example, in Chapter 18 it is suggested that employment level must be controlled by over-all system policy which recognizes sales rate, nature of the customer, inventories, and backlogs. The information sources and the policy-generating considerations are broader than the scope of the production manager. On the other hand, production efficiency (with presently available plant equipment) has its sources primarily within the production department, and policy should be established at the level of the production manager.

created by the pace of new scientific development and its effect on the rapid revision of designs and products. This is often interpreted as requiring a short-range view, justified by the observation that today's product is obsolete tomorrow. Whether or not there really is rapid change depends on our viewpoint and on the level of abstraction of our thinking. Two or three decades ago product design may indeed have been more constant. The required management policies were those governing the flow of materials and manpower into the continuous creation of the physical product. Planning needed to reach ahead only to the creation of plants and marketing organizations. Now product designs in many industries have a short life.[4] Policies must have a somewhat different emphasis. Rather than dealing primarily with the conversion of material into product, we deal more with the conversion of men, ideas, money, and laboratory space into designs. The continuous output of this system is a stream of new designs. Conceptually, this is still a conversion process with inputs flowing into output. At this level of abstraction, where our attention is focused on the conversion of resources into new designs, the time span of interest has actually lengthened, not shortened, compared with that for the production line. Whereas in the "good old days" the manager needed policies that governed the planning of factory capacity, he now needs policies that govern the creation and operation of a research organization, a much longer process. The horizon has receded, not approached. The continuous output was product, now it is new designs. The difference is only a slight change in emphasis in system structure and policies. We had new designs before, we still have production now. The system is the same but with a shift in some parameters.[5]

A common approach to levels, policy, decisions, and flow rates should erase the artificial distinctions that have grown up between studies of the company versus an industry, or between the teaching of microeconomics versus macroeconomics. The same fundamental kinds of decisions, the same importance of structure, the same need for some amount of aggregation according to the same principles, the same presence of uncertainty, the same perturbations from "noise" in decisions and flow rates are all to be found, whether one considers the corner drug store or international trade. These may differ in scope, in controversy, and in obscurity, but they all fit the same conceptual structure.

20.2 Principles of System Structure

Many places in this book have referred to the "orderly underlying principles" from which system behavior derives. It is too early in the study of system dynamics to attempt a comprehensive specific treatment of these principles, but their nature is beginning to emerge.

The principles to be discussed here all arise in the context of information-feedback systems. They are systems principles. They are not the principles of the management art such as have been taught in organization, production, and human relations courses.

---

[4] Perhaps artificially short to the detriment of company success.

[5] In fact, this apparent technological acceleration is a continuation of the slowly changing system emphasis that has carried us from an agricultural society (where men grew their own food), to a craft trade society (where men worked on special products directly for the consumer), to an early industrial society (where men used machines to make products for unknown men), to our late industrial society (where men made machines to be operated by other men to make consumer goods), to automation (where men design machines for others to make, which without men make products for men to consume), to an "artificial intelligence" society (where men will lay out the concepts that will be followed by machines, to make machines, to make products that men will consume). Economic evolution has followed a continuously deepening hierarchy of abstraction where human endeavor is more and more distant from the end goal of satisfying human needs. When the process finally bears fruit, a man-hour of effort produces an ever-increasing standard of living by having been multiplied through the intervening sectors that lie between human effort and consumption.

Because the principles apply to systems behavior, they do not fall into neat separate packages. We do not find here the equivalent of the separate laws of thermodynamics, or of Newton's laws of motion, or Einstein's relativity. Of course, even those are but incomplete and unrelated fragments of some broader future unifying concept.

The concept of a system implies interaction and interdependence. In attempting to identify factors that are common to all systems, we must keep the essential indivisibility in mind. In certain contexts, the principles discussed below are separable, at other times they may better be interpreted as different views of overlapping phenomena.

An information-feedback system derives its behavior from

- Structure
- Delays
- Amplification

It is a principle of dealing with such systems that we must look for the actual effective organizational structure and information sources, not merely those that appear in formal charts and operating procedures. We must watch closely for the presence of delays, especially those overlooked in informal information channels and in the higher levels of management decision making. Amplification is inherent in many of the system policies, especially in the overt[6] decisions.

To point the way toward the kinds of principles of system structure that will emerge, I shall list some of those that have been illustrated by the two dynamic models in Part III, Chapters 15 through 18:

- All system decisions implicitly or explicitly contain the concept of a desired state of affairs. Rates of flow in a system are determined by decisions, which are created by policies that ultimately rest on some attempt to bring actual system levels into agreement with a set of de-

[6] See Section 10.4.

sired levels. This process whereby decisions attempt to adjust actual conditions toward desired conditions defines an information-feedback system and sets the framework for model construction. A decision function is deficient unless it is working toward some objective.[7] This does not imply that the system must reach the equilibrium that is sought. The interacting efforts toward goals may keep the system continuously off balance.

- Many "worse-before-better" sequences are to be found where corrective forces first cause action opposite from the desired direction.[8] These can be highly unstabilizing. They arise as part of the frequent conflict between short- and long-term considerations.

- Amplification occurs at points where policies attempt to adjust levels in the system to values that change with varying flow rates associated with those levels. This was seen in the inventories proportional to sales rates in Chapters 15 and 17.

- Delays can create amplification by requiring that input flow rates exceed the output flow for an interval long enough so that the delay accumulates an internal content equal to the product of the rate of flow multiplied by the delay interval.[9]

- Variability of delay can create amplification. A variable delay can induce a variable inflow rate in an attempt to control the outflow.[10] Also, a variable delay can cause amplification by varying the outflow rate from a reservoir.[11]

- Averaging or smoothing of data inherently

[7] The outflow decision of a transportation system delay is working to adjust outflow rate to inflow rate. after taking into account (implicitly) the required level of goods in transit. A purchasing decision is working (usually explicitly) to adjust inventories to a desired level. The hiring action of a personnel office is adjusting employment to the authorized level of a factory work force.
[8] Sections 10.6 and 19.3.
[9] See Section 15.7.1.
[10] As in the variable-delay supply pipelines in Section 15.7.4.
[11] As in the prospective purchaser pool of Chapter 16 or in the release rate from the engineering department in Chapters 17 and 18.

causes a delay in the information.[12] Averaging is necessary to remove meaningless noise fluctuations from the data (as for example, sales information). The greater the noise in proportion to the desired information, the longer must be the time over which an average is taken; the longer the averaging time, the longer the information is delayed.

- When forecasting includes extrapolating past trends and cyclic fluctuations (as it usually does), it introduces a potential source of system amplification.[13] This amplification caused by an effort to anticipate the future can help produce a more undesirable future. Curves need to be smooth and well-behaved in order that they may be extrapolated into the future. But the curves are often derived from noisy data. The data are averaged to get smooth curves, but this averaging introduces a delay. The further into the future that the curve is to be extrapolated, the smoother it must be; but the more the data are averaged, the later is the result. Because of the erratic nature of much of the available industrial data, it is probable that only a few curves can be meaningfully extrapolated to the present, to say nothing of the future.

- All levels (inventories, bank balances, skill levels, etc.) are very similar in character to what we have otherwise isolated and called delays. As we have used "delay," it consists of a level or levels[14] whose input flow rate is controlled but whose output flow is uncontrolled and is generated by the internal structure of the delay. The other nondelay system levels have outflow rates that ordinarily are directly controlled and where only indirect control exists over inflow (as in the ordering of goods to replenish inventory). The policies for system control are usually designed to attempt to maintain some specified quantity in the level. All levels, whether in what we have called delays or elsewhere, serve to uncouple one rate of flow from another and thereby cause (or make it possible for) changes in one rate to lag

behind changes in another. In those level and rate combinations that are herein called delays, changes in an outflow rate lag behind changes in an inflow rate. It is typical of the other levels that changes in the inflow rates lag behind changes in the outflow rates.

- Delays and policies are frequency-sensitive. That is, they may amplify disturbances having a certain interval of periodicity and attenuate disturbances outside this range.[15] Also the phase shift and time delay between input and output will depend on the frequencies of the disturbance.[16]

- Delays are not inherently "bad." The effect of a delay depends on where it is in the system.[17] Delays may produce either amplification or attenuation of disturbances depending on where they appear.

The preceding principles identify some of the factors of consequence in system behavior. Principles for the formulation of models were discussed in Chapter 5. General principles relating component structure to specific over-all system behavior are ruled out at present by our inability to generalize about the nature of complex, nonlinear systems. Principles like the preceding can tell us what *may be* important. For the present, the discrimination between those factors that *may be* and those that *are* important must be obtained by putting the components into the context of the system and then determining the composite behavior and how that

---

[12] See Appendix E.
[13] See Appendix L. Also see Section 19.8.
[14] See Section 9.3.

[15] Section 15.7.3 discusses how the distribution system of Chapter 15 selects and amplifies from a noise input those frequencies that are near the natural period of the system. See also Appendix G.
[16] Appendix I shows how the output of a system can lag the input at some input periods and actually lead the input at other periods.
[17] In Section 15.7.7 increasing the delay in the adjustment of inventory was one of the most powerful parameters for reducing system fluctuation. In Chapter 18 the new policies for improved stability were based on longer rather than shorter delays in adjusting employment and inventories (in spite of the fact that in the original system the system was less stable as the delay in changing employment was increased).

over-all behavior depends on the separate components.

20.3 Academic Programs in Industrial Dynamics

The treatment of industrial management as the organization and control of a system, and the use of models to permit experiments in system behavior, can provide an integrating structure for the teaching of management. This can create motivation and effective "synthetic experience" in policy making. The study of system dynamics can be introduced at the beginning of an undergraduate academic program and continue through the doctoral and Executive Development Programs. At the Master's and doctoral thesis levels it provides a means for experimenting with new hypotheses and concepts of management organization and policy.

Purpose. Industrial dynamics in the management curriculum should unify the other subject matter of management. It should tie together functional areas and should add the time dimension. It should create an understanding of growth and of system-created change in industrial operations.

Gordon and Howell stress the importance of an integrating course in business policy.[18] They suggest that such a course must be of the case-study variety and must come in the senior year of an undergraduate curriculum. Section 20.1 discusses how the dynamic model can give rigor and precision to the discussion of a case. It seems clear, however, that such an approach should not be limited to a single term or even year. The student does not assimilate the feeling for systems behavior quickly. For full effect, the student must be steeped in a combination of theory, analysis, laboratory work, and design over a considerable period of time. There is certainly material here to be developed well beyond that which would be required to carry a continuous thread from the beginning of the

[18] Pages 206-208 of Reference 20.

sophomore year through the doctoral thesis level.

Enterprise design, through the use of dynamic models, provides an opportunity for a new kind of laboratory work in organizational theory and management policy. To a certain extent, the student can begin in his academic program to gain experience in the management of industrial systems.[19]

System dynamics and the associated models provide information feedback in the educational process itself. An idea can be developed, it can be set up as an experiment, the results can be evaluated, and the idea can be revised. This sequence of invention, experiment, evaluation, and review has been illustrated in connection with the models of Chapters 15 through 18.

Such work provides an environment in which it is possible to observe and evaluate certain of the personal characteristics of students which do not show up clearly in courses where material is studied and examinations test primarily the memory of facts. In converting descriptive knowledge to a dynamic model, the student is faced with a need to identify a problem, to visualize the dynamic concepts which are at work, to show initiative and judgment in selecting the factors to be incorporated, to cope with uncertainty and incomplete information, and to supply inventiveness in seeking system improvements.

A sequence of systems courses should attempt to instill in the student several attitudes:

- That there are fundamental characteristics of systems which apply whether these systems be electrical, mechanical, chemical, biological, corporate, or economic
- That there are important interactions between the parts of a system and that these systems

[19] In fact, I have seen this type of artificial experience carry a research team of graduate students so far into the understanding of the problems of one corporation that an executive of the company made the succinct comment, "It is clear that you already know more about the way this product is managed than anyone in the company."

interactions are often of greater consequence than the individual characteristics of the components

- That static analysis is an inadequate tool in dealing with the managerial environment

- That our intuition is unreliable in anticipating the behavior of complicated information-feedback systems

- That the experimental model-building approach is a powerful tool for handling situations that are beyond the grasp of intuitive or analytical mathematical solution

Teaching Procedure. Management, like engineering and medicine and architecture, is a practical profession dedicated to achieving specific goals. The successful practitioner must be highly motivated toward accomplishment. The end justifies the effort necessary. It therefore seems inappropriate to spend long years developing foundation subjects before giving the student an opportunity to view the objective.

The experimental study of system dynamics does not rest on any mathematics that cannot be taught in a few days any time after the high-school level. An education in industrial management could therefore begin with a study of quantitative system dynamics closely coordinated with verbal industrial dynamics, in the form of history and the reading and interpretation of the current daily and weekly business press. Such an approach would bring forth both the underlying fundamentals of business activity as well as the superficial manifestations of its problems. Such would then become a motivation for the study of science, engineering, finance, human relations, economic development, politics, and the other facets of management.

Based on some three years of teaching of industrial dynamics, several suggestions have developed:

- The subject is effectively taught as lecture and discussion combined with various forms of laboratory work. This laboratory work can include hand computation on simple examples, the study of more complicated systems by using digital computer simulation if a suitable machine is available, and individual and group projects in formulating models of industrial situations.

- Classroom simulation, with students playing the part of various components of the system, can be used to demonstrate the principles discussed in Section 20.2. Group simulation shows in a more personal and dramatic way than mere computed curves the behavior of larger systems. It can also show convincingly that the environment is a strong determinant of the "free-will" decisions and that the "obvious" best action is often so powerfully dictated by available information that intuitive decisions by different people are surprisingly alike.

- Even if a digital computer and automatic program compiler are available, emphasis should be placed on hand computation of simple dynamic sequences. A well-chosen example is highly informative. Furthermore, the time required to do the computation gives the student an opportunity to watch the progress of a disturbance as it is propagated through the system. Simple effects such as a step-function input to a supply pipeline are highly informative. By converting some of the delays from exponential to discrete delays, and by lengthening the solution interval to approximately a week, one of the stages of the distribution system of Chapter 15 comes within the range of practical hand computation. With proper coaching, the student can study the way in which rates integrate into levels, the way in which delays cause amplification, and the ways in which different decision policies affect the system. As an example of the latter, he can experiment with various omissions and additions in the purchasing equation of one distribution stage in Chapter 15.

- In teaching industrial dynamics the instructor must be careful to concentrate on the concepts of business and how to express and interrelate the business factors rather than to be carried away by the techniques and the methodology. There seems to be a tendency to elaborate the mathematics and the methodology beyond that justified by any real effect on the qualitative nature of the results. In so doing, the goal of business behavior can be easily submerged by the unnecessary refinement of technique.[20]

- A number of new staff members of management schools now have training in engineering. Some of these men show a strong tendency to cast management problems into the vocabulary of the engineering field. There is no more reason why management should be dealt with in terms of capacitors, resistors, vacuum-tube circuits, or hydraulic flows than there would be for teaching engineering in terms of inventories, average sales, and accounts receivable. The subject must not be handicapped by the insistence on an artificial conversion of vocabulary.

- The availability of a modern, large-scale digital computer lends drama, vitality, and scope to the teaching of experimental system dynamics. The teacher of such a course should have had experience in the building of models of rather complex systems and their simulation on a digital computer. However, for an instructor who has had this experience and who has justified confidence in what he is doing, I believe that a very effective course can be taught on the basis of hand-computed and group-simulated examples. The unavailability of a modern digital computer is to be regretted but need not prevent the teaching of industrial dynamics.

Related Subjects. The functional subjects in the separate management areas should treat the forces at work within the individual areas and show how decisions are influenced by informa-

---

[20] Students at M.I.T., even those having no previous experience with a digital computer, receive but one or two hours of lecture and description on the DYNAMO compiler described in Appendix A. Beyond this they are able and are expected to use the modern digital computer as a research tool by reading an instruction pamphlet and receiving some coaching from a laboratory assistant. Here is an example of a place where the instructor might be tempted to insert a lengthy interlude on digital computers and computer programming. This is not necessary. With reasonably adequate written material, it can be left to the student, just as is learning to use a slide rule. The class presentation should concentrate on management systems from the viewpoint of the manager, not the staff advisor.

tion arising from both within and without the functional specialty. An understanding of the separate functional areas is obviously essential to their inclusion in an over-all system. Some academic subjects whose pertinence to system dynamics is not so obvious justify special mention.

Good history of economic systems, of industries, of companies, and of special projects can serve as the descriptive raw material for the construction of dynamic models. Such history must of course contain the descriptive material relating to the important factors. Such history is hard to obtain. Often the reasons for decisions, which may have been evident at the time within the committee meeting or the board of directors meeting, are not recorded in the minutes and not available to the historian who comes on the scene late. The recollections and even many of the contemporary documents are apt to be biased in favor of the way the system should have operated and the decisions should have been made, rather than the way in which they were. History does not often enough treat the interacting forces of organizational stresses, personal interest, technological factors, reluctance to reach decisions, conflicting objectives, and the multitude of circumstances that eventually resolve themselves into a course of action. Likewise, history does not adequately record the alternative courses of action that might have been taken in an attempt to analyze what the ensuing consequences might have been.

A perceptive history of past managerial situations is one of the inputs to the better understanding of system dynamics. The ability to build models of complex systems and to study their behavior should make historians more sensitive to the important system variables. They should try better to search out and record the "policies" that tell how actions and pressures influence decision making.

In a quite different area, the courses taught in servomechanisms in many engineering departments provide an excellent background for the study of industrial systems. A good foundation in linear systems analysis develops an intuitive understanding and keen perception of the kinds of factors that give information-feedback systems their typical behavior characteristics. However, the problems of industrial dynamics are too complex for analytical treatment.

Although the fundamental principles of ordinary information-feedback systems are applicable, the quantitative nature of industrial and economic systems can be quite different from what the engineer ordinarily encounters. We do not have here the simple, single-loop control systems in which there is one point of error detection and one place where error-correcting control is exercised. Error detection and control are both distributed throughout the system. Every decision point constitutes both. Many of the practical design procedures devised for engineering systems are not applicable to social systems. Many of the methods that have been useful in engineering systems work because of the special characteristics of those systems. Engineering systems tend to be designed and used in such ways that the normal operating frequencies, the natural frequencies of the system, and the impressed noise frequencies tend to fall into different frequency bands. This does not seem to be true in our industrial systems. The disturbing noise, the natural frequencies of the systems, and the frequencies to which they are expected to respond all fall within the same frequency band. It is my belief, that the formal theory of servomechanisms will continue to be an important part of the educational background of those taking leadership in the field of industrial dynamics, but that servomechanisms theory will not be directly used as an experimental and operating tool.

M.I.T. in 1961. In 1961 industrial dynamics was included in each of the four programs of the School of Industrial Management at the Massachusetts Institute of Technology. At the undergraduate senior and the graduate student levels, two consecutive one-term courses were

available, making a full academic year as optional subjects. To the Sloan Fellows (men between 30 and 40 years old sent for a full calendar year by their companies to M.I.T.) the subject was taught for three class hours per week over a period of seven weeks, in addition to which there were nine hours of guided laboratory work. The Senior Executive Program (a ten-week series for industrial executives in the age bracket of 40 to 50 years) contained an introduction to the subject consisting of six lecture hours.

In addition to the academic work, theses at the undergraduate senior, the Master's, and the Ph.D. levels have been based on models of dynamic behavior. Several of these have been discussed in Chapter 19.

Future extension of the present two courses at the graduate level might take the form of four consecutive one-term courses (the typical M.I.T. Master's degree graduate program in management is of two years' duration). The first of these, as at present, would follow the subject matter of this book along with laboratory and classroom exercises as discussed above. The second term, as at present, would continue to be on advanced industrial applications, with experience in interpreting new situations into a mathematical model structure. A third course (now available as a series of subjects in the M.I.T. Department of Electrical Engineering and taken by staff members and doctoral students in industrial dynamics) would deal with Laplace transforms, transient response, frequency response, quantized-data systems, smoothing, data filtering, and random signals. A fourth subject might build on the first three, plus courses in economics, to treat the dynamics of economic systems.

As has been typical in other subjects in the past,[21] the material in industrial dynamics, as

[21] For example, courses on digital computers that were taught in the M.I.T. graduate school in 1949 and 1950 are a decade later available in a number of high schools.

developed first in the graduate school, is being adapted into the undergraduate teaching of management. Throughout the undergraduate program, systems studies could provide an integrating vehicle for the functional subjects and also bring in continuously the nature of industrial behavior, its problems, its managerial challenges, its history, and its current press.

20.4 People

In discussing how the characteristics of people fit them to make use of the methods of industrial dynamics discussed in this book, we must distinguish between two levels of interest and intent — for general student background on the one hand and on the other, for research, management, and educational leaders who expect to use the subject seriously.

For the ordinary student of management, industrial dynamics can be included in the curriculum to almost any extent from a few lectures upward. At any level of intensity it will give some enhancement of the concept of business as an integrated system. If properly taught in relationship to the backgrounds of the students, it can be fitted to almost any level of previous mathematical preparation. Herein lies a danger. The techniques and the methodology appear very simple. It is too easy to create the impression that a single graduate course or one academic year of study can cover the ground and can produce students who are expert in systems analysis.

The difficult part of industrial dynamics lies in the question of what problems to attack, and what factors to include as pertinent. The steps in an industrial dynamics study are

- Defining the objectives of the system under study
- Observing symptoms
- Detecting the real problem
- Visualizing the system at issue
- Estimating the boundaries within which lie the causes of the trouble

- Selecting the factors to be dealt with
- Constructing a formal model of the preceding
- Using the model to simulate system interactions under selected conditions
- Interpreting the significance of the simulation results
- Inventing system improvements
- Repeating all of these steps to move closer and closer to the true problems and to better management policies.

Starting the student on the techniques of the seventh and eighth items (constructing a model and using it) can be a very effective initial introduction to the subject of industrial dynamics. It gives an opportunity for the student to preview the more tangible aspects of the subject and to sample the nature of the goals. The real managerial content, however, lies in the first five items on problem definition and in the last three items on interpretation of results and system improvements.

This brings us to the second type of person — the man who is preparing himself to be a leader on either the academic or the industrial front.

The teacher or the enterprise designer, unless he is also an experienced and competent manager, may concentrate overly long on the techniques of the seventh and eighth items. Constructing a model and operating it are the most tangible of the steps and also the easiest. Unless great diligence is exercised, some simple and arbitrary assumptions will be made to substitute for the difficult perceptual and conceptual steps of the first six items. Furthermore, after a model is constructed and simulated, simple observation of results is likely to take the place of thoughtful interpretation of the reasons for the results. Invention of system improvements based on promising hypotheses that are then explored in the model may give way to mere trials of randomly chosen parameter variations.

The teacher or the corporate staff research man who is narrowly oriented toward methods can here fall into the same practices that were criticized in the Introduction with regard to operations research in business — the emphasis on the methodology may replace the emphasis on management. To maintain the proper perspective with regard to systems dynamics and the managerial objectives requires a man with breadth of viewpoint. He must be interested in the objective of improved systems, not in merely the method of analysis. Yet he must at the same time guard against breadth without depth, an approach that leads to a superficial treatment of subtle and complex problems.

So far, we are finding two types of men to be the most successful in this work. One type is the man whose interests are centering on the area of business management but who has had formal training in the analysis of information-feedback systems (obtained in the study of mechanical, electrical, or hydraulic equipment), and preferably has also had laboratory work and first-hand experience with such systems. The second type of successful man is the unusually competent and perceptive industrial manager. He also has dealt with complex information-feedback systems in the true-life environment. Even better, there are many men now becoming available with both kinds of capabilities.

To be successful, a man needs a good feeling for system dynamics, and also he must understand the factors that go into managerial decision making. He needs a strong motivation toward deeper insight into better management. By contrast, we have had much less success in the industrial dynamics work at M.I.T. with those trained primarily in mathematics, in conventional operations research, and in mathematical economics. Men who are narrowly interested in these latter areas seem often to lack one or all of the essential qualifications of dynamic systems intuition, motivation, and understanding of the real-life business processes.

After trying various shortcuts without notable success, we have at M.I.T. begun to develop some impressions of the kind of training that a man needs to become an effective leader in industrial dynamics. He must of course be interested in the subject. He must be interested not merely in gathering information and understanding systems, but he must also wish to move into system redesign and improvement. He needs high integrity. He should have the personality that makes it possible for him to delve into sensitive and embarrassing corporate circumstances and deficiencies without alienating people.

Regardless of his background he should work his way up through the formal theory of linear and sampled-data systems to the level of two or three graduate courses in the subject. If he lacks a background in differential equations and engineering, this may involve three or more preparatory undergraduate courses before he reaches the graduate-school level of systems study.[22] Such a sequence of courses in dynamic

[22] A sample sequence of related courses selected from the M.I.T. catalogue follows. Most students undertaking a study in depth of industrial dynamics will have at least part of the following material, and others may find it possible to abbreviate the sequence in various ways. This intensity of training in systems theory is appropriate for the doctoral candidate whose emphasis is in industrial dynamics. Courses in Electrical Engineering are indicated by the number 6; courses in Mathematics by the number 18. There is no reason why the subject matter of these courses might not be taught in the economics and business context, but so far it has not seemed necessary to duplicate material of the same underlying philosophy merely to get a change of vocabulary (so long as the more complex business systems are taught in the business vocabulary). In time, the systems theory represented in these courses could be adapted and somewhat shortened into a theory sequence for students of management.

6.01 *Introductory Circuit Theory*

Two-term sequence for students majoring in electrical engineering. Network topology and topological aspects of duality, Kirchhoff's laws, volt-ampere relations for the elements, and sources of excitation. Equilibrium equations and methods for their solution. Step function and impulse response of simple circuits, and their complex natural frequencies. Sinusoidal steady-state behavior. Impedance as a function of applied and natural frequencies

theory can be studied in parallel with material such as presented in this book. This book does not depend directly on such material, and yet a proper interpretation and understanding of the material discussed herein will be greatly enhanced by an understanding of the subject matter of linear dynamic theory.

Along with the preceding material, the student can study functional subjects of management, economic and industrial history, and read regularly the daily and weekly business press.

As the nature of management education changes, the place of doctoral level study in business will also change. At the present time

study to the Ph.D. degree is usually thought of as preparation for teaching. Such was the attitude toward advanced graduate study in engineering twenty-five years ago, but now, doctoral study is a common step in the route to top industrial technical positions. As the professional challenge in management becomes more demanding, the need will grow for greater depth of training in the academic program for management. The trends that we have seen in the other professions can be expected to develop in management. Advanced study will initially prepare men for teaching. But as the management profession evolves, advanced study will

---

and its geometrical interpretation in the complex frequency plane. Resonance phenomena, amplitude and frequency scaling, vector diagrams, reciprocal and complementary impedances. Energy storage and dissipation; active and reactive power. Mutual inductive coupling. Polyphase circuits. Usefulness of the duality principle given prominence throughout the discussions. Closely correlated laboratory work accompanying lectures and recitations.

6.02 *Electronic Devices and Circuits* Prerequisite: 6.01
Basic subject for students majoring in electrical engineering. Circuit models for electronic devices such as transistors, vacuum tubes, gas tubes, etc. Analysis of basic electronic circuits using nonlinear curves or piecewise-linear approximation for nonlinear operation and incremental models for linear operation. Associated laboratory.

6.05 *Electronic Circuits and Signals* Prerequisite: 6.02
Basic subject for students majoring in electrical engineering. Fourier representation of signals. Modulation, amplification, and generation of electrical signals. Limitations on signal transmission imposed by inherent energy storage, noise, and saturation in electronic systems. Associated laboratory.

18.05 *Advanced Calculus for Engineers* Prerequisite: Differential Equations. Graduate Subject.
Vector analysis; orthogonal curvilinear coordinates. Functions of a complex variable; calculus of residues; conformal mapping; ordinary differential equations; integration by power series; Bessel and Legendre functions. Characteristic-value problems; expansions in series of orthogonal functions, including Fourier series. Partial differential equations; characteristics;

solution of classical equations of mathematical physics. Sokolnikoff and Redheffer, *Mathematics of Physics and Modern Engineering.*

6.601 *Feedback Control Theory* Prerequisite: 6.05 and 18.05. Graduate Subject.
Models for physical systems, block diagrams, Laplace and Fourier transforms, stability criteria, transient response, frequency response, performance indices, and compensation schemes leading to trial-and-error design procedure for linear control systems. Properties and effects of stochastic signals. Introduction to nonlinear systems, sampled and quantized-data systems, and optimization theory. Newton, Gould, and Kaiser, *Analytical Design of Linear Feedback Controls.*

6.54 *Sampled and Quantized Systems* Prerequisite: 6.601. Graduate Subject.
Development and application of linear analysis techniques to design of discrete systems and to related problems. z-transform methods. Processes of sampling continuous data and of smoothing sampled data. Feedback systems employing sampled-error data, including stability studies of such systems. Study of continuous systems by numerical methods. Digital computer operations as data filtering, analysis and synthesis of linear digital computer programs in frequency domain. Development of a circuit theory for sampled and mixed systems making use of flow graphs. Comparison between techniques of numerical analysis and control system design. Response of sampled-data systems to random inputs. Study of quantization and round-off noise in numerical solution and real-time control. Analysis of errors in computation. Study of random Markov processes by sampled-data analogs. Design of statistically optimum discrete systems and adaptive systems.

become a normal channel to the higher managerial positions.[23]

The development of a stronger foundation of management science will not change the fact that management is an art. The better managers will still be the ones who have the judgment to use their tools in the proper places. There is not soon to be in management, any more than in engineering, law, or medicine, an orderly procedure for identifying the worthwhile goals or the road to them. Therefore, the full development of an expert in the field of industrial dynamics requires an opportunity for him to gain experience, to test judgment, and to build self-confidence. This seems to require some form of internship.

An effective internship has been provided at M.I.T. in the form of the full-time research assistantship for the graduate student. In the School of Industrial Management this is patterned after the program which has been highly effective in M.I.T. engineering departments during the last three decades. The full-time research assistant works at a new frontier of knowledge in an atmosphere of serious, timely importance. The work usually takes the form of developing and applying new methods to the solution of problems that are of intense interest to an outside cooperating organization. He works on new concepts and methods and on their conversion into use. The man is expected to be a responsible member of his research project, devoting full time to it except for the time actually in class. His graduate thesis is usually closely related to his staff work, so that the two mutually support each other. He can progress in his graduate study at a rate about half that of a full-time student.

A carefully selected full-time research assistant at the graduate school level can, with

proper faculty guidance, take a responsible part in the growth of industrial dynamics. He can help develop underlying methods and techniques. He can interpret industrial problems into systems structure. If he has the proper personality characteristics, he can work within an industrial organization to delve into the mode of operation and to interview the several management levels until the factors influencing decisions begin to clarify. The better of such men often come out of full-time research assistantships (after three or four years, or more if they have earned both S.M. and Ph.D. degrees) with published papers and a national or international reputation for their pioneering work. After a combination of study and practice, the man becomes capable of taking leadership in new academic programs or in setting up a new activity within an industrial company.

20.5 Management Games

Management games have received much attention, both in academic circles and as a corporate tool for executive training. Because of the publicity that management games have received, it is necessary to comment on them here. This is especially so because they superficially appear very similar to the dynamic models in this book, and yet educationally they move almost in the opposite direction.

As used here the term management game means a situation in which part of the industrial system is represented by a dynamic model, but certain of the information flows are presented to a person who makes decisions and inserts his decisions into the system as the simulation progresses. Most often there are a number of people involved who, along with the simulation carried out by a computer or a referee, make up the entire business process being represented.

There are some close similarities between the business game and the full dynamic model. In the dynamic model, as discussed in this book, all parts of the system and all policies are

[23] See Leavitt and Whisler, Reference 7, for additional discussion of how the future entry point to upper-management positions will more and more be through special training rather than by working up through the lower administrative levels of the corporation.

represented by decision rules that are mechanized by a digital computer. In the business game most of the system is represented by formal policies within a computer, but a few of the decisions are made by the human participants as the game progresses. In both there is a mathematical model, usually there is a computer, both concentrate on the business system, and both aspire to enhance understanding of the broader business scope. Here the similarities end.

The management game has the combined shortcomings of both the full mathematical dynamic model and also the learning process from real-life experience in an operating company. At the same time, it lacks the principal advantages of either of the other two. Like the real-life business system, the participant in the game views the remainder of the system as a black box that he attempts to poke and prod with his decisions to see how it responds. In a complex nonlinear system this is disappointingly unrewarding. Even if the model in the black box were a good one, and truly represented the particular business of interest, he still would not be able to fathom its complexities by this type of external teasing.

The emphasis in the business game is on the external manifestations, not on the internal structure and its implications. In fact, the internal structure gets relatively inadequate attention. The design emphasis is on achieving a superficial external appearance of reality, highly tempered by pressures to enhance the theatrical effect. We need only observe typical examples of such games to immediately detect misleading compromises that have been introduced in the interest of showmanship.

The most conspicuous warping of reality in game design is in the direction of shortening the pertinent time constants of the system. Most managers think they can get systems responses much faster than in fact can be accomplished. The management game, by shortening these time constants even more than the manager already believes them to be, can do nothing but accentuate already existing misconceptions. For example, at least one of the very popular early management games involved a research-and-development expenditure in which the profitable pay-off from research and development occurred within two or three quarter-years after the expenditure. This is so untypical as to be seriously misleading, especially to the participant who already underestimates the time required to achieve research results. In this particular case, the defense of the short time constant by the designer of the game was merely that longer time constants would take so long to react that the effects would not be visible within the one day of playing available to the participants.

The management game, if it teaches anything, is usually defended as illustrating good decision-making practice. This means good practice as viewed by the designer of the game. Game designers have not ordinarily been great managers skilled in innovative advancement of the management art. The game then becomes a vehicle for perpetuating the past clichés about what constitutes good management. The game is designed to reward the kinds of decisions that the game designer thinks are proper. The player is therefore being trained in the patterns that the game designer intuitively believes to be preferable. Any manager who plays the management game for the sake of learning better management should examine carefully whether or not the game judges his decisions in a way that is significantly related to the way his decisions react in his own company.

The management game has often been compared with devices like the Link cockpit trainer for the training of aircraft pilots. This is a good analogy and immediately shows up the shortcomings of the method. The Link Trainer is designed to enhance and ingrain automatic responses in the trainee. The kinds of situations to which he is to respond are designed in advance. The relationships are relatively simple from the standpoint of the system designer. The information to be presented to the trainee is predetermined. In fact, it is no great intellectual step to set down the precise formal rules that the trainee should follow to get the best possible response out of the system. This has been done in fully automatic autopilots for aircraft. In a sense, the pilot trainee is being processed through the Link Trainer because he is the most available and least expensive machine to carry out a predetermined set of policies regarding data analysis and decision making. By contrast, the Link Trainer is of practically no value in designing a new, different, and improved type of aircraft. The cockpit trainer has been designed and adjusted until it reacts sufficiently like a particular known actual aircraft so that it serves the purposes of obtaining emotional involvement from the pilot, presents to him the typical kinds of information, and receives and acts plausibly well on his reactions. It need not, however, contain any true appreciation for the underlying causal factors that make the behavior what it is. The aircraft designer goes instead to his wind-tunnel model, his mathematical models of aircraft behavior to be solved on a computer, and his considerations of performance objectives in order to bring out a new improved design. In this analogy, the manager of the future should be more in the position of a man seeking a new improved air-transport system than in the position of the man being trained as a robot operator of a piece of equipment.

On the other side, there are certain things that a management game can do well. Game designers stress the extent to which the management game achieves intensity of participation and emotional involvement from the players. This is certainly true but also applies to poker and football. Is emotional involvement the proper state of mind for objective, creative management? The management game usually leaves the player with a much stronger impression than he had before of the importance of the interacting parts of the management system. This is good.[24] Before, the player may have had only an intellectual belief in the importance of interactions between parts of the system. He may come away from the game with a deeper

---

[24] In connection with industrial dynamics studies, we have often devoted one day to a hand-computed game that is a simplification of the model described in Chapter 17. The game consists of putting the student in the position of the production manager of the company supplying components. He has but one decision to make: the number of people to be hired or released from the production work force each week. He is told the general structure of the system regarding the delays in the transit of orders from the customer to the production department. He is given the union contract conditions regarding termination notice to employees, and is told how long the training period is for new employees before they can become productive. He is also given an unusual piece of information not normally available to the production manager, to the effect that the customer is rigidly following a formal decision process for generating his orders for components. The forcefulness of the environment in predetermining the participant's decisions can be adjusted by changing such things as the length of the employee termination notice, the training delay, and the constant CISTC in Equation 17-89 that determines the extent to which customer inventory varies with delivery delay. The production manager makes one hiring or layoff decision each week (represented by perhaps 5 minutes of real time). By proper selection of coefficients, all within a reasonable range, it is possible to create a situation wherein almost every participant, whether he is a student or experienced corporate manager, will find himself tracing a highly unstable employment and sales pattern even when the customer is using the product at an absolutely constant rate in his own production process. Furthermore, this is true even when the participant has a monopoly (without knowing it) supplying the entire market demand. (As in Figures 18-5 and 18-19.) As soon as this history begins to unfold, the manager begins to see what he believes to be a seasonal pattern, and proceeds to make allowances for it in the future. This only makes matters worse, since the seasonal pattern is merely a reflection from the customer of the supplier's own varying delivery delay. The exercise drives home two points. First, that the interactions between the parts of the system can be tremendously powerful. Second, the assumed free-will managerial decision is strongly conditioned by the environment, and at least in the more unstable system structures it is highly predictable because it is so heavily forced by the available information sources.

conviction that the systems interactions are indeed of great importance. In general, he will not know what the nature of the interactions is, but he will be convinced that they are significant. The management game therefore can serve as an effective opening wedge into a true study of system dynamics wherein the manager goes behind the scenes and deals with the actual concepts of how the parts of the system relate to one another. One or two days of game playing are usually sufficient for this purpose.

The emphasis in the management game tends to endorse many of the actual weaknesses of present managerial practice. For example, the game emphasizes the short-term decision of the moment, rather than the importance of sound policy. It does not allow the experimental approach in which a given situation can be repeated with known changes in policy or structure to see what effect those changes would have. It emphasizes management by crisis decisions. It puts emphasis on outguessing competitors rather than on developing a plan for a strong independent course of action.

We see then the difference between industrial dynamics and management games. Industrial dynamics emphasizes the *basis* for reaching decisions. It emphasizes the study and finding of policies that should be used for guiding decisions. The attention is on the internal processes of the industrial system and how they interact. It is a study of the internal structure of cause-and-effect relationships. The goal is the design of a more successful system that may involve a new structure, new policies, and different sources of information.

On the other hand, the management game follows the pattern of the business world, with emphasis on intuitive decisions without formal attention to the consequences of various ways of reaching those decisions. Management games perpetuate the black-box nature of the business world. The game has a causal structure for its behind-the-scenes operation, but this structure is not itself the subject of study. The game stresses the short-range crisis and the immediate decision rather than the long-range planning of policies and organizations to avoid crises. If the management game tries to improve intuitive decision making, it runs the risk inherent in nonlinear systems of training for the particular combination of circumstances existing in the game but giving no basis for knowing when those conditions may cease. In short, the management game is a game.

20.6 Management Research

Much of academic research in management has been following the tradition of the liberal arts and the social sciences — gathering data about the past and seeking explanations of the present. The present can use more than explanation. It needs more effective concepts for industrial management and economic leadership.

University research in management has the same opportunity to assume aggressive, innovative leadership that we have seen in the great schools of medicine, science, and engineering. Breaking out of the old patterns must be based on new hypotheses supported by experimental verification. Daring experiments are necessary to find new concepts for enterprise design.

Experiments with actual organizations are too costly, too time consuming, and too difficult to interpret to be used as a routine experimental research method. Mathematical models can carry the principal burden of experiments in management systems. A limited number of real-life trials can be arranged to keep the model experiments attuned to reality.

Management is a profession of action. Management research should be attempting to develop new management viewpoints and new types of organizations and policies. Management research has the opportunity to lead the way in putting new ideas into practice. To do this, cooperative programs between university research and industrial organizations can provide contact with important problems, and also the opportunity to test the results of experiments with models by carrying recommendations into revisions of operating organizations.

Research opportunities exist all the way from inventory and production rate policies, as discussed in Part III of this book, to the problems of international trade and finance and economic development. In this latter area in particular, the world has much to learn. The newly emerging countries present a great challenge to the leadership of the Western democracies. Our record is not impressive in showing the way by which a primitive society can move rapidly to the level of the more advanced economies. Must they go through all of the stages through which we have gone? Or knowing now the forthcoming age of automation, can their societies perhaps skip the stage of large manual-labor and white-collar worker forces that are now on the decline in the more advanced countries? The successes of history have been infrequent, and furthermore there is grave doubt that countries today want to follow the patterns of development of those countries which have been successful in history. What then is the structure, the policy, the allocation of resources, the timing, the goals, and the aspirations that can lead to success? Here is an area for innovative model building. We need to combine the economic factors, the political, the educational, and the technological to obtain a better understanding of the dynamics of stable, rapid growth. What are to be the stages of economic development? What kinds of people will be required? Where are they to come from, and who will train them? The United States asserts to new countries the advantages of building in the democratic tradition, yet these countries do not have the foundation of an educated public that the United States had at its establishment. We hold ourselves up as an example of economic development, but we present no clear and coordinated plan through which others can harness education and political and economic forces for the common good. We have no plan because we lack an adequate understanding of the dynamics of growth. This is but one of the opportunities for management research.

The experimental approach should take the place of observation alone. Mathematical models should incorporate all the factors that our judgment tells us are essential to the solution of the problem at hand. No longer should we limit our attention to oversimplified analysis simply to achieve analytical solutions. We should abandon the quest for optimum solutions in the interest of attacking significant problems. More is to be gained by improving areas of major opportunity than by optimizing areas of minor importance.

CHAPTER 21

Industrial Dynamics
in Business

Industrial dynamics studies of the corporation should be started as a long-term program at an activity level low enough to avoid pressure for immediate results. Selection of proper men with the managerial viewpoint and the necessary technical skills is crucial. They must be given time to develop their understanding of the company's problems and of system dynamic behavior. Suitable men will be those also in demand to fill management positions in the company.

A frequently asked question regards the procedure for making industrial dynamics an effective tool in the management of a company. There are at least two important considerations. First a company must begin with the proper intent of continuity of effort and with plans for long-term, gradual development. Second, it must select suitable men who will undertake to analyze the major corporate problems and the factors on which improvement depends.

The rapidity with which results can be obtained in a particular company will depend greatly on the kind of people available for assignment to the industrial dynamics work and also on the environment within the company. Even under the most favorable circumstances where the proper kinds of men are initially available, a period of three or four years would usually elapse before the industrial dynamics studies have reached a point where definite managerial recommendations begin to emerge. This period of time is needed to establish the proper perspective and attitudes on the part of the study group, to allow them time to instill some confidence for their work in the operating managers of the company, to identify the truly important problems, and to clarify

the company and market factors on which the indicated courses of action may depend. After this, a first experiment in applying results to an actual operating situation may require two or three years or more to inaugurate and to evaluate. The whole undertaking can well require a decade before the approach is generally accepted as valid and powerful by the responsible managers of the company. During this decade, men who have acquired personal experience in corporation design will move into managerial positions and they can accelerate the acceptance of new methods of management.

If this interval seems unduly long, understanding corporate dynamics should be compared with other much simpler undertakings that have required at least as long. Electronic digital computers for the routine clerical processing of information have been known for over a decade, and yet we really should regard most of the computer installations in 1961 as providing experimental evaluation of things to come. A new product may take ten years from the beginning of research and development until it is successful in the market. In the use of industrial dynamics as a managerial tool, we are speaking of a process that will be inherently slower than the acceptance of isolated techniques of management science and slower than new-product development. We have come to accept the need for new-product development. It does not meet with emotional resistance. There is belief in its importance. New products are assisted by favorable management attitudes. On the other hand, new management methods may at first be viewed suspiciously, for a time they may be hindered rather than helped.

Industrial dynamics in the company will not evolve fast enough to bear on today's crisis. It should not be started with the expectation of showing the way out of a specific present difficulty, although of course it should help avoid similar difficulties at a future time. Industrial dynamics should not be undertaken as a method of forecasting specific future events or of guaranteeing the correctness of any specific decision. Instead it should be aimed at a better understanding of the management process and at improving the frequency of successful decisions rather than ensuring a particular decision.[1]

The executive ought also to bear in mind the typical enthusiasm profile of a new undertaking. At the time an activity is established, enthusiasm will be high, along with optimistic expecta-

tions for the future. This initial enthusiasm is apt to suffer when in spite of all warnings results do not come as rapidly as was hoped. In starting an industrial dynamics study, initial corporate models and their results will point the way toward what is needed in future models rather than toward immediately useful new concepts of management. The initial planning of an industrial dynamics study should recognize the probability that it must live through the typical secondary reaction of disappointment which so often accompanies new enterprises of any type. The hazard here may be greater than it is in the corresponding enthusiasm profile of a new product, because there is no history of success in other similar situations to support confidence. At this point the selection of an unsuitable leader of the industrial dynamics group may make its appearance. If so, the valley of despair will be greatly lengthened while a group reorganization takes place.

It is important, then, to avoid starting work in the area of industrial dynamics at a level of activity that cannot be sustained. It is much better to begin at a low level of activity, with plans for gradual growth, than to start with a larger program that from its very size must be committed to early success.

---

[1] The desire to treat the specific decision is usually stronger than the interest in the proper policy. This is a manifestation of the short-term view taking precedence over the long-term. An illustrative example occurred at a talk on industrial dynamics which I gave to executives of one company. During the ensuing discussion the executive in charge of a consumer-durable product line commented, "Very interesting. But how does industrial dynamics tell me how many to manufacture of our newly announced *abc* design?" The *abc* design was a convenience improvement in an otherwise mature and long-accepted kind of product. The question was whether or not to rush into full-scale production on the basis of initially promising sales to distributors. Some similar situations in the past had developed successfully, others had not supported at retail the initial enthusiasms of the distributors. Industrial dynamics should not be expected to enter at this late stage to answer the specific question. Instead, it could have been used to understand better the *nature* of this particular kind of product

innovation. Does company strength in the market really arise from innovation and being first with design changes? Or is the long-term position built on quality, performance, and trouble-free operation? Must the innovation be exploited rapidly, thereby making significant the original question of trying to establish production in anticipation of demand? Or is the risk primarily in the dissatisfied customers that might result from flooding the market prematurely with an unproved design? The policy that should govern production depends on the character and the time responses of the market. What kind of business are we in? Can production decisions be taken as demand develops and as sales information becomes available through normal channels, or does success depend on information not normally available? If the latter, are we justified in the cost of unusual information sources that can be timely in guiding the production decisions? These are questions that must be considered in advance; they cannot be answered soon enough after a crisis has developed.

The process of illuminating and improving the nature of the business system is not expensive. It does not require a large group. It is more dependent on quality than on quantity of personnel. In fact, a surprising amount can be done by one person given a period of three or four years. A one-man group is not advisable because it is of unstable size. Should the one man decide to leave the company or to take on some different responsibilities, the momentum and continuity of his work will be lost. Therefore a group of at least three persons seems to be indicated so that they can work with one another and act as an inspiration to each other and as a check on each other's work. A three-man group may be large enough so that new members can be fitted in to replace ones who leave, without a resulting break in momentum.

It has been stressed throughout this book that industrial dynamics is a tool for the professional, responsible, creative manager. Therefore the kind of man who undertakes the introduction of industrial dynamics into a company should be the kind of man who is qualified, or at least will be qualified, for an important managerial position. Because it is a long-term process, the company may start with a group of younger men, but they should be men who have the capability to develop into top-level executives. Their judgment, courage, integrity, and understanding of the business problems and objectives are paramount. Next, they must have, or they must develop, a real insight into the dynamic behavior of complicated systems. A desirable foundation of formal training is given in Footnote 22 on page 355. They should be men in whom the management has confidence, and men with whom the management enjoys working. They should be men with whom the innermost secrets and hopes of the organization can be discussed.

Men of these qualifications may seem very rare, but they are not. A large number of men between the ages of 25 and 40 who have a background of experience that makes them well-suited to this kind of understanding are now emerging into managerial positions in many companies. It is no doubt going to be easier for the companies who have had military systems and chemical-process and electrical aircraft control-system work in their technical departments to find qualified men for working in the dynamics of industrial organizations. Business schools will in time turn out qualified leaders. It will take them time, however, to develop their teaching programs and to provide the several years' sequence of training necessary to make experts. Until such an educational process has begun to be effective, the best-qualified people will come from within those industrial organizations that already have a tradition of working with dynamic systems.

The suitable candidate must have as his goal the building of a better company. He is not the mathematician interested in the analytical processes for their own sake. He is not the operations research specialist who looks upon himself as a scientist and an advisor instead of a leader with a personal responsibility for the success of the organization. He must have the courage to delve into the areas which are important, not merely the ones which lack political strength in the company.[2]

The man described here will be in great demand for other assignments in the company. Industrial dynamics is premature in a company until it is viewed with sufficient regard that it can compete for the most capable men. It may be as well to do nothing as casually to delegate the subject to a staff study just as a way to create the feeling of having done something

---

[2] I know one corporate group engaged in industrial dynamics research which seems to be extremely well-qualified technically but which at times appears to stumble on this point of courage. It has been their frank personal opinion, as well as mine, that the xyz area is crucial to the processes being studied. Yet they have shown great reluctance to delve into how that area is operated, how the decisions are arrived at, and how it ought to be represented in a system model because "it is not politically expedient in our company to question that department."

and to get the subject off the executive's desk.

The man who takes up industrial dynamics primarily to further his own personal cause within the company is apt to be foredoomed to failure. He may be looking for a gimmick to which to attach his own career. His competence and integrity are apt to be such that he is already under a cloud and not apt to enjoy the full confidence of his associates. It is of course essential that the man who undertakes such a new development be enthusiastic about it. Yet he must have qualifications well beyond that of enthusiasm.

There are no doubt exceptions, but industrial dynamics will not often fit happily into the conventional computer applications department nor into the usual staff advisory group of operations research experts. The former is busy with an overwhelming load of immediate detailed work; it is apt to approach industrial dynamics as a computer application rather than from a viewpoint of top management. The latter is often oriented more toward science and mathematics than toward management.

It appears that the most successful use of industrial dynamics will occur when it is undertaken on a long-term basis and placed in the hands of the line manager type of personality who approaches the subject as an opportunity for him to learn new management methods that he himself will have the future opportunity of applying as a responsible company executive.

Last, there is the question of the size of company to which industrial dynamics is most suitable. There is a common first presumption that industrial dynamics is a tool of use primarily to the largest corporations. As the field has thus far been developing, there seems to be little immediate support for this conclusion. In the largest organizations, the functional compartmentalization is apt to be stronger than in smaller companies, making it very much more difficult for any person actually to cut across all activities from research to marketing. It is beginning to appear that the aggressive, rapidly growing, medium- and small-size organizations may be the places where the methods discussed in this book will have their first important impact. Such organizations are often more flexible. They may be more responsive to the wishes of the company officers, so that if the officers want to explore a new management tool, the organization will indeed do so. In the newer companies, the management is often younger, expects to hold office longer, and takes a longer-range view of developing company strength than in more mature companies. The smaller organizations may be more fluid, so that the rigidities of functional subdivisions are not so much of a handicap. The costs of management systems research are low enough so that they present no great difficulty in an organization as small as one million dollars per year of business.

Because of the organizational hazards and the uncertainties of picking a suitably qualified man, there is no certainty of initial success in any particular industrial dynamics program. However, similar risks are taken in many other places where the potential gains are not as great. One is playing here for stakes comparable with the normally expected profit margins of the industry. The risks and the time and effort must be assessed against the possibility of solving some of today's most vexing problems of management.

APPENDICES

DYNAMO

D YNAMO is a special-purpose compiler for generating the running code to simulate the type of models discussed in this book on an IBM 704, 709, or 7090 digital computer.[1]

The DYNAMO compiler takes equations in the form that has been used in Part III of this book. It checks the equations for the kinds of logical errors that represent inconsistencies within the equation set itself, improper details in equation structure, and card-punching errors that represent impossible situations. An example of such an error response is shown in Figure A-1. Here a set of equations containing numerous logical errors was submitted to the 709 computer under the control of the DYNAMO compiler. Instead of returning a simulation run of the model, an analysis of the inconsistencies in the equations was printed out.

If the model passes the logical checks, the DYNAMO compiler then creates a computer running code. This code is run and the desired data tabulated and the desired graphical plots generated.

The DYNAMO compiler itself consists of about 10,000 instructions that guide the checking, running code generation, model simulation, and output data preparation. The DYNAMO compiler with a 32,000-word internal-memory IBM 709 computer can accept a dynamic model of some 1,500 variables.

DYNAMO generates various special functions and groups of equations in response to functional instructions such as STEP, DELAY3 and RAMP, that were seen in Chapters 15 through 18 as well as others not used in this book.[2]

It should be noted that the equation system specified in this book and mechanized by DYNAMO is suitable for the simulation of any information-feedback system, be it biological, chemical process, river basin control, economic, or industrial.

Figures A-2 through A-12 are photographic reproductions of selected pages of the computer run that produced Figure 18-5. The separate pages of the run are given separate figure numbers for identification. Explanations are in the figure captions.

---

[1] DYNAMO was created over a period of three years with an investment of some six man-years of skilled design by several of the country's experts in digital computer programming. A forerunner to the DYNAMO system was written by Mr. Richard K. Bennett, who developed the version used to produce the data for the computer runs that appeared in Reference 21 in July, 1958. DYNAMO itself was created by Dr. Phyllis Fox (Mrs. George Sternlieb) and Mr. Alexander L. Pugh, III, assisted by Mrs. Grace Duren and Mr. David J. Howard. The automatic curve-plotting features were designed primarily by Mr. Edward B. Roberts. DYNAMO was converted from the 704 to the 709 and 7090 computers by Mr. Pugh.

[2] See Reference 22 for the user's handbook for DYNAMO.

```
DYN   M637
RUN   1989JF/DT=0.1/LENGTH=10/PRTPER=1/PLTPER=1                        2/ 3   124.9 1961

      1. PROBLEM 92, NOTEBOOK 58JF57
      2. FOR FORRESTER BOOK ON INDUSTRIAL DYNAMICS
      3. APPENDIX
      4. SAMPLE OF DYNAMO ERROR COMMENTS

37B   BOX=BOXLIN(3,DT)                                                          JP0001
C     BOX*=1/2/3/4                                                              JP0002
18A   BOX*7.K=(A.K)(X.K+C.K)                                                    JP0003
13A   X.K=(A.K)(C.K)(XYZ)                                                       JP0004
5L    ABCDE.K=ABCDE.J+(DT)((1/30.0E+3)(PQRST.J-ABCDE.J)+(1/20.0E+3)(BOX*        JP0005

CARD NUMBER DOES NOT AGREE

      X1   3.J-ABCDE.J)                                                         JP0006

PQRST AND  PQRSU SHOULD BE THE SAME IN THIS EQN TYPE
      1L   PQRST.K=PQRSU.J+(DT)(IN.JK-OUT.JK)                                   JP0007

DYNAMO FOUND , WHERE IT EXPECTED (
      12A  A.K=(X.K)(C.K)                                                       JP0008

VARIABLE TYPE AND TIME-SUBSCRIPT ON VARIABLE DEFINED DO NOT AGREE
      20R  OUT.K=BAK.K/DEL                                                      JP0009

NO SUCH VARIABLE TYPE EXISTS
      8M   B.K=X.K+Y.K+Z.K                                                      JP0010

I HAS UNALLOWED SUBSCRIPT
      1L   C.K=C.J+(DT)(I.K-O.K)                                               JP0011

NO SUCH VARIABLE TYPE EXISTS
      21-  BOX*1.KL=(1/7)(A.K+B.K)                                              JP0012

BOX HAS DIFFERING LENGTH AND NUMBER OF I.C.
ABCDE REQUIRES INITIAL CONDITIONS
PQRST REQUIRES INITIAL CONDITIONS
C REQUIRES INITIAL CONDITIONS
BOX HAS BOXCAR NUMBER THAT EXCEEDS LENGTH OF TRAIN
XYZ IS UNDEFINED IN EQN FOR    X
NO OUTPUT REQUESTED
```

Figure A-1 Error print-out from DYNAMO compiler.

Statements indicating where logical errors have been found in the equations submitted. The first note refers to a continuation of an equation (an X1 card) in which the continuation card does not have the same equation number as the first card. Most of the other comments from the DYNAMO compiler are self-explanatory.

```
DYN   M637
RUN   1987JF/DT=0.25/LENGTH=0.0/PRTPER=2/PLTPER=2.                    2/ 3  118.7 1961

      1.  PROBLEM 91, NOTEBOOK 58JF55
      2.  FOR FORRESTER BOOK ON INDUSTRIAL DYNAMICS
      3.  CHAPTERS 17 AND 18, ELECTRONIC COMPONENT INDUSTRY
      4.  OLD POLICIES

      ORDER FILLING—FIGURE 17-6, SECTION 17.4.1

2L    RCF.K=RCF.J+(DT)(RRF.JK-RFIF.JK-RMOF.JK+0+0)   - - - - - -     17-  1
12N   RCF=(RRF)(DCPF)                       - - - - - - - - - - -     17-  2
1L    SOF.K=SOF.J+(DT)(RFIF.JK-SIF.JK)      - - - - - - - - - - -     17-  3
20R   SIF.KL=SOF.K/DSF                      - - - - - - - - - - -     17-  4
6N    SIF=RFIF                              - - - - - - - - - - -     17-  5
1L    IAF.K=IAF.J+(DT)(MIF.JK-SIF.JK)       - - - - - - - - - - -     17-  6
12N   IAF=(CIRF)(RRF)                       - - - - - - - - - - -     17-  7
44R   RFIF.KL=(FRFIF.K)(RCF.K)/DCPF         - - - - - - - - - - -     17-  8
12N   RFIF=(CNFIF)(RRF)                     - - - - - - - - - - -     17-  9
12N   RMOF.KL=(1/DCPF)((RCF.K)(1)+(-FRFIF.K)(RCF.K))  - - - - - -     17- 10
22R                                                                   17- 11

      INVENTORY SPLIT FRACTION—FIGURES 17-6 AND 17-7, SECTION 17.4.1

7A    FRFIF.K=CMFIF-AUX3.K                  - - - - - - - - - - -     A17- 12
28A   AUX3.K=(CMFIF)EXP(-AUX2.K)            - - - - - - - - - - -     B17- 12
44A   AUX2.K=(C1)(IAF.K)/CINF               - - - - - - - - - - -     A17- 13
29N   C1=(1)LOGN(AUX1)                      - - - - - - - - - - -     26N
26N   AUX1=(CMFIF+0+0)/(CMFIF-CNFIF+0)      - - - - - - - - - - -     B17- 13
6N    CINF=IAF                                                        17- 14

      INVENTORY REORDERING—FIGURE 17-8, SECTION 17.4.2

27A   MOITF.K=(AUX23.K/TIAF)+ASIF.K         - - - - - - - - - - -     A17- 15
9A    AUX23.K=IDF.K-IAF.K+OINF.K-OIAF.K     - - - - - - - - - - -     B17- 15
51R   MOIF.KL=CLIP(MOITF.K,0,MOITF.K,0)     - - - - - - - - - - -     17- 16
3L    ASIF.K=ASIF.J+(DT)(1/TASIF)(SIF.JK-ASIF.J)  - - - - - - - -     17- 17
6N    ASIF=RFIF                                                       17- 18
12A   IDF.K=(CIRF)(RSF.K)                   - - - - - - - - - - -     17- 19
3L    RSF.K=RSF.J+(DT)(1/TRSF)(RRF.JK-RSF.J)  - - - - - - - - - -     17- 20
6N    RSF=RRF                                                         17- 21
12A   OINF.K=(ASIF.K)(DMIF.K)               - - - - - - - - - - -     17- 22
7A    OIAF.K=BLIF.K-OPIF.K                                            17- 23

      MANUFACTURING—FIGURE 17-9, SECTION 17.4.3

1L    BLIF.K=BLIF.J+(DT)(MOIF.JK-BLIRF.JK)  - - - - - - - - - - -     17- 24
12N   BLIF=(RFIF)(DNBLF)                                              17- 25
1L    BLCF.K=BLCF.J+(DT)(RMOF.JK-PCOF.JK)   - - - - - - - - - - -     17- 26
18N   BLCF=(DNBLF)(RRF-RFIF)                - - - - - - - - - - -     17- 27
7A    BLTF.K=BLIF.K+BLCF.K                                            17- 28
42A   MMBLF.K=BLTF.K/((DMBLF)(CPLF))        - - - - - - - - - - -     17- 29
51A   MBLF.K=CLIP(MENPF.K,MMBLF.K,MMBLF.K,MENPF.K)  - - - - - - -     17- 30
20A   FLIF.K=BLIF.K/BLTF.K                                            17- 31
20A   FLCOF.K=BLCF.K/BLTF.K                 - - - - - - - - - - -     17- 32
13R   PCOF.KL=(FLCOF.K)(MBLF.K)(CPLF)       - - - - - - - - - - -     17- 33
7N    PCOF=RRF-RFIF                                                   17- 34
13A   PIOF.K=(FLIF.K)(MBLF.K)(CPLF)         - - - - - - - - - - -     17- 35
6R    BLIRF.KL=PIOF.K                                                 17- 36
```

Figure A-2 DYNAMO print-out. First page of model equations of Chapter 17.

The first line gives data needed by the computing center, and on the right-hand edge the date, clock time in hours and minutes to the nearest one-tenth minute, and the year. The computer began to process the run at 18.7 minutes after the hour. The second line specifies the conditions for the first machine run. The number 1987 indicates the run number. The solution interval DT is 0.25 week. The run length is given as 0.0, which means that the model itself will not be run but instead the computer will go immediately to the first rerun where operating parameters different from the basic model are specified. This technique is useful when it is desired to specify all input conditions as zero in the basic model, and then to designate as reruns the particular circumstances of interest. PRTPER gives the data print-out interval as once every 2 weeks. PLTPER gives the frequency with which information is to be plotted as once every 2 weeks.

The next four lines are notes giving information about the model. The equations then follow as given in Chapter 17.

Figure A-3 DYNAMO print-out. Second page of model equations of Chapter 17.

The left-hand margin indicates the equation composition and the equation type number. The letters are those used in Part III of this book, indicating the equation type (L for level, A for auxiliary, R for rate, N for initial condition, C for a constant, S for supplementary output information, and X for a continuation card).

DYNAMO in its present form does not contain an algebraic translator to interpret algebraic statements, but instead it uses an equation composition number that is supplied by the person preparing the input cards. From a list of some 60 equation types, the proper equation format is selected. The number indicates the equation format. Occasionally an equation format will not be available in the standard list to fit the desired equation, so that the equation may need to be separated into two or more parts. An example of this is seen in Equation 17-52, where the variable AUX4 substitutes into the equation for AUX5 and this in turn substitutes into Equation A17-52 for RMPF.

Figures A-2 through A-7 reproduce the information from the input cards describing the model and the desired simulation run.

```
PAGE 2      1987JF

 7A    ME1PF.K=MENPF.K-MBLF.K                                       17- 37
12A    PE1F.K=(ME1PF.K)(CPLF)                                       17- 38
 7R    PIF.KL=P1OF.K+PE1F.K                                         17- 39
 6N    PIF=RF1F                                                     17- 40
 1L    OPIF.K=OPIF.J+(DT)(PIF.JK-MIF.JK)                            17- 41
12N    OPIF=(DPF)(RF1F)                                             17- 42
39R    MIF.KL=DELAY3(PIF.JK,DPF)                                    17- 43
 1L    OPCF.K=OPCF.J+(DT)(PCOF.JK-SMOF.JK)                          17- 44
18N    OPCF=(DPF)(RRF+RF1F)                                         17- 45
39R    SMOF.KL=DELAY3(PCOF.JK,DPF)                                  17- 46
 2L    RMSF.K=RMSF.J+(DT)(RMRF.JK-PIF.JK-PCOF.JK+0+0+0)             17- 47
12N    RMSF=(RRF)(CRMSF)                                            17- 48
27A    DM1F.K=(BL1F.K/P1OF.K)+DPF                                   17- 49
27A    DMCOF.K=(BLCF.K/PCOF.JK)+DPF                                 17- 50

MATERIAL ORDERING--FIGURE 17-10, SECTION 17.4.4

12A    RMDF.K=(RSF.K)(CRMSF)                                        17- 51
 8R    RMPF.KL=PCOF.JK+PIF.JK+AUX5.K                               A17- 52
16A    AUX5.K=(AUX4.K)(RMDF.K)+(-AUX4.K)(RMSF.K)+(AUX4.K)(RMPNF.K)+(-AUX4  B17- 52
X1     .K)(RMPAF.K)                                                C17- 52
20A    AUX4.K=1./TRMAF                                             B17- 52
 6N    RMP=RRF                                                     C17- 52
12A    RMPNF.K=(RSF.K)(DRMF)                                        17- 53
 1L    RMPAF.K=RMPAF.J+(DT)(RMPF.JK-RMRF.JK)                        17- 54
12N    RMPAF=(RRF)(DRMF)                                            17- 55
39R    RMRF.KL=DELAY3(RMPF.JK,DRMF)                                 17- 57

LABOR SECTOR--FIGURE 17-11 AND 17-12, SECTION 17.4.5

 1L    LTF.K=LTF.J+(DT)(LHF.JK-LEPF.JK)                             17- 58
 6N    LTF=0                                                        17- 59
39R    LEPF.KL=DELAY3(LHF.JK,DLTF)                                  17- 60
 1L    MENPF.K=MENPF.J+(DT)(LEPF.JK-LDNF.JK)                        17- 61
20N    MENPF=RRF/CPLF                                               17- 62
 1L    LLF.K=LLF.J+(DT)(LDNF.JK-LTERF.JK)                           17- 63
 6N    LLF=0                                                        17- 64
39R    LTERF.KL=DELAY3(LDNF.JK,DLLF)                                17- 65
20A    LASF.K=RSF.K/CPLF                                            17- 66
12A    BLNF.K=(RSF.K)(DNBLF)                                        17- 67
42A    LBLAF.K=AUX6.K/((CPLF)(TBLAF))                              A17- 68
 7A    AUX6.K=LBTF.K-BLNF.K                                        B17- 68
 8A    LDF.K=LASF.K+LBLAF.K-MEIPF.K                                 17- 69
 7A    LCF.K=LDF.K-LAF.K                                            17- 70
 7A    LAF.K=LTF.K+MENPF.K                                          17- 71
22A    LCHF.K=(1/LLCF)(LCF.K)(1,0)+(-CLCTF)(MENPF.K)                17- 72
51R    LHF.KL=CLIP(LCHF.K,0,LCHF.K,0)                               17- 73
 6N    LHF=0                                                        17- 74
22A    LCDF.K=(1/LLCF)(LCF.K)(1,0)+(CLCTF)(MENPF.K)                 17- 75
51R    LDNF.KL=CLIP(0,LCDF.K,LCDF.K,0)                              17- 76
 6N    LDNF=0                                                       17- 77
 8A    MENTF.K=LTF.K+MENPF.K+LLF.K                                  17- 78
 1L    LCTF.K=LCTF.J+(DT)(LHF.JK+LDNF.JK)                           17- 79
 6N    LCTF=0                                                       17- 80

DELIVERY DELAY--FIGURE 17-13, SECTION 17.4.6

14A    DVF.K=AUX7.K+(DSF)(FRF1F.K)                                 A17- 81
18A    AUX7.K=(DMCOF.K)(1-FRF1F.K)                                 B17- 81
```

1987JF

```
7R   DFOF.KL=DCPF+DVF.K                                         17- 82
3L   DQDF.K=DQDF.J+(DT)(1/TAQDF)(DFOF.JK-DQDF.J)     -- -- --   17- 83
7N   DQDF=AUX8A+DCPF                                            A17- 84
14N  AUX8A=AUX8+(CNFIF)(DSF)                          -- -- --  B17- 84
16N  AUX8=(DPF)(1)+(-CNFIF)(DPF)+(DNBLF)(1)+(-CNFIF)(DNBLF)     C17- 84

     CUSTOMER ORDERING--FIGURE 17-14, SECTION 17.4.7

1L   EDPC.K=EDPC.J+(DT)(INPUT.JK-PDC.JK)                        17- 85
12N  EDPC=(INPUT)(DEEDC)                                        17- 86
1L   RCC.K=RCC.J+(DT)(PDC.JK-RRF.JK)                            17- 87
12N  RCC=(INPUT)(DRCC)                               -- -- --   17- 88
7A   DAERC.K=DRCC+AUX20.K                                       A17- 89
18A  AUX20.K=(DQDF.K)(1+CISTC)                                  B17- 89
7R   DDEDC.KL=DTC-DAERC.K                                       17- 90
8N   DTC=DNEDC+DRCC+AUX22                                       A17- 91
18N  AUX22=(DQDF)(1+CISTC)                                      B17- 91
3L   DEEDC.K=DEEDC.J+(DT)(1/TAEDC)(DDEDC.JK-DEEDC.J) -- -- --   A17- 92
8N   DEEDC=DTC-AUX21-DRCC                                       A17- 93
18N  AUX21=(DQDF)(1+CISTC)                                      B17- 93
44R  PDC.KL=(EDPC.K)(AUX9.K)/DEEDC.K                            A17- 94
7A   AUX9.K=1+NEDC.K                                            B17- 94
44R  RRF.KL=(RCC.K)(AUX10.K)/DRCC                               A17- 95
7A   AUX10.K=1+NPC.K                                            B17- 95
6N   RRF=INPUT                                                  17- 96

     CASH FLOW--FIGURE 17-15, SECTION 17.4.8

12R  RMIF.KL=(RMRF.JK)(CRMPF)                                   17- 97
1L   APF.K=APF.J+(DT)(RMIF.JK-RMCEF.JK)                         17- 98
13N  APF=(RRF)(CRMPF)(DAPF)                                     17- 99
20R  RMCEF.KL=APF.K/DAPF                                        17-100
18R  FGIF.KL=(CFGPF)(SIF.JK+SMOF.JK)                            17-101
12N  FGIF=(RRF)(CFGPF)                                          17-102
1L   ARF.K=ARF.J+(DT)(FGIF.JK-FGCRF.JK)              -- --      17-103
13N  ARF=(RRF)(CFGPF)(DARF)                                     17-104
39R  FGCRF.KL=DELAY3(FGIF.JK,DARF)                              17-105
12R  LCEF.KL=(MENTF.K)(CWRF)                                    17-106
12N  LCEF=(MENPF)(CWRF)                                         17-107
12R  ITAXF.KL=(0.5)(PBTRF.K)                                    17-108
6R   SDIVF.KL=SDLF.K                                            17-109
1L   SDTDF.K=SDTDF.J+(DT)(SDIVF.JK+0.0)                         17-110
6N   SDTDF=0                                                    17-111
4L   CASHF.K=CASHF.J+(DT)(1/1)(FGCRF.JK-RMCEF.JK-LCEF.JK-CCEF.JK-ITAXF.JK-  17-112
X1   SDIVF.JK)                                                  17-112
13N  CASHF=(CNCSF)(CFGPF)(RRF)                                  17-113

     PROFIT, DIVIDENDS--FIGURE 17-16, SECTION 17.4.9

27N  SICF=(CWRF/CPLF)+CRMPF                                     17-114
8A   PBTRF.K=AUX17.K-AUX18.K-CCEF                               A17-115
18A  AUX17.K=(AX17A.K)(CFGPF-SICF)                             B17-115
7A   AX17A.K=SIF.JK+SMOF.JK                                     C17-115
18A  AUX18.K=LCEF.JK-AUX16.K                                    D17-115
20N  AUX16.K=(AUX15)(MIF.JK+SMOF.JK)                            E17-115
18A  AUX15=CWRF/CPLF                                            F17-115
12R  NPRF.KL=(0.5)(PBTRF.K)                                     17-116
6N   NPRF=SDLF                                                  A17-EX
1L   NPTDF.K=NPTDF.J+(DT)(NPRF.JK+0.0)                          17-117
```

Figure A-4   DYNAMO print-out. Third page of model equations of Chapter 17.

198?JF

```
6N   NPTDF=0                                                                    17-118
3L   SDLF.K=SDLF.J+(DT)(1/TASDL)(NPRF.JK-SDLF.J)                                17-119
14N  SDLF=AUX13+(-0.5)(CCEF)                                                   A17-120
17N  AUX13=(RRF)(0.5)(CFGPF)+(RRF)(0.5)(-CRMPF)+(RRF)(0.5)(-AUX12)            B17-120
20N  AUX12=CWRF/CPLF                                                          C17-120

MODEL CONSTANTS----DEFINED AFTER EQUATION NUMBER

C    CEF=30000        17-112
C    CFGPF=100        17-101
C    CIRF=4           17- 19
C    CISTC=0.5        17- 89
C    CLCTF=0.0        17- 72
C    CMFIF=0.8        17- 12
C    CNCSF=1.0        17-113
C    CNFIF=0.7        17- 29
C    CPLF=2.66666     17- 97
C    CRMPF=20         17- 51
C    CRM5F=6          17-106
C    CWRF=80          17-105
C    DAPF=3           17-  9
C    DCPF=1           17- 65
C    DLLF=4           17- 60
C    DLTF=3           17- 29
C    DMBLF=1.0        17- 67
C    DNBLF=4          17- 91
C    DNEDC=30         17- 43
C    DPF=6            17- 95
C    DRCC=3           17- 57
C    DRMF=3           17-  5
C    DSF=1            17- 92
C    TAEDC=15         17- 83
C    TAQDF=4          17-119
C    TASDL=52         17- 17
C    TASIF=2          17- 68
C    TBLAF=20         17- 15
C    TIAF=6           17- 72
C    TLCF=10          17- 52
C    TRMAF=8          17- 20
C    TRSF=15          17- 20

SUPPLEMENTARY EQUATIONS FOR OUTPUT INFORMATION, SECTION 17.5

46S  BLTPC.K=(BLTF.K)(100.0)(1.0)/((CINPI)(DNBLF))(1.0)            17-121
46S  CASPC.K=(CASHF)(100.0)(1.0)/((CNCSF)(CFGPF)(CINPI))          17-122
12S  FRFPC.K=(FRFIF.K)(100.0)                                     17-123
46S  IAFPC.K=(IAF.K)(100.0)(1.0)/((CIRF)(CINPI)(1.0))            17-124
46S  IDFPC.K=(IDF.K)(100.0)(1.0)/((CIRF)(CINPI)(1.0))            17-125
46S  INPPC.K=(INPUT.JK)(100.0)(1.0)/((CINPI)(1.0)(1.0))          17-126
46S  MENPC.K=(MENPF.K)(CPLF)(100.0)/((CINPI)(1.0)(1.0))          17-127
46S  NPRPC.K=(NPRF.JK)(200.0)(1.0)/((AAA4)(1.0)(1.0))           A17-128
18N  AAA3=(CINPI)(CFGPF-SICF)                                    B17-128
7N   AAA4=AAA3-CCEF                                              C17-128
46S  RRFPC.K=(RRF.JK)(100.0)(1.0)/((CINPI)(1.0)(1.0))            17-129
7S   SCF.K=SIF.JK+SMOF.JK                                        17-130
12S  VIAF.K=(IAF.K)(SICF)                                        17-131
12S  VIPIF.K=(D132.K)(C132)                                     A17-132
42N  B132=CWRF/((2.0)(CPLF))                                    B17-132
```

Figure A-5 DYNAMO print-out. Fourth page of model equations of Chapter 17 and values of parameters. The notes after the values of parameters tell where in the text of Chapter 17 the parameter is discussed.

Figure A-6 DYNAMO print-out. Fifth page of model equations of Chapter 17 and print and plot instructions.

The first print card specifies, for example, that in the first column the variable BLTPC will be printed. The number in parentheses (0.2) species that the value will not be multiplied by a power of 10 for scaling and that two decimal places will be printed to the right of the decimal point. The star indicates that the second position in the first column is to be left blank. The remainders of the first and second lines are likewise specified. The cards with an initial X are continuations of the preceding card. The following print instructions do not specify scaling, and this will be automatically provided by DYNAMO to make the best use of the available five decimal digits that are allowed in the printing format.

The plot cards indicate several graphs of variables versus time, with up to 10 variables per plot. The name of the variable is specified and the symbol which is to be used for indicating that variable. If the plotting scales are not indicated, DYNAMO will automatically scale to ensure that the width of the page is effectively used and that the full range of the variable falls on the plotted page. This is done by scanning the computed data to find the maximum and minimum values before creating the control information for the plotting process.

```
7N    C132=CRMPF+B132                                    C17-132
7A    D132.K=OPIF.K+OPCF.K                               D17-132
12S   VRMSF.K=(RMSF.K)(CRMPF)                            17-133
8S    VTIF.K=VIAF.K+VIPF.K+VRMSF.K                       17-134
9S    SURPL.K=VTIF.K+ARF.K+CASHF.K-APF.K                 17-135

      INPUT TEST FUNCTIONS--FIGURE 17-17, SECTION 17.6

7R    INPUT.KL=CINPI+INPCH.K                      - - - - - - - - -    17-136
6N    INPUT=CINPI                                 - - - - - - - - -    17-137
9A    INPCH.K=STP1.K+GTH1.K+GTH2.K+SNE.K          - - - - - - - - -    17-138
45A   STP1.K=STEP(STH1,STT1)                      - - - - - - - - -    17-139
47A   GTH1.K=RAMP(GTHR1,GTHT1)                     - - - - - - - -    17-140
47A   GTH2.K=RAMP(GTHR2,GTHT2)                     - - - - - - - -    17-141
31A   SNE.K=(SIH)SIN(2PI)(TIME.K)/PER)            - - - - - - - - -    17-142
43A   NEDC.K=SAMPLE(NNEDC.K,CNSEC)                 - - - - - - -      17-143
31A   NNEDC.K=(1)NORMRN(0.0,CNAEC)                 - - - - - -        17-144
34A   NPC.K=SAMPLE(NNPC.K,CNSPC)                   - - - - - -        17-145
34A   NNPC.K=(1)NORMRN(0.0,CNAPC)                  - - - - - -        17-146

      TEST CONDITIONS-- DEFINED AFTER EQUATION NUMBER

C     CINPI=1000           17-137
C     CNAEC=0              17-144
C     CNAPC=0              17-146
C     CNSEC=0              17-145
C     CNSPC=0              17-143
C     GTHR1=0.0            17-140
C     GTHR2=0.0            17-141
C     GTHT1=0.0            17-140
C     GTHT2=0.0            17-141
C     PER=52               17-142
C     SIH=0                17-142
C     STH1=0               17-139
C     STT1=0               17-139

PRINT 1)BLTPC(0.2),*/2)CASPC(0.2),*/3)DQDF(0.2),*/4)FRRPC(0.2),*/5)IAFPC
X1    (0.2),*/6)IDFPC(0.2),*/7)INPPC(0.2),*/8)MENPC(0.2),*/9)NPRPC(0.2),*
X2    */10)RRFPC(0.2),*/11)NEDC(0.3),*/12)NPC(0.3),*/13)PDC(0.1),*/14)*,
X3    *
PRINT 1)APF/2)ARF/3)ASIF/4)BLCF/5)BLIF/6)BLIRF/7)BLTF/8)CASHF
PRINT 9)DDEDC/10)DEEDC/11)DFOF/12)DMCOF/13)DQDF/14)DVF/1)EDPC/2)FGCRF
PRINT 3)FRFIF/4)IAF/5)IDF/6)INPUT/7)LTAXF/8)LASF/9)LBLAF/10)LCEF
PRINT 11)LCF/12)LCTF/13)LDF/14)LDNF/1)LHF/2)MEIPF/3)MENPF/4)MENTF
PRINT 5)M1F/6)MMBLF/7)MOIF/8)NEDC/9)NPC/10)NPRF/11)NPTDF/12)OIAF
PRINT 13)OINF/14)PCOF/1)PDC/2)PEIF/3)PIF/4)PIOF/5)RCC/6)RCF
PRINT 7)RFIF/8)RMCEF/9)RMOF/10)RMPF/11)RMRF/12)RMSF/13)RRF
PRINT 14)RSF/1)SCF/2)SDIVF/3)SDLF/4)SDTDF/5)SIF/6)SMOF/7)SOF
PRINT 8)SURPL/9)VIAF/10)VIPIF/11)VRMSF/12)VTIF

PLOT  MOIF=0,PDC=P,RMOF=M,RMPF=S/NEDC=N1-0.5,0.5)/LBLAF=B/BLCF=L,BLIF=1/
X1    PIF=K,PCOF=G
PLOT  RMCEF=7(10000,30000.0)/SDIVF=D,ITAXF=T(0.0,20000.0)/FGCRF=R(90000,
X1    110000.0)/LCEF=W,CCEF=F(20000,40000.0)
PLOT  CASHF=C,VRMSF=M,APF=D(0,2000000)/ARF=R,VTIF=I(400000,600000)/VIAF=Q
X1    ,VIPIF=P(100000,300000)/SURPL=S(970000,1170000)
PLOT  IAF=Q/RRF=R/DQDF=L/MENPF=M/BLTF=B/LNF=H,LDNF=F/FRFIF=S/DEEDC=D/NPR
X1    F=P
PLOT  IAFPC=Q,IDFPC=T,INPPC=I,MENPC=M,RRFPC=R(50,150)/BLTPC=B,CASPC=C,NP
```

Figure A-7 DYNAMO print-out. End of model statement with no run as part of the basic model, followed by rerun instructions.

The last plot card calls for the grouping of variables into sets that are to be plotted to the same scales which are here numerically specified. Either or both ends of the plotting scale can be specified by the DYNAMO user or be left for automatic scaling on the basis of the actual data of the run.

At 19.1 minutes (0.5 minute after starting) model checking was complete, and code generation started.

The second group gives the information for a rerun with the new values of any parameters which are to be changed (here the noise-generation coefficients) with both new and old values. Running code had been created and then reset for the rerun in a total of 0.2 minute. With computer coding requiring such a short time, instruction codes for machine running are never saved. It is less expensive to re-create machine codes than to determine whether or not model changes have been made that would make obsolete the code last used.

```
PAGE 6    1987JF
   X1    RPC=P10.2001/DQDF=D10.8)/FRPPC=S(0.80)

STARTED GENERATING CODE AT    2/ 3  119.1
CODE STARTS IN / 13163.THERE ARE 1047   LINES OF CODING


PAGE 7    1988JF
   RUN   1988JF/DT=0.25/LENGTH=350/PRTPER=2/PLTPER=2

   1. PROBLEM 91, NOTEBOOK 58JF55
   2. FOR FORRESTER BOOK ON INDUSTRIAL DYNAMICS
   3. CHAPTERS 17 AND 18, ELECTRONIC COMPONENT INDUSTRY
   4. OLD POLICIES
   5. NOISE AT CUSTOMER ENGINEERING

        CNSEC   CNAEC
1988JF  5.000   .1000
1987JF   .000   .0000
STARTED TO RUN CODE AT   2/ 3   119.3
```

```
PAGE 8    1988JF    RATES .KL PRINTED.    STARTED PRINTING AT    2/ 3    120.7

TIME   BLTPC  CASPC  DQDF   FRFPC  IAFPC  IDFPC  INPPC  MENPC  NPRPC  RRFPC  NEDC   NPC    PDC    DVF

       APF    ARF    ASIF   BLCF   BLIF   BLIRF  BLTF   CASHF  DDEDC  DEEDC  DFOF   DMCOF  DQDF   DVF
       EDPC   FGCRF  FRFIF  IAF    IDF    INPUT  ITAXF  LASF   LBLAF  LCETF  LCF    LCTF   LDF    LDNF
       LHF    MEIPF  MENPF  MENTF  MIF    MMBLF  MOIF   NEDC   NPC    NPRF   NPTDF  OIAF   OINF   PCOF
       PDC    PEIF   PIF    PIOF   RCC    RCF    RFIF   RMCEF  RMOF   RMPF   RMPF   RMSF   RRF    RSF
       SCF    SDIVF  SDLF   SDTDF  SIF    SMF    SOF    SURFL  VIAF   VIPIF  VRMSF  VTIF

E+00   E+00   E+00   E+00   E+00   E+00   E+00   E+00   E+00   E+00   E+00   E+00   E+00   E+00   E+00

       E+03   E+03   E+00   E+00   E+00   E+00   E+00   E+03   E+00   E+00   E+00   E+00   E+00   E+00
       E+03   E+03   E+00   E+00   E+00   E+00   E+00   E+00   E+00   E+03   E+00   E+00   E+00   E+00
       E+00   E+00   E+00   E+00   E+00   E+00   E+00   E+00   E+00   E+00   E+00   E+03   E+00   E+00
       E+00   E+03   E+00   E+03   E+00   E+00   E+00   E+03   E+00   E+03   E+00   E+00   E+00   E+00
       E+00   E+03   E+03   E+03   E+00   E+00   E+00   E+03   E+03   E+03   E+03   E+03

.00    100.00 100.00 4.70   70.00  100.00 100.00 100.00 100.00 100.00 100.00 .000   .000   1000.0

       30.000 500.00 700.000 1200.00 2800.0 700.00 4000.0 100.00 30.000 30.000 4.7000 10.000 4.7000 3.7000
       30.000 100.00 .70000 4000.00 4000.0 1000.00 10.000 375.00 .000  30.000 .00   .00   375.00 .000
       .0000  .000   375.00 375.00  700.00 1500.00 700.0  .00000 0.    10.000 80    7.000 7000.00 300.00
       1000.0 10.000 700.00 700.00  3000.00 1000.0 700.00 20.000 300.00 1000.0 1000.00 6000.00 1000.0 1000.0
       1000.0        10.000 .00     700.00 300.00 700.00 1070.0 200.00 210.00 120.00 530.00

2.00   100.00 100.00 4.70   70.00  100.00 100.00 100.00 100.00 100.00 100.00 .000   .000   1000.0

       60.000 500.00 700.000 1200.00 2800.0 700.00 4000.0 100.00 30.000 30.000 4.7000 10.000 4.7000 3.7000
       30.000 100.00 .70000 4000.00 4000.0 1000.00 10.000 375.00 .000  30.000 .00   .00   375.00 .000
       .0000  .000   375.00 375.00  700.00 1500.00 700.0  .00000 0.    10.000 20    7.000 7000.0 300.00
       1000.0 10.000 700.00 700.00  3000.00 1000.0 700.00 20.000 300.00 1000.0 1000.00 6000.00 1000.0 1000.0
       1000.0        10.000 .00     700.00 300.00 700.00 1070.0 200.00 210.00 120.00 530.00

4.00   100.00 100.00 4.70   70.00  100.00 100.00 100.00 100.00 100.00 100.00 .000   .000   1000.0

       60.000 500.00 700.000 1200.00 2800.0 700.00 4000.0 100.00 30.000 30.000 4.7000 10.000 4.7000 3.7000
       30.000 100.00 .70000 4000.00 4000.0 1000.00 10.000 375.00 .000  30.000 .00   .00   375.00 .000
       .0000  .000   375.00 375.00  700.00 1500.00 700.0  .00000 0.    10.000 40    7.000 7000.0 300.00
       1000.0 10.000 700.00 700.00  3000.00 1000.0 700.00 20.000 300.00 1000.0 1000.00 6000.00 1000.0 1000.0
       1000.0        10.000 .00     700.00 300.00 700.00 1070.0 200.00 210.00 120.00 530.00

6.00   100.02 100.00 4.70   70.00  100.00 100.11 100.89 100.00 100.13 104.04 .139   .000   1134.2

       60.000 500.00 700.06  1200.8  2800.1 699.87 4000.9 100.00 30.000 30.000 4.7007 10.002 4.7000 3.7007
       29.863 100.00 .69999 3999.9  4004.3 1000.0 10.013 375.41 -.064 30.001 .34   .01   375.34 .000
       .0335  .000   375.01 375.01  700.00 1500.4 701.0  .13940 0.    10.013 60.0  7.000 7001.3 300.13
       1134.2 10.000 699.87 699.87  3121.1 1013.7 709.59 20.000 304.12 1000.3 1000.00 6000.0 1040.4 1001.1
       1001.7        10.000 60.0    701.73 300.00 701.73 1070.0 199.99 210.00 120.00 529.99

8.00   101.18 99.54  4.70   69.87  99.39  100.89 100.00 100.00 106.16 108.49 .139   .000   1124.3

       60.001 501.81 709.54  1222.0  2825.2 698.08 4047.2 99.54  29.995 30.000 4.7263 10.049 4.7035 3.7263
       29.603 100.04 .69872 3975.6  4035.7 1000.0 10.616 378.35 .215  30.023 3.28  .28   378.56 .000
       .3278  .000   375.28 375.28  699.97 1517.7 737.1  .13940 0.    10.616 80.5  7.023 7128.8 301.95
       1124.3 10.009 698.08 698.08  3254.8 1065.1 744.19 20.000 320.88 1002.5 1000.2 6000.1 1084.9 1008.9
       1028.9        10.009 80.0    728.91 300.03 728.91 1070.1 198.78 210.00 120.00 528.78
```

Figure A-8 DYNAMO print-out. First page of tabular data for Figure 18-5.

Printing started at 20.7 minutes, indicating that 1.4 minutes were required for evaluating the set of equations of Figures A-2 through A-7 each 0.25 week for 350 weeks, or 1,400 times. The first block gives the names of the variables in the arrangement for which values are printed in succeeding blocks. The second block gives the power of 10 by which the numerical values are to be multiplied. Succeeding blocks give numerical values at the specified printing interval.

PAGE 37 1988JF

TIME	BLTPC	CASPC	DQDF	FRFPC	IAFPC	IDFPC	INPPC	MENPC	NPRPC	RRFPC	NEDC	NPC	PDC	
	APF	ARF	ASIF	BLCF	BLIF	BLIRF	BLTF	CASHF	DDEDC	DEEDC	DFOF	DMCOF	DQDF	DVF
	EDPC	FGCRF	FRFIF	IAF	IDF	INPUT	ITAXF	LASF	LBLAF	LCEF	LCF	LCTF	LDF	LDNF
	LHF	MEIPF	MENPF	MENTF	MIF	MMBLF	MOIF	NEDC	NPC	NPRF	NPTDF	OIAF	OINF	PCOF
	PDC	PEIF	PIF	PIOF	RCC	RCF	RFIF	RMCEF	RMOF	RMPF	RMRF	RMSF	RRF	RSF
	SCF	SDIVF	SDLF	SDTDF	SIF	SMOF	SOF	SURPL	VIAF	VIPIF	VRMSF	VTIF		
346.00	85.65	97.16	4.37	70.77	103.84	101.40	100.00	98.20	94.30	89.47	-.240	.000	737.1	
	59.585	516.32	712.60	937.7	2488.5	713.28	3426.2	97.16	30.500	30.309	4.4842	9.498	4.3666	3.4842
	29.406	105.81	.70767	4153.5	4056.1	1000.0	9.430	380.26	-11.810	29.538	-.07	756.79	368.45	.007
	.0000	.000	368.27	369.23	707.31	1284.8	697.5	-.24026	0.	9.430	3390.3	6.755	6761.8	268.76
	737.1	.00	713.28	713.28	2684.0	942.3	666.87	19.862	275.48	980.8	984.6	6214.0	894.7	1014.0
	968.7	10.458	10.458	3366.4	686.24	282.48	686.24	1092.8	207.68	207.00	124.28	538.96		
348.00	82.36	108.94	4.40	71.18	106.04	99.41	100.00	98.08	79.66	81.87	-.240	.000	749.5	
	59.303	497.52	678.38	920.1	2374.1	706.86	3294.2	108.94	30.444	30.329	4.4118	9.368	4.4039	3.4118
	29.921	102.67	.71180	4241.5	3976.3	1000.0	7.966	372.78	-12.790	29.498	-7.92	757.39	359.99	.792
	.0000	.000	367.81	368.73	710.71	1235.3	584.9	-.24026	0.	7.966	3408.0	6.644	6348.7	273.96
	749.5	.00	706.86	706.86	2456.2	853.0	607.18	19.768	245.84	974.6	982.0	6217.4	818.7	994.1
	909.3	10.397	10.397	3387.3	634.07	275.19	634.07	1090.2	212.07	206.58	124.35	543.00		
350.00	71.22	117.11	4.37	72.03	110.94	96.99	100.00	97.40	67.18	78.70	-.041	.000	960.9	
	59.065	474.09	633.15	835.2	2013.6	688.42	2848.8	117.11	30.489	30.345	4.2222	8.946	4.3741	3.2222
	30.411	98.70	.72034	4437.5	3879.7	1000.0	6.718	363.72	-19.328	29.467	-20.87	760.05	344.39	2.087
	.0000	.000	365.24	368.34	711.28	1068.3	438.3	-.04116	0.	6.718	3422.7	6.262	5650.9	285.54
	960.9	.00	688.42	688.42	2361.0	802.4	578.01	19.688	224.40	962.4	977.2	6221.0	787.0	969.9
	863.9	10.282	10.282	3408.0	591.56	272.38	591.56	1084.6	221.87	206.12	124.42	552.42		

Figure A-9 DYNAMO print-out. Last page of tabular data for Figure 18-5.

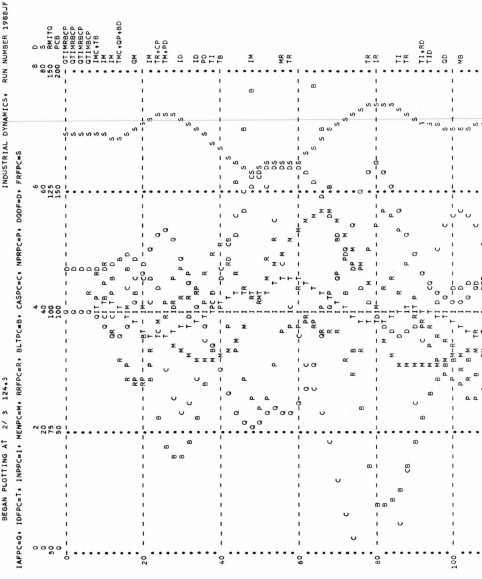

Figure A-10 DYNAMO print-out. First page of plot for Figure 18-5. This is the last plot as requested in Figures A-6 and A-7.

At the beginning of the plot are printed the designations of the variables and the symbols used to represent the variables on the graph. Next come the scales to which the variables are plotted. In the main body of the plot, time is printed along the bottom. The letter groups on the top margin indicate where curves have crossed on the plot. Within each group the first letter designates the letter that appears on the plot, and the following letters are those that are also represented by the first letter. A colored line can be run through each particular letter to make the plot easier to read.

379

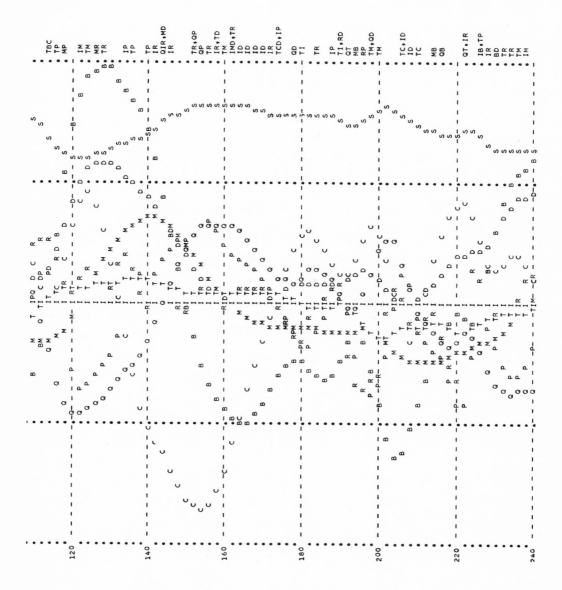

Figure A-11 DYNAMO print-out.
Second page of plot for
Figure 18-5.

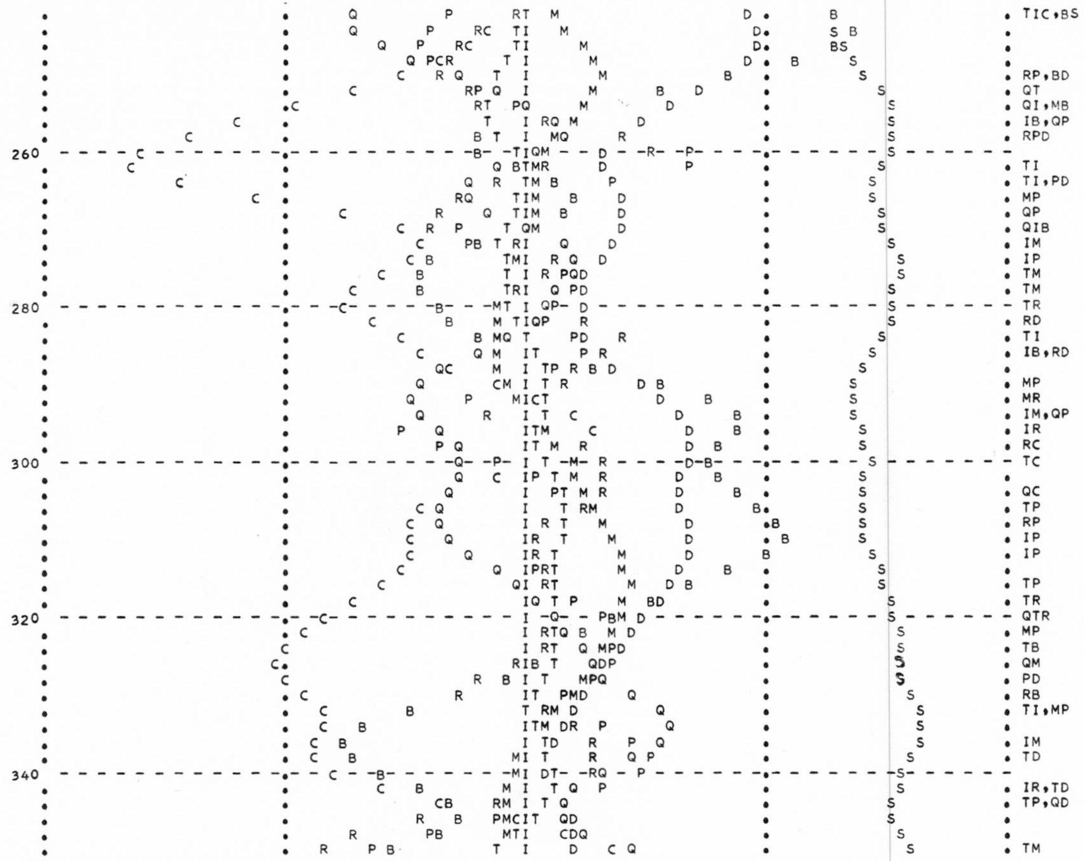

FINISHED RUNNING RUN NUMBER 1988JF AT 2/ 3 124.8

Figure A-12 DYNAMO print-out. Third page of plot for Figure 18-5.

The run and the preparation of the magnetic tape for off-line printing were completed at 24.8 minutes after the hour, 6.1 minutes after the reading of input information was started. The control cards could continue with groups for other parameter values, and these succeeding runs would be processed continuously as for the first rerun appearing in Figures A-7 through A-12.

Model Tabulations

FOR convenient reference, the models that were developed in Chapters 15 through 18 are here tabulated to give a consolidation of the applicable equations and parameters.

The production-distribution system of Chapters 2 and 15 follows in Figures B-1 through B-3.

The advertising addition and the production-distribution system of Chapters 2 and 16 are given in Figures B-4 through B-7.

The model of the electronic component industry of Chapters 17 and 18 with old policies was used in the example in Appendix A and appears in Figures A-2 through A-7.

The model of the electronic component industry with new policies of Chapter 18 is given in Figures B-8 through B-12.

DYN
RUN OO43
1991JF/DT=0.05/LENGTH=0.0/PRTPER=1/PLTPER=1

2/ 3 125.4 1961

```
1. PROBLEM 94, NOTEBOOK 58JF61
2. FOR FORRESTER BOOK ON INDUSTRIAL DYNAMICS
3. CHAPTER 15, PRODUCTION-DISTRIBUTION SYSTEM

RETAIL SECTOR—FIGURE 15-14, SECTION 15.5.1

1L   UOR.K=UOR.J+(DT)(RRR.JK-SSR.JK)                                  15- 1
1L   IAR.K=IAR.J+(DT)(SRR.JK-SSR.JK)                                  15- 2
20A  STRR.K=UOR.K/DFR.K                                               15- 3
20A  NIR.K=IAR.K/DT                                                   15- 4
51R  SSR.KL=CLIP(STR.K,NIR.K,NIR.K,STR.K)                             15- 5
27A  DFR.K=(MNR.K/IAR.K)+DMR                                         A15- 6
12A  MNR.K=(DUR)(IDR.K)                                              B15- 6
12A  IDR.K=(AIR)(RSR.K)                                               15- 7
3L   RSR.K=RSR.J+(DT)(1/DRR)(RRR.JK-RSR.J)                            15- 8
25R  PDR.KL=RRR.JK+(1/DIR)(IDR.K-IAR.K+LDR.K-LAR.K+UOR.K-UNR.K)       15- 9
16A  LDR.K=(RSR.K)(DCR)+(RSR.K)(DMR)+(RSR.K)(DFD.K)+(RSR.K)(DTR)      15-10
9A   LAR.K=CPR.K+PMR.K+UOD.K+MTR.K                                    15-11
18A  UNR.K=(RSR.K)(DHR+DUR)                                           15-12
1L   CPR.K=CPR.J+(DT)(PDR.JK-PSR.JK)                                  15-13
39R  PSR.KL=DELAY3(PDR.JK,DCR)                                        15-14
1L   PMR.K=PMR.J+(DT)(PSR.JK-RRD.JK)                                  15-15
39R  RRD.KL=DELAY3(PSR.JK,DMR)                                        15-16
1L   MTR.K=MTR.J+(DT)(SSD.JK-SRR.JK)                                  15-17
39R  SRR.KL=DELAY3(SSD.JK,DTR)                                        15-18

DISTRIBUTOR SECTOR—FIGURE 15-15, SECTION 15.5.2

1L   UOD.K=UOD.J+(DT)(RRD.JK-SSD.JK)                                  15-19
1L   IAD.K=IAD.J+(DT)(SRD.JK-SSD.JK)                                  15-20
20A  STD.K=UOD.K/DFD.K                                                15-21
20A  NID.K=IAD.K/DT                                                   15-22
51R  SSD.KL=CLIP(STD.K,NID.K,NID.K,STD.K)                             15-23
27A  DFD.K=(MOD.K/IAD.K)+DHD                                         A15-24
12A  MOD.K=(DUD)(IDD.K)                                              B15-24
12A  IDD.K=(AID)(RSD.K)                                               15-25
3L   RSD.K=RSD.J+(DT)(1/DRD)(RRD.JK-RSD.J)                            15-26
25R  PDD.KL=RRD.JK+(1/DID)(IDD.K-IAD.K+LDD.K-LAD.K+UOD.K-UND.K)       15-27
16A  LDD.K=(RSD.K)(DCD)+(RSD.K)(DMD)+(RSD.K)(DFF.K)+(RSD.K)(DTD)      15-28
9A   LAD.K=CPD.K+PMD.K+UOF.K+MTD.K                                    15-29
18A  UND.K=(RSD.K)(DHD+DUD)                                           15-30
1L   CPD.K=CPD.J+(DT)(PDD.JK-PSD.JK)                                  15-31
39R  PSD.KL=DELAY3(PDD.JK,DCD)                                        15-32
1L   PMD.K=PMD.J+(DT)(PSD.JK-RRF.JK)                                  15-33
39R  RRF.KL=DELAY3(PSD.JK,DMD)                                        15-34
1L   MTD.K=MTD.J+(DT)(SSF.JK-SRD.JK)                                  15-35
39R  SRD.KL=DELAY3(SSF.JK,DTD)                                        15-36

FACTORY SECTOR—FIGURE 15-16, SECTION 15.5.3

1L   UOF.K=UOF.J+(DT)(RRF.JK-SSF.JK)                                  15-37
1L   IAF.K=IAF.J+(DT)(SRF.JK-SSF.JK)                                  15-38
20A  STF.K=UOF.K/DFF.K                                                15-39
20A  NIF.K=IAF.K/DT                                                   15-40
51R  SSF.KL=CLIP(STF.K,NIF.K,NIF.K,STF.K)                             15-41
```

Figure B-1 Production-distribution model of Chapter 15, first page.

```
1991JF

27A   DFF.K=(MPF.K/IAF.K)+DHF                                    A15-42
12A   MPF.K=(DUF)(IDF.K)                                         B15-42
12A   IDF.K=(AIF)(RSF.K)                                         15-43
3L    RSF.K=RSF.J+(DT)(1/DRF)(RRF.JK-RSF.J)                      15-44
25A   MWF.K=RRF.JK+(1/DIF)(IDF.K-IAF.K+LDF.K-LAF.K+UOF.K-UNF.K)  15-45
51R   MDF.KL=CLIP(MWF.K,ALF,ALF,MWF.K)                           15-46
18A   LDF.K=(RSF.K)(DCF+DPF)                                     15-47
7A    LAF.K=CPF.K+OPF.K                                          15-48
18A   UNF.K=(RSF.K)(DHF+DUF)                                     15-49
1L    CPF.K=CPF.J+(DT)(MDF.JK-MOF.JK)                            15-50
39R   MOF.KL=DELAY3(MDF.JK,DCF)                                  15-51
1L    OPF.K=OPF.J+(DT)(MDF.JK-SRF.JK)                            15-52
39R   SRF.KL=DELAY3(MOF.JK,DPF)                                  15-53

      INITIAL CONDITIONS OF CHAPTER 15, SECTION 15.5.4

6N    RRR=RRI                                                    15-54
18N   UOR=(RRI)(DHR+DUR)                                         15-55
12N   IAR=(AIR)(RRI)                                             15-56
6N    RSR=RRR                                                    15-57
12N   CPR=(DCR)(RRI)                                             15-58
6N    PDR=RRR                                                    A15-59
12N   PMR=(DMR)(RRI)                                             15-60
12N   MTR=(DTR)(RRI)                                             B15-EX
6N    SSD=RRD                                                    15-62
18N   UOD=(RRI)(DHD+DUD)                                         15-63
12N   IAD=(AID)(RRI)                                             15-64
6N    RSD=RRI                                                    C15-EX
12N   CPD=(DCD)(RRI)                                             15-65
6N    PDD=RRD                                                    15-66
12N   PMD=(DMD)(RRI)                                             15-67
12N   MTD=(DTD)(RRI)                                             D15-EX
6N    SSF=RRF                                                    E15-EX
6N    MDF=RRF                                                    15-69
18N   UOF=(RRI)(DHF+DUF)                                         15-70
12N   IAF=(AIF)(RRI)                                             15-71
6N    RSF=RRI                                                    15-72
12N   CPF=(DCF)(RRI)                                             15-73
12N   OPF=(DPF)(RRI)

      PARAMETERS FOR CHAPTER 15, SECTION 15.5.5

C     AID=6
C     AIF=4
C     AIR=8
12N   ALF=(1000)(RRI)
C     DCD=2
C     DCF=1
C     DCR=3
C     DHD=1
C     DHF=1
C     DHR=1
C     DID=4
C     DIF=4
C     DIR=4
C     DMD=.5
C     DMR=.5
C     DRD=8
C     DRF=8
```

Figure B-2 Production-distribution model of Chapter 15, second page.

PAGE 3   1991JF

C   DRR=8
C   DPF=6
C   DTD=2
C   DTR=1
C   DUD=.6
C   DUF=1
C   DUR=.4

     INPUT TEST FUNCTIONS

7R   RRR.KL=RRI+RCR.K
7A   RCR.K=STP.K+SNE.K
45A STR.K=STEP(STH,STT)
31A SNE.K=(SIH)SIN((2PI)(TIME.K)/PER)

     TEST INPUT CONDITIONS

C   PER=52
C   RRI=1000
C   SIH=0.0
C   STH=0
C   STT=4

     SUPPLEMENTARY EQUATIONS

8S   TIS.K=IAR.K+IAD.K+IAF.K

PRINT 1)CPD/2)CPR/3)DFD/5)DFF/6)DFR/7)IAD/8)IAF/9)IAR/10)IDD/11)ID
X1   F/12)IDR/13)LAD/14)LAF/1)LAR/2)LDD/3)LDF/4)LDR/5)MDF/6)MNR/7)MOD/8
X2   )MOF/9)MPF/10)MTD/11)MTR/12)MWF/13)NID/14)NIF/1)NIR/2)OPF/3)PDD/4)
X3   PDR/5)PMD/6)PMR/7)PSD/8)PSR/9)RRD/10)RRF/11)RRR/12)RSD/13)RSF/14)R
X4   SR/1)SRD/2)SRF/3)SRR/4)SSD/5)SSF/6)SSR/7)STD/8)STF/9)STR/10)TIS/11
X5   )UND/12)UNF/13)UNR/14)UOD/1)UOF/2)UOR

PLOT RRR=R,RRD=D,RRF=F,MOF=M,SRF=S(0,2000)/IAF=A,UOF=U,IAD=C,IAR=B(0,10
X1   000)/DFF=W(0,10)

STARTED GENERATING CODE AT   2/ 3   125.6
CODE STARTS IN / 12556.THERE ARE   655   LINES OF CODING

A15-74
B15-74
C15-74
D15-74

Figure B-3 Production-distribution model of Chapter 15, third and last page.

386

```
DYN
RUN

0045
1993JF/DT=0.1/LENGTH=0.0/PRTPER=2/PLTPER=2

1. PROBLEM 95, NOTEBOOK 58JF63
2. FOR FORRESTER BOOK ON INDUSTRIAL DYNAMICS
3. CHAPTER 16, ADVERTISING MODEL.
4. MAIL DELAYS COMBINED WITH PURCHASING DELAY, ELIMINATING
   EQUATIONS 15-14, 15-15, 15-32, 15-33, 15-59, 15-66.

RETAIL SECTOR—FIGURE 15-14, SECTION 15.5.1

1L    UOR.K=UOR.J+(DT)(RRR.JK-SSR.JK)                          15- 1
1L    IAR.K=IAR.J+(DT)(SRR.JK-SSR.JK)                          15- 2
20A   STR.K=UOR.K/DFR.K                                        15- 3
20A   NIR.K=IAR.K/DT                                           15- 4
51R   SSR.KL=CLIP(STR.K,NIR.K,NIR.K,STR.K)                     A15- 5
27A   DFR.K=(MNR.K/IAR.K)+DHR                                  B15- 6
12A   MNR.K=(DUR)(IDR.K)                                       15- 7
12A   IDR.K=(AIR)(RSR.K)                                       15- 8
3L    RSR.K=RSR.J+(DT)(1/DRR)(RRR.JK-RSR.J)                    15- 9
25R   PDR.KL,RRR.JK+(1/DIR)(IDR.K-IAR.K+LDR.K-LAR.K+UOR.K-UNR.K) 15-10
16A   LDR.K=(RSR.K)(DCR)+(IO)(IO)+(RSR.K)(DFD.K)+(RSR.K)(DTR)  15-11
8A    LAR.K=CPR.K+UOD.K+MTR.K                                  15-12
18A   UNR.K=(RSR.K)(DHR+DUR)                                   15-13
1L    CPR.K=CPR.J+(DT)(PDR.JK-RRD.JK)                          15-14
39R   RRD.KL=DELAY3(PDR.JK,DCR)                                15-16
1L    MTR.K=MTR.J+(DT)(SSD.JK-SRR.JK)                          15-17
39R   SRR.KL=DELAY3(SSD.JK,DTR)                                15-18

DISTRIBUTOR SECTOR— FIGURE 15-15, SECTION 15.5.2

1L    UOD.K=UOD.J+(DT)(RRD.JK-SSD.JK)                          15-19
1L    IAD.K=IAD.J+(DT)(SRD.JK-SSD.JK)                          15-20
20A   STD.K=UOD.K/DFD.K                                        15-21
20A   NID.K=IAD.K/DT                                           15-22
51R   SSD.KL=CLIP(STD.K,NID.K,NID.K,STD.K)                     A15-23
27A   DFD.K=(MOD.K/IAD.K)+DHD                                  B15-24
12A   MOD.K=(DUD)(IDD.K)                                       15-25
12A   IDD.K=(AID)(RSD.K)                                       15-26
3L    RSD.K=RSD.J+(DT)(1/DRD)(RRD.JK-RSD.J)                    15-27
25R   PDD.KL,RRD.JK+(1/DID)(IDD.K-IAD.K+LDD.K-LAD.K+UOD.K-UND.K) 15-28
16A   LDD.K=(RSD.K)(DCD)+(IO)+(RSD.K)(DFF.K)+(RSD.K)(DTD)      15-29
8A    LAD.K=CPD.K+UOF.K+MTD.K                                  15-30
18A   UND.K=(RSD.K)(DHD+DUD)                                   15-31
1L    CPD.K=CPD.J+(DT)(PDD.JK-RRF.JK)                          15-34
39R   RRF.KL=DELAY3(PDD.JK,DCD)                                15-34
1L    MTD.K=MTD.J+(DT)(SSF.JK-SRD.JK)                          15-35
39R   SRD.KL=DELAY3(SSF.JK,DTD)                                15-36

FACTORY SECTOR— FIGURE 15-16, SECTION 15.5.3

1L    UOF.K=UOF.J+(DT)(RRF.JK-SSF.JK)                          15-37
1L    IAF.K=IAF.J+(DT)(SRF.JK-SSF.JK)                          15-38
20A   STF.K=UOF.K/DFF.K                                        15-39
20A   NIF.K=IAF.K/DT                                           15-40
51R   SSF.KL=CLIP(STF.K,NIF.K,NIF.K,STF.K)                     A15-41
27A   DFF.K=(MPF.K/IAF.K)+DHF                                  B15-42
12A   MPF.K=(DUF)(IDF.K)                                       15-42
12A   IDF.K=(AIF)(RSF.K)                                       15-43
```

2/ 3   127.5 1961

Figure B-4   Advertising model of Chapter 16, first page.

1993JF

```
3L    RSF.K=RSF.J+(DT)(1/DRF)(RRF.JK-RSF.J)                          15-44
25A   MWF.K=RRF.JK+(1/DIF)(IDF.K-IAF.K+LDF.K-LAF.K+UOF.K-UNF.K)      15-45
51R   MDF.KL=CLIP(MWF.K,ALF,ALF,MWF.K)                               15-46
18A   LDF.K=(RSF.K)(DCF+DPF)                                         15-47
7A    LAF.K=CPF.K+OPF.K                                              15-48
18A   UNF.K=(RSF.K)(DHF+DUF)                                         15-49
1L    CPF.K=CPF.J+(DT)(MDF.JK-MOF.JK)                                15-50
39R   MOF.KL=DELAY3(MDF.JK,DCF)                                      15-51
1L    OPF.K=OPF.J+(DT)(MOF.JK-SRF.JK)                                15-52
39R   SRF.KL=DELAY3(MOF.JK,DPF)                                      15-53
```

INITIAL CONDITIONS OF CHAPTER 15, SECTION 15.5.4

```
18N   UOR=(RSR)(DHR+DUR)          15-55
12N   IAR=(AIR)(RSR)              15-56
6N    RSR=RRR                     15-57
12N   CPR=(DCR)(RRI)              15-58
6N    PDR=RRR                     A15-EX
12N   MTR=(DTR)(RRI)              15-60
6N    SSD=RRD                     B15-EX
18N   UOD=(RSD)(DHD+DUD)          15-62
12N   IAD=(AID)(RSD)              15-63
6N    RSD=RRD                     15-64
12N   CPD=(DCD)(RRI)              15-65
6N    PDD=RRD                     C15-EX
12N   MTD=(DTD)(RRI)              15-67
6N    SSF=RRF                     D15-EX
6N    MDF=RRF                     E15-EX
18N   UOF=(RSF)(DHF+DUF)          15-69
12N   IAF=(AIF)(RSF)              15-70
6N    RSF=RRF                     15-71
12N   CPF=(DCF)(RRI)              15-72
12N   OPF=(DPF)(RRI)              15-73
```

PARAMETERS FOR CHAPTER 15, SECTION 15.5.5

```
C   AID=6
C   AIF=4
C   AIR=8
C   ALF=3000
C   DCD=2.5
C   DCF=1
C   DCR=3.5
C   DHD=1
C   DHF=1
C   DHR=1
C   DID=8
C   DIF=8
C   DIR=8
C   DRD=8
C   DRF=8
C   DRR=8
C   DTD=2
C   DPF=6
C   DTR=L
C   DUD=.6
C   DUF=1
C   DUR=.4
```

Figure B-5 Advertising model of Chapter 16, second page.

388

1993JF

```
ADVERTISING SECTOR— FIGURE 16-5, SECTION 16.1

17R    VDF.KL=(MAF.K)(UPF)(AVS)                          16— 1
3L     MAF.K=MAF.J+(DT)(1/DMS)(MOF.JK-MAF.J)             16— 2
39R    VCF.KL=DELAY3(VDF.JK,DVF)                         16— 3
39R    VMC.KL=DELAY3(VCF.JK,DVA)                         16— 4
3L     VAC.K=VAC.J+(DT)(1/DVC)(VMC.JK-VAC.J)             A16— 5
7A     DPC.K=DSV+AUX3.K                                  A16— 6
28A    AUX3.K=(AUX1.K)EXP(-AUX2.K)                       B16— 6
7A     AUX1.K=D2V-DSV                                    C16— 6
20A    AUX2.K=VAC.K/ASL                                  D16— 6
22R    RRR.KL=(1/DPC.K)((PPC.K)(1)+(PPC.K)(NPR.K))       16— 7
1L     PPC.K=PPC.J+(DT)(GNC.JK-RRR.JK)                   16— 8
7R     GNC.KL=GNI+CNC.K                                  16— 9

       INPUT TEST FUNCTIONS

43A    NPR.K=SAMPLE(NNR.K,ANL)                           16-10
34A    NNR.K=(1)NORMRN(0.0,ADN)                          16-11
7A     CNC.K=STP.K+SNE.K                                 16-12
45A    STP.K=STEP(STH,STT)                               16-13
31A    SNE.K=(SIH)SIN((2PI)(TIME.K)/PER)                 16-14

       INITIAL CONDITIONS FOR ADVERTISING EQUATIONS, SECTION 16.2

6N     MAF=RRF                                           16-15
13N    VDF=(MAF)(UPF)(AVS)                               16-16
6N     VAC=VMC                                           16-17
6N     ASL=VAC                                           A16-18
7N     DPC=DSV+A3                                        A16-19
28N    A3=(A1)EXP(-A2)                                   B16-19
7N     A1=D2V-DSV                                        C16-19
20N    A2=VAC/ASL                                        D16-19
12N    PPC=(RRR)(DPC)                                    16-20
6N     RRR=GNI                                           16-21
6N     RRI=GNI                                           A16-EX

       SUPPLEMENTARY EQUATIONS

75     PLT1.K=IAF.K-IDF.K
85     TIS.K=IAR.K+IAD.K+IAF.K

       PARAMETERS FOR ADVERTISING-MARKET, SECTION 16.3

C      UPF=100
C      AVI=0
C      AVS=0.06
C      DMS=4
C      DSV=15
C      DVA=4
C      DVC=6
C      DVF=6
C      D2V=70
C      GNI=1000

       TEST INPUT CONDITIONS

C      ADN=0.0
C      ANL=1.0
```

Figure B-6 Advertising model of Chapter 16, third page.

```
C     PER=52
C     SIH=0
C     STH=0.0
C     STT=4
PRINT 1)CPD/2)CPF/3)CPR/4)DFD/5)DFF/6)DFR/7)IAD/8)IAF/9)IAR/10)IDD/11)ID
X1   F/12)IDR/13)LAD/14)LAF/1)LAR/2)LDD/3)LDF/4)LDR/5)MDF/6)MNR/7)MOD/8
X2   )MOF/9)MPF/10)MTD/11)MTR/12)MWF/13)NID/14)NIF/1)NIR/2)OPF/3)PDD/4)
X3   PDR/5)VAC/6)VCF/7)VDF/8)VMC/9)RRD/10)RRF/11)RSD/12)RSF/13)RSR/14)S
X4   RD/1)SRF/2)SRR/3)SSD/4)SSF/5)SSR/6)STD/7)STF/8)STR/9)UND/10)UNF/11
X5   )UNR/12)UOD/13)UOF/14)UOR
PRINT 1)DPC/2)GNC/3)MAF/4)NPR/5)PLT1/6)PPC/7)RRR/8)TIS

PLOT  RRR=R,RRD=D,RRF=F,MOF=M,SRF=S(0,2000)/IAF=A,UOF=U,IAD=C,IAR=B(0,10
X1   000)/DFF=W(0,10)
PLOT  VAC=X,VDF=V/DPC=Y/PPC=Z/PLT1=P/RRR=R,RRD=D,RRF=F,MAF=G,GNC=I(0,200
X1   0)

STARTED GENERATING CODE AT    2/ 3   127.7
CODE STARTS IN  / 12666.THERE ARE  753   LINES OF CODING
```

Figure B-7 Advertising model of Chapter 16, fourth and last page.

390

```
DYN   M637
RUN   2017JF/DT=.25/LENGTH=0.0/PRTPER=2/PLTPER=2

      1. PROBLEM 100, NOTEBOOK 58JF77
      2. FOR FORRESTER BOOK ON INDUSTRIAL DYNAMICS
      3. CHAPTER 18, ELECTRONIC COMPONENT INDUSTRY
      4. NEW POLICIES
      5. NEW PARAMETER VALUES FOR TRSF, TRSFI, AND TBLAF

      ORDER FILLING--FIGURE 17-6, SECTION 17.4.1

2L    RCF.K=RCF.J+(DT)(RRF.JK-RFIF.JK-RMOF.JK+0+0+0)      17- 1
12N   RCF=(RRF)(DCPF)                                     17- 2
1L    SOF.K=SOF.J+(DT)(RFIF.JK-SIF.JK)                    17- 3
12N   SOF=(RFIF)(DSF)                                     17- 4
20R   SIF.KL=SOF.K/DSF                                    17- 5
6N    SIF=RFIF                                            17- 6
1L    IAF.K=IAF.J+(DT)(MIF.JK-SIF.JK)                     17- 7
12N   IAF=(CIRF)(RRF)                                     17- 8
44R   RFIF.KL=(FRFIF.K)(RCF.K)/DCPF                       17- 9
12N   RFIF=(CNFIF)(RRF)                                   17- 10
22R   RMOF.KL=(1/DCPF)((RCF.K)(1)+(-FRFIF.K)(RCF.K))      17- 11

      INVENTORY SPLIT FRACTION--FIGURES 17-6 AND 17-7, SECTION 17.4.1

7A    FRFIF.K=CMFIF-AUX3.K                                A17- 12
28A   AUX3.K=(CMFIF)EXP(-AUX2.K)                          B17- 12
44A   AUX2.K=(C1)(IAF.K)/CINF                             C17- 12
29N   C1=(1)LOGN(AUX1)                                    A17- 13
26N   AUX1=(CMFIF+0+0)/(CMFIF-CNFIF+0)                    B17- 13
6N    CINF=IAF                                            17- 14

      INVENTORY REORDERING--FIGURE 18-11, SECTION 18.3.1

3L    ASIF.K=ASIF.J+(DT)(1/TASIF)(SIF.JK-ASIF.J)          17- 17
6N    ASIF=RFIF                                           17- 18
3L    RSF.K=RSF.J+(DT)(1/TRSF)(RRF.JK-RSF.J)              17- 20
6N    RSF=RRF                                             17- 21
3L    RSF1.K=RSF1.J+(DT)(1/TRSF1)(RRF.JK-RSF1.J)          18- 1
6N    RSF1=RRF                                            18- 2
12A   IDF.K=(CIRF)(RSF1.K)                                18- 3
24A   LIAF.K=(1/AAL1)(IDF.K-IAF.K-OINF.K-OPIF.K+0+0)      A18- 4
12N   AAL1=(TBLAF)(CPLF)                                  B18- 4
12A   OINF.K=(ASIF.K)(DPF)                                18- 5

      MANUFACTURING--FIGURE 18-12, SECTION 18.3.2

42A   LDCOF.K=BLCF.K/((DNBLF)(CPLF))                      18- 6
51A   LCOF.K=CLIP(LDCOF.K,MENPF.K,MENPF.K,LDCOF.K)        18- 7
12R   PCOF.KL=(LCOF.K)(CPLF)                              18- 8
7A    LIF.K=MENPF.K-LCOF.K                                18- 9
12R   PIF.KL=(LIF.K)(CPLF)                                18- 10
1L    BLCF.K=BLCF.J+(DT)(RMOF.JK-PCOF.JK)                 17- 26
18N   BLCF=(DNBLF)(RRF-RFIF)                              17- 27
7N    PCOF=RRF-RFIF                                       17- 34
6N    PIF=RFIF                                            17- 40
1L    OPIF.K=OPIF.J+(DT)(PIF.JK-MIF.JK)                   17- 41
12N   OPIF=(DPF)(RFIF)                                    17- 42
39R   MIF.KL=DELAY3(PIF.JK,DPF)                           17- 43
```

Figure B-8 Model of electronic components industry, new policies, Chapter 18, first page.

391

2017JF

```
1L    OPCF.K=OPCF.J+(DT)(PCOF.JK-SMOF.JK)                              17- 44
18N   OPCF=(DPF)(RRF-RFIF)                                             17- 45
39R   SMOF.KL=DELAY3(PCOF.JK,DPF)                                      17- 46
2L    RMSF.K=RMSF.J+(DT)(RMRF.JK-PIF.JK-PCOF.JK+0+0+0)                 17- 47
12N   RMSF=(RRF)(CRMSF)                                                17- 48
27A   DMCOF.K=(BLCF.K/PCOF.K)+DPF                                      17- 50

MATERIAL ORDERING--FIGURE 17-10, SECTION 17.4.4

12A   RMDF.K=(RSF.K)(CRMSF)                                            17- 51
8R    RMPF.KL=PCOF.JK+PIF.JK+AUX5.K                                   A17- 52
16A   AUX5.K=(AUX4.K)(RMDF.K)+(-AUX4.K)(RMSF.K)+(AUX4.K)(RMPNF.K)+(-AUX4   B17- 52
X1    .K)(RMPAF.K)                                                    C17- 52
20A   AUX4.K=1./TRMAF                                                  17- 52
6N    RMPF=RRF                                                         17- 53
12A   RMPNF.K=(RSF.K)(DRMF)                                            17- 54
1L    RMPAF.K=RMPAF.J+(DT)(RMPF.JK-RMRF.JK)                            17- 55
12N   RMPAF=(RRF)(DRMF)                                                17- 56
39R   RMRF.KL=DELAY3(RMPF.JK,DRMF)                                     17- 57

LABOR SECTOR--FIGURE 18-13, SECTION 18.3.3

1L    LTF.K=LTF.J+(DT)(LHF.JK-LEPF.JK)                                 17- 58
6N    LTF=n                                                            17- 59
39R   LEPF.KL=DELAY3(LHF.JK,DLTF)                                      17- 60
1L    MENPF.K=MENPF.J+(DT)(LEPF.JK-LDNF.JK)                            17- 61
20N   MENPF=RRF/CPLF                                                   17- 62
1L    LLF.K=LLF.J+(DT)(LDNF.JK-LTERF.JK)                               17- 63
6N    LLF=0                                                            17- 64
39R   LTERF.KL=DELAY3(LDNF.JK,DLLF)                                    17- 65
20A   LASF.K=RSF.K/CPLF                                                17- 66
13A   BLNF.K=(AUX23)(DNBLF)(RSF.K)                                    A18- 11
7N    AUX23=1-CNFIF                                                   B18- 11
42A   LBLAF.K=AUX24.K/((CPLF)(TBLAF))                                 A18- 12
7A    AUX24.K=BLCF.K-BLNF.K                                           B18- 12
8A    LDF.K=LASF.K+LBLAF.K+LIAF.K                                      18- 13
7A    LCF.K=LDF.K-LAF.K                                                17- 70
7A    LAF.K=LTF.K+MENPF.K                                              17- 71
22A   LCHF.K=(1/TLCF)(LCF.K)(1,0)+(-CLCTF)(MENPF.K))                   17- 72
51R   LHF.KL=CLIP(LCHF.K,0,LCHF.K,n)                                   17- 73
6N    LHF=0                                                            17- 74
22A   LCDF.K=(1/TLCF)(LCF.K)(1,0)+(CLCTF)(MENPF.K))                    17- 75
51R   LDNF.KL=CLIP(0,-LCDF.K,LCDF.K,n)                                 17- 76
6N    LDNF=0                                                           17- 77
8A    MENTF.K=LTF.K+MENPF.K+LLF.K                                      17- 78
1L    LCTF.K=LCTF.J+(DT)(LHF.JK+LDNF.JK)                               17- 79
6N    LCTF=0                                                           17- 80

DELIVERY DELAY--FIGURE 17-13, SECTION 17.4.6

14A   DVF.K=AUX7.K+(DSF)(FRFIF.K)                                     A17- 81
1RA   AUX7.K=(DMCOF.K)(1-FRFIF.K)                                     B17- 81
7R    DFOF.KL=DCPF+DVF.K                                              17- 82
3L    DQDF.K=DQDF.J+(DT)(1/TAQDF)(DFOF.JK-DQDF.J)                      17- 83
7N    DQDF=AUX8A+DCPF                                                 A17- 84
14N   AUX8A=AUX8+(CNFIF)(DSF)                                         B17- 84
16N   AUX8=(DPF)(1)+(-CNFIF)(DPF)+(DNBLF)(1)+(-CNFIF)(DNBLF)          C17- 84

CUSTOMER ORDERING--FIGURE 17-14, SECTION 17.4.7
```

Figure B-9 Model of electronic components industry, new policies, Chapter 18, second page.

```
1L    FDPC.K=FDPC.J+(DT)(INPUT.JK-PDC.JK)                                    17- 85
12N   FDPC=(INPUT)(DEEDC)                                                    17- 86
1L    RCC.K=RCC.J+(DT)(PDC.JK-RRF.JK)                                        17- 87
12N   RCC=(INPUT)(DRCC)                                                      17- 88
7A    DAERC.K=DRCC+AUX20.K                                                  A17- 89
18A   AUX20.K=(DQDF.K)(1+C1STC)                                            B17- 89
7R    DDEDC.KL=DTC-DAERC.K                                                   17- 90
8N    DTC=DNEDC+DRCC+AUX22                                                 A17- 91
18N   AUX22=(DQDF)(1+C1STC)                                               B17- 91
3L    DEEDC.K=DEEDC.J+(DT)(1/TAEDC)(DDEDC.JK-DEEDC.J)                        17- 92
8N    DEEDC=DTC-AUX21-DRCC                                                 A17- 93
18N   AUX21=(DQDF)(1+C1STC)                                               B17- 93
44R   PDC.KL=(EDPC.K)(AUX9.K)/DEEDC.K                                      A17- 94
7A    AUX9.K=1+NEDC.K                                                      B17- 94
44R   RRF.KL=(RCC.K)(AUX10.K)/DRCC                                         A17- 95
7A    AUX10.K=1+NPC.K                                                      B17- 95
6N    RRF=INPUT                                                             17- 96
```

CASH FLOW--FIGURE 17-15, SECTION 17.4.8

```
12R   RMIF.KL=(RMRF.JK)(CRMPF)                                              17- 97
1L    APF.K=APF.J+(DT)(RMIF.JK-RMCEF.JK)                                    17- 98
13N   APF=(RRF)(CRMPF)(DAPF)                                                17- 99
2oR   RMCEF.KL=APF.K/DAPF                                                   17-100
18R   FGIF.KL=(CFGPF)(SIF.JK+SMOF.JK)                                       17-101
12N   FGIF=(RRF)(CFGPF)                                                     17-102
1L    ARF.K=ARF.J+(DT)(FGIF.JK-FGCRF.JK)                                    17-103
13N   ARF=(RRF)(CFGPF)(DARF)                                                17-104
39R   FGCRF.KL=DELAY3(FGIF.JK,DARF)                                         17-105
12R   LCEF.KL=(MENF.K)(CWRF)                                                17-106
12N   LCEF=(MENPF)(CWRF)                                                    17-107
6R    ITAXF.KL=(0.5)(PBTRF.K)                                               17-108
1L    SD1VF.KL=SDLF.K                                                       17-109
6N    SDTDF=0                                                               17-110
      SDTDF.K=SDTDF.J+(DT)(SD1VF.JK+0.0)
4L    CASHF.K=CASHF.J+(DT)(1/1)(FGCRF.JK-RMCEF.JK-LCEF.JK-CCEF-ITAXF.JK-   17-111
X1    SD1VF.JK)                                                            17-112
13N   CASHF=(CNCSF)(CFGPF)(RRF)                                             17-113
```

PROFIT, DIVIDENDS--FIGURE 17-16, SECTION 17.4.9

```
27N   SICF=(CWRF/CPLF)+CRMPF                                                17-114
8A    PBTRF.K=AUX17.K-AUX18.K-CCEF                                         A17-115
18A   AUX17.K=(AUX17A.K)(CFGPF-SICF)                                       B17-115
7A    AUX17A.K=SIF.JK+MOF.JK                                               C17-115
18A   AUX18.K=LCEF.JK-AUX16.K                                              D17-115
7A    AUX16.K=(AUX15)(MIF.JK+SMOF.JK)                                      E17-115
2oN   AUX15=CWRF/CPLF                                                      F17-115
12R   NPRF.KL=(0.5)(PBTRF.K)                                               17-116
6N    NPRF=SDLF                                                            A17-EX
1L    NPTDF.K=NPTDF.J+(DT)(NPRF.JK+0.0)                                    17-117
6N    NPTDF=0                                                              17-118
3L    SDLF.K=SDLF.J+(DT)(1/TASDL)(NPRF.JK-SDLF.J)                          17-119
14N   SDLF=AUX13-(0.5)(CCEF)                                               A17-120
17N   AUX13=(RRF)(0.5)(-CRMPF)+(RRF)(0.5)(-AUX12)                          B17-120
2oN   AUX12=CWRF/CPLF                                                      C17-120
```

MODEL CONSTANTS---DEFINED AFTER EQUATION NUMBER

Figure B-10 Model of electronic components industry, new policies, Chapter 18, third page.

```
C    CCFF=30000          17-112
C    CFGPF=100           17-101
C    CIRF=4              17- 19
C    CISTF=0.5           17- 89
C    CLCTF=0.0           17- 72
C    CMFIF=0.8           17- 12
C    CNCSF=1.0           17-113
C    CNFIF=0.7           17- 13
C    CPLF=2.66666        17- 29
C    CRMPF=20            17- 97
C    CRMSF=6             17- 51
C    CWRF=80             17-106
C    DAPF=3              17-100
C    DARF=5              17-105
C    DCPF=1              17-  9
C    DLLF=4              17- 65
C    DLTF=3              17- 60
C    DNBLF=4             17- 67
C    DNEDC=30            17- 91
C    DPF=6               17- 43
C    DRCC=3              17- 95
C    DRMF=3              17- 57
C    DSF=1               17-  5
C    TAEDC=15            17- 92
C    TAQDF=4             17- 83
C    TASDL=52            17-119
C    TASIF=2             17- 17
C    TBLAF=50            17- 68
C    TLCF=10             17- 72
C    TRMAF=8             17- 52
C    TRSF=26             17- 20
C    TRSF1=52            18-  1
```

SUPPLEMENTARY EQUATIONS FOR OUTPUT INFORMATION, SECTION 17.5

```
46S    BLCPC.K=(BLCF.K)(100.0)(1.0)/((CINPI)((AAA2)(DNBLF))              17-122
7N     AAA2=1.0-CNFIF                                                    17-123
46S    CASPC.K=(CASHF)(100.0)(1.0)/((CNCSF)(CFGPF)(CINPI))               17-124
12S    FRFPC.K=(FRFTF.K)(100.0)  - - - - - - - - - - - -  -  -  -        17-124
46S    IAFPC.K=(IAF.K)(100.0)(1.0)/((CIRF)(CINPI)(1.0))   -  -  -  -     17-125
46S    IDFPC.K=(IDF.K)(100.0)(1.0)/((CIRF)(CINPI)(1.0))   -  -  -  -     17-125
46S    INPPC.K=(INPUT.JK)(100.0)(1.0)/((CINPI)(1.0)(1.0))    -  -  -     17-126
46S    MFNPC.K=(MFNPF.K)(CPLF)(100.0)/((CINPI)(1.0)(1.0))       -  -     17-127
46S    NPRPC.K=(NPRF.JK)(200.0)(1.0)/((AAA4)(1.0)(1.0))            -     A17-128
18N    AAA3=(CINPI)(CFGPF-SICF)          -  -  -  -  -  -  -  -  -  -    B17-128
7N     AAA4=AAA3-CCEF                                                    C17-128
46S    RRFPC.K=(RRF.JK)(100.0)(1.0)/((CINPI)(1.0)(1.0))   -  -  -  -     17-129
7S     SCF.K=TIF.JK+SMOF.JK                                              17-130
12S    VIAF.K=(IAF.K)(SICF)                                              17-131
12S    VIPIF.K=(D132.K)(C132)                                            A17-132
42N    R132=CWRF/((2.0)(CPLF))                                           B17-132
7N     C132=CRMPF+B132                                                   C17-132
7A     D132.K=OPIF.K+OPCF.K                                              D17-132
12S    VRMSF.K=(RMSF.K)(CRMPF)                                           17-133
8S     VTIF.K=VIAF.K+VIPIF.K+VRMSF.K                                     17-134
9S     SURPL.K=VTIF.K+CASHF.K-APF.K                                      17-135
```

INPUT TEST FUNCTIONS--FIGURE 17-17, SECTION 17.6

Figure B-11 Model of electronic components industry, new policies, Chapter 18, fourth page.

```
7R    INPUT.KL=CINPI+INPCH.K                                    17-136
6N    INPUT=CINPI                                               17-137
9A    INPCH.K=STP1.K+GTH1.K+GTH2.K+SNE.K                        17-138
45A   STP1.K=STEP(STH1,STT1)                                    17-139
47A   GTH1.K=RAMP(GTHR1,GTHT1)                                  17-140
47A   GTH2.K=RAMP(GTHR2,GTHT2)                                  17-141
31A   SNE.K=(S1H)SIN((2PI)(TIME.K)/PER)                         17-142
43A   NEDC.K=SAMPLE(NNEDC.K,CN<EC)                              17-143
34A   NNEDC.K=(1)NORMRN(0.0,CNAEC)                              17-144
43A   NPC.K=SAMPLE(NNPC.K,CNSPC)                                17-145
36A   NNPC.K=(1)NORMRN(0.0,CNAPC)                               17-146

C       TEST CONDITIONS-- DEFINED AFTER EQUATION NUMBER

C       CINPI=1000        17-137
C       CNAEC=0           17-144
C       CNAPC=0           17-146
C       CN.EC=0           17-143
C       CN.PC=0           17-145
C       GTHR1=0.0         17-140
C       GTHR2=0.0         17-140
C       GTHT1=0.0         17-141
C       GTHT2=0.0         17-141
C       PER=52            17-142
C       S1H=0             17-142
C       STH1=0            17-139
C       STT1=0            17-139

PRINT 1)BLCPC(0.2),*/2)CASPC(0.2),*/3)DQDF(0.2),*/4)FRFPC(0.2),*/5)IAFPC
X1    (0.2),*/6)IDFPC(0.2),*/7)INPPC(0.2),*/8)MENPC(0.2),*/9)NPRPC(0.2),
X2    */10)RRFPC(0.2),*/11)NEDC(0.3),*/12)NPC(0.3),**/13)PDC(0.1),**/14)**,
X3    *
PRINT 1)APF/2)ARF/3)ASIF/4)BLCF/5)BLNF/6)CASHF/7)DEEDC/8)DEEDC
PRINT 9)DFOF/10)DMCOF/11)DQDF/12)DVF/13)EDPC/14)FGCRF/1)FGCRF/2)IAF
PRINT 3)IDF/4)INPUT/5)ITAXF/6)LASF/7)LBLAF/8)LCEF/9)LCEF/10)LCOF
PRINT 11)LCTF/12)LDCOF/13)LDF/14)LDNF/1)LHF/2)LIAF/3)LIF/4)MENPF
PRINT 5)MENTF/6)MIF/7)NEDC/8)NPC/9)NPRF/10)NPTDF/11)OINF/12)PCOF
PRINT 13)PDC/14)PIF/1)RCC/2)RCF/3)RFIF/4)RMCEF/5)RMOF/6)RMPF
PRINT 7)RMRF/8)RMSF/9)RRF/10)RSF/11)RSF/12)SCF/13)SDIVF
PRINT 14)SDLF/1)SDTDF/2)S1F/3)SMOF/4)SOF
PRINT 8)SURPL/9)VIAF/10)VIPIF/11)VRMSF/12)VTIF

PLOT  SCF=C,PDC=P,RMOF=M,RMPF=S/NEDC=N(-0.5,0.5)/LBLAF=B,LIAF=I/BLCF=L/P
X1    IF=K,PCOF=G
PLOT  RMCEF=7(10000,30000,0)/SDIVF=D,ITAXF=T(0,0.20000,0)/FGCRF=R(90000,
X1    110000,0)/LCEF=W,CCEF=F(20000,40000,0)
PLOT  CASHF=C,VRMSF=M,APF=D(0,2000000)/ARF=R,VTIF=I(400000,600000)/VIAF=Q
X1    /VIPIF=P(100000,300000)/SURPL=S(970000,117000)
PLOT  IAF=Q,IDF=T/RRF=R,PCOF=G,PIF=K/NEDC=N(-0.5,0.5)/PDC=P(500,1500)
PLOT  IAF=Q/RRF=R/DQDF=L/MENPF=M/BLCF=B/LHF=H,LDNF=F/FRFIF=S/DEEDC=D/NPR
X1    F=P
PLOT  IAFPC=Q,IDFPC=T,INPPC=I,MENPC=M,RRFPC=R(50,150)/BLCPC=B,CASPC=C,NP
X1    RPC=P(0,200)/DQDF=D(0,8)/FRFPC=S(0,80)

STARTED GENERATING CODE AT   2/13 2357.2
CODE STARTS IN  / 13133.THERE ARE 1013   LINES OF CODING
```

Figure B-12 Model of electronic components industry, new policies, Chapter 18, fifth and last page.

Example of Computation Sequence

TO clarify the computation procedure used for the dynamic models in this book, a very simplified example may be helpful. Figure C-1 shows a flow diagram (see Chapter 8 for symbols) of a simplified version of the equations of the retail sector in Chapter 15. Figure C-2 is a tabulation of the corresponding equations. Figures C-3 through C-6

give the computed numerical values and plotted results.

The definitions of terms and reasons for choice of equations are not given here, because the example is not included for the sake of the system itself but to show how the equations give rise to a computing sequence.

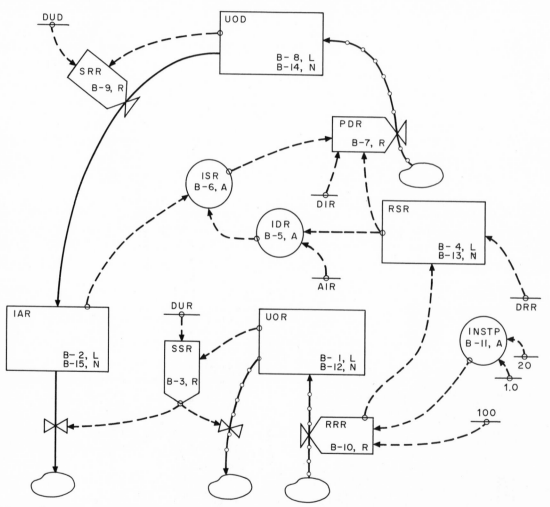

Figure C-1   Flow diagram for example of computing sequence.

See Chapter 8 for symbols and Section 15.5.1 for a similar set of equations.

```
DYN    M637
RUN    1990JF/DT=0.50/LENGTH=18.5/PRTPER=0.5/PLTPER=0.5
```

2/ 3   125.0 1961

```
       1. PROBLEM 93, NOTEBOOK 58JF59
       2. FOR FORRESTER BOOK ON INDUSTRIAL DYNAMICS
       3. EXAMPLE FOR APPENDIX TO SHOW COMPUTATION STEPS
       4. IN THE PRINTED RESULTS, LEVELS ARE GIVEN ON THE FIRST LINE,
          AUXILIARY VARIABLES ON THE SECOND, AND RATES ON THE THIRD

1L     UOR.K=UOR.J+(DT)(RRR.JK-SSR.JK)               B- 1, L
1L     IAR.K=IAR.J+(DT)(SRR.JK-SSR.JK)               B- 2, L
20R    SSR.KL=UOR.K/DUR                              B- 3, R
3L     RSR.K=RSR.J+(DT)(1/DRR)(RRR.JK-RSR.J)         B- 4, L
12A    IDR.K=(AIR)(RSR.K)                            B- 5, A
7A     ISR.K=IDR.K-IAR.K                             B- 6, A
27R    PDR.KL=(ISR.K/DIR)+RSR.K                      B- 7, R
1L     UOD.K=UOD.J+(DT)(PDR.K-SRR.JK)                B- 8, L
20R    SRR.KL=UOD.K/DUD                              B- 9, R
7R     RRR.KL=100.0+INSTP.K                          B-10, R
45A    INSTP.K=STEP(20.1)                            B-11, A
12N    UOR=(DUR)(100.0)                              B-12, N
6N     RSR=100.0                                     B-13, N
12N    UOD=(DUD)(100.0)                              B-14, N
12N    IAR=(AIR)(100.0)                              B-15, N

C      DUR=1.0
C      DRR=2.0
C      AIR=3.0
C      DIR=2.0
C      DUD=2.0

PRINT  1)**,SRR(0,2)/2)IAR(0,2)/3)**,SSR(0,2)/4)UOR(0,2)/5)**,RRR(0,2)
X1     /6)**,INSTP(0,2)/7)RSR(0,2)/8)**IDR(0,2)/9)**ISR(0,2)/10)**,PDR(0,
X2     2)/11)UOD(0,2)

PLOT   RSR=A,SSR=S,PDR=P,SRR=D,RRR=R(0,200)/UOR=B,IAR=I,IDR=T,UOD=U(0,400
X1     )/ISR=N(-100,100)

STARTED GENERATING CODE AT     2/ 3   125.1
CODE STARTS IN  / 12110.THERE ARE 76    LINES OF CODING
STARTED TO RUN CODE AT 2/ 3  125.1
```

Figure C-2 Tabulation of Equations for example of computing sequence.

These equations are in standard notation and should be self-explanatory except for Equation B-11. INSTP is a step function of 20 units amplitude occurring when time is equal to 1 week. This is an external disturbance, the propagation of which will be observed through the system. The equations are somewhat simplified from the retail stage in Chapter 15. The initial conditions are for a steady-state initial flow of 100 units per week through the system.

PAGE 2   1990JF   RATES .KL PRINTED.   STARTED PRINTING AT   2/ 3   125.1

TIME	SRR	IAR	SSR	UOR	RRR	INSTP	RSR	IDR	ISR	PDR	UO3
E+00	E+00	E+00	E+00	E+00	E+00	E+00	E+00	E+00	E+00	E+00	E+00
.000	100.00	300.00	100.00	100.00	100.00	.00	100.00	300.00	.00	100.00	200.00
.500	100.00	300.00	100.00	100.00	100.00	.00	100.00	300.00	.00	100.00	200.00
1.000	100.00	300.00	100.00	100.00	120.00	.00	100.00	300.00	.00	100.00	200.00
1.500	100.00	300.00	110.00	110.00	120.00	20.00	105.00	315.00	15.00	112.50	200.00
2.000	103.12	295.00	115.00	115.00	120.00	20.00	108.75	326.25	31.25	124.37	206.25
2.500	108.44	289.06	117.50	117.50	120.00	20.00	111.56	334.69	45.62	134.37	216.87
3.000	114.92	284.53	118.75	118.75	120.00	20.00	113.67	341.02	56.48	141.91	229.84
3.500	121.67	282.62	119.37	119.37	120.00	20.00	115.25	345.76	63.14	146.83	243.34
4.000	127.96	283.76	119.69	119.69	120.00	20.00	116.44	349.32	65.56	149.22	255.92
4.500	133.27	287.90	119.84	119.84	120.00	20.00	117.33	351.99	64.09	149.38	266.55
5.000	137.30	294.62	119.92	119.92	120.00	20.00	118.00	353.99	59.38	147.69	274.60
5.500	139.90	303.30	119.96	119.96	120.00	20.00	118.50	355.49	52.19	144.59	279.79

Figure C-3 Example of computing sequence. First page of numerical tabulation.

This is the print-out of computed values resulting from the equations of Figure C-2. A solution interval of DT = 0.5 week is used, and all computations are printed. The step input (in the box) first occurs when time is 1 week. The arrows show the propagation of the numerical changes. Levels are given on the first line of each time step and depend on levels and rates from the preceding time step. Auxiliary variables appear on the second line and depend on levels or on other auxiliaries computed earlier in the same time step. Rates are on the last line and do not depend on each other but only on levels and auxiliaries already computed in the same time step.

TIME	SRR	IAR	SSR	UOR	RRR	INSTP	RSR	IDR	ISR	PDR	UOD
6.000	141.07	313.27	119.98	119.98	120.00	20.00	118.87	356.62	43.35	140.55	282.14
6.500	140.94	323.82	119.99	119.99	120.00	20.00	119.16	357.47	33.65	135.98	281.83
7.000	139.70	334.29	120.00	120.00	120.00	20.00	119.37	358.10	23.81	131.27	279.40
7.500	137.59	344.14	120.00	120.00	120.00	20.00	119.52	358.57	14.43	126.74	275.19
8.000	134.88	352.94	120.00	120.00	120.00	20.00	119.64	358.93	5.99	122.64	269.76
8.500	131.82	360.38	120.00	120.00	120.00	20.00	119.73	359.20	-1.18	119.14	263.64
9.000	128.65	366.29	120.00	120.00	120.00	20.00	119.80	359.40	-6.89	116.35	257.30
9.500	125.58	370.62	120.00	120.00	120.00	20.00	119.85	359.55	-11.07	114.32	251.15
10.000	122.76	373.40	120.00	120.00	120.00	20.00	119.89	359.66	-13.74	113.02	245.52
10.500	120.32	374.78	120.00	120.00	120.00	20.00	119.92	359.75	-15.04	112.40	240.65
11.000	118.34	374.95	120.00	120.00	120.00	20.00	119.94	359.81	-15.14	112.37	236.68
11.500	116.85	374.12	120.00	120.00	120.00	20.00	119.95	359.86	-14.26	112.82	233.70
12.000	115.84	372.54	120.00	120.00	120.00	20.00	119.96	359.89	-12.65	113.64	231.68

Figure C-4 Example of computing sequence. Second page of numerical tabulation.

TIME	SRR	IAR	SSR	UOR	RRR	INSTP	RSR	IDR	ISR	PDR	UOD
12.500	115.29	370.46	120.00	120.00	120.00	20.00	119.97	359.92	-10.54	114.70	230.58
13.000	115.14	368.11	120.00	120.00	120.00	20.00	119.98	359.94	-8.17	115.90	230.29
13.500	115.33	365.68	120.00	120.00	120.00	20.00	119.98	359.95	-5.73	117.12	230.66
14.000	115.78	363.35	120.00	120.00	120.00	20.00	119.99	359.97	-3.38	118.30	231.56
14.500	116.41	361.24	120.00	120.00	120.00	20.00	119.99	359.97	-1.26	119.36	232.32
15.000	117.15	359.44	120.00	120.00	120.00	20.00	119.99	359.98	.54	120.26	234.29
15.500	117.93	358.01	120.00	120.00	120.00	20.00	120.00	359.99	1.97	120.98	235.85
16.000	118.69	356.98	120.00	120.00	120.00	20.00	120.00	359.99	3.01	121.50	237.38
16.500	119.39	356.32	120.00	120.00	120.00	20.00	120.00	359.99	3.67	121.83	238.79
17.000	120.30	356.02	120.00	120.00	120.00	20.00	120.00	359.99	3.98	121.99	240.01
17.500	120.50	356.02	120.00	120.00	120.00	20.00	120.00	360.00	3.98	121.99	241.00
18.000	120.87	356.27	120.00	120.00	120.00	20.00	120.00	360.00	3.73	121.86	241.74
18.500	121.12	356.70	120.00	120.00	120.00	20.00	120.00	360.00	3.29	121.65	242.24

Figure C-5 Example of computing sequence. Third and last page of numerical tabulation.

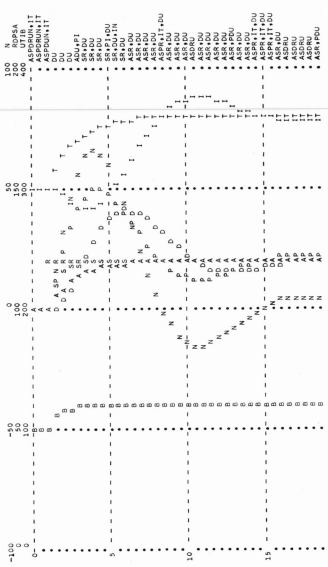

Figure C-6 Example of computing sequence. Plotted results.

The second line shows the letter used to plot each variable. All variables are plotted except **INSTP**, which causes the step change in **RRR**. The letters beside the scales show the variables plotted to each scale. The letter groups on the upper margin show those variables that coincide on the plot, with the first letter in a group standing for all letters in the group.

Solution Interval for Equations

IN Section 7.5 is a discussion of a rule of thumb for selecting the proper solution interval DT to use with a model of a dynamic system. That section should be reread before continuing here.

The choice of interval depends on the relationship between the levels and flow rates in the system. Levels are related to the inflows and outflows by the average delay that an item experiences in the level. This is true for all levels, not just those appearing in configurations we have been calling delays. As the solution interval becomes too long, the content of a level becomes comparable with the quantity that is transmitted in or out during one solution interval. As this begins to happen, either the solution interval is too long, or the level in question has lost its meaning in the system and can be eliminated (the mail delays were eliminated in going from the production-distribution model of Chapter 15 to the advertising model of Chapter 16).

The effect of changing the solution interval can be seen by considering the equations of a first-order delay (see Chapter 9, Equations 9-1 and 9-2):

LEV.K = LEV.J + (DT)(IN.JK − OUT.JK)

$$\textbf{OUT.KL} = \frac{\textbf{LEV.K}}{\textbf{DEL}}$$

Suppose that the delay is initially empty, with zero in and out rates, and that a step input rate of one unit per unit of time occurs at time = 0. Figure D-1 shows the resulting output for various ratios of solution interval over delay, DT/DEL. The horizontal scale is nondimensionalized in terms of time divided by the delay DEL.

If the solution interval is vanishingly small, the true exponential curve is obtained as shown for a zero interval. When DT is half of DEL, the level rises to half of its final value, and the output rate

likewise at the time of the first computation. Half of the remaining difference between output and input rates is closed at each time interval.

When the solution interval equals the delay time, the level and the output rate reach their final values at the time of the first computing step. The exponential delay has taken on some of the characteristics of a pipeline delay but, of course, has an input defined only at the computing intervals. It is not a method for creating a general pipeline delay.

For still longer solution intervals, the first level computed (as for the curve for DT $= \frac{3}{2}$ of the delay) will exceed the steady-state value. The output rate will exceed the input rate. The next computation of the level will produce a value less than the steady state. If the solution interval lies between DEL and 2(DEL), a decaying oscillation will result in the output.

At DT equal to 2(DEL), a sustained oscillation in output rate (either zero or twice the input) will result from a step input. When the solution inter-

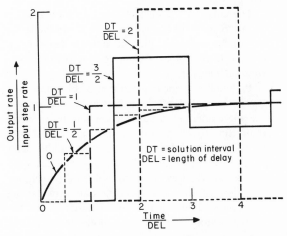

Figure D-1 Unit-step response of first-order delay with various ratios of solution interval DT to delay DEL.

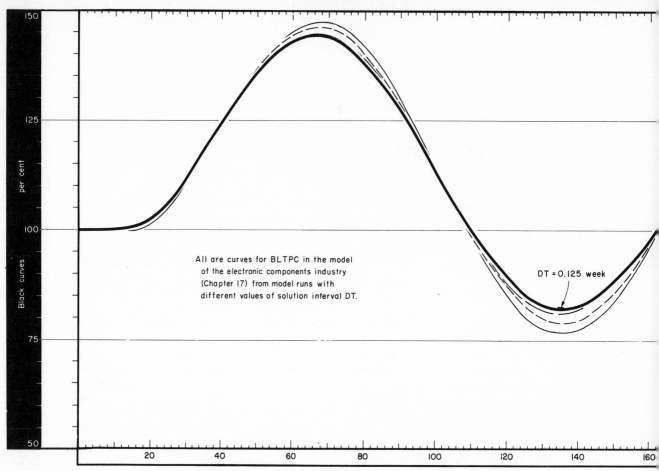

Figure D-2 Effect of changing solution interval.

val DT exceeds 2(DEL), the output-rate fluctuation will grow.

The curve in Figure D-1 for a solution interval equal to half the delay constant is probably an acceptable approximation if only a few of the system delays come as close as this to the chosen solution interval.

It should be kept in mind that a third-order delay is composed of three first-order sections. If the approximation in each section is to be no worse than the DT/DEL $= \frac{1}{2}$ curve in Figure D-1, the solution interval must be $\frac{1}{6}$ or less of the delay constant of any third-order exponential delay.

The criteria used here for selecting a solution interval arise from the structural nature of the system and its internal dynamic considerations. The proper solution interval for a model is not related

to such extraneous matters as the frequency with which data may have been collected in the real system that is being modeled. The solution intervals chosen as herein suggested will be much shorter than have often appeared in the literature of economic models, where intervals as great as a year have sometimes been attempted, even where short-term annual system variations were under study.

The effect of solution interval can be tested empirically with different model runs to see if operation is in a region where solution interval is affecting the results. This was done with the model used for Figure 18-9, which is the model of the electronic components industry, old policies, with TBLAF equal to 40 weeks. Figure D-2 shows the results. Runs were made with DT equal to 0.125

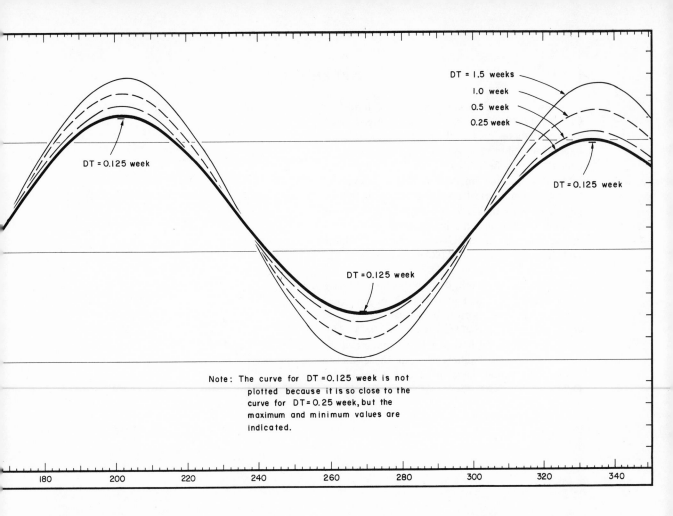

Note: The curve for DT = 0.125 week is not plotted because it is so close to the curve for DT = 0.25 week, but the maximum and minimum values are indicated.

week, 0.25 week (as used in Chapter 18), 0.5 week, 1.0 week, 1.5 weeks, and 2.0 weeks. The 0.125- and 0.25-week values of DT were too close together to be distinguished on the figure. The run for DT = 2.0 weeks was numerically unstable, and at 76 weeks numbers exceeded the permissible register lengths of the computer.

This particular computer run should be especially sensitive to the effect of solution interval because the step input was used, and the system oscillation is "free running" with no driving function to control the periodicity. Even so, the time of the third peak is within 1 week of 335 weeks for each of the curves.

The amplitudes accumulate a little more variation than does the period over the two full cycles of disturbance, as in Table D-1.

Table D-1 Effect of solution interval on amplitude

Solution Interval (weeks)	Amplitude of Third Peak (% of initial value)
0.125	148
0.25	149.9
0.5	153.9
1.0	163.3
1.5	174.8

The ratio of the third peak to the first is 0.79 for a solution interval of 0.125 week, 0.85 for an interval of 1.0 week, and 0.90 for an interval of 1.5 weeks. These are unimportant differences, in view of the accuracy needed and the wide variations of model results for the different conditions in Chapter 18.

A few of the time constants in the model of Chapter 17 (DCPF, DMBLF, and DSF) are 1 week long, and these are less than the longest solution interval in Table D-1. As seen there, the error in computation is beginning to build up rather rapidly. At a solution interval of 2 weeks, some of the internal numerical interactions lead to an explosive computation. The relationship between the solution interval and the height of the third peak of the backlog curve BLTPC is plotted in Figure D-3. The amplitude of BLTPC is measured in per cent of its initial value.

By our rule of thumb, DT would ordinarily have been selected as equal to 0.5 week or less (less than half of the shortest first-order delay, here equal to 1 week for DCPF, DMBLF, and DSF). This choice of DT would have produced numerical results insignificantly different from those that a smaller DT would have produced (see Figure D-2).

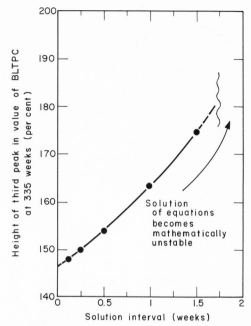

Figure D-3 Computed result versus solution interval.

Smoothing of Information

THE flow rates in an industrial or economic system are usually irregular. The decisions that generate the flows are influenced by a multitude of minor events. Irregularity is caused by the individuality of human actions, the existence of a week end in the otherwise smooth flow of the week's activities, impending labor strikes, the weather, holidays, errors in the assembly and analysis of data, decisions made on an undefined and varying basis, quantity-discount prices that encourage large and intermittent orders, setup costs and the technological nature of production processes that cause separate production lots rather than continuous production, nonuniform accounting periods such as irregular lengths of months that introduce fluctuation into summarized data, international and national news events that alter public optimism, the sampling of data sources periodically, and the intermittent making of decisions weekly, monthly, quarterly or annually.

On the other hand, many of our managerial actions that depend on these irregular flows must be restrained to react only smoothly. Factory production rates cannot be permitted to fluctuate as rapidly as do the day-by-day incoming orders. Inventories are built up and decreased slowly. The manager is in the position of trying to filter out and ignore the fluctuations introduced by other decision makers, while at the same time trying to detect as soon as possible any enduring change to which he wants to respond.

The effort to detect underlying, significant changes in data, while ignoring the superimposed, meaningless fluctuations, is called smoothing or averaging. Smoothing of incoming information occurs to some extent at every decision-making point in the system. In turn, each of these same decisions also contributes its source of noise to the rates being controlled.

We should recognize two separate kinds of mechanisms for data smoothing. The most obvious, but the least frequent, is the formal, numerical processing of data into averages. The more frequent, although more illusive, is the intuitive smoothing introduced by a wait-and-see attitude in reaching decisions.

The existence of a formal, mathematical smoothing procedure is easy to locate in a system. It is a practice based on an established policy. Weekly, monthly, quarterly, and annual summaries of sales, or production, or costs are averages for the periods specified. These formal averaging processes are found in many information flow channels.

In addition, further smoothing is usually introduced at the decision points in a system. We might refer to this as "psychological smoothing." Full and immediate action is seldom taken on a change of incoming information, even though that information may already be the result of numerical averaging processes applied to the raw data. This tendency to delay action until the change in information is insistent and until many indicators point in the same direction is smoothing introduced by judgment, or procrastination, or indecisiveness.

Smoothing processes are fundamental to a proper treatment of system dynamics. Smoothing is essential for filtering out short-period noise. But smoothing inherently introduces delays in information channels and in decisions. Smoothing changes the sensitivity of the system to different periodicities that may exist in data fluctuations.

This is its purpose. Consequently, smoothing distorts for better or worse the information flows in a system.

Smoothing is always a compromise. One is caught in the dilemma between more smoothing to reduce meaningless noise and less smoothing to reduce the time delay in extracting the desired, meaningful information. Smoothing is characterized by these two effects — the attenuation of rapid fluctuations, and the creation of a time delay. These will be examined after discussing two common kinds of smoothing.

A discussion of smoothing methods can become a complex subject in its own right. We shall deal here with only the simple concepts.

Smoothing is a process of taking a series of *past* information values and attempting to form an estimate of the *present* value of the underlying significant content of the data. Smoothing methods can vary widely in the weight (or importance) attached to various ages of the past data. We shall discuss two smoothing methods — the moving average and exponential smoothing.

The moving average attaches the same weight or importance to each past value of a time series back to a cutoff time before which no data are included. "Average sales for the last 8 weeks" is a moving average. Sales for 8 weeks are added together (equal importance) and the total divided by 8. The average would be computed as follows:

$$\text{Moving average} = \frac{1}{8}(S_1 + S_2 + S_3 + S_4 + S_5 + S_6 + S_7 + S_8)$$

where S_1 is the most recent full week, S_2 is the next earlier week, etc. For the general case of a moving average taken over T periods

$$\text{Moving average} = \frac{1}{T}(S_1 + S_2 + \cdots + S_T)$$

The moving average is customarily found in business where formal, numerical averages are obtained. It is simple to explain. The process lends itself easily to the form in which data have usually been kept manually and to the manual clerical computing methods. Weekly, monthly, and annual levels of average activity are moving averages in the sense that they give equal weight to all the data within the interval, but they create additional information distortion by virtue of being computed at widely separated intervals equal to the averaging period, and should be considered also as intermittent or sampled values.

The other common smoothing procedure to be considered here is exponential smoothing. In exponential smoothing, data points are given progressively less weight as they become older. The weighting of past values of data is exponential, meaning that the importance of each successive past value falls by the same ratio. For example, exponential smoothing with an 8-week time constant as applied to weekly sales totals would be computed as follows:

$$\text{Exponential average} = \frac{1}{8}\left[S_1 + \left(1 - \frac{1}{8}\right)S_2 \right.$$
$$+ \left(1 - \frac{1}{8}\right)^2 S_3$$
$$+ \left(1 - \frac{1}{8}\right)^3 S_4$$
$$\left. + \left(1 - \frac{1}{8}\right)^n S_{n+1} + \cdots \right]$$

where S_1 is the sales figure for the most recent full week, S_2 for the next earlier week, etc. For the general case of a smoothing time constant T weeks long, the average would be indicated as follows:

$$\text{Exponential average} = \frac{1}{T}\left[S_1 + \left(1 - \frac{1}{T}\right)S_2 \right.$$
$$+ \left(1 - \frac{1}{T}\right)^2 S_3 + \cdots$$
$$\left. + \left(1 - \frac{1}{T}\right)^n S_{n+1} + \cdots \right]$$

It will be noted that in principle every past term makes a contribution to the average. As a practical matter, the weighting coefficients become so small after the number of terms have reached several times the value of T that the contribution is inconsequential. The sum of the coefficients

$$1 + \left(1 - \frac{1}{T}\right) + \left(1 - \frac{1}{T}\right)^2 + \cdots$$
$$+ \left(1 - \frac{1}{T}\right)^n + \cdots$$

in the series approaches the value of T as the number of terms taken becomes larger and larger. Therefore, for a constant value of sales in each past period, the average approaches, as it should, that same value.

The exponential smoothing gives the greatest weight to the most recent value and attaches progressively less significance to older information. This is more nearly the process of intuitive averaging than is the moving average that attaches equal importance to information back to a sharp cutoff time. It seems like a good form of smoothing to use in representing attitudes and impressions in a system model.

The exponential average has a technical advantage over the moving average for use with computing machines. Contrary to what one might suppose from the preceding equations, the exponential average is easier to calculate. The average at time $= 1$ as given above is

$$A_1 = \frac{1}{T}\left[S_1 + \left(1 - \frac{1}{T}\right)S_2 + \left(1 - \frac{1}{T}\right)^2 S_3 + \cdots \right.$$
$$\left. + \left(1 - \frac{1}{T}\right)^n S_{n+1} + \cdots \right]$$

where the S's are the progressively older values of the variable, for example sales, being averaged The subscripts indicate time into the past. At the next time interval, when time $= 0$,

$$A_0 = \frac{1}{T}\left[S_0 + \left(1 - \frac{1}{T}\right)S_1 + \left(1 - \frac{1}{T}\right)^2 S_2 + \cdots \right.$$
$$\left. + \left(1 - \frac{1}{T}\right)^{n+1} S_{n+1} + \cdots \right]$$

This is equal to a new term plus the old average A_1 multiplied by the exponential weighting factor; therefore

$$A_0 = \frac{1}{T}S_0 + \left(1 - \frac{1}{T}\right)A_1$$
$$= A_1 + \frac{1}{T}\left(S_0 - A_1\right)$$

Each new value of the average can be calculated from the previous value of the average and the new value of the variable being smoothed. The old value of the average can then be discarded, and it is necessary to carry forward only the single numerical value of the average and not a long sequence of earlier data. The last form above is the same as used in Equation 15-8, when we recognize that the interval between solutions need not be the same as the unit of time used for the specification of the exponential time constant T. The amount of correction to the old average in each solution interval will be the above correction multiplied by the length of the solution interval. The smoothing time constant T was previously nondimensional as a multiple of the interval between data points. Generalized to any solution interval, the exponential average becomes:

$$A.K = A.J + \frac{DT}{T}(S.JK - A.J) \qquad \text{E-1, L}$$

A average value of S (same units as S)
DT solution interval (units of time)
T exponential smoothing time constant (units of time)
S variable being smoothed (in its own units of measure)

Diagrammatically, the exponential smoothing is shown in Figure E-1. At the beginning of the computation at time K the old value of the average A.J has been carried forward from time J. The noisy variable that is being smoothed is shown as S.JK. The difference (S.JK − A.J) in Equation E-1 is labeled x. When this is multiplied by $1/T$ it gives the correction to be made in each full unit of time, and when further multiplied by DT it gives y, which is the correction to be made in this solution period.

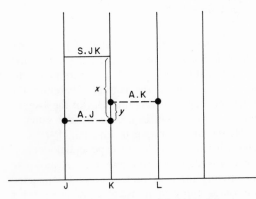

Figure E-1 Exponential smoothing.

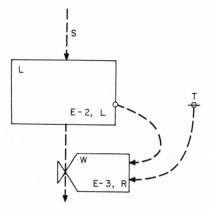

Figure E-2 First-order exponential smoothing and delay.

We shall now discuss the inherent fact that averaging creates a delay in information. This can first be shown by the similarity between Equation E-1 and the usual pair of equations that have been used for a first-order exponential delay. Suppose the S in Equation E-1 be designated as the input to a delay with W as the output, as in Figure E-2. The first-order exponential delay equations would be

$$L.K = L.J + (DT)(S.JK - W.JK) \qquad \text{E-2, L}$$

$$W.KL = \frac{L.K}{T} \qquad \text{E-3, R}$$

L Level in delay (units of S multiplied by time)
S input information flow (in its own units of measure)
W output flow from the delay (in the same units as S)
T exponential-smoothing time constant (units of time)

Equation E-3 can be rewritten to its value one time period earlier as

$$W.JK = \frac{L.J}{T}$$

and this can be substituted into Equation E-2 to obtain

$$L.K = L.J + (DT)\left(S.JK - \frac{L.J}{T}\right)$$

If we now define

L.K = (T)(A.K)

because we note that L in Equation E-2 is T times as large as A in Equation E-1, and substitute this into the preceding equation

(T)(A.K) = (T)(A.J) + (DT)(S.JK − A.J)

and dividing by T

$$A.K = A.J + \frac{(DT)}{T}(S.JK - A.J)$$

which is the same as Equation E-1. The exponential-smoothing and first-order-delay equations are equivalent.

The first-order exponential average causes a delay in information that is of the same amount and transient form as a first-order exponential delay. The smoothing time constant is equivalent to the delay constant that was discussed in Chapter 9. The transient-response curves in Chapter 9 and Appendix G for first-order delays also apply to first-order smoothing.

The delay created by smoothing can be shown in a diagram. Figure E-3 is for a moving average. A uniformly rising true value of the variable· is shown. At any time the average value is that which the true value had at half the averaging period earlier, or stated differently, the average

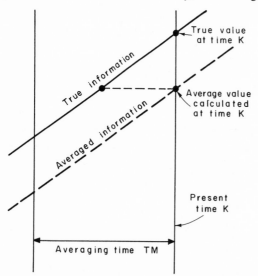

Figure E-3 Delay caused by a moving average.

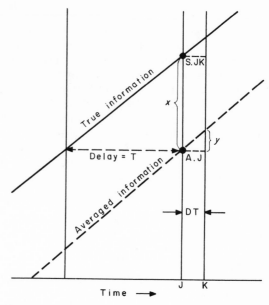

Figure E-4 Delay caused by an exponential average.

value is the same as the true information delayed by half the averaging interval.

Figure E-4 shows the delay in terms of an exponential-smoothing time constant as applied to a steadily rising trend. We can see that the delay must be equal to the time constant T, as follows. By similar triangles,

$$\frac{y}{(DT)} = \frac{x}{T} = \frac{S.JK - A.J}{T}$$

$$y = \frac{(DT)}{T}(S.JK - A.J)$$

where y is the change in the value of the average in the figure and is the same as the right-hand term in Equation E-1, which also represents the change in the value of the average. Therefore, the T in the figure representing the delay of the average behind the true value is necessarily the same in value as the time constant in Equation E-1.

The constant delay due to exponential smoothing as in Figure E-4 is true only for a steady-state ramp (trend) input as shown. More complicated relationships apply in transient conditions or for steady-state signals of other types. Appendix G shows that for a sinusoidal input the delay never exceeds a quarter of a period of the input fluctuation.

A flow of information is distorted in both amplitude and time by being smoothed. The nature of the distortion depends on the character of the changes that occur in the input information, the kind of smoothing, and the amount of smoothing that is required by the kind and degree of unwanted disturbances that exist in the information. Almost all information flows are smoothed by formal mathematical processes, or by psychological processes, or both before being used in deci-

sions. Delays and amplification (here the frequency-sensitive reduction in amplitude) are caused by smoothing, and as we have seen in Part III, these have important effects on the dynamic behavior of systems.

Even in noise-free runs of a model (such as many of the figures in Part III) the smoothing processes must be represented in the model. The smoothing, which the noise imposes, unavoidably acts as a filter to distort the desired information; the smoothing distortion must be present even in the noise-free runs if the system is to be properly represented.

Noise

$\mathbf{A}$N understanding of noise is essential in working with models of information-feedback systems. The decision functions that we can formulate will account for only the major factors influencing an action stream. Multitudes of effects will impinge from outside the system under study. As discussed in Appendix E, the presence of noise, that is, random events, makes data smoothing necessary, and with smoothing come resulting delays. As seen in Figures 15-20 and 18-5, noise provides the excitation to induce modes of disturbance to which a system is sensitive. As will be discussed in Appendix K, noise limits the ability to predict the future of a system.[1]

In this book we have chosen to formulate models first around the continuous, noise-free flows of information, decisions, and action. After the noise-free dynamic character of the system is observed, noise is then added to see what randomness contributes to system operation. This is a different study sequence from the approach taken in stochastic models in which decisions are formulated to create sequences of separate events whose statistical probabilities of occurrence may be biased by the state of the system. I believe that starting first with an exploration of the noise-free system leads more quickly to understanding the way that the underlying system structure affects operation.

When we are ready to introduce noise components into system decisions, we must concern ourselves with the details of methodology for doing this. How is noise to be specified? What characteristics of noise interest us? A noise signal is one containing power (where power measures the integral of the square of the signal) over a wide range of frequencies.

There are many possible views of a noise signal. In the terminology of the physical sciences the term "white noise" is used to describe a continuous waveform whose *power density* spectrum is constant at all frequencies from zero to infinity and whose amplitude probability density is Gaussian. It is a continuous signal that contains an infinite amount of power, can have instantaneous values infinitely large, and its present value tells nothing about a particular future value, even an infinitesimal time interval away.

By having a constant power density spectrum, as in white noise, we mean that the power is the same in any frequency band of a specified width, regardless of where that band is centered. See Figure F-1. For the example of a wide-band electronic noise source, the same amount of power would be measured after we pass the noise through a filter 1,000 cycles per second wide that includes the range from 1,000 cycles per second to 2,000 cycles per second, as would be measured after a filter that passes the frequency range 1,111,000 cycles per second to 1,112,000 cycles per second. It should be noted that by definition the noise power in white noise depends on the arithmetic width of the passband, not on the ratio of the lower to upper frequency. In the first example the

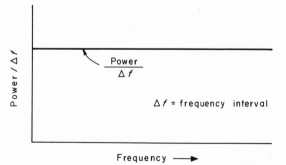

Figure F-1 White noise showing its constant power per frequency interval.

[1] Discussions of noise in information-feedback systems will be found in Seifert and Steeg in Reference 5, Laning and Battin in Reference 23, and Davenport and Root in Reference 24.

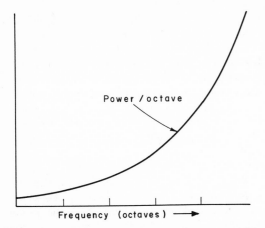

Figure F-2 White noise showing its exponential rise in power per octave.

upper frequency bound of the band is twice the lower. In the second example it is less than 0.1% higher. The noise power is the same in every frequency band of the same absolute width but is very different in frequency bands measured in octaves (an octave is a band bounded by an upper frequency that is twice the lower). In a white-noise source, the noise power per octave doubles with each higher octave. See Figure F-2. For example, assume that 1 unit of power is found in the octave from 1,000 cycles per second to 2,000 cycles per second. Then, if the noise is from a white-noise source, the octave from 2,000,000 cycles per second to 4,000,000 cycles per second will contain 2,000 units of power.

White noise specifies a particular power distribution over a frequency range, but this particular specification does not necessarily describe the kind of noise that we want to use. We shall now relate the ideas of white noise and noise power more specifically to the use of noise in models of social systems.

A true white-noise generator, since it implies a range that extends to infinite power and infinite frequencies, clearly does not exist. However, excellent approximations exist over certain frequency ranges. One view of a noise signal is to consider it a series of discrete random numbers. These numbers might be equally spaced in time. With such a set we can think of a continuous noise signal as the curve that connects these values, as shown in

Figure F-3. A waveform as in Figure F-3 is a good approximation to white noise up to the vicinity of frequencies whose periods are twice the interval between noise samples. In other words, the highest frequency that need be present in the curve passing through a series of random values uniformly spaced in time is half the frequency with which the random samples themselves appear.

A series of equally spaced random numbers is easy to use as a noise source in work with models of social systems. But will it represent the decision-making perturbations that we wish to incorporate? The problem here is the same as in the selection of other model relationships and parameters. We are interested in noise sources that represent the character of the disturbances that we believe exist in the actual system. Merely taking a series of random numbers does not give any assurance of serving the purpose. What should be the average deviation? What should be the noise power versus frequency distribution? How often should noise samples be selected? How vulnerable are our conclusions to the noise specifications?

Fortunately, the conclusions we wish to draw from models do not seem to be particularly sensitive to the specifications of the noise signals that are used. However, some general observations and guides should be mentioned.

A noise signal described by a sequence of random numbers as in Figure F-3 approximates a white-noise source at frequencies that are low compared with the sampling frequency. Such a signal has uniform noise power per increment in frequency but not per octave. When we observe a waveform, we are probably most sensitive visually to power per octave. We can see in Figure F-3

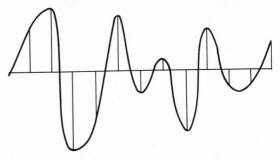

Figure F-3 Uniformly spaced random numbers and continuous noise curve.

the predominant noise power at a frequency of half the sampling frequency. We do not see or sense the lower-frequency components because they are so small on a power-per-octave basis.

A noise signal cannot simply be chosen as a sequence of random numbers because this procedure completely and arbitrarily specifies the entire power spectrum, and the specification may not be suitable to our purpose. As an example, consider a random variable representing weather in a model of an economic system or a commodity market. Suppose, for reasons otherwise determined, that the model is to be evaluated daily. We might then select a random number daily for the amount of rainfall. Its random character would then need to reproduce the daily variation to be expected in rainfall. This is not sufficient. The daily random numbers must be filtered to control the weekly, monthly, annual, and even longer-term rainfall variation, because rainfall is not a purely random day-by-day phenomenon but has long-term correlations as well. This can be seen by what would happen if the daily random numbers were summed to create annual rainfall. A proper distribution on the assumption that each day were an independent event might not yield, in groups of 365, a variation that would represent what we expect to find from year to year. From the viewpoint of noise power, an annual frequency is some seven octaves removed from a daily frequency. The power in a white-noise signal drops by a factor of 2 per octave. The power in the octave between 0.75 and 1.5 cycles per year will be some 300 times less than that in the daily variation. This is another way of saying that in rainfall successive minutes or hours or days or months are not independent uncorrelated events. Even successive years may not be randomly related. This being so, we must give some attention to the character of a noise signal over the entire frequency band of interest.

A simple control of noise power density was used in Chapters 15 through 18, where noise was sampled and held for an interval longer than the solution interval for the equation. This had the effect of suppressing the higher frequencies near the solution-interval frequency in order that more low-frequency power could be generated without excessive high-frequency power values. In Chap-

ter 15, Equation 15-79, noise was sampled and held 1 week in a model whose equations were being evaluated each 0.05 week. Suppose that we had attempted to create a week-by-week random variation in sales covering a 2-to-1 range (not as great as this in Chapter 15) by adding groups of random numbers taken 20 at a time. The hour-by-hour and day-by-day variation might have to be absurdly high, sometimes requiring canceling more orders than were already placed, to make the long-period variation large enough.

When we consider the character of a random-noise signal, the frequency selectivity of systems is especially striking. In Figure 18-5 the random signal is sampled and held for 5 weeks. Its strongest power content on a power-per-octave basis is in the highest frequency range represented. This is the range of periods around 10 weeks, or a frequency of some 5 cycles per year. Yet this high-frequency power is almost completely absorbed by the smoothing and delays of the system. The system as a whole responds to the much smaller amount of noise energy representing periods of two years or about a half a cycle per year. This is the frequency range that the system amplifies to magnitudes greater than represented in the input noise signal.

It should be noted that smoothing suppresses the higher frequencies in a noise source but will pass the lower frequencies. These lower frequencies are the components in the noise that autocorrelate over longer time periods.

In estimating noise variables we must be alert to distinguish low-frequency disturbances from outside the system (by definition, noise) from the internal natural frequencies of the system. It is probably impossible to determine by observation how much of any low-frequency randomness is to be attributed to the actual amplitude of external disturbance and how much is to be attributed to the internal amplification of that input by the system. We shall usually rely on the knowledge of details of system structure to create the model sensitivity versus frequency and shall then choose (as discussed in Chapter 13) an amplitude of random signal that yields observed system behavior amplitudes. Only if the required noise is of implausible magnitude will independent knowledge of real-system noise be useful in determining model validity.

One should observe that the selection of the noise power spectrum should be related to the time constants of the part of the system with which the noise is associated. In Figure 17-14 and Equations 17-94 and 17-95 noise signals were introduced in two places — the output flow of the engineering design pool EDPC and the output of the orders in process in customer purchasing RCC. The noise at the engineering department output acts on a pool of much longer time constant (the average value of transit time through engineering DEEDC is about 30 weeks). This is a large pool in terms of the flow rate through it. It can support long-term noise components in the outflow deci-

sion. Were we to try to introduce long-term noise at some pool in the system where the delay is short, the pool would decline at a rapid rate compared with the period of the noise component and would attenuate the noise effect. Noise at large pools is of special importance. The large pools can support varying flows that have periods which are comparable with those of the system itself. The prospective purchaser pool PPC of Figures 16-1 and 16-5 is also an example. These are points where long-period noise can affect the system. They are the points that in actual systems are apt to be affected by long-period external disturbances.

The use of noise in dynamic models needs much more study. This appendix has attempted to point out some of the important considerations.

APPENDIX G

Phase and Gain Relationships

EARLIER sections have commented on the way in which data smoothing and the delays in flow channels cause amplitude and timing distortions in the flow rates. Chapter 9 discussed the transient characteristics of exponential delays in response to step and impulse inputs. We also are often interested in what smoothing and delays will do to sinusoidally varying signals that pass through them.

A sinusoidal signal passing through any kind of passive filter such as the exponential smoothing and the delays that have been used in this book will of course emerge with the same frequency.

This is for a steady-state continuing sinusoidal signal. The transient, sudden beginning of such a signal may, like a step function, contain components of other frequencies. A linear filter does not cause new frequencies to be generated within the filter system. On the other hand, a nonlinear filter can create harmonics of an input variation.

In dealing with simple exponential filters, we are interested in how the output is related to the input both in phase and in amplitude. Both of these relationships will depend on the way in which the period of the sinusoidal input is related to the delay time constant of the filter.

Figure G-1 shows the ratio of the amplitude of the output signal to the amplitude of the input signal as it depends on the ratio of the filter delay to the period of the input signal. Three curves are shown. One is for a first-order exponential delay or smoothing equation, another is for a third-order exponential delay, and the horizontal line is for an infinite-order delay. All represent the same delay. In other words, for a given steady-state continuous input they would each have the same total amount stored in the system. On the left-hand side of the diagram we note that for the first-order and third-order delays the output signal is approximately equal to the input when the period of the sinusoid is very long compared with the delay. As the period becomes shorter, the output amplitude begins to decrease. When the time taken for the period of one cycle of the signal becomes equal to the delay, the amplitude of the output fluctuations has fallen to around 0.1 of the amplitude of the input sinusoid. This shows how the process of smoothing tends to eliminate the high-frequency disturbances. The high frequencies are absorbed by the filter while the longer period disturbances are passed through it. We see this filtering action taking place throughout our industrial systems. Inventories exist to absorb short-term fluctuations of incoming sales. Production rates are not controlled to vary with the week-by-week changes in sales rate, but instead respond to longer average values of sales. Managerial decisions are delayed while confirmation is sought for apparent changes in conditions.

Beside changing the amplitude of a sinusoidal signal passing through the exponential filter, the time relationship between the input and the output will also be altered. This is shown in Figure G-2. The first-order exponential delay creates a maximum phase difference of $\frac{1}{4}$ period of lag of the peaks and valleys in the output behind the corresponding peaks and valleys in the input signal. Each such section of a filter can produce another $\frac{1}{4}$-period lag. We see, therefore, that the maximum lag created by the third-order exponential delay is $\frac{3}{4}$ of a period. On the other hand,

the infinite-order exponential delay that is the discrete or pipeline delay, as discussed in Chapter 9, creates a delay that is continuously proportional to frequency or inversely proportional to period.

It should be noted in Figures G-1 and G-2 that the total delay is the same for each of the curves. Therefore, the third-order exponential delay is made up of three first-order sections, each of which is one-third as long a delay as that in the first-order delay.

In Figure G-2 the time delay differs very little between the three curves until the period of the sinusoidal disturbance decreases to ten times that of the filter delay. As the period becomes shorter and approaches that of the delay, the phase shift between the three curves changes rapidly.

In Figure G-2 we can see that for a given time delay the phase relationship of the output compared with the input depends on the frequency of the signal. For example, a signal having a period of ten times the delay of the filter will be shifted about $\frac{3}{16}$ of a period. On the other hand, we see that a signal with a shorter period that is equal to the delay will be shifted by approximately an additional half period when passing through a third-order delay. This means, as will be seen in Appendix M, that a system which contains sufficient amplification can find that period of oscillation at which critical feedback loops in the systems have an unstable positive-feedback relationship. The attempt to suppress a fluctuation by countercyclical policy may simply cause the natural period of the system to shift until a new mode is found where positive feedback occurs.

When several exponential delays are cascaded one after another, the total effect is easy to determine for a steady-state sinusoidal signal traversing the series. The amplitude ratios of the individual filters should be multiplied together. The phase-shift angles of the separate filters should be added.

The curves of Figures G-1 and G-2 are convenient in estimating the effect to be expected on signals of various periodicities, depending on the delay or smoothing time constant.

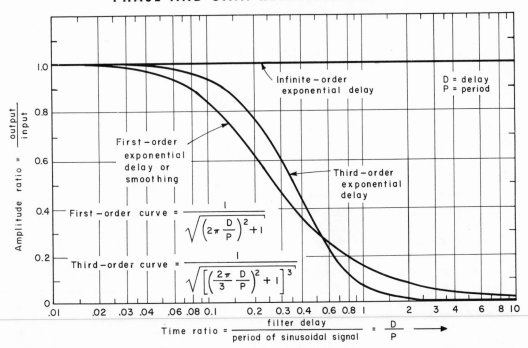

Figure G-1 Ratio of output amplitude to input amplitude versus ratio of delay time to input period.

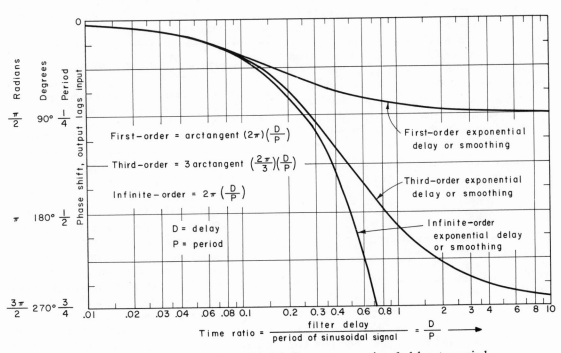

Figure G-2 Phase shift of exponential filter versus ratio of delay to period.

Delays

THREE points regarding delays that have not appeared in earlier sections are included here.

H.1 Information versus Material Delays

A distinction needs to be made between the delay of an information flow and the delay of the flow of specific physical items. In earlier sections delay equations have been used to delay the transit of material and orders. Smoothing equations have been used to delay the flow of information. It has already been stated in Appendix E that the transient behavior of the two is the same. Nevertheless, they differ somewhat in concept, and they differ in the way they will behave if the time constant of the delay is no longer constant but becomes a variable.

A material delay should not lose or create any units in the flow that is traveling through it. This means that in a material delay with a constant inflow rate the outflow will exhibit a transient change if the delay constant is changed. It is necessary that the output should differ from the input for a time long enough for the internal level stored in the delay to be adjusted.

The following are the equations for a first-order exponential delay in which the delay is variable:

$$\text{LEV.K} = \text{LEV.J} + (\text{DT})(\text{IN.JK} - \text{OUT.JK}) \qquad \text{H-1, L}$$

$$\text{OUT.KL} = \frac{\text{LEV.K}}{\text{DEL.K}} \qquad \text{H-2, R}$$

LEV	LEVel stored in delay (units)
DT	Solution interval (time)
IN	INflow rate (units/time)
OUT	OUTflow rate (units/time)
DEL	DELay, variable (time)

Equation H-1 accumulates the difference between the inflow and the outflow. Equation H-2 bases the outflow rate on the level generated by the preceding equation. For a constant inflow rate and steady-state conditions existing initially, the content of the level would be equal to the inflow rate multiplied by the delay. If the delay were now reduced to half its previous value, the quantity in the level must necessarily be reduced even though the inflow rate is constant. This will require that the outflow rate be higher than the inflow for a period of time. The higher outflow will result from Equation H-2 until the stored level has fallen to the proper value to make the outflow equal the inflow.

On the other hand, in delaying the flow of a constantly repeated value of information, this value should not change merely because the transmission delay is changed. This transient independence of changes in delay can be examined in the following smoothing equation:

$$\text{INS.K} = \text{INS.J} + \frac{\text{DT}}{\text{DEL.K}}(\text{IN.JK} - \text{INS.J}) \qquad \text{H-3, L}$$

INS	INput Smoothed (in the units of measure of IN)
DT	Solution interval (time)
DEL	DELay, variable (time)
IN	INput informaton (in its own units of measure)

In the steady state when there is a constant inflow rate IN, the smoothed value INS will have the same value. The difference in the right-hand term of Equation H-3 will be zero. The value of the delay DEL can therefore change without changing the value of the smoothed output. This is as it should be for the delayed transmission of information. Furthermore, Equation H-3 does not change the units of measure between the input and the smoothed value that is the output. In Equation H-2 the output of the material delay is measured in the same units as the input. However, the content of the level is measured in units of input multiplied by time. This turns out to be awkward and meaningless in many information channels.

For the delayed transmission of physical quantities, delay equations such as H-1 and H-2 or those in Chapter 9 should be used. For the delay of information, smoothing equations as discussed in Appendix E and as given in Equation H-3 should be used.

H.2 Alternative Equations for Exponential Delays

The difference Equations H-1 and H-2 can be written in various forms, depending on convenience. The form given in Section H.1 above is somewhat inefficient, since it requires storing the two quantities LEV and OUT from one time period until the next. In principle, a first-order exponential delay should require the storage of only one numerical value between time periods.

In representing a third-order exponential delay, the DYNAMO compiler uses a single equation for each first-order delay stage, which can be derived as follows. Write Equation H-2 in its form for one time period earlier (and return to a constant delay)

$$\text{OUT.JK} = \frac{\text{LEV.J}}{\text{DEL}}$$

Rearrange to become

$$\text{LEV.J} = (\text{DEL})(\text{OUT.JK})$$

Substitute into Equation H-1 to get

$$\text{LEV.K} = (\text{DEL})(\text{OUT.JK}) + (\text{DT})\ (\text{IN.JK} - \text{OUT.JK})$$

Substitute into Equation H-2 to get

$$\text{OUT.KL} = \text{OUT.JK} + \frac{\text{DT}}{\text{DEL}}(\text{IN.JK} - \text{OUT.JK})$$

This last equation is of the same form as the smoothing equation except that it bases the new outflow rate on the new inflow rate and the old outflow rate. This means that only the old outflow rate OUT need be carried over from the previous time period. Three equations of this form constitute a third-order exponential delay. The outflow of the first is defined as the variable flowing into the second, and the outflow of the second becomes the inflow to the third. The outflow from

the third is the outflow from the delay. Each section would contain one-third of the total delay. This would yield the following set of equations that is automatically supplied by the DYNAMO compiler in response to the functional notation that a third-order delay is wanted:

$$\text{R1.KL} = \text{R1.JK} + \frac{\text{DT}}{(\text{DEL})/3}(\text{IN.JK} - \text{R1.JK})$$

$$\text{R2.KL} = \text{R2.JK} + \frac{\text{DT}}{(\text{DEL})/3}(\text{R1.JK} - \text{R2.JK})$$

$$\text{OUT.KL} = \text{OUT.JK} + \frac{\text{DT}}{(\text{DEL})/3}(\text{R2.JK} - \text{OUT.JK})$$

IN	INflow rate (units/time)
DEL	total DELay for the three delay sections (time)
R1	intermediate rate variable created by DYNAMO (units/time)
R2	intermediate rate variable created by DYNAMO (units/time)
OUT	OUTflow rate (units/time)

The preceding group of equations would be created by the DYNAMO compiler in response to the following functional equation form in the system equations:

$$\text{OUT.KL} = \text{DELAY3}(\text{IN.JK, DEL})$$

H.3 System Effect from Time Response of Delays

Section 9.4 stressed the time response of various exponential delays and related these to physical phenomena. Throughout the rest of the book third-order delays appear most places in flow channels. Actually it seems that systems will usually not be very sensitive to the time response of the delay representation.[1]

To illustrate the effect of changing the time response of delays, the delays in the advertising model of Chapter 16 were changed. Figure H-1 shows the results. All three curves are for the factory output. All are for the same step input of generation of need at the customer used in Figure 16-7. The line for factory output in Figure 16-7

[1] This is contrary to the findings of Phillips in Reference 25. I believe that Phillips observed a high dependence on the "time-forms of lagged responses," primarily because of the very simple systems with which he was working.

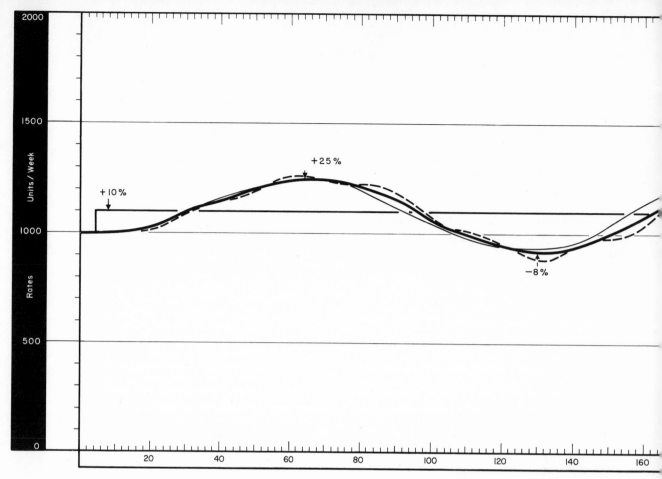

Figure H-1 Effect of changing order of the exponential delays in the advertising model of Chapter 16.

is reproduced here for the system containing the third-order delays as formulated in Chapters 15 and 16.

All of the third-order delays of the system of Chapter 16 were then changed to first-order delays, yielding the curve shown. This curve is very slightly more stable and of a very slightly shorter period. The differences are not significant.

The system of Chapter 16 was then changed so that all of the third-order delays were replaced with pipeline delays. For the pipeline delays a time resolution of 0.1 week was used, which is the solution interval of the equations. In other words, a 2-week delay would be replaced by 20 boxcars (see Appendix N) in a boxcar-delay chain that is indexed once each solution interval. The contents emerging at a particular solution interval are those stored 20 intervals earlier. The time responses of first-order, third-order, and

pipeline delays are shown in Figures 9-3 through 9-8. Figure H-1 illustrates the effect of introducing the pipeline delays into a system. The gross overall behavior of the system is not greatly changed. The period of fluctuation is a little longer. The amplitude of fluctuation is growing somewhat more rapidly. The most significant difference is the appearance of the natural period of the distribution chain in the total system response. The pipeline delays do not absorb the higher-frequency disturbances in the way that a first- or third-order exponential delay will do. In fact, the true pipeline delay does not attenuate any frequencies that are passed through it. This tends to permit the inventory fluctuations reaching the factory to be transmitted down the advertising channel to affect the input to the distribution chain.

The effect of the time response of delays may depend on where the delay is in the system. One

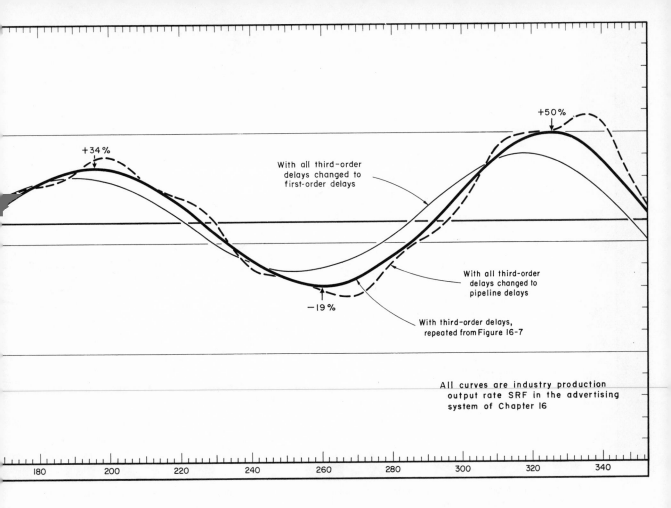

+34%

With all third-order
delays changed to
first-order delays

+50%

With all third-order
delays changed to
pipeline delays

With third-order delays,
repeated from Figure 16-7

−19%

All curves are industry production
output rate SRF in the advertising
system of Chapter 16

180 200 220 240 260 280 300 320 340

should not draw the general conclusion that the time response does not matter. However, for most of the incidental delays within a large system we can expect to find that the time response of the delay is not a critical factor. Third-order exponential delays are usually a good compromise.

APPENDIX I

Phase Shift and Turning Points

SOME attempts to forecast changes in economic activity are based on a search for leading and lagging time series. Certain aspects of economic activity tend toward leading or lagging in time-phase relationship with respect to other aspects. Turning points are sought in one time series in hopes that they will be an advance indication of a similar change in other time series. In Chapter 17 appeared a typical time-phase relationship between production and inventories. This implied a leading time-series relationship of employment to inventories. Nevertheless, efforts to base economic pre-

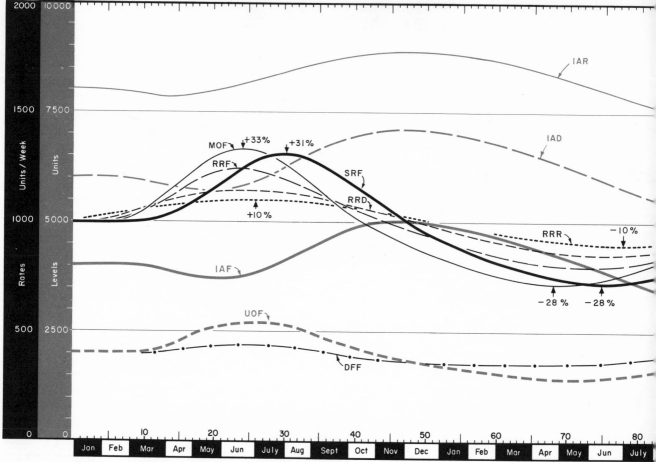

Figure AI-1 Response of production-distribution system of Chapter 15 to a 10% sinusoidal input of two-year period.

dictions on leading and lagging time series have not been conspicuously successful.

The search for leading and lagging time series rests on the hope that a fixed time-lag relationship between different system variables will tend to exist. Even the very simple system described in Chapters 15 and 16 can be used to shed some light on the nature of this problem. As already seen in Appendix G, the phase shift between two variables can depend on the frequency of the disturbance traversing the system. In Appendix G the discussion related only to delay and smoothing functions that contain no amplification. In the absence of amplification, the output of a system will lag the input. The output, however, need no longer lag the input in a system containing frequency-sensitive amplification. A dependent variable may either lead or lag the independent input disturbance, depending on the frequency of the

disturbance and the system characteristics.

In Figure 15-19 is the simple production-distribution system of Chapter 15 upon which is impressed a sinusoidal disturbance with a one-year period. Under these circumstances there is a first maximum in retail sales that is 10% above average and occurs at 13 weeks. The first peak in manufacturing orders at the factory MOF reaches a peak 7 weeks later at the 20th week. This seems fairly reasonable and to be expected in view of the delays existing in the distribution system. However, the second peak in retail sales occurs at the 65th week, and the corresponding peak in the factory manufacturing orders occurs only 1 week later. Suppose now that we examine the same system with an input disturbance having a longer period.

Figure AI-1 shows the same production-distribution system with a 10% sinusoidal variation

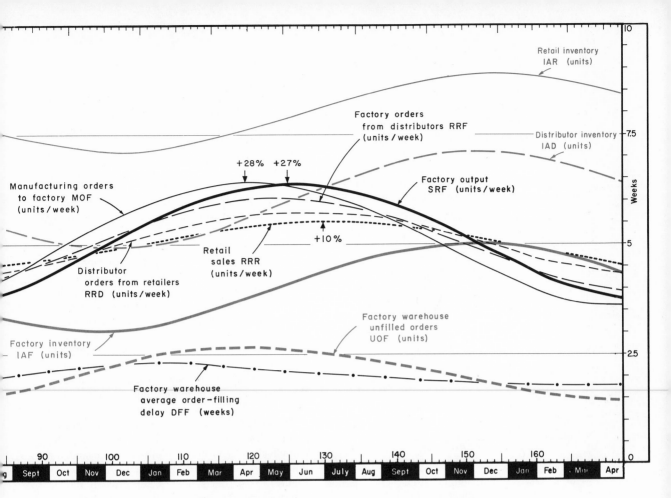

at retail sales having a period of two years. Retail sales RRR reach a first maximum at 26 weeks. Manufacturing orders MOF reach a first maximum 2 weeks earlier. The second peak in retail sales occurs at 130 weeks. The corresponding peak in manufacturing orders at the factory reaches a peak 11 weeks before the peak in retail sales. Clearly, in this system, manufacturing orders are dependent on retail sales, yet manufacturing orders reach their peak-and-valley conditions before the corresponding turning points in retail sales. This is an example of the "acceleration principle" of economics as it applies to inventory and pipeline accumulation. The effect of inventories is evident in the recession period in Figure AI-1 beginning at the 40th week and following through into the recovery that starts after the 80th week. Beginning at the 38th week, manufacturing orders are below retail sales. Factory production falls below retail

sales at the 46th week. After that time inventories are declining along with retail sales. If the managerial policies relate inventories to the sales level, then inventories would continue to decline as sales decline. In fact the rate of decline of inventories would depend on the rate at which sales are declining, and therefore, the amount that factory production falls below retail sales depends on the rate of fall of retail sales. The maximum rate of fall of retail sales occurs where the slope of the retail sales line is the greatest, which is at the 52nd week. After this point, retail sales begin to fall less and less rapidly. Soon the managerial policies begin to call for slowing down and eventually ceasing the liquidation of inventories. As this happens, the gap between manufacturing rate and retail sales must close. Conditions for an upturn in manufacturing begin to take shape after the 52nd week, and actually result in an upturn in

423

manufacturing orders by the 68th week. This occurs well ahead of the upturn in retail sales that, when it occurs, calls for an expanding inventory and pipeline investment and a rapid rise in manufacturing.

We can plot for the system as a whole the same types of curves that were given in Appendix G for the simple delay or smoothing equations. The

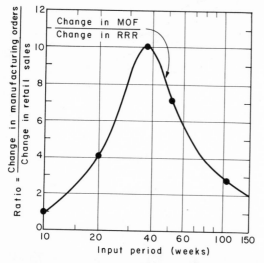

Figure AI-2 System amplification versus period of sine input at retail sales.

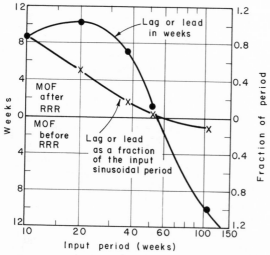

Figure AI-3 Lag or lead of manufacturing orders relative to retail sales versus period of sine input at retail sales.

system-amplification and the time-delay relationships between retail sales and manufacturing orders can both be plotted as a function of the period of the input sinusoidal retail sales.

Figure AI-2 shows how the distribution system amplifies sinusoidal inputs of different periods. Each point on this curve has been obtained from a computer run similar to Figure AI-1. For a very rapid fluctuation at retail having a period of 10 weeks, the system shows no amplification. The fluctuation caused at the factory is the same amplitude as that at retail sales. At even more rapid variation in retail sales, there is attenuation in the system, and the factory would respond less than the variation at retail. The very short term fluctuations do not propagate through the system. However, as the periodicity of the input lengthens, the system amplification increases. In Figure 15-18 the system had a natural period of a little under 40 weeks. When the input disturbance is near this frequency, a maximum amplification occurs. Here the fluctuation in factory manufacturing orders is ten times as great as the fluctuation at retail sales. As has been noted elsewhere, this amplification is probably unusually high and not typical of many industrial systems. As shown in Figure 15-24, the system is sensitive to the inventory and the pipeline adjustment times DIR, DID, and DIF, and in the example system these are probably shorter than would be typical.

Beyond the peak value of amplification, the ratio of output to input falls and eventually begins to approach a ratio of 1. This occurs as the period of the disturbance becomes long enough for inventory and pipeline changes to take place slowly over a long enough period so that they do not add appreciably to the retail sales demands.[1]

At the same time that amplification is changing with frequency, interesting changes are taking place in the time-phase relationships between the variables. Figure AI-3 shows the lag or lead relationship between factory manufacturing orders and retail sales. One curve shows the relationship in weeks and the other as a fraction of the period of

[1] The computer runs for Figure AI-2 and AI-3 were taken with a 5% input variation of retail sales in order that the peak amplification at the resonant frequency not cause entry into the highly nonlinear region of zero production rate.

the input sinusoidal disturbance. For the very short period inputs, the time delay is in the 8- to 10-week interval, with manufacturing orders lagging behind changes in retail sales. In fact in this region a complete phase reversal can occur with factory manufacturing orders at a peak at the same time that retail sales are at a minimum. As the periodicities of the input become longer, the inventory and pipeline accumulation effects can take place to create a phase displacement that causes manufacturing orders first to approach retail sales in time and then actually to occur in advance of retail sales. The most rapid rate of change in the lag or lead relationship occurs near the peak in Figure AI-2. In Figure AI-3 manufacturing orders lag behind retail sales by about 7 weeks, when the input disturbance has a 38-week period. By the time the input disturbance has reached a period of 55 weeks, which is only slightly more than a year, the manufacturing orders are occurring simultaneously with retail sales. By the time the disturbance to which the system is being subjected has reached a period that is 2 years long, the manufacturing orders are leading some 10 weeks ahead of retail sales.

Much more investigation of this phenomenon is called for. However, I believe that many of our industrial systems will be found to be operating generally in a region comparable with the zone between 40 weeks and 100 weeks in Figure AI-3. A fuller development of a model with the inclusion of factors not appearing in this simplified example will tend to lengthen the natural period of the system. This means that the region of rapidly changing lag or lead relationships will probably lie in a periodicity range comparable to that exhibited by business cycle fluctuation. Small variations in the rate of change of system variables can cause the lag or lead relationships to shift rather rapidly as indicated by Figure AI-3. The multitude of noise sources containing periodicities in this range will create similar effects. The lag or lead relationships are dependent in a very complicated way on the past history of the pertinent variables and their rates of change. It is hardly to be expected that decisive indicators of future events can be found in these relationships. Certain variables will tend, however, to be related in specific ways, and on the average certain directions of time relationship will predominate over the reverse directions.

The problem of the leading and lagging series can be examined in Figure 18-5. In that system of Chapters 17 and 18 the situation is very simple, and the cause-and-effect relationships are free of the complicated interlocking loops that would be found in a complete economic system. As a result we should expect in Figure 18-5 that the leading and lagging phase relationships would be more consistent and have more decisiveness than they would in a system of greater complication. By inspection, it appears that the backlog curve leads the employment curve and the employment curve leads the inventory curve most of the time. This is especially true for the large and long-period disturbances. The lead-lag relationship begins to break down, however, for the smaller disturbances. We should expect the time relationships to become less decisive in a complex system with more interlocking relationships between many industries.

Value of Information

INFORMATION is the substance from which managerial decisions are made. As with other products the quality of the raw material partially determines the quality of the output. The manager is well aware that information sources are important. But does he know with any assurance the measures of information quality?

Information sources, like other flows in the industrial system, are subject to being distorted. The distortion can occur in amplitude and in timing. The distortion will depend on the time-varying character of the information. Amplification can exist in information channels. The kinds of distortions seen in Appendices E, G, and I occur in the information flows.

The manager is interested in more useful information on which to base decisions. Greater utility can be obtained both by improving existing information sources and by basing decisions on new and different sources that have not been used in the past. The cost of obtaining improved information can usually be estimated with fair accuracy. But what is better information worth? This is not so easily answered. Better information is worth the value that we attach to the improved industrial performance which results when the better information is available. Unless we can determine the change in system performance that will result from a changed information flow, we cannot determine its value.

The value of information has usually been determined by highly subjective means that necessarily include an estimate of what the information will do to the dynamic behavior of the system. Our ability to estimate the characteristics of information-feedback systems is poor. It is to be expected that one of the weakest areas of managerial judgment is in placing a dollar value on an information source.

The United States has a tremendous data-processing industry that caters to the managerial conviction that better information means better management. Machine processing of business data is usually justified on the basis of machine-processing costs relative to costs in the previously existing system. For lack of any real measure of the value of information, this justification is almost never made on the basis of the relationships between information cost and information value.

It is my belief, based on experiences with industrial organizations, that some of the most important and useful information is going unused and untouched. At the same time, great efforts are devoted to attempting to acquire information that even were it available would do little for the success of the organization. As an example, some organizations have succeeded in speeding up the flow of sales information and production scheduling to the point where the random-noise variations in the market can now be directly imposed on the production process. This tends to ignore the proper use of inventories for absorbing such variations. Carried to an extreme, the result of more timely information can be harmful. The effect can be to cause the manager to put more and more stress on short-range decisions. In Chapter 18 the system improvements did not result so much from changing the type of information available or its quality nearly so much as from changing the sources of information used and the nature of the decision based on the information.

A system model can be used to see how changed information flows will affect the system. Doing this requires a firm hypothesis about how the information affects decisions. Herein would be the first test of much of the information generated in industrial organizations. For much of the information selected, it would be nearly impossible to suggest a hypothesis about the way such information affects the flow of business activity. Inability to suggest a use for an information source raises the possibility that it is not useful. Given the as-

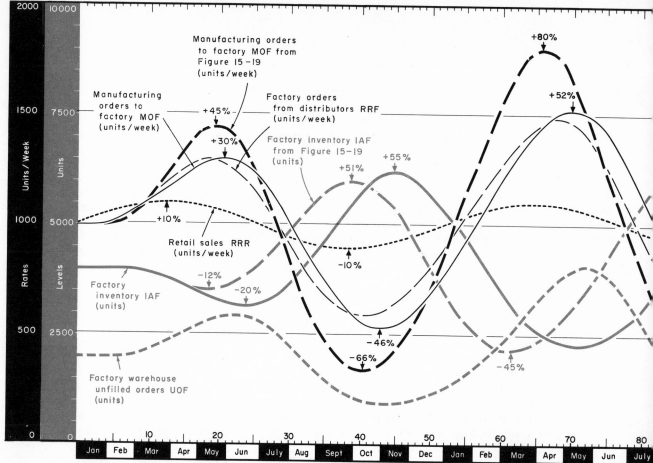

Figure J-1 Production-distribution system of Chapter 15 with retail sales information available at factory.

sumption of some new and presumably better information source, in setting up a model we are required to state how this information flow is to affect the decision streams.

An illustrative example can be based on the production-distribution system of Chapter 15. It is sometimes assumed that the problems of production planning would disappear if full knowledge were available about the behavior of retail sales. In the model of Chapter 15 we can assume the availability of retail sales at the factory level and see what this would do in the presence of various kinds of market patterns.

Figure J-1 shows such an altered system. Here the *average* sales at retail are assumed to be available immediately at the factory and to be incorporated into the manufacturing rate decision. (The exact changes in system equations are given later

in Equations J-1 to J-5.) Here the weight or confidence given to the retail sales information has been made five times the weight given to factory incoming orders from distributors. This represents a very high confidence placed in the data about retail activity. Improvement has occurred but not as much as might have been hoped. The fluctuation in manufacturing orders is about five times as great as at retail compared to eight times as great in Figure 15-19. The change in the manufacturing decision has caused some shifting of the position of the inventory curve. The rapidly rising inventory still causes a depression in the manufacturing schedule even though desired inventory is now based on the information regarding retail sales rather than the information on incoming factory orders.

The need for great confidence in the retail sales

information is apparent if we examine conditions between the 30th and 35th weeks. Factory production schedules are still greater than incoming factory orders because of the rapidly falling incoming orders. The backlog of unfilled orders is declining, and the factory inventory is shooting upward. Here is a condition where the information available directly at the factory seems to contradict the information available about retail sales. When faced with falling orders, rising inventory, and falling backlog of unfilled orders, the factory manager would need the confidence and courage necessary to continue a production rate higher than his own locally available information indicates.

The manager must also be able to provide both the required investment money and the warehouse capacity for the fluctuating inventory. The availability of retail sales information at the factory does nothing about the inventory policies of the system. A detailed study of such a system might lead to the conclusion that the solution to better system behavior lies not in more information at the factory but rather in a change in the operating policies of the distribution system.

The following changes in the equations of Chapter 15 were made to yield Figure J-1. Equations 15-43, 15-45, 15-47, and 15-49 are replaced by

$$IDF.K = (AIF)(WAS.K) \qquad \text{J-1, A}$$

$$MWF.K = \frac{1}{1 + CRWF}[RRF.JK + (CRWF)(RSR.K)]$$
$$+ \frac{1}{DIF}(IDF.K - IAF.K + LDF.K$$
$$- LAF.K + UOF.K - UNF.K) \qquad \text{J-2, A}$$

$$LDF.K = (WAS.K)(DCF + DPF) \qquad \text{J-3, A}$$

$$UNF.K = (WAS.K)(DHF + DUF) \qquad \text{J-4, A}$$

$$WAS.K = \frac{1}{1 + CRWF}[RSF.K + (CRWF)(RSR.K)]$$
$$\text{J-5, A}$$

IDF	Inventory Desired at Factory (units)
AIF	proportionality constAnt for Inventory at Factory (weeks)
WAS	Weighted Average of Sales at retail and factory (units/week)
MWF	Manufacturing rate Wanted at Factory (units/week)
CRWF	Constant, Retail Weighting at Factory (dimensionless ratio of retail to factory sales)
RRF	Requisitions (orders) Received at Factory (units/week)
RSR	Requisitions (orders) Smoothed at Factory (units/week)
DIF	Delay in Inventory (and pipeline) adjustment at Factory (weeks)
IAF	Inventory Actual at Factory (units)
LDF	pipeLine orders Desired in transit through Factory (units)
LAF	pipeLine orders Actual in transit through Factory (units)
UOF	Unfilled Orders at Factory (units)
UNF	Unfilled orders, Normal, at Factory (units)
DCF	Delay in Clerical processing of manufacturing orders at Factory (weeks)
DPF	Delay in Production lead time at Factory (weeks)
DHF	Delay due to minimum Handling time required at Factory (weeks)
DUF	Delay, average, in Unfilled orders at Factory caused by out-of-stock items when inventory is "normal" (weeks)
RSF	Requisitions (orders) Smoothed at Factory (units/week)

The weighted average of sales WAS in Equation J-5 takes the place of the smoothed factory sales RSF that was previously used. This places a weight on average retail sales that is CRWF times as great as the weight placed on smoothed orders at the factory. For Figure J-1, CRWF = 5.

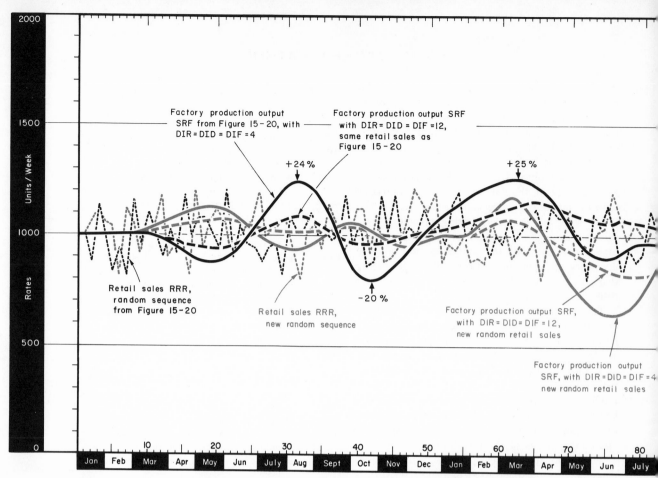

Figure K-1 Noise and prediction in a system model.

APPENDIX K

Prediction of Time Series

IN Chapter 13 the point was made that the exact path followed by an industrial or economic system depends on two separate considerations. First is the orderly system structure for which we can hypothesize approximate rules of behavior. The second is the noise that occurs in all decision functions and about which we know only its existence and perhaps its approximate magnitude and statistical character. Noise is that part of the decision flow for which we have no satisfactory causal hypothesis.

Our social systems are "broadband," by which we mean that they do not have the structural character that will cause present conditions to determine the exact state of the system far into the future.

We find then that the general character of the behavior of a system can be largely attributed to the orderly system structure and policies but that the instantaneous state of the system is also dependent on the random-noise disturbances. This is indicated by the experiment illustrated in Figure K-1, which is a composite figure from four different model runs.

The solid black line is the factory production output SRF from Figure 15-20. Also shown is the random retail sales input RRR from Figure 15-20. As discussed in Chapter 15, the system has selected

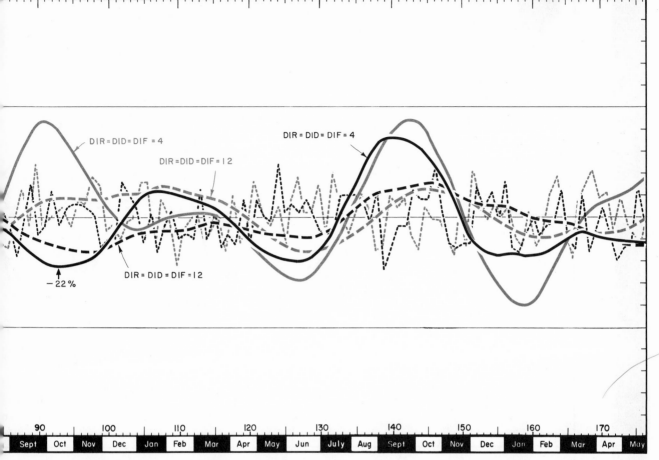

DIR = DID = DIF = 4

DIR = DID = DIF = 12

DIR = DID = DIF = 4

DIR = DID = DIF = 12

−22%

| 90 | 100 | 110 | 120 | 130 | 140 | 150 | 160 | 170 |

| Sept | Oct | Nov | Dec | Jan | Feb | Mar | Apr | May | Jun | July | Aug | Sept | Oct | Nov | Dec | Jan | Feb | Mar | Apr | May |

from the noise certain components to which the system itself is particularly sensitive.

In solid red is the factory production output for the same identical system again started in the same steady-state initial condition but now with a different random sequence at retail sales, as shown in red.[1] We note that the factory production with the new noise sequence is different in detail from factory production with the first noise sequence. However, it appears to belong to the same "family." Its amplitudes of excursion are about the same. The abruptness with which it changes is similar. The two factory production curves appear to be qualitatively similar. We note, however, that the solid red curve does not serve as a useful predictor of what the solid black curve

[1] These random numbers come from a pseudorandom-number generator that can be caused to repeat the same sequence of random numbers or if started at a new beginning point will generate a different sequence. The new sequence will be different in detail but will have the same general statistical character as the first sequence.

will do. Both systems begin at the same initial point. By the 15th week one curve has started up, and the other has started down. The initial period of 10 weeks is the same for both and is indicative of the time taken for a disturbance to propagate itself from the retail level to the factory production rate. The factory production lead time itself represents 6 weeks of the interval before the curves begin to diverge. Therefore, the factory manufacturing schedules MOF, if plotted, would begin to diverge even sooner.

The two solid curves can be viewed as a situation in which we have a "real system" and a "model" that are identical. The structure of the model is identical in every detail with the real system. Only the noise sequences differ. If the presence of noise is admitted, we must necessarily come to the conclusion that even the perfect model may not be a useful predictor of the specific future state of the system it represents.

This does not keep the model from being a useful predictor of system improvement that will

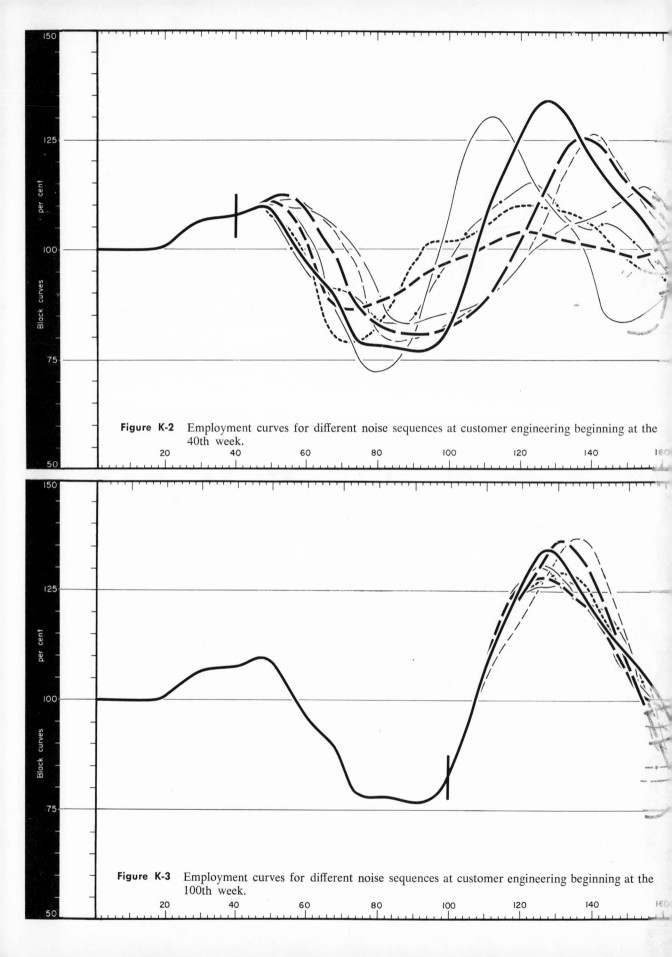

Figure K-2 Employment curves for different noise sequences at customer engineering beginning at the 40th week.

Figure K-3 Employment curves for different noise sequences at customer engineering beginning at the 100th week.

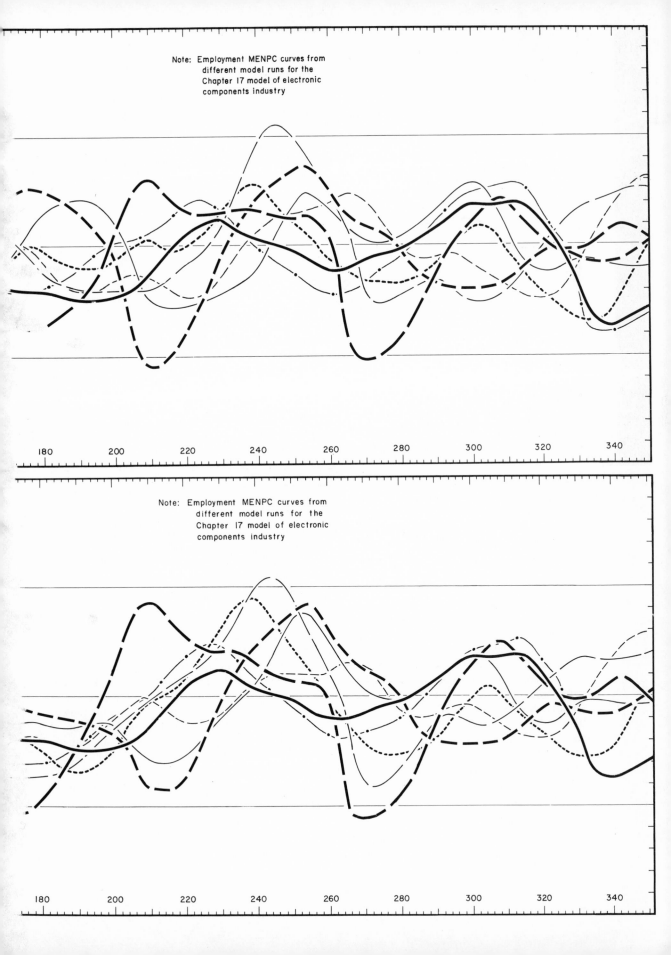

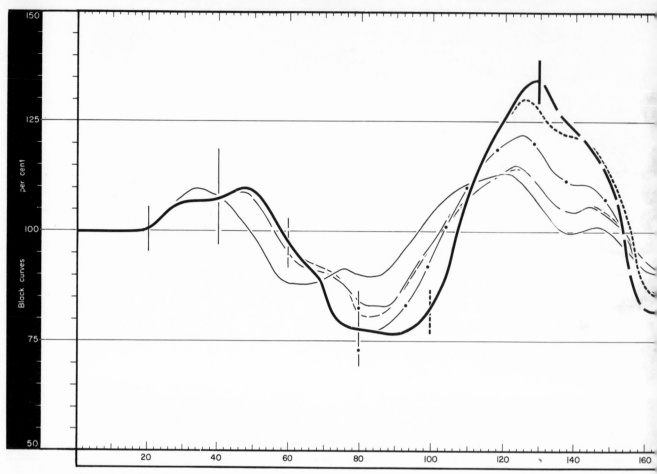

Figure K-4 Employment curves for noise at customer engineering transferred at different times from one noise sequence to another.

come from design changes. In Chapter 15 a more stable system resulted from more gradual adjustment of inventory and pipeline contents. In Figure K-1 the reduced tendency toward fluctuation is evident for either noise-sequence input to the system when the inventory and pipeline adjustment time is increased. The black dashed line indicates the production curve that results from the black random input and should be compared with the solid black curve. The red dashed line results from the red random input and should be compared with the solid red line. In both cases we see that the new system is improved after the policy change, regardless of the particular noise sequence. The dashed lines are similar to one another, and both show a lower disturbance amplitude than the solid lines. The dashed lines result from setting the inventory adjustment time DIR, DID and DIF all at 12 weeks. The solid lines use a value of 4 weeks, the same as was used in Chap-

ter 15 (except for Figure 15-24, where was shown the effect of the inventory adjustment time on the system response to a step input).

Some related tests to demonstrate the effect of noise on a system now follow, using the model of the electronic components industry as developed in Chapter 17.

In Section 13.7 it was suggested that the ability of a model to predict the future state of a system might depend on the condition of the system at the time that the prediction is undertaken. This can be seen by comparing Figures K-2 and K-3. Each curve in these two figures represents the employment MENPC from a different model run. A number of different noise generators, all with the same statistical characteristics, were established in the model, and each of these began at its own same point at the start of each model run. In Equation 17-94 noise was inserted into the system only at the output of the customer engineering

434

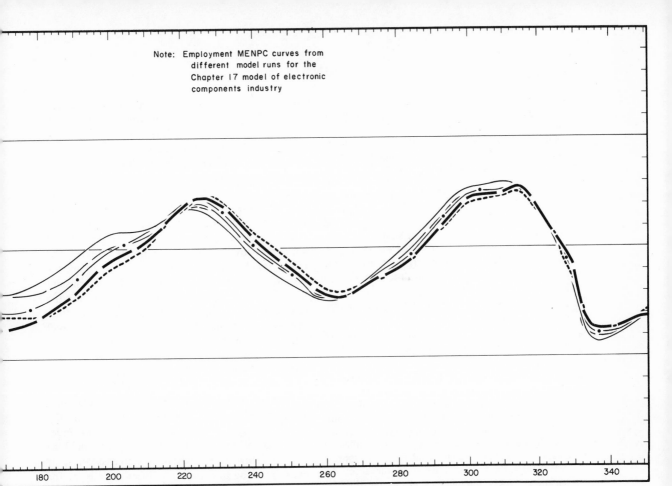

Note: Employment MENPC curves from different model runs for the Chapter 17 model of electronic components industry

180   200   220   240   260   280   300   320   340

department. At the start of each model run, Equation 17-94 was connected to the same noise source. In Figure K-2 at the 40th week in each run the noise input was shifted to a different generator. In other words, the figure shows the system history after the 40th week for different noise sequences. The curves all have the same qualitative behavior. The rapidity with which they diverge after the 40th week in Figure K-2 illustrates how long the conditions existing at the 40th week persist in determining the actual instantaneous state of the system.

Figure K-3 represents the same experiment, in which the transfer from the initial noise source to one of the other noise sources occurs at the 100th week. At the 100th week the internal stresses in the system are considerably greater than they were at the 40th week. The system has been operating below average production for a long period of time, and backlogs have risen and inventories have fallen. Conditions have been established for a rapid recovery, which then takes place nearly in-

dependently of the particular sequence of random events. All of the curves peak around the 130th week and rebound to a region of underproduction before beginning to diverge in response to the separate noise sequences that are the driving forces.

Examination of Figures K-2 and K-3 might lead to the conclusion that substantial prediction for as much as 40 or 60 weeks into the future is possible. However, it is necessary to remember that the experiment illustrated in these figures is highly idealized. For the following reasons the conditions illustrated in these figures are more favorable than would be encountered in real life:

• Except for the different noise sources, the response curves illustrated all come from identically the same model. If we think of one of the curves in Figure K-2 or K-3 as representing the real system and another as representing the prediction obtained from a model, it implies that we have a model which is a perfect representation of the real system.

435

- In Figure K-2 all model runs begin under identically the same conditions at the 40th week, and in Figure K-3 all model runs are identical at the 100th week. The similar situation under real-life circumstances of using a model to predict the future of a real system would require knowledge of all of the conditions of the real system at the time the prediction is being made. This would require knowledge of many more variables than ordinarily are measured and, furthermore, would require knowledge of their present values. Actually, much of the data on industrial and economic systems are received late by nearly as much as the interval of useful prediction illustrated in Figures K-2 and K-3.

- The system illustrated in these two figures is very simple.

- The system illustrated has but one major mode of behavior, represented by the fluctuation having about a two-year period.

- The system illustrated in Figures K-2 and K-3 is highly oscillatory. As shown in Figure 18-1, any disturbance tends to persist for a substantial period of time. This indicates that any set of initial conditions tends to transmit its implications rather far into the future. After this system was modified in Chapter 18, it had stronger tendencies to suppress disturbances. In a system where disturbances die out more quickly, the influence of any set of initial conditions will likewise persist for a shorter time.

- Only one noise source was used in Figures K-2 and K-3. A multiplicity of noise sources, perhaps one at each decision point, could be expected to degrade still further the rather inadequate prediction ability illustrated in the figures.

- All of the model runs in Figure K-2 and Figure K-3 are for constant external conditions represented by the constant flow of orders INPUT into the customer sector. Any variation in this flow rate would further degrade the prediction ability of the model.

Figures K-2 and K-3 show that the conditions existing at a particular point in time do not determine for very long the exact future state of a system. This can also be illustrated in the inverse manner by showing that for a given noise sequence the system will converge to the same behavior, regardless of the initial conditions. Such an experiment is shown in Figure K-4. Here again each curve represents the employment level MENPC from a different model run. For each model run

the same two noise sources were used. One source was used at the start of the run, and Equation 17-94 was at some point shifted to the other noise source. Each curve represents a different time of making the transfer. Runs are illustrated for the transfer point between the first and second noise sources occurring at 20, 40, 60, 80, 100, and 130 weeks. These give six different initial conditions at the time of transfer to the second noise sequence. Figure K-4 shows the convergence of all six curves toward the same ultimate pattern. The initial conditions contribute to the state of the system only for the time that it takes a transient to decay. In Figure 18-1, this decay rate for the electronic components industry model was 50% per cycle. In Figure K-4 we see that only a small influence from any initial condition persists after the 260th week. For systems that attenuate disturbances rapidly, the initial conditions do not long influence system condition. This means that conditions within the system at a particular moment do not long determine conditions in the future. (Under extreme circumstances, the independence of initial conditions may not hold as we move toward models of unstable systems, or ones operating in certain nonlinear regions, or for systems whose principal characteristic is growth.)

The conclusions from these figures are interpreted as supporting the views in Chapter 13, where it was proposed that a dynamic model should be used for determining the behavior character of a system but not its specific future state. A model should predict how the character of the system will be changed by changes in policy and organizational form. Many factors will combine to keep the model from predicting the future state of a system beyond the length of time that the existing conditions and trends will resist being influenced by system noise. These limitations on prediction include the fact that a model will not be perfect, the present state of the real system is unknown and only past data are available, and social systems do not have a structure giving them strong momentum characteristics that carry the implications of present conditions forward independently of the random events which we will not be able to include among the model variables.

APPENDIX L

Forecasting

IN Appendix K prediction was used as meaning the employment of formal models for predicting the specific future values of time series of interest. Forecasting is distinguished in this appendix as meaning any of the more informal processes by which estimates are made about the future. In particular, as used here, forecasting applies to such activities as the short-range sales forecast.

Forecasting is especially subject to misunderstanding when considered outside the context of the system of which it is a part. In most organizations that take sales forecasting seriously, the process is highly subjective and is based on skill, experience, and judgment. If any evaluation of the forecasting results is attempted, it is to view in retrospect the similarity between forecast and actual events. Almost never is it studied for its influence on the dynamics of the total industrial system.

Forecasters always stress the wide scope of factors that they take into account in arriving at their forecast. However, an examination of the entire forecasting procedure will usually reveal that extrapolation is one of the strongest basic processes in a forecast. The forecast is based very largely on an assumption of persistence. This may include a belief in the persistence of present values of variables, and a belief in the persistence of a recent trend, and a belief in the persistence of past cyclical fluctuation.

In Chapters 15 through 18 no explicit forecasting procedures were incorporated, and therefore the system might be interpreted as one in which the decision makers merely anticipated the continuation of the present state of affairs. Appendix N deals with a forecast that assumes the continuation of a past cyclical seasonal behavior. The present appendix is intended to raise questions about that component of forecasting which assumes a continuation of recent trends.

We cannot avoid taking actions that imply some assumption about the future. However, the form of this assumption can be one of the very important decision-making policies in determining total system behavior. In many situations forecasts can be either self-validating or self-defeating.

An example of a self-validating forecast arises in some situations involving the growth of new products, where sales growth is erroneously assumed to reflect growth in market demand, whereas in fact it reflects growth in production capacity. This can happen in a situation where demand continuously exceeds supply, where a persistent backlog of unfilled orders exists, and where this delay in the filling of orders is sufficiently long to depress the natural ordering rate. Under these circumstances exponential growth can often persist for several years, with demand consistently running ahead of supply and with market growth rate determined primarily by supply growth rate.

Another form of self-validating forecast arises where seasonal sales are forecast, followed by managerial actions that tend to create seasonality. This is the mechanism illustrated in Appendix N.

In the example that follows, the self-defeating type of forecast is illustrated, where the forecast can actually accentuate the difficulty that it is intended to alleviate. This example takes the production-distribution system of Chapter 15 and adds to the ordering decision a forecast that assumes the persistence of recent trends.

A sales forecasting procedure based on trend extrapolation has been added at the retail, distributor, and factory manufacturing order stages. It is assumed, as would usually be true, that the current actual sales figures are fluctuating from random events by such an amount that they must be smoothed before being meaningful. Sales smoothing has already been provided in Chapter 15.

437

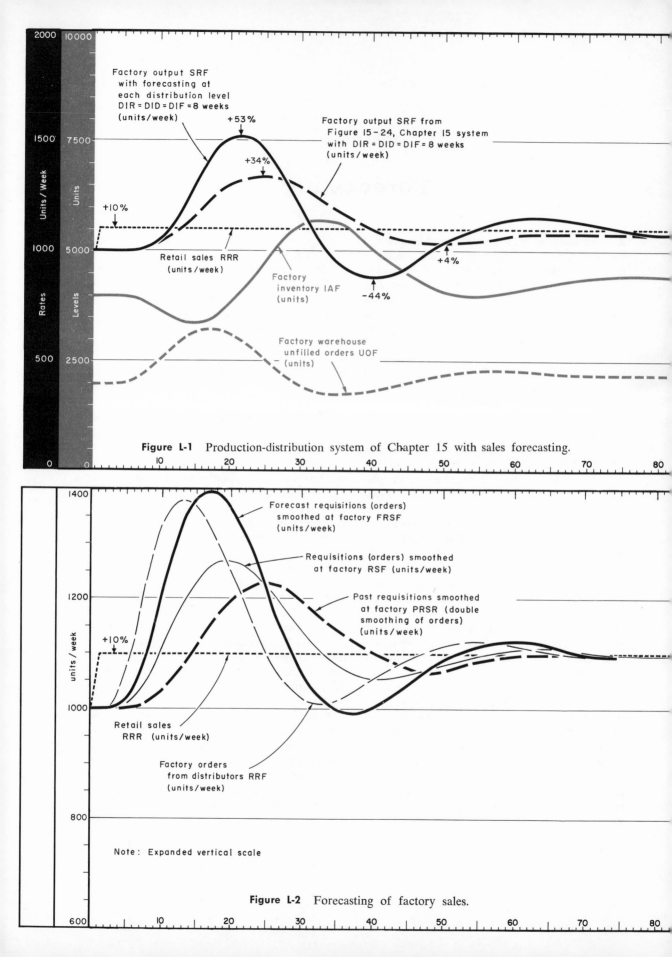

Figure L-1 Production-distribution system of Chapter 15 with sales forecasting.

Factory output SRF
with forecasting at
each distribution level
DIR = DID = DIF = 8 weeks
(units/week)

+53%

+34%

Factory output SRF from
Figure 15-24, Chapter 15 system
with DIR = DID = DIF = 8 weeks
(units/week)

+10%

Retail sales RRR
(units/week)

Factory
inventory IAF
(units)

-44%

+4%

Factory warehouse
unfilled orders UOF
(units)

Forecast requisitions (orders)
smoothed at factory FRSF
(units/week)

Requisitions (orders) smoothed
at factory RSF (units/week)

Past requisitions smoothed
at factory PRSR (double
smoothing of orders)
(units/week)

+10%

Retail sales
RRR (units/week)

Factory orders
from distributors RRF
(units/week)

Note: Expanded vertical scale

Figure L-2 Forecasting of factory sales.

Smoothed sales are then examined for trend and are extended into the future. The effect of doing this is shown in Figure L-1.

Because trend extrapolation here introduces a highly unstabilizing influence into the system, the reference system to which to add the forecasting was taken as the more stable system in which DIR = DID = DIF = 8, as shown in Figure 15-24. With the trend extrapolation used in the ordering decisions, the factory production rate peaks earlier and at a higher value than before. On the initial upswing, factory production reaches 53% above the beginning value instead of 34%. The secondary undershoot is also greater.

Trend extrapolation here has approximately the opposite effect to that caused by making more gradual adjustments in inventories and supply pipeline content. It yields a system that is more "excitable." It exhibits more vulnerability to random events.

Figure L-2 shows the relationships between factory orders and smoothed and forecast orders. Forecast orders still lag behind actual orders but not nearly so much as smoothed orders. The forecast-orders curve is a better approximation to the true-orders curve than is the smoothed-orders curve.

The system instability that arises here from extrapolating smoothed orders to get forecast orders depends on the amount of forecasting that is attempted. The further into the future that forecasting is attempted, the greater will be the resulting system instability. Figure L-2 is best understood after examining the equations for forecasting that were incorporated in the model.

Forecasting is to be accomplished by extrapolating forward the curve for smoothed orders. This is illustrated in Figure L-3 for the relationships that would exist in the factory sector if the incoming orders to the factory RRF were a uniformly rising trend. The straight-line curve for RRF is shown and its present value RRF.JK. This trend line is exponentially smoothed to obtain the smoothed value of sales RSF.K. As discussed in Appendix E, this is equivalent to a delayed value of the factory orders RRF. The delay is shown as DRF. To extrapolate forward the smoothed value of sales, we need to know the slope of the smoothed-orders curve RSF. This can be done if we know a past point on the curve as at b. If the smoothed-orders curve

is again smoothed, we can generate the value of a trailing point as at b. The presently known value of the point at b is given as PRSF.K. The smoothing time chosen for obtaining this trailing point is shown as DPRSF. Given the point RSF.K and b, which is calculated as PRSF.K, the straight line can now be projected forward as to c. The forward forecasting time is shown as FTF. The presently known value of the forecast is shown as FRSF.K. We should note that the forecast value does not necessarily give us a value in advance of the present true value of factory orders RRF.JK. The delays necessary to reduce noise by smoothing and to determine the slope of the smoothed curve have given values equivalent to those at e and f. When these are extrapolated forward, they yield a value as at d. Whether or not this extrapolation can be carried up to the present value RRF.JK or into the future beyond depends on the amount of noise that is present in the factory-orders curve and the amount of smoothing that is required. Whether or not any forecasting should be done depends on the effect it will have in the system as a whole.

The model equations necessary to accomplish the steps in Figure L-3 will now be developed for the factory sector. Smoothed factory orders RSF are already available from Equation 15-44.

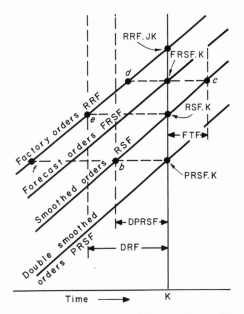

Figure L-3 Time relationships in forecasting equations.

To make available a trailing point on the curve for RSF, this value will be smoothed again as in Equation L-1:

$$PRSF.K = PRSF.J + \frac{DT}{DPRSF}(RSF.J - PRSF.J) \qquad \text{L-1, L}$$

$$FRSF.K = RSF.K + (FTF)\left(\frac{RSF.K - PRSF.K}{DPRSF}\right) \qquad \text{L-2, A}$$

PRSF	Past Requisitions (orders) Smoothed at Factory (units/week)
DT	Delta Time, equation solution interval (weeks)
DPRSF	Delay in Past Requisition Smoothing at Factory (weeks)
RSF	Requisitions (orders) Smoothed at Factory (units/week)
FRSF	Forecast Requisitions Smoothed at Factory (units/week)
FTF	Forecast Time at Factory (weeks)

Equation L-2 gives the forecast value at c in Figure L-3. In the right-hand parentheses is the slope of the line for smoothed orders. This is multiplied by the time into the future FTF, and the product is added to the present value of smoothed orders RSF.

Some related changes are necessary in the other equations. The old value of desired inventory IDF in Equation 15-43 is retained for use in determining the ability to fill orders from inventory. A new equation similar to Equation 15-43 is based on forecast sales FRSF instead of smoothed sales RSF, to give the desired inventory that is the basis for inventory and ordering decisions. Equation 15-45 for the manufacturing decision is changed to use the new value of desired inventory. Equations 15-47 for LDF and 15-49 for UNF should be altered to be based on forecast sales rather than smoothed sales.

A similar pair of equations is also required to generate forecast values at the distributor and retailer levels, along with alterations in related equations.

Figure L-2 shows the week-by-week time relationships between these new variables at the factory. The forecast value is much closer to the actual value than either of the smoothed curves. The effect of this, however, is to make the system more responsive to changes and to counteract the effect of the sales averaging. It makes the system more responsive to noise and to high-frequency disturbances and increases the amplification of the distribution chain and the tendencies toward instability. These should not be taken as generalities but show the dangers that may exist in a forecasting practice whose interactions with the remainder of the industrial system have not been explored.

Countercyclical Policies

THERE is often reference to countercyclical policies in economics. By this is meant a policy that is to have such a direction as to counteract any fluctuating tendencies of the system. Such discussions are usually on a very general plane. A verbal model is quite inadequate to test the effect that might follow the inauguration of such a proposed policy.

In Figures 16-7 and 16-8 appeared a cyclical disturbance of about two years' duration that arose because of an advertising policy. The policy was to authorize advertising expenditures proportional to scheduled production. After examining that policy and the results, many persons suggest that the system would be stabilized by the inverse policy. They suggest that reducing advertising expenditure as sales rise and raising advertising expenditure as sales fall would obviously be the wise and reasonable policy. Such a proposal can be tested by inverting the relationships originally used in Chapter 16.

To do this the model will need a long-term reference against which to judge whether or not manufacturing rate is high or low. This will be done by generating a very long term average of manufacturing rate as a reference level, as is done in Equation M-1:

$$\text{LMSF.K} = \text{LMSF.J} + \frac{\text{DT}}{\text{DLMSF}} \, (\text{MOF.JK} - \text{LMSF.J})$$

$$\text{M-1, L}$$

LMSF   Long-term Manufacturing rate Smoothed at Factory   (units/week)

DT   Delta Time, equation solution interval   (weeks)

DLMSF   Delay in Long-term Manufacturing rate Smoothing at Factory   (weeks)

MOF   Manufacturing Orders into Factory (units/week)

Equation M-1 is a standard information-smoothing equation. The length of the smoothing time constant might be 100 weeks or longer so that the value will not rise and fall appreciably during the periodicities to which the system is subject.

With the long-term reference level available from Equation M-1, the current average manufacturing rate can be compared with it, as in Equation M-2:

$$\text{VDF.KL} = (\text{UPF})(\text{AVS})[\text{LMSF.K} + \text{ACV}(\text{MAF.K} - \text{LMSF.K})] \quad \text{M-2, R}$$

VDF   adVertising Decision at Factory   (dollars/week of advertising authorization)

UPF   Unit Price of goods at Factory   (dollars/unit)

AVS   constAnt, adVertising as fraction of Sales revenue   (dimensionless)

LMSF   Long-term Manufacturing rate Smoothed at Factory   (units/week)

ACV   constAnt for Change in adVertising   (dimensionless multiplier of difference between short- and long-term manufacturing rate average)

MAF   Manufacturing short-term Average rate at Factory   (units/week)

With a negative value for ACV, Equation M-2 gives an advertising expenditure rate that is proportional to long-term sales but is inversely proportional to short-term sales. Equation M-2 replaces Equation 16-1 in Chapter 16. The coefficient ACV determines the amount and the direction of advertising change relative to the short-term changes in manufacturing rate. If ACV is +1, the equation reduces to the previous form found in Equation 16-1. If ACV is −1, then the direction of the short-term advertising change is inverted in relationship to changes in manufacturing level. Advertising rises by the same percentage that manufacturing falls, and vice versa.

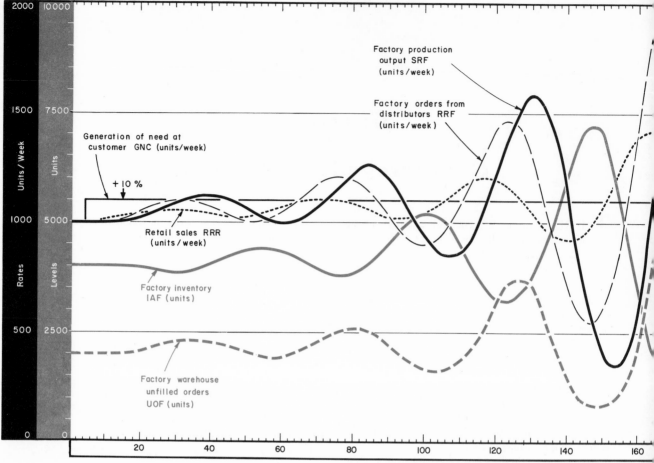

Figure M-1 Countercyclical advertising policy in Chapter 16, production-distribution-market model.

The results are shown in Figure M-1. The system is more unstable than before. An instability now appears with a period of about 45 weeks instead of the previous period of 130 weeks. The amplitude is growing rapidly and is more than doubling each cycle.

This illustrates a natural characteristic of information-feedback systems. Appendices E, G, and I have shown that phase shift and amplification are both critical functions of periodicity. A system will be oscillatory at those frequencies where the inphase component of amplification around any loop begins to approach unity. When the amplification is greater than unity, oscillations tend to grow. Any energetic and decisive policy may be a source of amplification. The reversal of

a policy, as in going from Figure 16-7 to Figure M-1, may simply cause the system to shift to a new periodicity where the phase shifts are again of such a kind that the feedback signal amplifies and reinforces an initial disturbance.

This appendix has not been included because managers might be apt to follow an inverse advertising policy. Instead, it is included to illustrate the way in which information-feedback systems automatically seek to generate internal fluctuation at that frequency at which the inphase gain around the feedback loop is near a maximum. The so-called countercyclical policies in the management of economic systems will not necessarily lead to stability. Any effect may be only to shift the periodicities to which the economic system is sensitive.

Self-generated Seasonal Cycles

INDUSTRIAL policies adopted in recognition of seasonal sales patterns may often accentuate the very seasonality from which they arise. A seasonal forecast can lead to action that may cause fulfillment of the forecast. In closed-loop systems this is a likely possibility. The analysis of sales data in search of seasonality is fraught with many dangers. As discussed in Appendix F, random-noise disturbances contain a broad band of component frequencies. This means that any effort toward statistical isolation of a seasonal sales component will find some seasonality in the random disturbances. Should the seasonality so located lead to decisions that create actual seasonality, the process can become self-regenerative.

Self-induced seasonality appears to occur many places in American industry. Sometimes it is obvious and clearly recognized, and perhaps little can be done about it. An example of the obvious is the strong seasonality in items such as cameras sold in the Christmas trade. By bringing out new models and by advertising and other sales promotion in anticipation of Christmas purchases, the industry tends to concentrate its sales at this particular time of year.

Other kinds of seasonality are much less clear. Almost always when seasonality is expected, explanations can be found to justify whatever one believes to be true. A tradition can be established that a particular item sells better at a certain time of year. As this "fact" becomes more and more widely believed, it may tend to concentrate sales effort at the time when the customers are believed to wish to buy. This in turn still further heightens the sales at that particular time.

An example is shown in Figure N-1. Here the production-distribution-market-advertising system of Chapter 16 has been modified to include an analysis of seasonal sales. Advertising is based on the forecast level of seasonal sales and is initiated in time to become effective at the market at the time of the forecasted sales level. The year is divided into 2-week intervals. The sales during corresponding 2-week intervals from preceding years are exponentially averaged. These averages form the estimates of future sales during corresponding intervals.

443

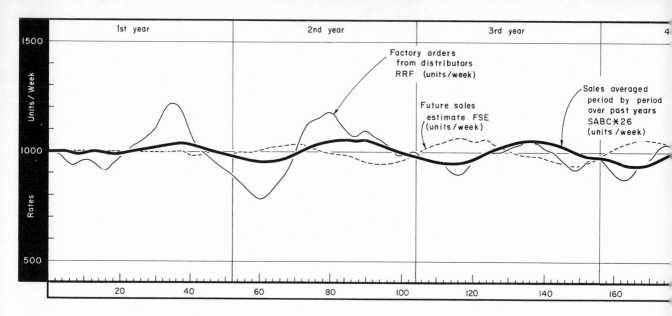

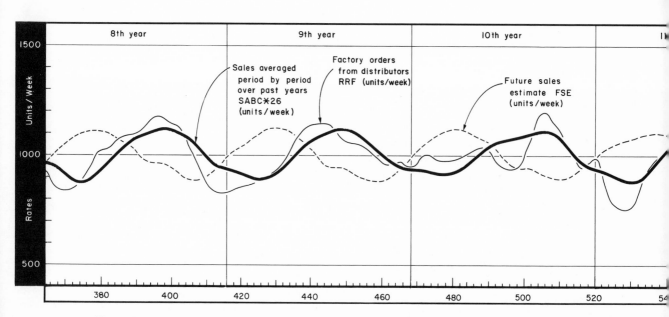

Figure N-1 Seasonality caused by advertising policies based on seasonal analysis of sales data.

Random noise is introduced as was done before at the retail sales point. The figure shows the actual factory orders, the sales averaged period by period over succeeding years to derive seasonality, and the anticipated sales (derived from earlier years) being used at any time in the advertising decision. In the first year random events cause a dip in sales during the first part of the year and a rise during the last part. This is the kind of behavior shown in Figure 15-20. The randomness causes a very small distortion in the average that had started in steady state as a constant value. In the second year the seasonality tends to repeat, again partly for random reasons, partly as a result of the fact that the distribution system has a natural period of about one year, and partly be-

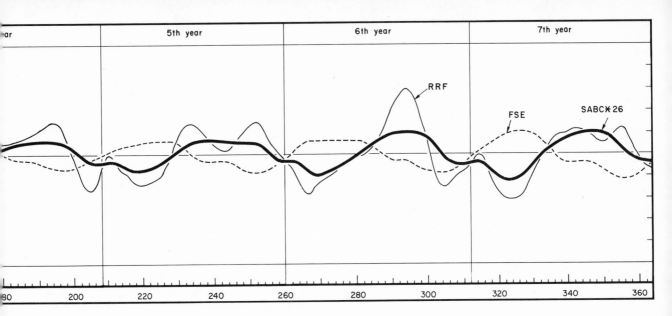

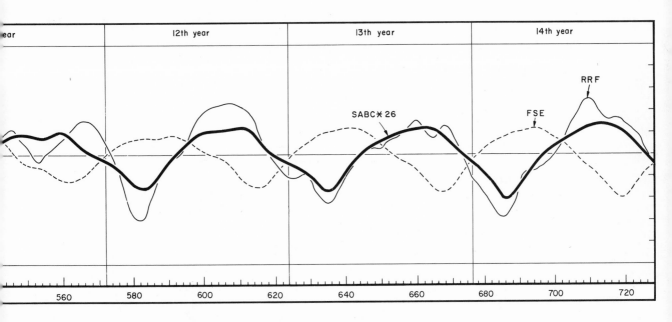

cause the small indication of seasonality appearing in the first year has distorted the advertising rate in the second year. These cause a further deepening of the seasonal sales average. As the observed market seasonality increases it distorts advertising and marketing efforts, and this creates still more seasonality. Following the curve year by year, we see that a strong seasonality has devel-

oped by the fourteenth year. The seasonal average shows a strong dip in the first half of the year and a rise in the second half. This causes sufficient effect on the market through advertising that it depresses sales in the first half of the year and raises them in the second half. Random noise continues to cause actual sales to fluctuate around the seasonal average.

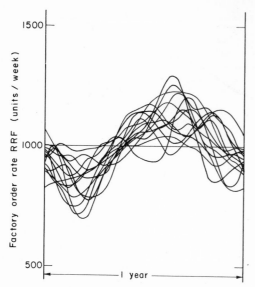

Figure N-2 Superimposed factory orders from distributors RRF for fourteen years.

Figure N-2 shows the incoming factory orders from distributors RRF superimposed for the fourteen years. Without any tie to the calendar whatsoever except for the seasonal analysis procedure, this particular model run continues to hold a fixed relationship between the average sales curve and the beginning of each year. In an actual industrial system such a result would quickly become associated with particular calendar months, so that inventory policies, sales promotion, vacations, and other influences would tend to become associated with calendar dates and thereby permanently freeze the annual pattern.

The equations will now be developed to show how the seasonal analysis was produced in the model.

A cyclic boxcar train is used to store the seasonal average values for each 2-week period of the year. Each 2 weeks the boxcar train will shift, bringing into place the past average value for the

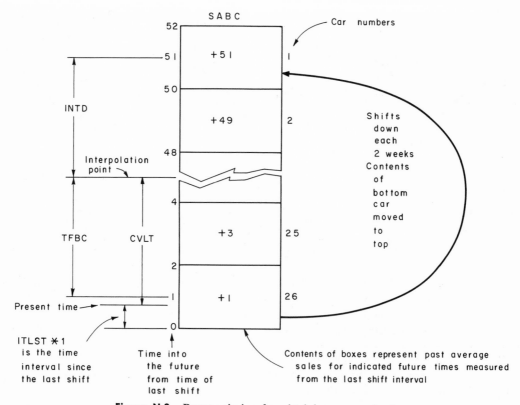

Figure N-3 Boxcar timing for obtaining seasonal sales.

current period. Into this old average new sales are averaged.

Figure N-3 will help to clarify the timing of the cyclic boxcar, the averaging, and the interpolation for a future value in the table of averages. The following equation is used to designate a cyclic boxcar:

$$SABC = BOXCYC \ (26, 2) \qquad \text{N-1, B}$$

SABC Seasonal Average Boxcar Cycle, the designation of this particular boxcar train, any one car of which is designated by adding an asterisk and the number of the car in the train

BOXCYC a functional designation instructing the DYNAMO compiler that a cyclic boxcar train be set up

26 the number of cars in the train

2 the time interval between shifts of the train (weeks)

In Figure N-3 the bottom (or 26th) box will be designated as the one representing the current 2-week time period. Equation N-2 averages present factory orders RRF into the previous value of this box. The time constant TSD (here taken as four years) is the exponential-smoothing time constant controlling the influence of the current sales in changing the year-to-year average.

$$SABC{*}26.K = SABC{*}26.J + \frac{DT}{2(TSD)} \ (RRF.JK \\ - SABC{*}26.J) \qquad \text{N-2, L}$$

SABC*26 the variable content of car No. 26 in the cyclic train SABC, sales in the current 2-week period averaged over past years (units/week). (A car designation permits the asterisk and car number beyond the standard five symbols normally allowed in naming a variable.)

DT Delta Time, equation solution interval (weeks)

2 the weeks per year that this box is active in the averaging process (weeks/year)

TSD Time constant for Seasonal Detector for averaging sales data (years)

RRF Requisitions (orders) Received at Factory (units/week)

The flow diagram for the new equations appears in Figure N-4.

For interpolation purposes, it will be necessary to know at any given time how much of the 2-week interval has already passed. This can be accomplished by accumulating time during the 2-week interval and then discarding the accumulated value at the same time that the boxcar train is shifted. This can be done with a linear boxcar train, the last value of which is discarded each shifting interval. Equation N-3 sets up a one-car boxcar train for this purpose:

$$ITLST = BOXLIN \ (1, 2) \qquad \text{N-3, B}$$

ITLST Interpolation Time Linear Shift Train used to accumulate the time elapsed since the last shift

BOXLIN functional notation instructing the DYNAMO compiler to establish a linear boxcar train

1 the number of cars in the train

2 the shift interval (weeks)

Equation N-4 accumulates time in the boxcar, set up by Equation N-3, during any one 2-week interval:

$$ITLST{*}1.K = ITLST{*}1.J + DT \qquad \text{N-4, L}$$

ITLST*1 the variable in the first box of train ITLST, the amount of time since the last shift of the boxcar trains (weeks)

DT Delta Time, the equation solution interval (weeks)

It is now possible to interpolate in the table of seasonal values for the future sales estimate that is to be used in the advertising decision. In Figure N-3 the constant CVLT represents the interval of time required to act on advertising decisions. It is here taken as 18 weeks. Immediately after a boxcar shift, time can be thought of as rising from the bottom up to the 2-week level, at which time the entire boxcar train is shifted down again. The interpolation point in the table rises gradually over a 2-week period, drops with the shifting of the boxcar train so that it retains its relative position in the table, and continues to progress smoothly through the table. Since the interpolation process happens to start at boxcar number 1, it is necessary to know the value of the variable INTD. This is first done by finding TFBC in Equation N-5, which represents the geometrical relationships in Figure N-3:

$$\text{TFBC.K} = \text{CVLT} + \text{ITLST} * 1.\text{K} - 1 \qquad \text{N-5, A}$$

TFBC	Time From center of Bottom Car (see Figure N-3), used for interpolation in the boxcar train (weeks)
CVLT	Constant, adVertising Lead Time (weeks)
ITLST∗1	time since the last boxcar shifting (weeks)
1	Constant, to adjust table interpolation designation to mid-point (in time) of boxcars (week)

Equation N-6 gives the interpolation distance INTD from the top of the table:

$$\text{INTD.K} = 50 - \text{TFBC.K} \qquad \text{N-6, A}$$

INTD	INTerpolation Distance from top of box car train (weeks)
50	time interval between mid-points of top and bottom cars (weeks)
TFBC	Time From center of Bottom Car for interpolation (weeks)

Equation N-7 specifies to the DYNAMO compiler the information necessary to do a linear interpolation in the table. As a result, the future sales estimate will be available. This is a value

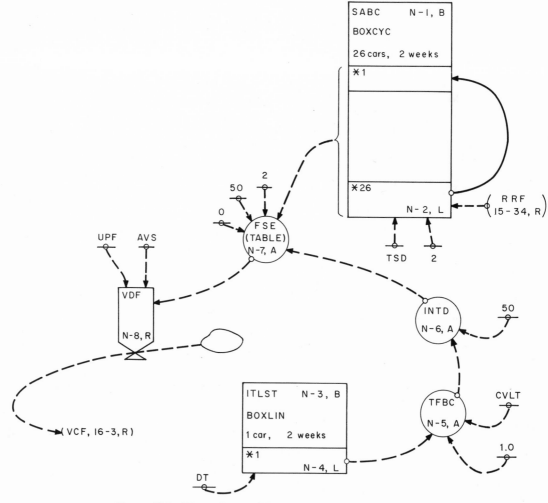

Figure N-4 Flow diagram for averaging of seasonal sales.

CVLT (18 weeks in this example) in advance of the current date taken from the average seasonal sales analysis.

FSE.K = TABLE(SABC, INTD.K, 0, 50, 2)   N-7, A

FSE	Future Sales Estimate (units/week)
TABLE	functional designation instructing DYNAMO to interpolate linearly for a value in a table
SABC	the name designating the table from which interpolated value is to be drawn, a boxcar train can be used as a table
INTD	the interpolating variable (weeks)
0	the first value of the interpolating variable for which a number is stored (weeks)
50	the last value of the interpolating variable for which a number is stored (weeks)
2	the interval between stored values (weeks)

Equation N-8 replaces Equation 16-1 and bases advertising on the estimate of future seasonal sales:

VDF.KL = (UPF)(AVS)(FSE.K)   N-8, R

VDF	adVertising Decision at Factory (dollars/week of advertising authorization)
UPF	Unit Price of goods at Factory (dollars/unit)
AVS	constAnt, adVertising as fraction of Sales revenue (dimensionless)
FSE	Future Sales Estimate (units/week)

These eight equations generate a seasonal average and a seasonal sales forecast based on it.

APPENDIX O

Beginners' Difficulties

IT seems that a discussion of what should be done in a new activity does not constitute adequate instruction. The positive approach alone does not provide forewarning against the pitfalls that lie along the road. Therefore, I shall try in this section to explain some of the difficulties experienced by students and staff members who have undertaken the kinds of systems studies discussed in this book.

The following are arranged as a collection of isolated topics somewhat in the sequence in which they arise in a systems analysis.

Courage

Industrial dynamics is an approach that should help in important top-management problems. Very few workers in the area of management science have had their aspirations conditioned to the expectation of major successes in the more difficult and challenging fields. The solutions to small problems yield small rewards. Very often the most important problems are but little more difficult to handle than the unimportant. Many men predetermine mediocre results by setting initial goals too low. The attitude must be one of enterprise design. The expectation should be for major improvement in the systems. The attitude that the goal is to explain behavior, which is fairly common in academic circles, is not sufficient. The goal should be to find management policies and organizational structures that lead to greater success.

Defining the Questions

A model should be designed to answer specific questions. A systems study must be for a purpose if it is to be productive. The questions must be meaningful and tangible and specific if they are to serve to guide a program. Determining the problems and the goals is the most critical part of almost any undertaking. The beginner tends to forge ahead into detailed construction of a model before its purpose has been adequately defined.

Scope of System

An important decision is to determine the boundaries of the system under study. The beginner can easily err on either side. On the one hand, he may lack boldness to be comprehensive enough and thereby limit his attentions to a subsystem within which the answers to the questions do not lie. On the other hand, he may define questions so vaguely that the objectives cannot be used to limit the scope of what must be included initially. The key to success will lie in clear questions which are broad enough to encompass matters of major consequence but which initially limit a system to proportions that fit the skill, time, and experience of the investigator.

Automatic Formulation and Evaluation

Many beginners in the study of system dynamics seem to feel that there must be a "scientific method" that will ensure their ability to create a proper model. They look for statistical procedures that can be applied to the scanty data ordinarily available, in the hope of having an orderly procedure that will create a system model by objective methods. They seem unwilling to express a professional opinion and to take risks on the basis of personal knowledge and judgment about the character of the system being improved. For the present, there seem to be no such objective methods that are effective. The attitude of requiring such safety and assurance of accepted methodology is not the attitude of the risk-taking manager. One does not achieve innovation and creativity by being timid.

Similar attitudes arise with respect to model evalution. The insecure system designer looks for "objective" criteria for the evaluation of the pertinence of a model. The compulsion to use such objective measures is sometimes so strong that refuge is taken in procedures that lack a sound foundation.

The world has thus far been run on the basis of judgment applied to the making of individual decisions. We are now on the threshold of moving the decision-making procedure back by one level of abstraction. When this has been done, judgment must cope with policy detection and policy design.

Hypothesis about Dynamic Behavior

The beginner usually fails to realize the importance of an initial hypothesis about dynamic behavior. There is often a feeling that to propose modes of dynamic behavior before a system model is constructed is to prejudge the answers. In a sense this is exactly what is needed. We start with a hypothesis for behavior. We build a model to see if the mode of behavior could exist and whether or not it can result from the initial assumptions. The experimental work is designed to prove or disprove the initial hypothesis.

The initial hypothesis is part of the establishment of the initial questions and goals for the study. Without this initial mental and verbal model of the dynamic behavior being studied, there is no basis for deciding what factors might be important and which ones could be neglected. It is very unlikely that a meaningful model will result from wandering through an organization and incorporating into the model whatever happens to come to one's attention.

A model of a chemical pilot plant serves as an analogy to what we are attempting here. It is made after a very tangible hypothesis about the plant's probable method of operation and the product it is to make. A pilot plant is designed and built to see if the hypothesis is correct and to see what new problems and insights are developed as one goes from a proposed method of operation to the experimental testing of the ideas. This is the way we use our dynamic model of an enterprise. A pilot plant is not constructed before a statement of objectives and probable mode of operation. We do not build a plant at random and then wait to see what kind of product it might produce.

Verbal Descriptive Model

Following closely after the hypothesis about dynamic behavior is the verbal model or verbal description of the system with which one is dealing. In general, the verbal model should come before the mathematical model. The verbal model should treat the description of how the parts behave and how they interact with one another. The verbal model is the support and the rationale for the hypothesis about dynamic behavior. If the verbal model is sufficiently complete, the mathematical model will follow as a consequence. The beginner is apt to fail to recognize the essential importance of the descriptive model as a step toward the creation of the equivalent formal mathematical model. The latter is a translation and clarification from the initial verbal language.

System Perspective

It is easy to view the system under consideration from the wrong distance. Students of economics tend to view from too great a distance, and they fail to see the essential decision points, nonlinearities, and interconnections of the system.

Managers and those having personal, intimate, first-hand knowledge of the system tend to view it from too close a distance. They want to put in too much detail. Separate, individual decisions take on too much importance. Apparent exceptions, which can be treated as noise disturbances, blind them to average behavior.

Each decision point should be represented as it might be seen by the next higher or the second higher administrative level. Very often the proper viewpoint from which to model a corporation is that of the perceptive banker or the skilled management consultant. At the levels of these men, it is the broad sweep of the system that catches the attention. They are aware of technical progress, of managerial and employee skills, of system objectives, and of markets. Yet, these viewpoints do not get enmeshed in the hour-by-hour and day-by-day details of the separate parts of the system.

Use of Descriptive Knowledge

The beginner usually neglects information that is in descriptive form and is inclined to look too long and futilely for sufficient information in the form of numerical data. He fails to appreciate that the overwhelming preponderance of our information about social systems is now in descriptive form. There is often an unwillingness to translate this descriptive knowledge into quantitative form. Descriptive information almost always implies relative magnitudes and differing importance of factors and assumptions. Attaching arbitrary scales and quantitative values to these concepts can provide orderliness without carrying any im-

plications about accuracy. Once the system has been cast in the terms of formal equations and numerical values of parameters, one can then determine much about the required accuracy. Creating a quantitative description of a system is a separate matter from achieving accuracy. These two different considerations are often confused by the person who does not think of mathematical and numerical notation as being merely another language for expressing ideas.

Our descriptive knowledge is rich in information about the probable form of decision functions under extreme limiting conditions. The use of the full nonlinear breadth of all the information that we have available makes the task of successfully describing system components much easier.

Creating Definitions

A companion to the reluctance to use the available descriptive information is the reluctance to create precise definitions and arbitrary scales of measurement for the quantities that exist in the descriptive knowledge. Many of the concepts and the so-called intangibles do not have generally recognized scales of measurement. This does not keep the investigator from creating his own precise definitions and his own arbitrary scale of measurement. It does require that he crystallize and clarify what he means by the terms which he uses. Here again, courage becomes important. The man must take the leadership in creating a field of knowledge, a vocabulary, a precision in descriptive terms, and a method of ranking and measurement. He must be willing to accept criticism that can be leveled only at the person who steps forward with a proposal. He must decide whether to defend himself against the criticism as being unwarranted or to use it to strengthen further the structure that he is building.

Insensitivity to Cause-and-Effect Mechanisms

The beginner may often fail to look at the system closely enough to observe the factors that create the dynamic behavior in which he is interested.

There is often a tendency to want to disaggregate into too fine a structure within any given flow channel, and at the same time to neglect to separate the flow channels well enough from each other.

In very many places, as discussed in earlier chapters, the "worse-before-better" sequences occur. These can often be very important to system dynamics but can be overlooked if one is unperceptive of system interactions. For example, the expansion and training of an engineering department does not just happen. It requires the diversion of managerial and technical time that might otherwise be put to different purposes. The whole structure of skills and the development of people and organizations must take place before they can produce an output. Many of the actual practical working mechanisms of a system tend to create the conflict between short-term and long-term results.

The aggregate flow channels must represent the sequences that are traversed by individual items following the channels. Very often a single event must be traced in order to discover the information channels of the system. But one must not let this study of the individual event trap him into an unduly close view of the system and a desire to model separate decisions rather than policy.

In all of these warnings against the hazards of system analysis we see that the investigator is faced by the usual form of decision wherein he must not do too much or too little of any of the things that are a part of success. As in the systems he is modeling, his own decisions are balancing the amount of his action against the goal which he has set.

Perceived versus Actual

We have said earlier that the decision maker in a system acts on what he perceives to be the state of affairs. This perception is usually not identical with the actual state of affairs but depends on the sources of information used and the amount of prejudice and distortion with which the information is viewed. The system analyst faces the same problem. He comes with ideas already formed about the system with which he is to deal. If he does not have sufficient firmness in his approach, he may never reach decisions and effective action. If his prejudices lead to blindness, the system he analyzes may be merely the one that he initially

thought to be present or which he wished were present. The man with no experience with actual organizations may come with a strong idealized but unrealistic concept of their behavior acquired through academic study of simplified abstractions. On the other hand, the man who has been immersed in a particular system too long may be unable to distinguish the actual system from what he wishes it were. A system model is apt to become a reflection of how the participants hope the system operates rather than a picture of how it does operate. Both of these, of course, are probably different from how it should operate. Wishful thinking and strongly formed past prejudices are both hazards to successful dynamic analysis.

Decision Points in System Context

The novice at system modeling may underestimate the importance of extensive, detailed, first-hand observation of the system with which he is dealing. He may rely too much on what he is told are the factors and objectives at a certain place in the system. He may not take the time to discover that the pressures on people at that point come from quite unexpected places in the organization. He may not observe that the sources of information that are assumed to be necessary may simply not exist. He may not discover the extent to which the goals and objectives are purely local ones designed to keep the participants out of difficulty rather than goals that constructively promote the welfare of the whole organization. A model of system behavior depends on an adequate representation of the decision-making policies within the system. The beginner usually does not realize the depth to which he must probe to distinguish the superficial veneer of rationalization from the deeper effective motivations.

Model Permanence

The beginner often looks upon a model as something that he should build to last. He thinks of *the* model that can be constructed and then used to answer all questions. As he progresses in his understanding of enterprise design, he comes to realize that each new question may carry with it an extension and a variation of the system previously defined. A model evolves with one's understanding of the system with which he is dealing. It can start with a very simple structure aimed at the essence of one particular phenomenon. It can then be extended to bring in new facets and to study new questions. The typical life of a model may be but a day or week before new extensions give it new form and meaning.

Decisions versus Policy

Management and management education have given such emphasis to decision making that the inexperienced systems analyst usually has a hard time grasping the full significance of dealing instead with the policy that creates the flow of decisions. A new orientation in attitude must take place. Time is required to develop an understanding of the full meaning of policy. It takes time to gain confidence that our existing knowledge of policy is sufficient for at least a major start in understanding system behavior.

Excessive Detail

The beginner typically includes too much detail in model formulation. This is a natural result of the fact that it is easier in the kinds of models discussed in this book to include a factor than it is to present a convincing argument that a factor does not matter. Junk is added in the model structure to avoid discriminating thinking about whether or not various factors are necessary. This is understandable and probably will always happen. Again it is a matter of degree. The beginner is more conspicuously vulnerable to including too much than he will be after some experience in building models and after the discovery of how much simplification is possible.

Some detail, even when it does not affect system performance, is justified in order to provide apparent reality and easier communication with others less skilled in model building. For this reason the models in Part III of this book are more detailed than would be necessary to capture the essence of the dynamic performance exhibited in the various figures of Chapters 15 through 18.

The clarity of the initial dynamic hypothesis and the initial verbal description of the system will largely determine one's vulnerability to wandering off into unnecessary complexity and detail.

Underestimating Time Constants

The beginner almost invariably underestimates the lengths of delays and time constants that will exist in our social systems. He overlooks the long educational delays. He fails to appreciate the persistence of prejudice and past personal experience. He fails to examine the sequential steps through which an action must go, and may estimate a total over-all time so short that it is a physical impossibility in the actual system. Time estimates by students are often so short that they do not even fall within the limits of plausibility, once there has been a thorough discussion of the actual real-life mechanisms involved.

Discontinuous Functions

The beginner tends to be carried away by his knowledge of the discreteness of various decisions and actions. This arises partly from viewing the system from too close a range. He wants budgets established once a year, production quotas each quarter, inventories reported each month, individual model years of product handled separately, and so forth. These are all factors that may be important in some of the more subtle questions that may be asked as a study progresses. However, they seldom involve the broad interactions that lead to conspicuous success or failure.

System dynamics can best be visualized if cast first in terms of continuous flows until the interactions of the major time constants, decisions, and levels can be understood.

Another amateur tendency in the area of discontinuous functions comes from carelessness in thinking about the nonlinear forms of decision functions. There is a tendency to make up decision functions from a straight-line linear section that is abruptly terminated when some limiting threshold is reached. In general, decision functions are not discontinuous. Much of the essential character of our industrial organizations arises because of the changes that begin to occur progressively as various limiting conditions are approached. In actual systems, pressures build up gradually, leading progressively over a period of time to awareness and action. Spurious system behavior is very apt to be introduced by artificial breaks and discontinuities in decision functions. One can very seldom argue that life continues as normal up to some very particular point at which the entire character of an organization changes.

Coefficients without Meaning

The beginner may not take the time to be certain that every parameter and variable in his system has a solidly defined meaning that can be related to the actual system with which he is working. Coefficients are often inserted to make dimensional units correct without thinking through whether or not these coefficients arise out of the actual structure and practices of the system. I believe it is possible to make every variable and every coefficient relate directly to its counterpart in the real system and be discussable with and understandable by the practical operating people in the organization. Forcing this degree of reality in equations and parameters helps to clarify thinking and to lead to greater model validity.

Dimensional Units of Parameters and Variables

The careless analyst may fail to define the dimensional units of measure of variables and parameters. This is apt to merge into a lack of a clear understanding of their meaning, and from this there rapidly follows a departure of the model system from the relationships of real life and from those intended in the initial verbal description. The units of measure of every factor should be carefully defined, preferably in terms that are as close as possible to those that have meaning in the verbal description and practical use of that factor.

Defective Decision Functions

Very often the beginner develops decision functions which he recognizes as having major defects but which he defends on the basis that the information-feedback loops threaded through the remainder of the system will prevent the occurrence of those circumstances under which the particular decision function would be wrong. This is a dangerous practice. One is now dealing with a system that fails to pass the test of "not being obviously wrong." It contains relationships that can be established as wrong on the basis of a logical argument. In general this is unnecessary. A little thought and care can make the relationships compatible with at least the obvious tests and challenges. The most powerful defense of a dynamic model lies in the extent to which all of the components are acceptable. The complex information-feedback behavior is usually not subject to strong intuitive argument. To depend on the information-feedback network in order to avoid a state of affairs wherein defective decision functions would have the opportunity to create unexpected difficulty is to compromise in that region where soundness can be achieved and confidence can be greatest. One's trust is relegated to that area of the model about which he knows the least. The dynamic behavior is what we are trying to learn about the system. We want to know the dynamic behavior that will result from the best knowledge we have about the component parts and how they must contribute to the total system.

Simultaneous Initial-Condition Equations

The beginner usually has a great deal of difficulty in the establishment of initial-condition equations for a model and very often finds himself in the position of having to solve a system of simultaneous algebraic equations to get a consistent set of system parameters. This often arises from a poor choice of what is to be independently specified and which parameters may necessarily have values dependent on others already picked.

It is my feeling that a skillful handling of the way in which parameters are defined and numerical values are chosen can always lead to a direct determination of initial conditions without solution of simultaneous equations for the initial condition of the system. This was true in the models of Chapters 15 through 18, although it must be recognized that those systems are not complex. A thorough treatment of the methodology for handling initial conditions is beyond the scope of this book, and the matter will be left at present simply with the assertion that I have always found it possible to achieve a straightforward determination of initial conditions. This, however, is not the usual experience of the person first undertaking model construction.

Infinite Ratios

Sooner or later the model builder usually encounters a situation where he has set up a functional relationship as the ratio of two variables where the denominator can take on a small or zero value. Since real life does not contain infinite values for significant variables, this can be looked on as a defect in the equation formulation. In fact, some of these do exist in Chapters 15 through 18, and the difficulties would become apparent were the model called upon to create properly system activity all the way from zero sales rate upward. The difficulty is especially likely in transient growth models, where all of the variables may have initial values of zero. In general, the difficulty is surmounted by an adequately realistic and wide-ranging statement of decision functions that cannot under any circumstances take on impossible values.

Loops without Levels

In the discussion of model structure in Chapter 6, it was pointed out that rates create levels and that levels are the only proper inputs to the determination of rates. It was, however, further noted that as a practical matter one sometimes uses a rate instead of a short-term average as an input to another rate. This must be watched very carefully, because it is rather easy to create a loop whereby a chain of auxiliary equations may arise from one rate to control another rate that may eventually lead to a third which appears as an input to one of the first equations. In the flow diagram of the system this leads to a loop in which no level equation exists. Such a situation is contrary to the concepts on which the model structure was built. This kind of erroneous structure can lead to a high-frequency numerical instability in the model structure, where the instability is a characteristic of the model defect and not of the system being modeled.

Engineers

A number of the people entering the field of dynamic analysis of business systems have a background in engineering. They may wish to carry over familiar methodology which may not be appropriate in the industrial context. They may have strong tendencies to want to convert the structure and the terminology of the business system into an electrical or a hydraulic network. This greatly increases the communications difficulty between the systems analyst and those in the actual business system. It is also a handicap because the engineering systems are not well equipped with examples of highly nonlinear phenomena. This conversion to, for example, an electrical network tends to initiate a very inappropriate line of thinking that restricts consideration to linear systems.

Those coming from the field of servomechanisms may have difficulty in distinguishing fundamental concepts from the practical art of the engineering field. The fundamental concepts of information-feedback systems are universally applicable. The practical art is apt to depend on the particular classes of systems that are encountered. Most especially, the physical systems are usually characterized by having their own internal natural periods well separated from the frequencies that are imposed upon them in their ordinary operation. This leads to certain practical design procedures that may not carry over into nonlinear systems in which the noise frequencies, the natural frequencies of the system, and the frequencies to which the system should respond all lie within approximately the same frequency band.

The engineer is more accustomed to thinking in terms of differential equations than he is in terms of integral equations. The difference equations used in this book are all essentially integrations of rates. It is common in engineering to speak of velocity as the derivative of position and acceleration as the derivative of velocity. However, nature does not usually take derivatives. Actually the inverse sequence is a more natural viewpoint. Forces create acceleration. Acceleration is integrated to get velocity, and velocity is integrated to get position. Mathematically these are equivalent. Conceptually they give one a somewhat different viewpoint of the system and its structure.

Scientists

Students coming in from the areas of science, whether it be physics, mathematics, or the social sciences, are apt to have their attention focused too exclusively on methodology and technique. They are accustomed to looking for analytic solutions to problems. This habit has in turn forced them to the consideration of only linear systems.

They may have been trained so exclusively in linear systems and linear approximations to systems that they lack an awareness of the extent to which our life and our world depend for their very existence on nonlinear behavior.

A fascination with methodology is apt to lead to a quite disproportionate amount of attention to peripheral questions of technique. The mathematician may want to substitute some more elegant methodology for the first-order integration such as used in this book, even though there can probably be no objective demonstration of the necessity. He may become involved in the discontinuities and the computing phenomena that re-volve around the selection of a solution interval DT. He may become involved in trying to determine how large this interval can be made, rather than merely making sure that it is small enough to raise no questions.

The science student may lack any first-hand awareness of the political factors that enter into even the so-called scientific and technical decisions. He is apt to feel that rigorous and objective methods must of necessity be found for the construction of models, even though in his home field there are no rigorous and objective methods that ensure the success of the design of engineering systems and models of engineering devices.

REFERENCES

[Including place in this book where cited]

1. Drucker, Peter F., "Thinking Ahead: Potentials of Management Science," *Harvard Business Review,* Vol. 37, No. 1, pp. 25-30, 146-150 (January-February 1959). [Section I.2.]

2. Porter, A., *An Introduction to Servomechanisms,* Methuen & Co., London, and John Wiley & Sons, New York, 1950. [Section 1.1.]

3. MacMillan, R. H., *An Introduction to the Theory of Control in Mechanical Engineering,* The University Press, Cambridge, 1951. [Section 1.1.]

4. Brown, Gordon S., and Campbell, Donald P., *Principles of Servomechanisms,* John Wiley & Sons, New York, 1948. [Section 1.1.]

5. Jury, Eliahu I., *Sampled-Data Control Systems,* John Wiley & Sons, New York, 1958. [Section 1.1.]

   Ragazzini, John R., and Franklin, Gene F., *Sampled-Data Control Systems,* McGraw-Hill Book Company, New York, 1958. [Section 1.1.]

   Seifert, William W., and Steeg, Carl W., Jr., Editors, *Control Systems Engineering,* McGraw-Hill Book Company, New York, 1960. [Section 1.1, Appendix F.]

6. Tustin, Arnold, *The Mechanism of Economic Systems,* Harvard University Press, Cambridge, Mass., 1953. [Section 1.4.]

7. Leavitt, Harold J., and Whisler, Thomas L., "Management in the 1980's," *Harvard Business Review,* Vol. 36, No. 6, pp. 41-48 (November-December 1958). [Sections 3.3, 20.4.]

8. Kovach, Ladis D., "Life Can Be So Nonlinear," *American Scientist,* Vol. 48, No. 2, pp. 218-225 (June 1960), published by the Society of the Sigma Xi. [Section 4.1.]

9. Klein, Lawrence R., *Economic Fluctuations in the United States 1921-1941,* Cowles Commission Monograph No. 11, John Wiley & Sons, New York, 1950. [Chapter 12.]

10. Koopmans, Tjalling C., Editor, *Statistical Inference in Dynamic Economic Models,* Cowles Commission Monograph No. 10, John Wiley & Sons, New York, 1950, Third Printing, July 1958. [Chapter 12.]

11. Churchman, C. West, *Theory of Experimental Inference,* The Macmillan Company, New York, 1948. [Section 13.7.]

12. Vidale, M. L., and Wolfe, H. B., "An Operations-Research Study of Sales Response to Advertising," *Operations Research,* Vol. 5, No. 3, pp. 370-381 (June 1957). [Section 16.1.]

13. The following S.M. theses have been completed at the M.I.T. School of Industrial Management under the supervision of the author. Being S.M. theses, only part time for some six months was available to the students for their completion. Time did not permit as many revised drafts as may have been desirable, nor was there time for the student to interpret and take full advantage of the system model that he formulated. However, each of the following has made a substantial contribution in opening new facets of the field of industrial dynamics. Ph.D. theses and staff research are now under way to exploit interesting areas opened by several of the following theses. Microfilm or photocopies of theses can be purchased from the Library, Massachusetts Institute of Technology, Cambridge, Mass.

   Ballmer, Ray W., Sloan Fellow at M.I.T., 1959-1960, from the Kennecott Copper Corporation, S.M. Thesis, School of Industrial Management, *Copper Market Fluctuations: An Industrial Dynamics Study,* 1960, i-ix plus 125 pages. [Section 19.3.]

   Fey, Willard R., *The Stability and Transient Response of Industrial Organizations,* S.M. Thesis, Department of Electrical Engineering, 1961, M.I.T., 8 unnumbered front pages, 132 pages plus appendix of 22 pages. [Section 17.4.]

   Hurford, Walter J., Sloan Fellow at M.I.T., 1959-60, from the Westinghouse Electric Corporation, S.M. Thesis, School of Industrial Management, *Application of Industrial Dynamics to the Growth of the Fuel Manufacturing Industry for Nuclear Thermal Electric Power Plants,* 1960, i-vii plus 123 pages. [Section 19.8.]

   Katz, Abraham, Sloan Fellow at M.I.T., 1957-58, from the Radio Corporation of America, S.M. Thesis, School of Industrial Management, *An Operations Analysis of an Electronic Systems Firm,* 1958, 7 front plus 109 pages. [Section 19.4.]

   Kinsley, Edward R., Sloan Fellow at M.I.T., 1958-59, from the Texas Instruments Company,

REFERENCES

S.M. Thesis, School of Industrial Management, *The Managerial Use of Industrial Dynamics as Illustrated by a Company Growth Model,* 1959, i-vii plus 180 pages. [Section 19.2.]

Raff, Alfred I., *Dynamics of the Tankship Industry,* S.M. Thesis, Department of Naval Architecture and Marine Engineering, M.I.T., 1960, 5 unnumbered front pages and 100 pages. [Section 19.9.]

Schlager, Kenneth J., Sloan Fellow at M.I.T., 1960-61, from the AC Spark Plug Division of General Motors Corporation, S.M. Thesis, School of Industrial Management, *Systems Analysis of the Copper and Aluminum Industries: An Industrial Dynamics Study,* 1961, i-viii plus 264 pages. [Section 19.3.]

Walter, Franklin, Sloan Fellow at M.I.T., 1958-59 from the Chrysler Corporation, S.M. Thesis, School of Industrial Management, *An Analysis Relating Lead Time and Market Penetration in the Auto Industry,* 1959, i-vii plus 125 pages. [Section 19.1.]

14. Robinson, Dwight E., "Fashion Theory and Product Design," *Harvard Business Review,* Vol. 36, No. 6, pp. 126-138 (November-December 1958). [Section 19.1.]

15. Katz, Abraham, "An Industrial Dynamic Approach to the Management of Research and Development," *IRE Transactions on Engineering Management,* Vol. EM-6, No. 3, pp. 75-80 (September 1959), Institute of Radio Engineers, New York. [Section 19.4.]

16. Brown, Robert G., *Statistical Forecasting for Inventory Control,* McGraw-Hill Book Company, New York, 1959. [Section 18.8.]

17. Winters, Peter R., Forecasting Sales by Exponentially Weighted Moving Averages," *Management Science,* Vol. 6, No. 3, pp. 324-342 (April 1960). [Section 19.8.]

18. Zannetos, Zenon S., *The Theory of Oil Tankship Rates,* Ph.D. Thesis, Department of Economics and Social Science, Massachusetts Institute of Technology, September 1959, 12 unnumbered front pages plus 299 pages. [Section 19.9.]

19. Pierson, Frank C., and others, *The Education of American Businessmen,* The Carnegie Series in American Education, McGraw-Hill Book Company, New York, 1959. [Section 20.1.]

20. Gordon, Robert Aaron, and Howell, James Edwin, *Higher Education for Business,* Columbia University Press, New York, 1959. [Section 20.3].

21. Forrester, Jay W., "Industrial Dynamics — A Major Breakthrough for Decision Makers," *Harvard Business Review,* Vol. 36, No. 4, pp. 37-66 July-August 1958). [Appendix A.]

22. Pugh, Alexander L., III, *DYNAMO User's Manual,* The M.I.T. Press, Cambridge, Mass., 1961. Available through The Technology Store, 40 Massachusetts Avenue, Cambridge, Mass. [Appendix A.]

23. Laning, J. Halcombe, Jr., and Battin, Richard H., *Random Processes in Automatic Control,* McGraw-Hill Book Company, New York, 1956. [Appendix F.]

24. Davenport, Wilbur B., Jr., and Root, William L., *An Introduction to the Theory of Random Signals and Noise,* McGraw-Hill Book Company, New York, 1958. [Appendix F.]

25. Phillips, A. W., "Stabilisation Policy and the Time-Forms of Lagged Responses," *Economic Journal* (London), Vol. 67, No. 266, pp. 265-277 (June 1957). [Appendix H.]

Index

Academic program, for freshman, vii
  in industrial dynamics, 350-354
  at M.I.T., 353-354
Accounting information, 335-336
Accuracy, defined, 57
AC Spark Plug Division, 324, 458
Adequacy, *see* Validity
Advertising, added to production-distribution system, 36-42, 187-207
  causing production instability, 36-42, 187-207
  proportional to sales, 36, 190-192
  time response to, 194-195
Aggregation, into continuous flows, 65
  of whole industry, 187, 211, 340-341
  see also Delays, aggregation, and Variables, aggregation
Alfred P. Sloan Foundation, ix
Amplification, from adjusting levels, 348
  affected by decay rate, 202-203
  as characteristic of information-feedback system, 62-63
  from delays, 348
  from forecasting, 138, 349
  frequency-sensitive, 202, 349, 442
  from inventory policies, 138, 173, 209
  from ordering procedures, 22, 173-175
  in pipelines, 138, 173
  stability by reducing, 204
  of system, 254, 260, 424
  from variable delay, 182, 202, 212, 348
  see also Delays, Noise, amplification of, Smoothing
Analog computer, 18-19, 68
Assumptions, *see* Models, assumptions about system
Atomic fuel, 339
Automobile, *see* Market
Auxiliary equations, *see* Equations, auxiliary, Flow diagram, auxiliary-variable symbol
Averaging of data, *see* Smoothing, Equations, smoothing

Balance sheet, continuous, 263, 296
Ballmer, Ray W., 322, 457
Bandwidth, 54
  described, 54, 125
  in nonlinear system, 125
  prediction related to, 125, 126, 430
  solution interval related to, 80
  as system characteristic, 54

Battin, Richard H., 412, 458
Beef, production dynamics, 104
Bennett, Richard K., ix, 369
Boundaries of models, *see* Models, scope
Bowles, Edward L., viii
Bowman, Edward H., ix
Boxcar train, flow-diagram symbols, 84-85
  see also Delays, boxcar train
Boyd, Constance D., ix
Brooks, E. P., viii
Brown, Gordon S., viii, 14, 457
Brown, Robert G., ix, 338, 458
Bryant, Lynwood S., ix
Buoncristiani, John F., ix
Bursk, Edward C., ix
Business games, *see* Management games
Business Week, 98

Campbell, Donald P., 14, 457
Capital equipment, flow symbol for, 82
  flow system, described, 71
  introduced, 13
Carlson, Bruce R., ix
Cash flow, 240-245, 263, 296
Chapman, John F., ix
Chrysler Corporation, 313, 458
Churchman, C. West, 123, 457
Combining variables, *see* Variables, aggregation
Commodity dynamics, 321-324
  inventories, 322
  price changes, 322
Competition, 336-337
Computer, *see* Analog computer, Digital computers
Computing, sequence, 73-75, 396-401
  time required, *see* Digital computers, time required
Constants, *see* Parameters
Controlled experiments, 43
  models for, 55
Correlated data, 118
Countercyclical policy, *see* Policy, countercyclical
Cycles created by forecasts, 443-449

Damping, defined, 113
Data, collection of, 141
  see also Information for constructing models, data, Parameters, Statistical estimation
Davenport, Wilbur B., Jr., 412, 458

Dead-zone instability, 233-234, 274
Decision functions, to control rates, 69
  determining form, 103-107
  factors to include, 104
  inputs to, 103
  noise in, 107-108
  nonlinearity in, 106-107
  omission of feedback paths, 107
  parameters, model correction of inaccuracy, 105
  as policy statement, 69
  positive feedback, 104
  short-term vs. long-term effects, 104
  see also Decision making, Equations, rate, Flows, Information, for constructing models, Policy, Rates
Decision making, as continuous process, 96
  controlling flows, 95
  delays in, 151
  determining basis of, 17
  discussion of, 93-108
  as foundation for industrial dynamics, 17
  implicit, 102, 151, 164
  information-feedback context, 94
  levels as input, 95
  in military tactics, 17
  nature of process, 95-96
  overt, 102, 151, 162-164
  three parts, 95
    action, 96
    apparent state, 96, 452
    desired state, 95-96, 148, 348
  threshold in, 233-234, 274
  viewing distance, 96
  see also Decision functions, Policy
Deferrability of purchase, 36, 187-189
Delays, 86-92, 418-421
  in advertising response, 194-195
  aggregation, effect of, 110-111
  boxcar train, examples, 447
    flow-diagram symbols, 84-85
  as characteristic of information-feedback-system, 15
  characteristics, length determining steady-state, 87
  transient response, 87, 89-92
  communications, 155
  at conversion points, 64
  decision-making, representation, 151
  equations for, 87-89, 418-419
  exponential, equations for third-order, 89, 419
  higher-order, 88

Delays, frequency-sensitive, 415-417
  importance in models, 62
  impulse response, 90-92, 144
  information, 418
  initial value, 167
  inventory, *see* Inventory, delivery delay
  material, 418
  period of oscillation related to, 202
  in production-distribution example, 22-23, 140
  quantity in transit, 86, 89
  representation of, 86-92
  smoothing, equivalence to, 418
  smoothing as cause, *see* Smoothing, creating delay
  step response, 90-92
  structure of, 86-87
  system effect of, changing value, 32-33, 182-183, 268-276, 349
    time response, 419-421
  transient response, curves for, 90, 92
  transportation, 155
  underestimating, 453
  see also Amplification, from variable delay, Flow diagram, delay symbols
Descriptive information, *see* Information, for constructing models, descriptive
Design of enterprises, *see* Enterprise design
Diagrams, *see* Flow diagram
Differential analyzer, 68
Digital computers, 369
  capacity, 68
  cost reduction, 18
  as foundation for industrial dynamics, 18-19
  technical advancement, 19
  time required, 199, 381
  see also Simulation, cost
Digital Equipment Corporation, ix
Dimensions of measure, *see* Units of measure
Discrete events, 64-66
Distribution system, 21-42, 137-186
Douty, Arthur L., ix
Drucker, Peter F., 3, 457
DT, defined, 73-74
  see also Solution interval
Durand, David, ix
Duren, Grace, ix, 369
Dynamics, *see* Commodity dynamics, Growth dynamics, Market, dynamics of, Mining, Research and development
DYNAMO computer compiler, 369-381
  checking defined variables, 80
  creating sets of delay equations, 89
  delay equations used in, 419
  error print-out, 370
  growth ramp, 248
  initial conditions for delays, 197
  ordering auxiliary equations, 78

DYNAMO computer compiler, ordering initial-value equations, 166
  teaching of, 352

Economic development, academic research, 361
  as challenge to management, 6
  dynamics of, viii, 361
Education, liberal-arts, 2
  see also Management education
Electric generating industry, 341-343
Engineering, precedent for management, 4-6
  see also Men, engineers
Enterprise design, goals, 44, 449
  steps in, 13, 43-45
Equations, auxiliary, dependent on evaluation sequence, 78
  described, 78-79
  example, 78
  as subdivision of rate equation, 78
  basis for, 140-141
  classes of, 76
  constants, *see* Parameters
  initial-value, for delays, 167
    described, 79
    ordered by DYNAMO, 166
    production-distribution model, 165-168
    for rates, 217
    simultaneous, 454
    steady-state start, 166
  level, described, 76-77
    example, 76
    independence of evaluation sequence, 77
    independent of solution interval, 77
    as integrations, 68, 76
    loops without, 455
  parameters, *see* Parameters
  rate, auxiliary as subdivisions of, 77-78
    described, 77-78
    example, 78
    formulation of, 143
    independence of evaluation sequence, 77
  smoothing, 150-151
    see also Smoothing
  symbols for variables in, 75
  system of, 73-80
    computing sequence, 73-75, 396-401
  supplementary, described, 79
    examples, 246-248
  tabulation of, 383-395
  test input generation, 248-251
  see also Decision functions, Delays, equations for, Information for constructing models, Integration, Levels, Rates, Solution interval
Equilibrium, *see* Equations, initial-value
Error function, 61

Exogenous variables, *see* Input to model, exogenous variable as, Variables, exogenous
Exponential averaging, *see* Smoothing, exponential
Exponential delays, *see* Delays, exponential
Extreme conditions, *see* Models, non-linear

Feedback, *see* Information-feedback systems
Fey, Willard R., ix, 215, 457
Flow diagram, auxiliary variable symbol, 83
  delay symbols, boxcar, 84
    exponential, 84
  described, 81-85
  flow symbols, 82
  level symbol, 81-82
  parameter symbol, 83
  rate symbol, 82
  source-and-sink symbol, 82
  variable on other diagrams, symbols, 83
Flows, capital equipment, 71
  continuous treatment, 64-66
  defined, 69
  determined by levels, 69
  flow-diagram symbols, 82
  information, 71-72
  material, 70
  money, 70-71
  orders, 70
  personnel, 71
  six, described, 70-72
    introduced, 13
  see also Decision functions, Flow diagram, Rates
Forbes Magazine, 98
Forced oscillation, 200
Forecasting, 437-440
  amplification from, 138, 349, 437
  as decision equations, 337
  defined, 437
  discussion of, 337-339
  in electric generating industry, 342
  extrapolation in, 337, 437
  in ordering decisions, 138
  self-defeating, 437
  self-validating, 437
  system independence of, 337
  see also Prediction
Ford Foundation, ix
Forrester, Ethel W., viii
Forrester, Jay W., 369, 458
Forrester, M. M., viii
Foundations for industrial dynamics, 14-19
Fox, Phyllis (Mrs. George Sternlieb), ix, 369
Franklin, Gene F., 14, 457
Frequency-sensitive amplication, *see* Amplification, frequency-sensitive, Delays, frequency-sensitive, Smoothing, frequency selectivity

Future, *see* Industrial dynamics, interval to develop

Gain, *see* Amplification, Delays, Smoothing
General Motors Corporation, 324, 458
Goals of enterprise design, 44, 449
Goetz, Billy E., ix
Gordon, Robert Aaron, 350, 458
Growth dynamics, viii, 317-321, 331-334

Haberstroh, Chadwick J., ix
Harvard Business Review, ix
History, importance, 352
Howard, David J., ix, 369
Howell, James Edwin, 350, 458
Hunting in servomechanisms, 61
Hurford, Walter J., 339, 457

Implicit decision, *see* Decision making, implicit
Impulse response, *see* Delays, impulse response
Independent variable, *see* Variables, exogenous
Infinite values, 454-455
Institute of Management Sciences, 345
Internship, 357
Industrial dynamics, approach to a problem, 21
  defined, 13
  effect on manager, 45-46
  foundations, 14-19
  integrating academic program, 350
  interval to develop, 46, 362-363
  premises, 13-14
  size of company for, 365
  steps, 13, 43-45
  time required to study system, 213, 276, 362
  to study system, 213, 276, 362
Information, as basis for decisions, 69
  for constructing models, 54, 57-59, 101
  data, 57, 59, 101, 117, 118, 141
  for decision functions, 103
  descriptive, 57-59, 101, 117, 451
  example of, 218-219
  see also Equations, basis for
  distortion, 63
  about extreme conditions, 107
  flow symbol for, 82, 83
  flow system, described, 71-72
  introduced, 13
  harmful results, 427
  interconnecting other flows, 71-72
  retail, at factory, 428
  unavailable about future, 74, 337
  value of, 427-429
  see also Variables, true vs. indicated
Information-feedback systems, amplification in, 14-15, 62-63, 442
  characteristics, source of, 15-16
  closed, 51-52
  counterbalancing influences, 172
  defined, 14

Information-feedback systems, delays as characteristic of, 14-15
  history, 16
  information in, 94
  progress in understanding, 16
  social systems as, 15, 53, 61-63
  structure, as characteristic of, 14-15
  theory, as foundation of industrial dynamics, 14-17, 455
Initial conditions, *see* Equations, initial-value
Initial-value equations, *see* Equations, initial-value
Input to model, equations to generate, 248-251
  exogenous variable as, 112-114
  historical time series, 172
  periodic, one-year, to electronic component model, 258-259, 286, 292
  one-year, to production-distribution system, 26-27, 175-177
  two-year, to electronic component model, 260-261, 294
  random, to advertising-market system, 40-41
  to electronic component model, 262, 264-265, 296
  to production-distribution system, 27-29, 177-180
  related to exogenous variables, 114
  see also Noise
  step, to advertising-market system, 38-41, 200-203
  to electronic component model, 253-257, 291-293
  frequency content, 172
  to production-distribution system, 25, 172-175
Instability caused by advertising, 37-41
Intangibles in model, 328
Integration, first-order vs. higher-order, 80
  in level equations, 68
International Business Machines Corporation, ix
Interpretation of model results, 45
Inventory, adjustment rapidity of, 34-35, 152, 170, 186
  amplification from, 138, 173-175
  in commodity systems, 322
  delivery delay, 146-148, 169
  desired, 148-149, 222
  sales related to, 217, 222
  theory, 149
  time relation to production, 205-206, 209, 254-255, 290
  turnover, 149, 170
  see also Pipelines

J, as previous time step, defined, 74
JK, as past time interval, defined, 74
Jarmain, W. Edwin, ix
Johnson, Howard W., viii
Judgment in selecting model objective, 116

Jury, Eliahu I., 14, 457

K, as present time step, defined, 74
Katz, Abraham, 325, 326, 457, 458
Kennecott Copper Corporation, 322, 457
Kinsley, Edward R., 320, 457
KL, as following time interval, defined, 74
Klein, Lawrence R., 112, 457
Koopmans, Tjalling C., 113, 457
Kovach, Ladis D., 51, 457

L, as next time step, defined, 74
Labor, *see* Personnel
Lagging time series, *see* Leading-lagging time series
Lags, *see* Delays
Laning, J. Halcombe, Jr., 412, 458
Laplace transforms, 215
Leading-lagging time series, 421-425
  frequency-dependent, 202
  variability, 425
Leavitt, Harold J., 45, 457
Level equations, *see* Equations, level
Levels, average rates as, 68, 140
  defined, 68-69
  of distribution, 32-33, 183-185
  flow-diagram symbol, 81-82
  in six flow networks, 68
  test for, 68
  to uncouple rates, 349
  variables in distribution system, 140
  see also Equations, level
Life cycle, growth phase, 317-321
  mature phase, 139
Limiting conditions, *see* Models, nonlinear
Line functions of management, 46
Linear model, defined, 50
  see also Models, linear, Systems, linear
Long-range planning, *see* Planning

MacMillan, R. H., 14, 457
Management, as an art, 1-3, 357
  challenge to, 6-8
  converting information to action, 93
  effectiveness, 7
  frontier, 1
  laboratory, 43
  line functions, 46
  professional status, 2
  staff functions, 46
  see also Management science
Management education, case study, 2, 345
  fragments, 2
  history, basis of models, 352
  industrial dynamics in, 45, 344-361
  servomechanisms courses, 352, 355-356
  unifying, 350
Management games, 357-360
  defined, 357
  dynamic models related to, 357-358

Management science, art vs., 1-4
  policy detection in, 99
  see also Operations research
Manager, development, 45-46
  effect of industrial dynamics on, 45-46
  his task, 1
  related to management science, 8-9
Market, added to production-distribution system, 36-42, 187-207
  advertising, 36-42, 187-191
  automobile, 313-317
  differentiation by dynamic character, 336
  dynamics of, 311-317
Massachusetts Institute of Technology, academic program, 353
  Computation Center, ix
  Lincoln Laboratory, viii
  Servomechanisms Laboratory, viii
Materials, flow symbols for, 82
  flow system, described, 70
  introduced, 13
Mathematical models, *see* Models
Mathematical training for simulation, 18
Mathematics, in experimental dynamics, 351
  linear analysis, 50
  optimum solutions, 3
Men, engineers, 455
  qualifications for industrial dynamics, 364-365, 449
  scientists, 455-456
Mental models, *see* Models, abstract
Mining, 104, 322-324
Minute Maid Corporation, ix
Model building, hypothesis, importance, 450
  poor practices, 145, 146, 449-456
  see also Models, nonlinear
Models, 49-59
  abstract, 49-50
    mental, 49-50, 116, 117
    verbal, 49-50, 450
  accuracy vs. precision, 57
  assumptions about system, 141
  basic structure, 67-70
  as business cases, 345
  boundaries, *see* Models, scope
  characteristics to represent industrial systems, 52
  classification, 49-52
  closed, 51
  complexity for industrial representation, 52, 61
  content, 60-61
  continuous flows in, 64-66
  correspondence to real system, 54, 63-64
  delays in, 62
  dynamic, 50
  in engineering, 53
  information feedback in, 61
  information for, *see* Information
  as "law of behavior," 123
  linear, defined, 50

Models, mathematical, 44
  for clarity, 44
  term distinguished from economic usage, 52
  translation from English, 44, 50
  mechanizing, 55
  nonlinear, 50
    formulation easier than linear, 107, 143
  objectives, 53, 56, 137, 213
  open, 51
  physical, 49-50
  results, interpretation, 45
  in science, 53
  scope, 55-56, 138, 210-211, 450
  significance, *see* Validity
  of solar system, 123
  stable, 51
  static, 50
  steady-state, defined, 51
  structure, 67-70
  as system description, 18
  testing, *see* Validity
  time relationships in, 62
  transient, defined, 51
  unstable, 51
  validity of, *see* Validity
  see also Model building, Systems
Money, balance sheet, continuous, 263, 296
  cash flow, 263, 296
  flow symbol for, 82
  flow system, described, 70-71
    effect on decisions, 138-139
    example, 240-246
    vs. information flow, 335
    introduced, 13
Moving average, *see* Smoothing, moving-average

Natural period, *see* Period
Networks, 70-72
Noise, 412-415
  amplification of, 175, 180, 414
  converted to annual cycles, 22, 27-29, 443-449
  in decision functions, 107-108
  in economic systems, 124
  forecasting, effect on, 443
  frequency spectrum, 108, 179, 262, 264, 269, 412-415
    control of, 250, 262, 414
  power in, 412
  prediction, effect on, 124, 430-436
  random numbers related to, 413
  white, 108
    defined, 412
  see also Smoothing
Nonlinear models, *see* Models, nonlinear
Nonlinear policies, *see* Policy, nonlinearity in
Nonlinear systems, *see* Systems, nonlinearity in
Nord, Ole C., ix
Notation, *see* Symbols
Nuclear fuel, 339

Operations research, 3, 148
  see also Management science
Orders, flow symbol for, 82
  flow system, described, 70
    introduced, 13
  policy governing, 152-156, 220-222
Organization, functional, 329
  project, 330
  see also Structure
Oscillation, forced, 200
Overt decision, *see* Decision making, overt

Parameters, flow-diagram symbol, 83
  model insensitivity to values, 105, 171, 268
  plausible range, estimate, 171
  selecting values, 171-172
  sensitivity, test in model, 172, 268-276
  see also Information
Performance of system, defined, 116
Period, natural, maximum amplification related to, 255
  of production-distribution system, 175
  of oscillation related to delays, 202
Periodic input, *see* Input, periodic
Personnel, flow symbol for, 82
  flow system, described, 71
    introduced, 13
  policies, effect, 269-276
  representation in model, 229-235
Phase, *see* Leading-lagging time series
Phase shift in delays, 415-417
Ph.D. degree, as training for management, 356
Phillips, A. W., 419, 458
Pierson, Frank C., 345, 458
Pipelines, amplification in, 138, 173
  content adjustment, 152
  in-transit quantity, 152, 155, 173
Planning, long-range, models for, 339-340
  challenge to management, 7
Policy, consistency in, 102
  countercyclical, instability from, 205, 441-442
  defined, 93, 96-97
  detection of, 97-102
  discussion of, 93-108
  frequency-sensitive, 349
  intuitive judgment vs., 98
    dynamic variables in, 99
  knowledge of, basis of civilization, 98
  nonlinearity in, 305-307, 453
    infinite ratios, 454-455
  ordering, 152-156, 220-222
  three levels of abstraction, 97, 100
  as transfer function, 97
  see also Amplification, Decision making
Porter, A., 14, 457
Precision, defined, 57

INDEX

Prediction, as decision-making rule, 127
  dependence on system condition, 434
  by leading-lagging series, 421-422
  noise effect on, 430-436
  system independence of, 127
  time series, 430-436
  unstable system related to, 125
  see also Forecasting, Validity
Price, in commodities, 322
Procedure, *see* Model building
Product growth, *see* Growth dynamics
Production, capacity formulation, 164
  limit, effect of, 30-31, 40-41, 181, 204-206
  fluctuation, 21-22
  representation of, 162-164, 223-228
  see also Inventory, time relation to production
Production-distribution system dynamics, 21-42
Pugh, Alexander L., III, ix, 341, 369, 458
Purpose of model, *see* Models, objectives

Radio Corporation of America, 325, 457
Raff, Alfred I., 341, 458
Ragazzini, John R., 14, 457
Random input, *see* Input, random, Noise
Random numbers, characteristics of, 108, 412-415
  modification of, 108
  pseudorandom generator, 431
  see also Input, random, Noise
Rate equations, *see* Decision functions, Equations, rate, Rates
Rates, defined, 69
  determined by levels, 69
  in distribution system, 140, 143
  equations for, 77-78
  flow-diagram symbol, 82
  used in other rate equations, 69, 77, 152
  see also Decision functions, Equations, rate, Flows
Reading instructions, 11, 48, 136, 309
Reis, Wendyl A., Jr., ix, 215
Research, organization for, 5
Research and development, dynamic analysis of, 324-329
Richards, Faith, ix
Roberts, Edward B., ix, 325, 369
Robinson, Dwight E., 312, 458
Root, William L., 412, 458

SAGE air defense system, viii
Sage, Nathaniel McL., Sr., viii
Schlager, Kenneth J., 324, 458
Scientific method, 450
Scope of models, *see* Models, scope
Seasonality, created by forecasts, 338, 443-449
  from random disturbances, 28

Seasonality, self-induced, 29, 443-449
Seifert, William W., 14, 412, 457
Sensitivity tests, *see* Parameters, sensitivity, test in model
Sequence, computing, 73-75, 396-401
Servomechanisms, *see* Information-feedback systems
Servomechanisms courses, background for industrial dynamics, 352-353, 355-356
Shapiro, Eli, viii
Significance of a model, *see* Validity
Simulation, applied to management problems, 18
  cost, 44, *see also* Digital computers, time required
  defined, 18, 23
  mathematical training for, 18
  in military systems, 18
  procedure, 24
  using digital computer, 45
Sinusoidal input, *see* Input, periodic
Sloan Foundation, Alfred P., ix
Smoothing, 406-411
  creating delay, 150, 348-349, 407, 409-411
  equation, *see* Equations, smoothing
  exponential, 408-411
  formal, 407
  frequency selectivity, 414, 415-416
  intuitive, 407
  moving-average, 407
  noisy data, 150, 406
  by system, 262
Solution interval, 403-406
  data unrelated to, 62, 404
  defined, 73-74
  delays related to, 79, 403-404
  empirical test, 404-406
  length of, 79-80, 168, 199, 403-406
  levels related to, 403
  rule of thumb for, 79-80, 406
  symbol for, 73
  see also Equations, level, independent of solution interval
Sprague, Robert C., ix
Sprague Electric Company, ix
Stability, *see* Models, stable, Systems, stable
Staff management, 46
Statistical estimation, after need demonstrated, 171
Statistics, forecasting by, 337-338
  not to measure policy, 100-101
  see also Information, for constructing models
Steady state, *see* Models, steady-state
Steeg, Carl W., Jr., 14, 412, 457
Step input, *see* Input, step
Sternlieb, Mrs. George, ix, 369
Structure, as characteristic of information-feedback systems, 14-15
  see also Information-feedback system, structure, Organization
Sum of the squares, *see* Validity, judging, by least-squares tests
Superposition, 50

Supplementary equations, *see* Equations, supplementary
Supply pipelines, *see* Pipelines
Swanson, Carl V., ix, 215
Symbols, for terms in equations, 75
  for time notation, 75-76
Systems, experimental analysis as foundation of industrial dynamics, 17-18
  frequency selectivity, 28-29
  improvement, as step in enterprise design, 45, 276-277
  instability in, 66
  linear, defined, 50
  linearity in, 66
  nonlinearity in, 66, 318
  parameters, insensitivity to, 171
  performance, defined, 116
  stability, by reducing amplification, 204
  significance of, 202-203
  stable, 51
  study effect on, 128
  unstable, 51, 203
  see also Amplification, of system, Models
Systems engineering, as precedent for management science, 5-6

Tabulation of model equations, 383-395
Tanker, oil shipping, 341
Testing, *see* Validity
Texas Instruments Company, 320, 457
Time delays, *see* Delays
Time notation, 75-76
Time-phase relationships, *see* Amplification, Delays, Leading-lagging time series, Smoothing, Systems
Time required, *see* Digital computers, time required, Industrial dynamics, time required
Time series, *see* Input to model, Prediction, time series
Top management, decision-making, 329
  functional, 329
  project organization, 330
  time allocation, 332-335
Transfer function, defined, 52
Transient model, *see* Models, transient
Transient response, *see* Delays, transient response
Transportation delays, 155
Turning points, *see* Leading-lagging time series
Tustin, Arnold, 19, 457

Units of measure, 451, 454
  dimensional correctness, 64, 142
  not equivalent dollars, 63
Unstable models, *see* Models, unstable
Unstable systems, *see* Systems, instability in

463

Validity, behavior of model related to, 119, 120
  boundaries of system, 117
  by controlling system to model, 122
  correspondence of detailed parameters and structure, 116
  defense of model details, 117, 263
  differing views, 122-129
  engineering and economic models contrasted, 53-54
  formal test, lack of, 116
  foundations, two, 117
  judging, 115-129
    as applied to managers, 120
    by leading-lagging time series, 120, 263
    by least-squares tests, 56, 126-127
    by qualitative results, 121, 263-268
    by real-system results, 115
  noise, related to prediction, 124
  nonquantitative, 128-129
  not-implausible test, 119-120, 263
  objective tests, 117, 118, 123, 450
    basis, 123
  parameter values, influence on, 118-119, 121

Validity, in predicting behavior characteristics, 125, 126
  in predicting change, 116
  prediction related to, 54, 56, 123-128
    limitation on, 125
  related to purpose, 115-122
  similarity of problem symptoms, 120
  statistical analysis, vs. noise, 118
    significance tests, 123
  subjective evaluation, 115-116, 123
  test over wide range, 119
  value of objectives, related to, 116
  variables, choice of, determines, 118
  see also Prediction
Variables, aggregation, 109-111
    same channels as individual items, 109
    of similar decision functions, 110
  correspondence in model and real system, 63-64
  defining all, 80
  exogenous, 112-114
    dangers in use, 113
    defined, 112

Variables, exogenous, destruction of feedback paths, 113
    independence of, 112
    noise inputs contrasted, 114
    as single test input, 113
    as test input, 112
    see also Input to model
  independent, see Variables, exogenous
  true vs. indicated, 63-64, 103
  see also Equations, Flow diagram
Verbal models, 49-50
Vidale, M. L., 194, 457

Wall Street Journal, 98
Walter, Franklin, 313, 458
Ward, Ernest L., ix
Westinghouse Electric Corporation, 339, 457
Weymar, Helmut, ix
Whisler, Thomas L., 45, 357, 457
Winters, Peter R., 338, 458
Wolfe, H. B., 194, 457
Worse-before-better sequences, 348

Zannetos, Zenon S., 341, 458
Zero denominators, 455